F. Kremer

HANDBOOK OF POLYETHYLENE

PLASTICS ENGINEERING

Founding Editor

Donald E. Hudgin

Professor
Clemson University
Clemson, South Carolina

1. Plastics Waste: Recovery of Economic Value, *Jacob Leidner*
2. Polyester Molding Compounds, *Robert Burns*
3. Carbon Black-Polymer Composites: The Physics of Electrically Conducting Composites, *edited by Enid Keil Sichel*
4. The Strength and Stiffness of Polymers, *edited by Anagnostis E. Zachariades and Roger S. Porter*
5. Selecting Thermoplastics for Engineering Applications, *Charles P. Mac-Dermott*
6. Engineering with Rigid PVC: Processability and Applications, *edited by I. Luis Gomez*
7. Computer-Aided Design of Polymers and Composites, *D. H. Kaelble*
8. Engineering Thermoplastics: Properties and Applications, *edited by James M. Margolis*
9. Structural Foam: A Purchasing and Design Guide, *Bruce C. Wendle*
10. Plastics in Architecture: A Guide to Acrylic and Polycarbonate, *Ralph Montella*
11. Metal-Filled Polymers: Properties and Applications, *edited by Swapan K. Bhattacharya*
12. Plastics Technology Handbook, *Manas Chanda and Salil K. Roy*
13. Reaction Injection Molding Machinery and Processes, *F. Melvin Sweeney*
14. Practical Thermoforming: Principles and Applications, *John Florian*
15. Injection and Compression Molding Fundamentals, *edited by Avraam I. Isayev*
16. Polymer Mixing and Extrusion Technology, *Nicholas P. Cheremisinoff*
17. High Modulus Polymers: Approaches to Design and Development, *edited by Anagnostis E. Zachariades and Roger S. Porter*
18. Corrosion-Resistant Plastic Composites in Chemical Plant Design, *John H. Mallinson*
19. Handbook of Elastomers: New Developments and Technology, *edited by Anil K. Bhowmick and Howard L. Stephens*
20. Rubber Compounding: Principles, Materials, and Techniques, *Fred W. Barlow*

21. Thermoplastic Polymer Additives: Theory and Practice, edited by John T. Lutz, Jr.
22. Emulsion Polymer Technology, Robert D. Athey, Jr.
23. Mixing in Polymer Processing, edited by Chris Rauwendaal
24. Handbook of Polymer Synthesis, Parts A and B, edited by Hans R. Kricheldorf
25. Computational Modeling of Polymers, edited by Jozef Bicerano
26. Plastics Technology Handbook: Second Edition, Revised and Expanded, Manas Chanda and Salil K. Roy
27. Prediction of Polymer Properties, Jozef Bicerano
28. Ferroelectric Polymers: Chemistry, Physics, and Applications, edited by Hari Singh Nalwa
29. Degradable Polymers, Recycling, and Plastics Waste Management, edited by Ann-Christine Albertsson and Samuel J. Huang
30. Polymer Toughening, edited by Charles B. Arends
31. Handbook of Applied Polymer Processing Technology, edited by Nicholas P. Cheremisinoff and Paul N. Cheremisinoff
32. Diffusion in Polymers, edited by P. Neogi
33. Polymer Devolatilization, edited by Ramon J. Albalak
34. Anionic Polymerization: Principles and Practical Applications, Henry L. Hsieh and Roderic P. Quirk
35. Cationic Polymerizations: Mechanisms, Synthesis, and Applications, edited by Krzysztof Matyjaszewski
36. Polyimides: Fundamentals and Applications, edited by Malay K. Ghosh and K. L. Mittal
37. Thermoplastic Melt Rheology and Processing, A. V. Shenoy and D. R. Saini
38. Prediction of Polymer Properties: Second Edition, Revised and Expanded, Jozef Bicerano
39. Practical Thermoforming: Principles and Applications, Second Edition, Revised and Expanded, John Florian
40. Macromolecular Design of Polymeric Materials, edited by Koichi Hatada, Tatsuki Kitayama, and Otto Vogl
41. Handbook of Thermoplastics, edited by Olagoke Olabisi
42. Selecting Thermoplastics for Engineering Applications: Second Edition, Revised and Expanded, Charles P. MacDermott and Aroon V. Shenoy
43. Metallized Plastics: Fundamentals and Applications, edited by K. L. Mittal
44. Oligomer Technology and Applications, Constantin V. Uglea
45. Electrical and Optical Polymer Systems: Fundamentals, Methods, and Applications, edited by Donald L. Wise, Gary E. Wnek, Debra J. Trantolo, Thomas M. Cooper, and Joseph D. Gresser
46. Structure and Properties of Multiphase Polymeric Materials, edited by Takeo Araki, Qui Tran-Cong, and Mitsuhiro Shibayama
47. Plastics Technology Handbook: Third Edition, Revised and Expanded, Manas Chanda and Salil K. Roy
48. Handbook of Radical Vinyl Polymerization, Munmaya K. Mishra and Yusuf Yagci
49. Photonic Polymer Systems: Fundamentals, Methods, and Applications, edited by Donald L. Wise, Gary E. Wnek, Debra J. Trantolo, Thomas M. Cooper, and Joseph D. Gresser
50. Handbook of Polymer Testing: Physical Methods, edited by Roger Brown

51. Handbook of Polypropylene and Polypropylene Composites, *edited by Harutun G. Karian*
52. Polymer Blends and Alloys, *edited by Gabriel O. Shonaike and George P. Simon*
53. Star and Hyperbranched Polymers, *edited by Munmaya K. Mishra and Shiro Kobayashi*
54. Practical Extrusion Blow Molding, *edited by Samuel L. Belcher*
55. Polymer Viscoelasticity: Stress and Strain in Practice, *Evaristo Riande, Ricardo Díaz-Calleja, Margarita G. Prolongo, Rosa M. Masegosa, and Catalina Salom*
56. Handbook of Polycarbonate Science and Technology, *edited by Donald G. LeGrand and John T. Bendler*
57. Handbook of Polyethylene: Structures, Properties, and Applications, *Andrew J. Peacock*

Additional Volumes in Preparation

Handbook of Polyolefins, Second Edition, Revised and Expanded, *edited by Cornelia Vasile*

Supramolecular Polymers, *edited by Alberto Ciferri*

Polymer and Composite Rheology, Second Edition, Revised and Expanded, *edited by Rakesh K. Gupta*

Handbook of Elastomers, Second Edition, Revised and Expanded, *edited by Anil K. Bhowmick and Howard L. Stephens*

HANDBOOK OF POLYETHYLENE
Structures, Properties, and Applications

ANDREW J. PEACOCK
Exxon Chemical Company
Baytown, Texas

MARCEL DEKKER, INC.   NEW YORK · BASEL

To Shavon

ISBN: 0-8247-9546-6

This book is printed on acid-free paper.

Headquarters
Marcel Dekker, Inc.
270 Madison Avenue, New York, NY 10016
tel: 212-696-9000; fax: 212-685-4540

Eastern Hemisphere Distribution
Marcel Dekker AG
Hutgasse 4, Postfach 812, CH-4001 Basel, Switzerland
tel: 44-61-261-8482; fax: 44-61-261-8896

World Wide Web
http://www.dekker.com

The publisher offers discounts on this book when ordered in bulk quantities. For more information, write to Special Sales/Professional Marketing at the headquarters address above.

Copyright © 2000 by Marcel Dekker, Inc. All Rights Reserved.

Neither this book nor any part may be reproduced or transmitted in any form or by any means, electronic or mechanical, including photocopying, microfilming, and recording, or by any information storage and retrieval system, without permission in writing from the publisher.

Current printing (last digit):
10 9 8 7 6 5 4 3 2 1

PRINTED IN THE UNITED STATES OF AMERICA

Preface

The aim of this book is to provide a comprehensive introduction to the field of polyethylene in all its aspects as it applies to production, properties, and applications. Specifically, it correlates molecular structure with morphological features and thus with properties and end-use applications. Starting from a molecular description of the principal variants of polyethylene, it constructs a unified picture of polyethylene's melt structure and solid-state morphology and explains how this relates to processing variables and end-use applications.

An introductory chapter acquaints the reader with the field of polyethylene and provides an outline of polyethylene's molecular structure, morphology, properties, markets, and uses. Subsequently, the body of the book enlarges upon these themes. A chapter devoted to the history of polyethylene describes the development of the field from 1933 to the present day. Market development is explained in terms of the innovations that permitted molecular tailoring and expansion into new applications. Current catalysis and production processes are surveyed to explain the formation of the molecular features that distinguish the different types of polyethylene. The relationship between molecular structure and end-use properties begins with an examination of polyethylene's semicrystalline morphology and how this is formed from the molten state during crystallization. A complete range of physical attributes is discussed, encompassing solid-state mechanical, chemical, thermal, optical, and electrical characteristics and melt rheological properties. Methods of characterizing molecular characteristics and physical properties are described in the context of end-use applications. Chemical degradation, oxidation, and stabilization are described, as well as the deliberate chemical modification of surfaces. The molecular processes active during deformation are described in order to explain the properties of oriented structures, including high-modulus fibers and billets. The commercial processing techniques used to convert raw polyethylene to products are discussed, with emphasis on properties and end-use applications. The markets of polyethylene are broken down by use and molecular type. Finally, emerging trends in polyethylene production and usage are described to indicate the future trends of the industry.

The intended audience of this book includes chemists, engineers, physicists, and supervisory personnel who wish to expand their knowledge of the field of polyethylene. It would also serve as an introduction for graduate students or others considering a career in polymers. In order to reach as wide an audience as possible, no prior knowledge of the field of polymers is assumed. All relevant terms and background are explained prior to detailed discussion.

This book could not have been written without the help, cooperation, and encouragement of many people. I am indebted to various colleagues who read parts or all of the manuscript during its preparation, and who offered many critical and useful observations. Professor Leo Mandelkern was most helpful with the chapters dealing with morphology, crystallization, and properties. Gary Brown reviewed several chapters and offered suggestions, especially with regard to microscopic analysis. In particular I must express my utmost gratitude to Dr. Ferdinand Stehling, a retired colleague, who spent much time and energy reviewing the entire work during its preparation. Ferd's insight and encouragement were invaluable and added immeasurably to the quality of the book as a whole. Last, but not least, I must thank my wife, Shavon, who for more than half of our married life has had to tolerate my spending evenings and weekends closeted with books, papers, and a computer.

Andrew J. Peacock

Contents

Preface   *v*

1. Introduction   1
2. Commercial Development of Polyethylene   27
3. Production Processes   43
4. Morphology and Crystallization of Polyethylene   67
5. Properties of Polyethylene   123
6. Characterization and Testing   241
7. The Chemistry of Polyethylene   375
8. Orientation of Polyethylene   415
9. Use and Fabrication of Polyethylene Products   459
10. The Future of Polyethylene   509

Index   *523*

1
Introduction

I. THE ESSENCE OF POLYETHYLENE

A. Molecular Structure

In its simplest form a polyethylene molecule consists of a long backbone of an even number of covalently linked carbon atoms with a pair of hydrogen atoms attached to each carbon; chain ends are terminated by methyl groups. This structure is shown schematically in Figure 1.

Chemically pure polyethylene resins consist of alkanes with the formula $C_{2n}H_{4n+2}$, where n is the degree of polymerization, i.e., the number of ethylene monomers polymerized to form the chain. Unlike conventional organic materials, polyethylene does not consist of identical molecules. Polyethylene resins comprise chains with a range of backbone lengths. Typically the degree of polymerization is well in excess of 100 and can be as high as 250,000 or more, equating to molecular weights varying from 1400 to more than 3,500,000. Low molecular weight polyethylenes (oligomers) with a degree of polymerization between 8 and 100 are waxy solids that do not possess the properties generally associated with a plastic. When the degree of polymerization is less than 8, alkanes are gases or liquids at ordinary temperatures and pressures. Polyethylene molecules can be branched to various degrees and contain small amounts of unsaturation.

1. Variations on a Theme

Many types of polyethylene exist, all having essentially the same backbone of covalently linked carbon atoms with pendant hydrogens; variations arise chiefly from branches that modify the nature of the material. There are many types of branches, ranging from simple alkyl groups to acid and ester functionalities. To a lesser extent, variations arise from defects in the polymer backbone; these consist principally of vinyl groups, which are often associated with chain ends. In the solid state, branches and other defects in the regular chain structure limit a sample's crystallinity level. Chains that have few defects have a higher degree of

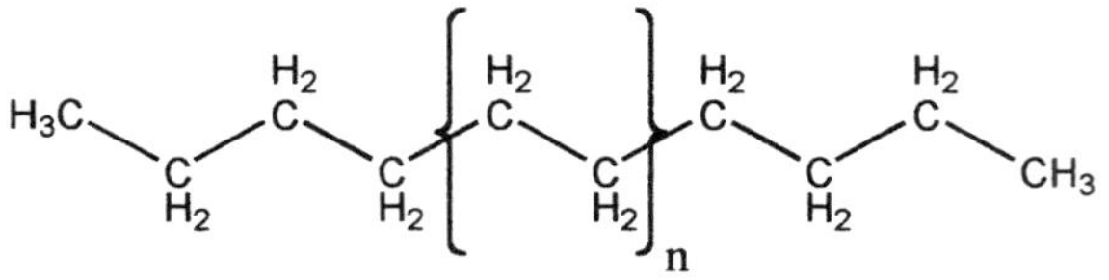

Figure 1 Chemical structure of pure polyethylene.

crystallinity than those that have many. As the packing of crystalline regions is better than that of noncrystalline regions, the overall density of a polyethylene resin will increase as the degree of crystallinity rises. Generally, the higher the concentration of branches, the lower the density of the solid. The principal classes of polyethylene are illustrated schematically in Figure 2.

a. High Density Polyethylene. High density polyethylene (HDPE) is chemically the closest in structure to pure polyethylene. It consists primarily of unbranched molecules with very few flaws to mar its linearity. The general form of high density polyethylene is shown in Figure 2a. With an extremely low level of defects to hinder organization, a high degree of crystallinity can be achieved, resulting in resins that have a high density (relative to other types of polyethylene). Some resins of this type are copolymerized with a very small concentration of 1-alkenes in order to reduce the crystallinity level slightly. High density polyethylene resins typically have densities falling in the range of approximately 0.94–0.97 g/cm^3. Due to its very low level of branching, high density polyethylene is sometimes referred to as linear polyethylene (LPE).

b. Low Density Polyethylene. Low density polyethylene (LDPE) is so named because such polymers contain substantial concentrations of branches that hinder the crystallization process, resulting in relatively low densities. The branches primarily consist of ethyl and butyl groups together with some long-chain branches. A simplified representation of the structure of low density polyethylene is shown in Figure 2b. Due to the nature of the high pressure polymerization process by which low density polyethylene is produced, the ethyl and butyl branches are frequently clustered together, separated by lengthy runs of unbranched backbone. Long-chain branches occur at random intervals along the length of the main chain. The long-chain branches can themselves in turn be branched. The mechanisms involved in the production of branches are discussed in Chapter 3. The numerous branches characteristic of low density polyethylene molecules inhibit their ability to crystallize, reducing resin density relative to high density polyethylene. Low density polyethylene resins typically have densities falling in the range of approximately 0.90–0.94 g/cm^3.

c. Linear Low Density Polyethylene. Linear low density polyethylene (LLDPE) resins consist of molecules with linear polyethylene backbones to which are attached short alkyl groups at random intervals. These materials are produced by the copolymerization of ethylene with 1-alkenes. The general structure of linear low density polyethylene resins is shown schematically in Figure 2c. The branches most commonly encountered are ethyl, butyl, or hexyl groups but can be a variety of other alkyl groups, both linear and branched. A typical average separation of branches along the main chain is 25–100 carbon atoms. Linear low density polyethylene resins may also contain small levels of long-chain branching, but there is not the same degree of branching complexity as is found in low density polyethylene. Chemically these resins can be thought of as a compromise between linear polyethylene and low density polyethylene, hence the name. The branches hinder crystallization to some extent, reducing density relative to high density polyethylene. The result is a density range of approximately 0.90–0.94 g/cm^3.

d. Very Low Density Polyethylene. Very low density polyethylene (VLDPE)—also known as ultralow density polyethylene (ULDPE)—is a specialized form of linear low density polyethylene that has a much higher concentration of short-chain branches. The general structure of very low density polyethylene is shown in Figure 2d. A typical separation of branches would fall in the range of 7–25 backbone carbon atoms. The high level of branching inhibits crystallization very effectively, resulting in a material that is predominantly noncrystalline. The high levels of disorder are reflected in the very low densities, which fall in the range of 0.86–0.90 g/cm^3.

e. Ethylene-Vinyl Ester Copolymers. By far the most commonly encountered ethylene-vinyl ester copolymer is ethylene-vinyl acetate (EVA). These copolymers are made by the same high pressure process as low density polyethylene and therefore contain both short- and long-chain branches in addition to acetate groups. The general structure of ethylene-vinyl acetate resins is shown schematically in Figure 2e (in which "VA" indicates an acetate group). The acetate groups interact with one another via dispersive forces, tending to cluster. The inclusion of polar groups endows such copolymers with greater chemical reactivity than high density, low density, or linear low density polyethylene. The acetate branches hinder crystallization in proportion to their incorporation level; at low levels these copolymers have physical properties similar to those of low density polyethylene, but at high levels of incorporation they are elastomeric. Due to the incorporation of oxygen, ethylene-vinyl acetate copolymers exhibit higher densities at a given crystallinity level than polyethylene resins comprising only carbon and hydrogen.

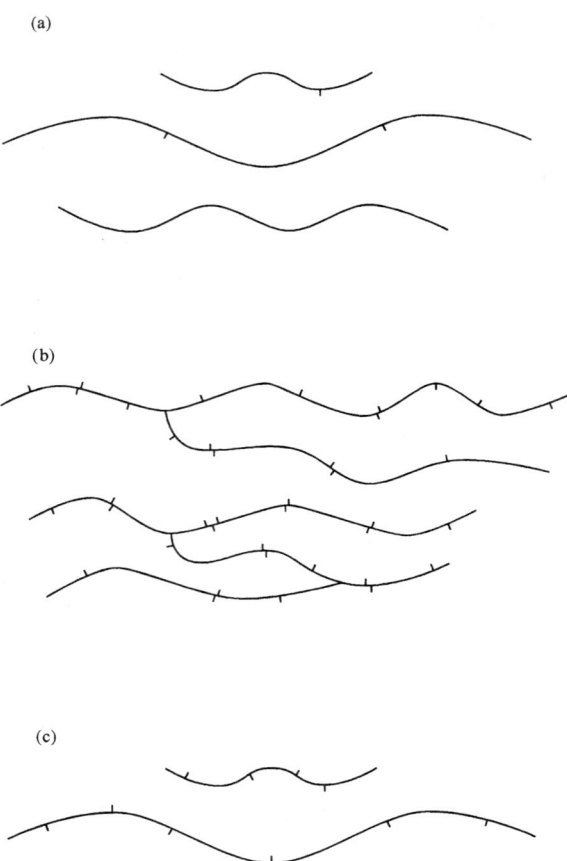

Figure 2 Schematic representations of the different classes of polyethylene. (a) High density polyethylene; (b) low density polyethylene; (c) linear low density polyethylene; (d) very low density polyethylene; (e) ethylene-vinyl acetate copolymer; (f) cross-linked polyethylene.

f. Ionomers. Ionomers are copolymers of ethylene and acrylic acids that have been neutralized (wholly or partially) to form metal salts. The copolymerization of these molecules takes place under conditions similar to those under which low density polyethylene is made; thus, in addition to polar groups, ionomers contain all the branches normally associated with low density polyethylene. The

(d)

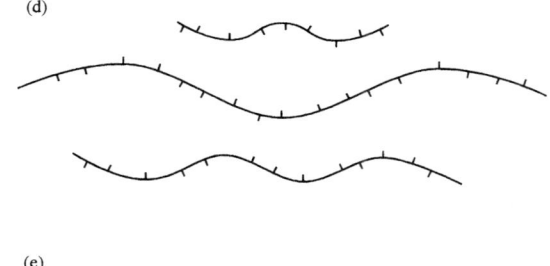

(e)

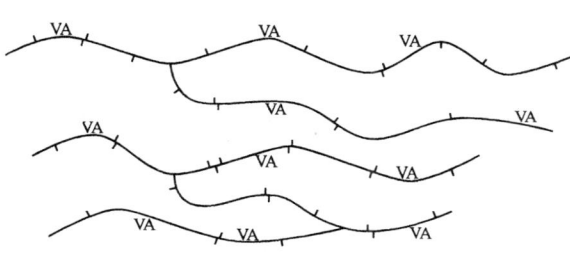

(f)

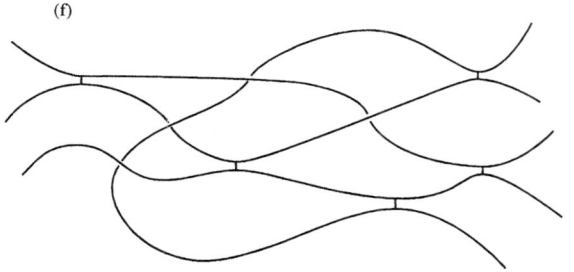

neutralized acid functionalities from adjacent chains interact with the associated metal cations to form clusters that bind neighboring chains together. A two-dimensional representation of an ionomer cluster is shown in Figure 3. The complex branching structure of ionomers and the existence of polar clusters drastically reduce their ability to crystallize. Despite their low levels of crystallinity, the density of ionomers is normally the highest of all polyethylenes due to the relatively high atomic weight of the oxygen and metal atoms in the ionic clusters.

    g. Cross-Linked Polyethylene. Cross-linked polyethylene (XLPE) consists of polyethylene that has been chemically modified to covalently link adjacent chains. A schematic representation of cross-linked polyethylene is shown

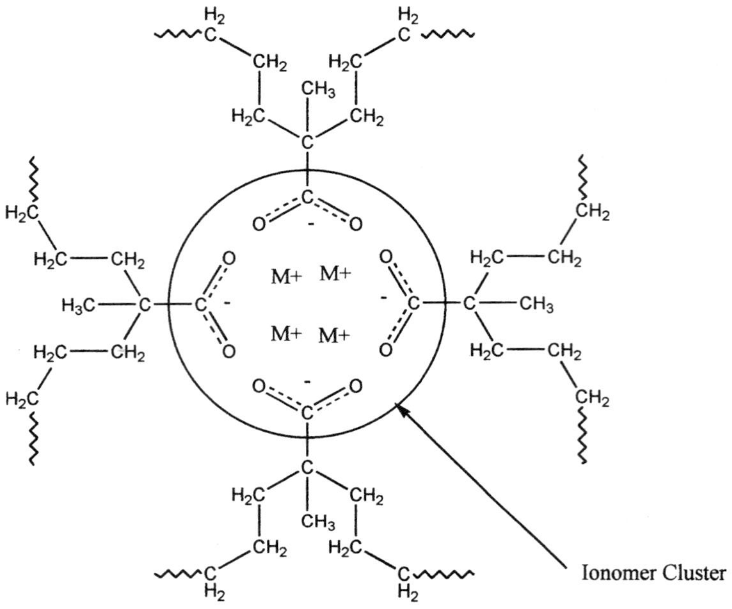

Figure 3 Schematic representation of an ionomer cluster.

in Figure 2f. Cross-links may comprise either direct carbon–carbon bonds or bridging species such as siloxanes. Cross-links occur at random intervals along chains; the concentration can vary widely, from an average of only one per several thousand carbon atoms to one per few dozen carbon atoms. The effect of cross-linking is to create a gel-like network of interconnected chains. The network is essentially insoluble, although it can be swollen by various organic solvents. This is in direct contrast to the non-cross-linked varieties of polyethylene that are soluble in appropriate solvents at high temperature. Cross-links greatly hinder crystallization, limiting the free movement of chains required to organize into crystallites. Thus the density of a cross-linked polyethylene is lower than that of the polyethylene resin on which it is based.

B. Molecular Composition

Polyethylene resins consist of molecules that exhibit a distribution of molecular lengths and branching characteristics. The characteristics of a polyethylene resin could be uniquely described if each of its component molecules were defined in terms of its exact backbone length and the type and placement of each branch. This cannot be achieved, because separative techniques are not adequate to divide any resin into its myriad constituent molecules, nor could the molecules be characterized with sufficient precision even if homogeneous fractions could be obtained. In practice one must settle for determining various average characteristics that are representative of the molecular weight and branching distribution.

The size of a polyethylene molecule is normally described in terms of its molecular weight. All polyethylene resins consist of a mixture of molecules with a range of molecular weights. The average molecular weight and the distribution of chain lengths comprising a polyethylene resin profoundly affect is properties. The molecular weights of molecules found in commercial resins may range from a few hundred up to 10 million.

1. Molecular Weight Distribution

The distribution of molecular sizes within a polyethylene resin can be described in terms of various molecular weight averages. The molecular weight averages are calculated as the moments of the distribution of molecular masses. The molecular weight distribution (MWD) of a polyethylene resin is normally plotted on a semilogarithmic scale, with the molecular weight on the abscissa and the fractional mass on the ordinate. Such a plot (derived from size elution chromatography) is shown in Figure 4, indicating various molecular weight averages. The molecular weight distribution may be (and often is) simplistically defined in terms of the ratio of two of the molecular weight averages. The breadth and shape of the molecular weight distribution curve can vary greatly; distribution plots can exhibit multiple peaks, shoulders, and tails. Molecular weight characteristics have a profound effect on the physical properties of polyethylene resins, affecting such properties as viscosity, environmental stress cracking, and impact strength. The relationship between properties and molecular weight distribution is discussed in Chapter 5.

a. Number-Average Molecular Weight. The number-average molecular weight ($\overline{M}_n$) of a polyethylene resin is defined in terms of the number of molecules and molecular weight of the chains making up a series of fractions that account for the molecular weight distribution. Thus, a molecular weight distribution plot is divided into 50 or more fractions, the characteristics of which are used to calculate the number-average molecular weight.

The number-average molecular weight is calculated according to

$$\overline{M}_n = \frac{\Sigma M_i N_i}{\Sigma N_i} = \frac{\Sigma W_i}{\Sigma N_i}$$

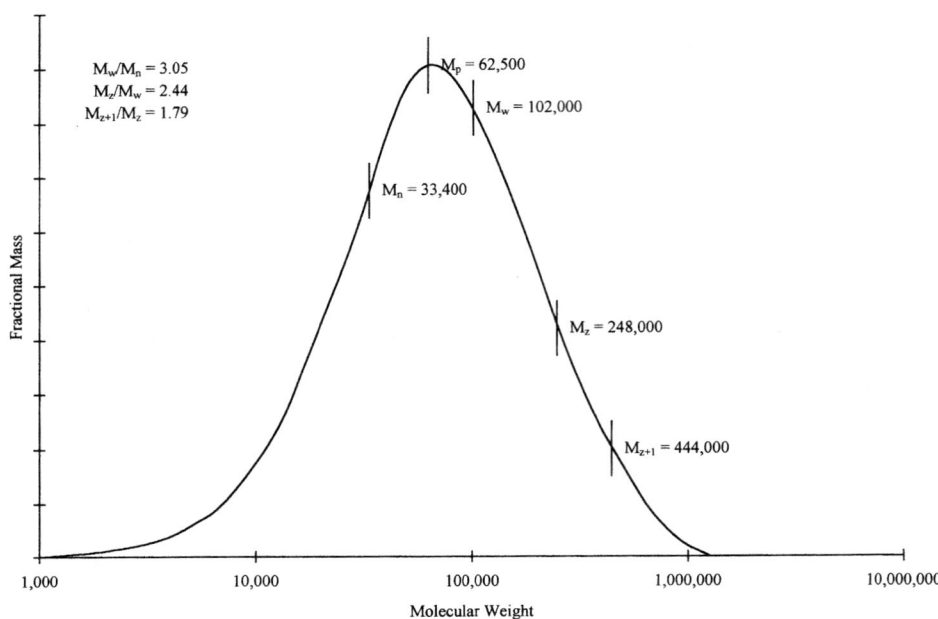

Figure 4 Typical molecular weight distribution plot of polyethylene.

where:

M_i = molecular weight of chains in fraction i
N_i = number of chains in fraction i
W_i = weight of chains in fraction i

The number-average molecular weight is a function of all the molecular weight species present, but it is most sensitive to the lower molecular weight fractions, which generally contain the largest numbers of molecules. Thus a low molecular weight tail will reduce the number-average molecular weight to a much greater extent than a high molecular weight tail will increase it.

b. *Weight-Average Molecular Weight.* The weight-average molecular weight ($\overline{M}_w$) is calculated from the same parameters used to calculate the number-average molecular weight, but a greater emphasis is placed on the higher molecular weight species.

The weight average molecular weight is calculated according to

$$\overline{M}_w = \frac{\sum M_i^2 N_i}{\sum M_i N_i} = \frac{\sum M_i W_i}{\sum W_i}$$

For a typical polyethylene resin, the weight-average molecular weight is particularly sensitive to the central portion of the molecular weight distribution, where the mass of the fractions is greatest. High and low molecular weight tails on the molecular weight distribution generally have only a small effect on the weight-average molecular weight.

   c. z-Average Molecular Weight. The z-average molecular weight ($\overline{M}_z$) is calculated in a similar manner to weight-average molecular weight, with even greater emphasis placed on the role of the higher molecular weight species.

The z-average molecular weight is calculated according to

$$\overline{M}_z = \frac{\Sigma M_i^3 N_i}{\Sigma M_i^2 N_i} = \frac{\Sigma M_i^2 W_i}{\Sigma W_i}$$

The z-average molecular weight is sensitive to the higher molecular weight species in a polyethylene resin. Changes in the central portion of the molecular weight distribution have a minor effect on the z-average molecular weight, and changes in low molecular weight tails are generally inconsequential. On the face of it, this molecular weight average may appear to be a rather strange way of characterizing a polyethylene resin, but there are many properties that are related to it, such as melt elasticity and shear thinning behavior.

   d. (z + 1)-Average Molecular Weight. Following the trend of the weight- and z-average molecular weights, the $(z + 1)$-average molecular weight ($\overline{M}_{z+1}$) is extremely sensitive to the highest molecular weight fractions.

The $(z + 1)$-average molecular weight is calculated according to

$$\overline{M}_{z+1} = \frac{\Sigma M_i^4 N_i}{\Sigma M_i^3 N_i} = \frac{\Sigma M_i^3 W_i}{\Sigma W_i^2}$$

The $(z + 1)$-average molecular weight is not routinely quoted when describing a polyethylene resin's molecular weight distribution. Its greatest use is when a resin contains an extended tail of high molecular weight material.

   e. Peak Molecular Weight. The peak molecular weight (M_p) is simply the molecular weight at the maximum of a conventional molecular weight distribution plot. For a normally distributed molecular weight distribution curve, the molecular weight of the peak falls between the number- and weight-average molecular weight values.

   f. Viscosity-Average Molecular Weight. The viscosity-average molecular weight ($\overline{M}_v$) depends upon the complete molecular weight distribution of a resin. For a normally distributed resin it falls between the number- and weight-average molecular weights. It can be precisely measured from the viscosities of a series of very dilute polymer solutions. More commonly it is estimated from the molecular weight distribution obtained from size exclusion chromatography.

g. Breadth of Molecular Weight Distribution. The value most frequently used to describe the breadth of a polyethylene resin's molecular weight distribution is the ratio of its weight- to number-average molecular weights ($\overline{M}_w/\overline{M}_n$). The $\overline{M}_w/\overline{M}_n$ ratio is often imprecisely referred to as the "molecular weight distribution" or the dispersity (Q). However, $\overline{M}_w/\overline{M}_n$ is not a unique identifier of a molecular weight distribution; it is possible to envisage an infinite number of molecular weight distributions that would exhibit a given $\overline{M}_w/\overline{M}_n$ ratio. Values of $\overline{M}_w/\overline{M}_n$ for commercial resins can vary from 2.0 to 25 or more. When used in conjunction with the molecular weight averages, the breadth of distribution can be used to predict various resin properties in both the solid and molten states.

Other measures of the breadth of a molecular weight distribution include the ratio of the *z*- to weight-average molecular weights ($\overline{M}_z/\overline{M}_w$) and that of the (*z* + 1)- to weight-average molecular weights ($\overline{M}_{z+1}/\overline{M}_w$). These values can give an indication of the skewness of a distribution when compared to $\overline{M}_w/\overline{M}_n$. The larger the value of $\overline{M}_z/\overline{M}_w$ in comparison to $\overline{M}_w/\overline{M}_n$, the more pronounced is the high molecular weight tail.

2. Composition Distribution

The term "composition distribution" (CD) refers to the distribution of branches among the molecules that comprise a polyethylene resin. It is principally used when discussing the characteristics of linear low density polyethylene. As comonomers are incorporated by mechanisms that are to a greater or lesser extent statistically random, the concentration of branches will vary along the length of a molecule and from molecule to molecule. Due to the nature of the polymerization process it is frequently the case that the average concentration of branches on a molecule is related to its molecular weight. Often it is found that those molecules making up the higher molecular weight fractions also display the lowest levels of branching. It is possible to represent the overall molecular composition of a resin as a three-dimensional plot in which weight fraction is plotted as a function of average concentration of branches and molecular weight. Such a plot is illustrated in Figure 5.

C. Morphology

The term "morphology" is used to describe the organization of polyethylene molecules in the solid or molten state. A complete structural description of the morphology of a polyethylene sample should include terms defining the levels of ordering on all scales, ranging from angstroms up to millimeters. In its solid state, polyethylene exists in a semicrystalline morphology; that is, the material

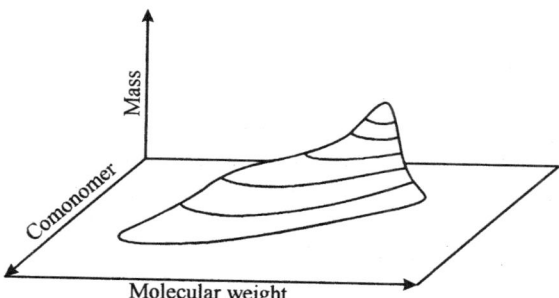

Figure 5 Molecular composition plotted as fractional mass as a function of average branch concentration and molecular weight.

contains some regions that exhibit short-range order normally associated with crystals, interspersed with regions having little or no short-range order. A generic semicrystalline structure is illustrated schematically in Figure 6. The morphology of polyethylene is discussed in depth in Chapter 4; in this introduction only a brief outline of the most important states of order is given.

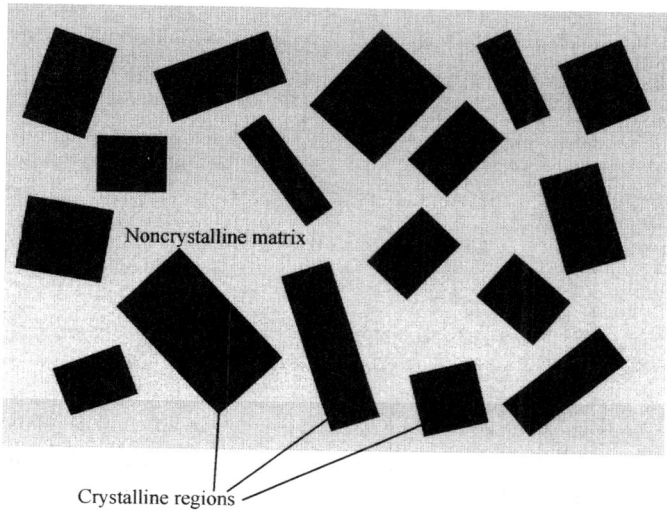

Figure 6 Generic illustration of semicrystalline morphology.

1. Noncrystalline Structure

When a freely jointed molecular chain is allowed to equilibrate with no external forces acting upon it, it will adopt a configuration known as a random coil. In this state the molecule possesses maximum entropy. A polymer random coil can be envisaged if the molecular chain is built up one monomer at a time, the angle between successive monomers being chosen arbitrarily. Thus the backbone describes a random trajectory in three dimensions. In practice, steric hindrance and the requirement that no two chain segments occupy the same space limit the available configurations.

Polyethylene chains adopt a random coil configuration when allowed to equilibrate in the molten state or when dissolved in an ideal solvent. In the molten state, and to a lesser extent in solution, the random coils of adjacent molecules overlap, resulting in various degrees of chain entanglement, depending primarily on chain length and concentration in solution. Molten polyethylene and polyethylene solutions have much higher viscosities than conventional low molecular weight organic materials, primarily due to the entanglements between chains.

When molten polyethylene solidifies, the chains in some regions become organized into small crystals known as crystallites. Disordered chains surround the crystallites; this is the essence of semicrystallinity. A typical polyethylene molecule has a length many times the average dimensions of the crystalline and noncrystalline phases; as such, various parts of it can be incorporated into different crystallites, linking them together via intervening disordered segments. The disordered molecular segments do not correspond to short lengths of random coil because of constraints placed upon them by connections to crystallites. Thus, the noncrystalline regions cannot be described as truly random, because some degree of preferential alignment is inevitably present. In addition, chain segments in the noncrystalline regions of a sample can be preferentially aligned by deformation associated with preparation procedures. In this volume the term ''amorphous'' is reserved for regions with no discernible ordering (such as the equilibrated molten state); regions between crystallites are referred to as ''noncrystalline'' or ''disordered.''

2. Crystal Unit Cell

When polyethylene is cooled from the melt, certain portions of it crystallize. The building block of crystalline structures is the unit cell, which is the smallest arrangement of chain segments that can be repeated in three dimensions to form a crystalline matrix. Thus the unit cell contains all the crystallographic data pertinent to the complete crystallite. The chain segments in a crystal are extended to their maximum length, the backbone taking up a configuration referred to as a ''planar zigzag.'' Under all but the most exceptional circumstances polyethylene chains pack to form orthorhombic crystals. The orthorhombic crystal structure of polyethylene is shown from two viewing angles in Figure 7. The orthorhombic

Introduction

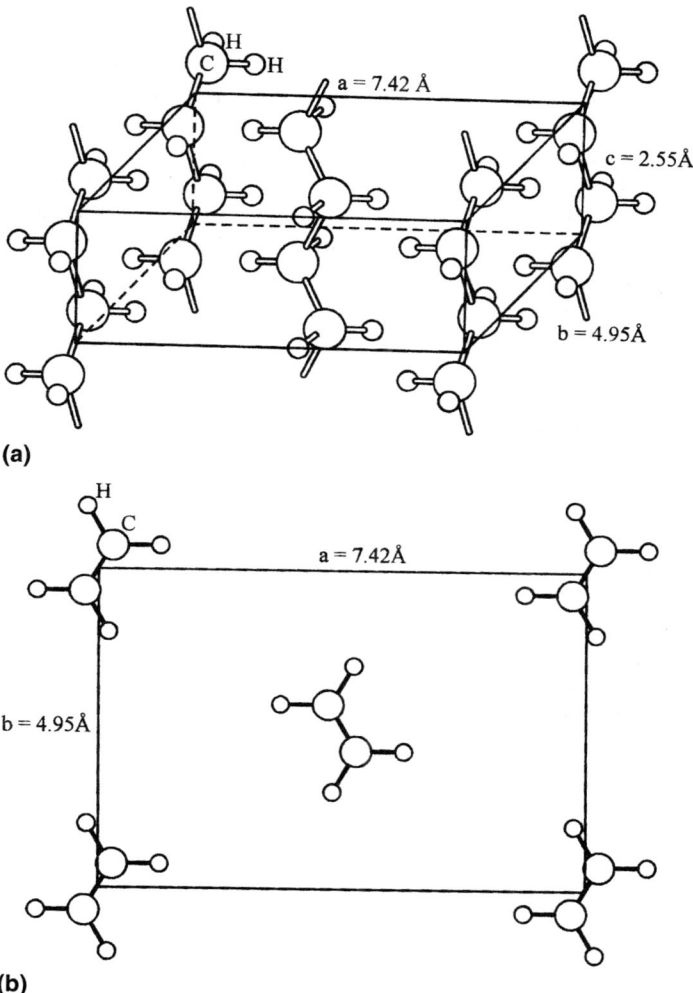

Figure 7 Polyethylene orthorhombic crystal habit. (a) Orthogonal view; (b) view along the c axis.

packing habit is characterized by unit cells whose faces make angles of 90° to one another, with the lengths of the a, b, and c axes being unequal.

As can be seen from Figure 7b, each polyethylene unit cell consists of one complete ethylene unit and parts of four others, for a total of two per unit cell. When a series of unit cells are packed together in a three-dimensional array, a crystal is formed.

3. Crystallite Structure

When polyethylene crystallizes, it does so only to a limited extent therefore the crystals are of finite size. The small crystals that make up the crystalline regions of solid polyethylene are known as crystallites. The most common crystal growth habit of polyethylene is such that a crystallite's a and b dimensions are much greater than its c dimension. Such crystallites, with two dimensions being very much greater than the third, are termed "lamellae." An idealized representation of a lamella is shown in Figure 8. Polyethylene lamellae are typically from 50 to 200 Å thick. Their lateral dimensions can vary over several orders of magnitude, from a few hundred angstroms up to several millimeters for crystals grown from solution. Lamellae can adopt a variety of formats, including curved, fragmented, and bifurcating. The chain axes of molecular segments making up the lamellae are rarely normal to the basal plane of the crystal; chains can exhibit tilt angles of up to 30° from the perpendicular.

4. Spherulite Structure

Semicrystalline polyethylene is made up of crystallites, between which are found disordered regions. The most common large-scale structures composed of crystalline and noncrystalline regions are called "spherulites." Spherulites are so named because their growth habit is approximately spherical, lamellae growing outward radially from nucleation sites. A schematic representation of a spherulite is shown in Figure 9. As spherulites grow they impinge on one another to form irregular polyhedrons. The bundles of lamellae making up a spherulite are arranged in such a way that their b axes (the direction in which growth occurs) are preferentially aligned with the radii of the spherulite. The lamellae comprising spherulites often twist and bifurcate.

Depending upon the concentration of nucleation sites, spherulites can vary in size from a few nanometers up to several millimeters across. Because they are

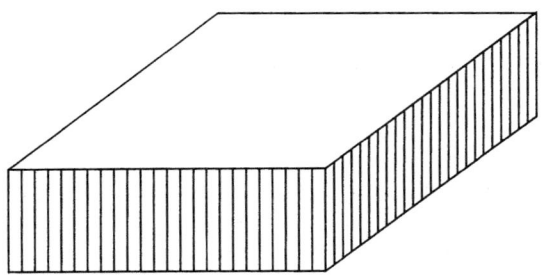

Figure 8 Idealized representation of a polyethylene lamella.

Introduction

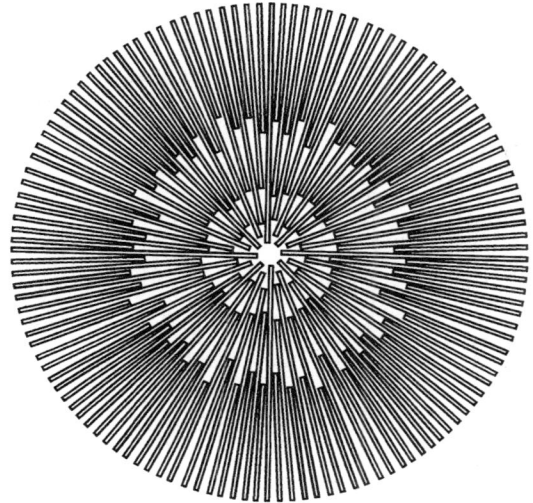

Figure 9 Schematic representation of a spherulite.

composed of lamellae arranged parallel to their radii, spherulites exhibit anisotropy; that is, the properties of individual sections vary as a function of testing direction. The size and perfection of spherulites influence certain physical properties.

II. POLYETHYLENE ATTRIBUTES

A. Intrinsic Properties

The various types of polyethylene exhibit a wide range of properties, the specific attributes depending on the molecular and morphological characteristics of the polyethylene resin. Each variant of polyethylene has its own characteristics, and within each type there is a spectrum of properties. There is much overlap between the ranges of properties available for the different variants of polyethylene. The relationships linking molecular structure and physical properties are discussed in Chapter 5.

A numerical comparison of the different types of polyethylene, highlighting the typical ranges of some key solid-state properties, is presented in Table 1. Figures 10–14 illustrate some of these data graphically. None of these data should be considered absolute; specific preparation conditions and testing configurations,

Table 1 Principal Properties of Different Types of Polyethylene

Property	HDPE	LDPE	LLDPE	VLDPE	EVA	Ionomer
Density (g/cm³)	0.94–0.97	0.91–0.94	0.90–0.94	0.86–0.90	0.92–0.94	0.93–0.96
Degree of crystallinity (% from density)	62–82	42–62	34–62	4–34	—	—
Degree of crystallinity (% from calorimetry)	55–77	30–54	22–55	0–22	10–50	20–45
Flexural modulus (psi @ 73°F)	145,000–225,000	35,000–48,000	40,000–160,000	<40,000	10,000–40,000	3,000–55,000
Tensile modulus (psi)	155,000–200,000	25,000–50,000	38,000–130,000	<38,000	7,000–29,000	<60,000
Tensile yield stress (psi)	2,600–4,500	1,300–2,800	1,100–2,800	<1,100	5,000–2,400	—
Tensile strength at break (psi)	3,200–4,500	1,200–4,500	1,900–6,500	2,500–5,000	2,200–4,000	2,500–5,400
Tensile elongation at break (%)	10–1,500	100–650	100–950	100–600	200–750	300–700
Shore hardness Type D	66–73	44–50	55–70	25–55	27–38	25–66
Izod impact strength (ft-lb/in. of notch)	0.4–4.0	No break	0.35–No break	No break	No break	7.0–No break
Melting temperature (°C)	125–132	98–115	100–125	60–100	103–110	81–96
Heat distortion temperature (°C @ 66 psi)	80–90	40–44	55–80	—	—	113–125
Heat of fusion (cal/g)	38–53	21–37	15–43	0–15	7–35	14–31
Thermal expansivity (10^{-6} in/in/°C)	60–110	100–220	70–150	150–270	160–200	100–170

Introduction

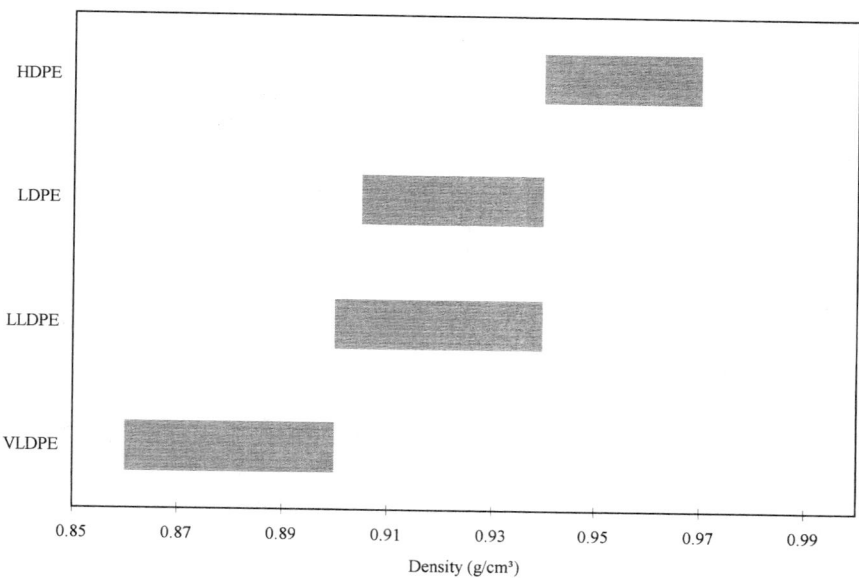

Figure 10 Typical density ranges of various classes of polyethylene.

particularly with respect to oriented specimens, can result in samples whose properties fall outside the ranges indicated.

The following subsections describe some of the characteristics of the various types of polyethylene that are directly manifest to the human senses.

1. High Density Polyethylene

Molded parts made from high density polyethylene are opaque white materials. To the touch they feel slightly waxy. Unless there has been thermal degradation during molding, high density polyethylene has no discernible taste or smell. High density polyethylene is the stiffest of all polyethylenes; a 1/8 in. thick molded plaque can be flexed slightly by hand. Aggressive manipulation can produce permanent deformation, with some whitening in the bend region. Thin films have a distinctive crisp sound when handled and readily take on permanent creases. When stretched, films deform substantially by necking, certain portions deforming more than others, becoming white in the process. Once punctured, thin films of high density polyethylene tear readily.

2. Low Density Polyethylene

Items molded from low density polyethylene are generally translucent; at thicknesses up to 1/8 in., newsprint laid directly in contact is readable through the

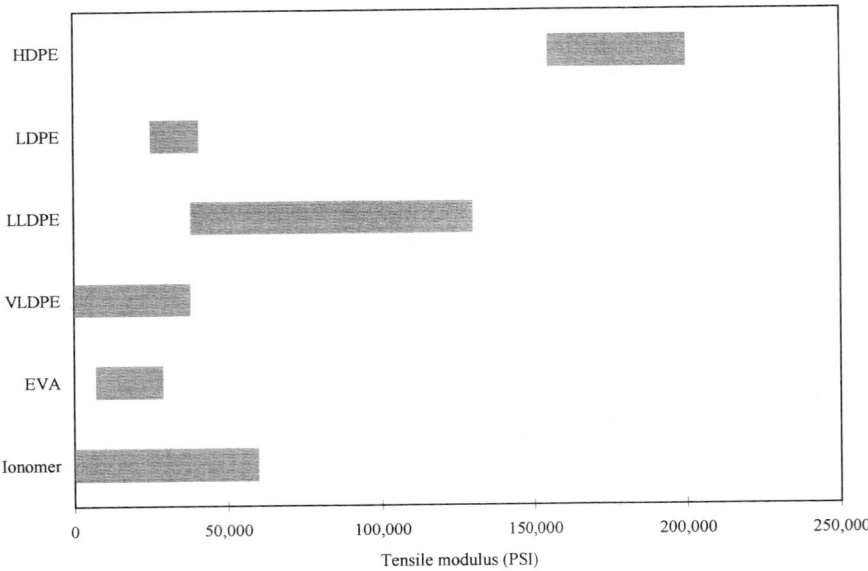

Figure 11 Typical tensile moduli of various classes of polyethylene.

low density polyethylene (LDP). They feel somewhat waxy, and there may be a trace of surface bloom. Low density polyethylene is quite pliable; it is readily flexed by hand at thicknesses up to 1/8 in. Samples show much resilience, rarely taking on a permanent set unless deformed substantially. In common with most other polyethylene resins, they have no taste or odor unless chemically altered by degradation or some other process. Thin films of low density polyethylene deform uniformly when stretched, with little if any whitening in the strained regions. They show substantial deformation before the onset of tearing, which does not proceed readily.

3. Linear Low Density Polyethylene

Items molded from linear low density polyethylene resins are generally somewhat hazy white materials. Surfaces feel slightly waxy and have little if any surface bloom. They exhibit no discernible taste or odor. Depending on the comonomer content, they can vary from being quite pliable to being stiff materials that flex only slightly before a permanent set is achieved. The maximum stiffness exhibited is only slightly less than that of the softest high density polyethylene samples. Thin films of linear low density polyethylene appear quite clear. Films are highly resistant to being punctured or torn. Film deformation proceeds by necking, the deformed region becoming hazy.

Introduction

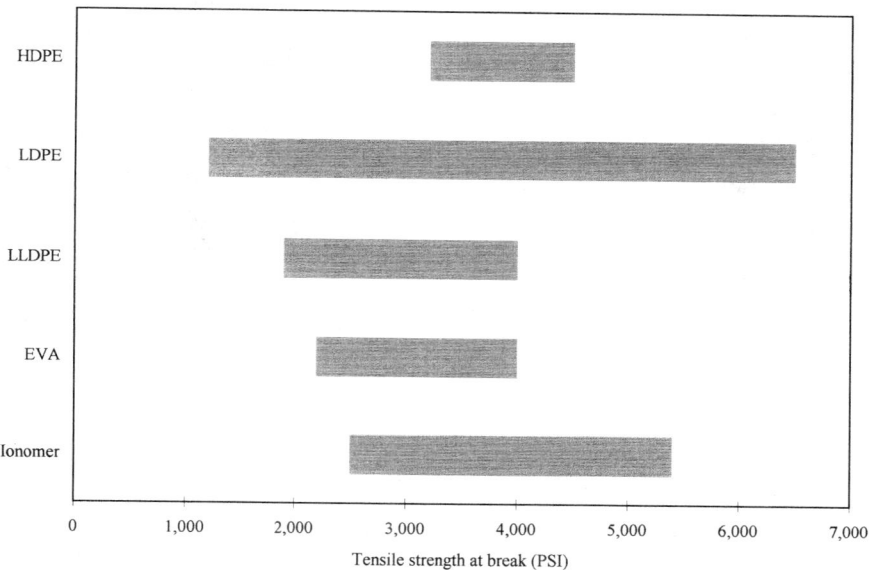

Figure 12 Typical tensile strengths of various classes of polyethylene.

4. Very Low Density Polyethylene

Very low density polyethylene is seldom molded into thick parts. Films are very soft and flexible and are readily deformed. Surfaces often have a somewhat tacky feel and exhibit a slight surface bloom. They should not have any taste or odor. Films are resilient, much of the deformation being recoverable if strain does not exceed 100%. Films are not readily torn or punctured. Very low density polyethylene is quite clear, with haze being negligible in thin films.

5. Ethylene-Vinyl Acetate Copolymer

Ethylene-vinyl acetate copolymers vary in stiffness depending upon the level of comonomer incorporation. At their stiffest they are comparable to low density polyethylene. At the other end of the spectrum they are as flexible as very low density polyethylene.

6. Ionomers

Ionomers make very flexible films with a somewhat rubbery feel. Deformation is recoverable to a large extent even at extensions in excess of 100%. Ionomer

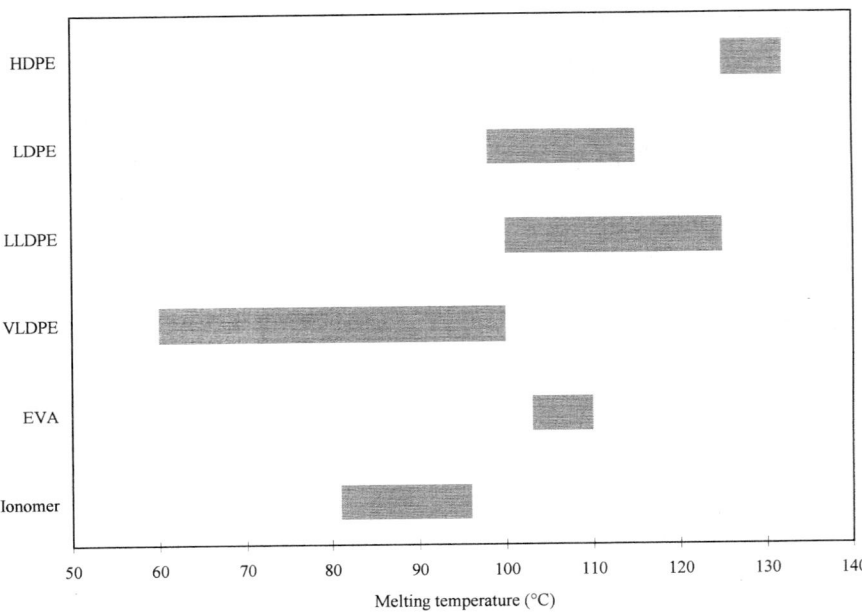

Figure 13 Typical melting temperatures of various classes of polyethylene.

films generally have negligible haze. Films are highly resistant to being punctured, cut, or torn. Certain types of ionomers can exhibit a noticeable taste and odor.

7. Cross-Linked Polyethylene

The properties of cross-linked polyethylene depend very much on the base resin and the degree of cross-linking. In general they exhibit properties smilar to those of the resin from which they are derived but may be somewhat more flexible. Certain types of chemical cross-linking impart a distinctive odor.

B. Comparative Properties

Polyethylene is used to fabricate many items that can be manufactured from a wide range of competing materials, both polymeric and nonpolymeric. Each raw material confers specific properties on the final article that may or may not be required for it to be functional. The choice of material is often very complex,

Introduction

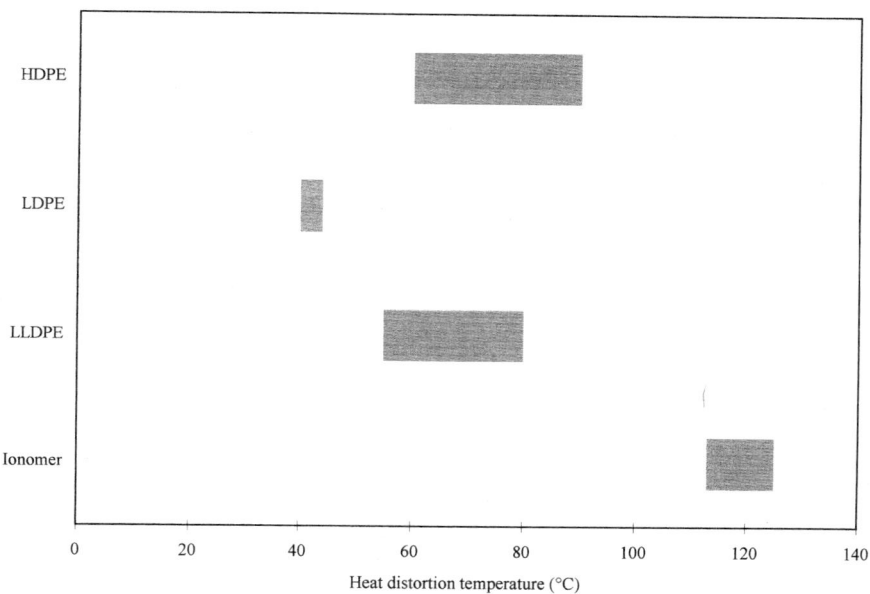

Figure 14 Typical heat distortion temperatures of various classes of polyethylene.

involving factors other than the required product attributes, such as material cost, ease of fabrication, and aesthetic appeal. Polyethylene has relatively modest physical attributes in comparison to many other materials, but its overall balance of properties may be the deciding factor in its favor. Polyethylene has few outstandingly good attributes, but it has few outstandingly bad ones either.

1. Polyethylene Versus Other Synthetic Polymers

Polymeric items can generally be placed into one of two categories: (1) nondurable applications that do not require that a product withstand large loads for extended periods of time and (2) durable applications, which often involve the transmission or support of considerable loads. An example of the first is fresh produce packaging film, in which a clear view of the contents is more desirable than great load-bearing capabilities. The second category is exemplified by such items as drums and pails, which are designed to withstand various forms of physical abuse. A distinction is drawn between commodity polymer resins, which are used in non-stress-critical applications, and engineering resins, which are capable of bearing loads.

a. In Comparison to Other Commodity Polymers. Commodity polymers are generally considered to be those thermoplastics (a thermoplastic resin is one that melts and flows when heated) that are reasonably cheap (less than ~$0.50/lb), are used in large quantities (tens of millions of pounds per year), and have relatively modest physical properties. Polyethylene falls into this category along with the likes of polypropylene, polystyrene, and poly(ethylene terephthalate). The properties of various competing resins are listed in Table 2.

b. In Comparison to Engineering Resins. The term "engineering resin" covers a wide range of materials that have properties that are particularly desirable from the point of view of structural engineering. Such properties may include high elastic modulus, low creep, and heat distortion temperatures in excess of 200°C. Such resins typically cost many times as much as polyethylene. Commodity and engineering resins do not compete directly with one another except in a very limited range of applications.

II. PRINCIPAL MARKETS AND USES

Polyethylene with its broad spectrum of physical properties is employed in a multitude of applications. The key to its adaptability lies in its tunable semicrystalline morphology, which can be controlled by manipulating molecular and processing variables. Toughness, hardness, clarity, and other physical characteristics can be regulated by altering its molecular weight, comonomer type, and comonomer content. Resins suited to most commercial thermoplastic fabrication processes can be created by controlling molecular weight, molecular weight distribution, and branching characteristics. Manipulation of polyethylene prior to and during crystallization also influences its solid-state properties. Polyethylene resins can thus be adapted to many end uses by virtue of both their physical properties and processing characteristics. This section briefly outlines some of the relationships between the properties of and principal uses for the various types of polyethylene. These relationships are addressed at length in Chapter 9.

Worldwide, the annual consumption of polyethylene exceeds 80 billion pounds, of which approximately 35% is used in the United States. High density, low density, and linear low density polyethylene fill the vast majority of this demand, with ethylene-vinyl acetate copolymer, very low density polyethylene, and ionomers being used in much lesser amounts.

A. High Density Polyethylene

The linear nature of high density polyethylene permits the development of high degrees of crystallinity, which endow it with the highest stiffness and lowest

Table 2 Selected Material Properties of Various Commodity Resins

Property	Poly(ethylene terephthalate)	Polypropylene	Polystyrene	Polyurethane (thermoplastic)	PVC (unplasticized)	PVC (plasticized)
Density (g/cm³)	1.29–1.40	0.90–0.91	1.04–1.05	1.12–1.24	1.30–1.58	1.16–1.35
Flexural modulus (psi @ 73°F)	350,000–450,000	170,000–250,000	380,000–500,000	—	300,000–500,000	—
Tensile modulus (psi)	400,000–600,000	165,000–225,000	330,000–485,000	—	350,000–600,000	—
Tensile strength at break (psi)	7,000–10,500	4,500–6,000	5,200–8,200	4,500–9,000	5,900–7,500	1,500–3,500
Tensile elongation at break (%)	30–300	100–600	1.2–3.6	60–550	40–80	200–450
Izod impact strength (ft-lb/in. of notch)	0.25–0.7	0.1–1.4	0.35–0.45	1.5–No break	0.4–22	Varies greatly
Melting temperature (°C)	212–265	160–175	74–110*	75–137	75–105*	75–105*
Heat distortion temperature (°C @66 psi)	75	107–121	68–107	46–135	57–82	—
Thermal expansivity (10⁻⁶ in/in/°C)	65	81–100	50–85	0.5–0.8	50–100	70–250

* Glass transition temperature.

permeability of all the types of polyethylene. This combination makes it suitable for many small, medium, and large liquid containment applications, such as milk and detergent bottles, pails, drums, and chemical storage tanks. Its low permeability, corrosion resistance, and stiffness are desirable pipe attributes, water, sewer, and natural gas transportation being the principal outlets. High density polyethylene's good tensile strength makes it fit for short-term load-bearing film applications, such as grocery sacks and trash can liners. Other household and commercial low load capacity applications include food storage containers, crates, pallets, trash cans, and toys. An added advantage in such applications is its high abrasion resistance. The chemical resistance and low permeability of high density polyethylene sheeting are exploited in its use as a liner sheet for liquid and solid waste containment pits. Fabricated items may be cross-linked to further improve their resistance to chemical and physical abuse in such applications as chemical storage tanks and small water craft.

B. Low Density Polyethylene

The numerous short-chain branches found in low density polyethylene reduce its degree of crystallinity well below that of high density polyethylene, resulting in a flexible product with a low melting point. Long-chain branches confer desirable processing characteristics, high melt strengths coupled with relatively low viscosities. Such characteristics eminently suit it to the film-blowing process, products of which are its principal outlet, accounting for more than half of all its use. Major applications include low load commercial and retail packaging applications and trash bags. Other uses include diaper backing, shrink-wrap, vapor barriers, agricultural ground cover, and greenhouse covers. Low density polyethylene can be coated onto cardboard to create a waterproof and heat-sealable composite widely used in fruit juice and milk cartons. Minor uses include wire and cable insulation and flexible pipe. Injection- and blow-molded items made from this resin are flexible and reasonably tough, suiting them for such applications as squeeze bottles and food storage containers.

C. Linear Low Density and Very Low Density Polyethylene

The generic classification linear low density and very low density polyethylene covers a broad spectrum of resins, ranging from transparent elastomers that are essentially noncrystalline to rigid opaque materials that share many of the characteristics of high density polyethylene. The majority of linear low density polyethylene falls within the density range encompassed by low density polyethylene and thus shares many of the same markets. In the realm of film—which is its largest outlet—linear low density polyethylene distinguishes itself by superior toughness. Such films are used in many packaging and nonpackaging applications, including grocery sacks, fresh produce packages, stretch-wrap, domestic trash can liners, and scientific balloons. Linear low density polyethylene is also extruded to form wire and cable insulation, pipes, and sheet for use in applications where the stiffness of high density polyethylene is not required. Items such as food container lids and toys, where flexibility combined with toughness is needed, are injection molded. On a larger scale, linear low density polyethylene is used for food processing containers, storage tanks, and highway construction barriers.

At high levels of comonomer incorporation, where crystallinity is largely suppressed, very low density polyethylene is used where clarity, softness, strain recovery, and toughness are at a premium. Such applications including medical tubing, meat packaging, and diaper backing.

D. Ethylene-Vinyl Acetate Copolymer

The numerous short-chain alkyl and acetate branches of ethylene-vinyl acetate copolymer limit its ability to crystallize. The resulting materials have low modulus and good clarity. In addition, the bulky acetate side groups inhibit the sliding of chains past one another during deformation, resulting in good strain recovery relative to other classes of polyethylene. Their high branch content results in low lamellar thicknesses, which translates into low melting and processing temperatures. Long-chain branches endow these copolymers with melt characteristics similar to those of low density polyethylene. Ethylene-vinyl acetate copolymers are used primarily in packaging films, where their flexibility, toughness, elasticity, and clarity are desirable attributes. Outlets for such products include meat packaging and stretch-wrap. Ethylene-vinyl acetate is also used for coating cardboard and as wire and cable insulation. The other main use of ethylene-vinyl acetate copolymer is as a component of adhesives.

2
Commercial Development of Polyethylene

I. OVERVIEW

The development of polyethylene as an integral part of everyday life did not proceed smoothly. It was produced serendipitously several times before the utility of synthetic polymers was appreciated. It was not until the 1930s that chemists, attempting to produce an entirely different product, inadvertently created polyethylene and recognized its potential. Initially polyethylene was a highly branched low density material with a limited range of physical properties. Its commercialization was accelerated by the need for war materiel. In the 1950s new polymerization processes were developed that produced essentially linear polymers with higher densities, thus extending the range of polyethylenes available. In the 1960s the copolymerization of ethylene with small amounts of other alkenes extended the range of products even further. Currently, development is aimed at synthesizing polyethylene resins that have properties particularly suited for specific applications.

II. PRE-1933 INCIDENTAL PRODUCTION

The first to record the preparation of polyethylene was von Pechmann in 1898 [1], followed shortly thereafter by Bamberger and Tschirner [2]. In both cases polyethylene was produced by the decomposition of diazomethane, but the commercial significance of the discovery went unappreciated. Strictly speaking, the decomposition of diazomethane yields polymethylene, the only difference between this and linear polyethylene being that polymethylene molecules can have any number of carbon atoms, whereas polyethylene must have an even number. Friedrich and Marvel, in a 1930 paper [3], reported the unexpected polymerization of ethylene to a "non-gaseous" product. They did not appreciate the significance of their observation and did not investigate the polymerization reaction. Also in 1930, Carothers et al. [4] reported the production of paraffin waxes by the action of sodium on decamethylene bromide. The maximum molecular weight achievable by this method was approximately 1400. The production of paraffin waxes by the Fischer–Tropsch reduction of carbon dioxide by hydrogen was reported in 1935 by Koch and Ibing [5]. The highest molecular weight produced was in the vicinity of 2000, but once again the importance of the discovery was missed.

III. DISCOVERY

A. Inadvertent Polymerization

In the early 1930s the British company Imperial Chemical Industries (ICI) established a research program with the goal of investigating the high pressure chemistry of selected organic compounds, including ethylene. On 29 March 1933, Eric Fawcett and Reginald Gibbon were investigating the high pressure reaction of ethylene with benzaldehyde. After an experiment that failed in its intended purpose—the benzaldehyde having been recovered unchanged—a subgram quantity of a white waxy solid was found lining the reaction vessel. The product was correctly identified as a polymer of ethylene, the first time its existence was recognized.

This reaction was not reproducible; attempts to repeat it sometimes led to uncontrollable exothermic reactions with accompanying excessive pressure that damaged equipment. It was not untill December 1935 that Michael Perrin established a set of conditions that could be used to polymerize ethylene consistently. His first successful experiment yielded approximately 8g of polyethylene. The key to reproducibility lay in the contamination of the ethylene by trace levels of oxygen. Oxygen reacted with ethylene to yield peroxides that subsequently decomposed to yield free radicals that initiated the polymerization process.

The polyethylene made by Perrin was a ductile material with a melting temperature of about 115°C. This material was what we know today as low density polyethylene. In 1936 ICI took out the first patent on the manufacture of polyethylene [6].

B. Early Exploitation

The properties of the new material produced in Perrin's experiment were investigated, and its potential as an electrical insulator was soon recognized, along with its chemical inertness and inherent flexibility. Work continued on the project, with the aim of developing the apparatus necessary for commercial production.

The successful development of the required plant was no small achievement, involving the design of a reaction vessel capable of withstanding a pressure of 22,500 psi. A pilot plant was established in 1937, and by the outbreak of World War II, ICI was producing polyethylene commercially. Even before the first commercial unit came on-stream, it was recognized that it would not meet the expected demand. A newer and bigger line was commissioned, which went into production in 1942.

The first polyethylene output was slated for use as an insulator of submarine communication cables. Priorities changed with the outbreak of hostilities, and the earliest production was used almost exclusively as an electrical insulator employed in the newly developed technology of radar. In this application its high dielectric strength and low loss factor proved invaluable. The use of polyethylene as an insulator enabled components to be made much smaller than those insulated with traditional materials, which facilitated the mounting of such equipment on airplanes and in other confined locations. It was not until the last year of the war that polyethylene was used as an insulator for communication cables linking England and France. Some early formulations of polyethylene included 12.5–15% polyisobutylene, which increased the plastic range of the molten material and improved its low temperature flexibility. As production methods improved, higher molecular weight grades of polyethylene became available, making such compounding unnecessary.

The advantages of polyethylene over existing insulators were so great that Union Carbide and du Pont quickly recognized a need for greater production. Both of these American companies obtained licenses from ICI and rapidly went into production. Commercial output of polyethylene in the United States began in 1943, soon overtaking production in the United Kingdom. As in Britain, initial uses were largely determined by the needs of the war industry.

IV. POST WORLD WAR II MARKET DEVELOPMENT

After the conclusion of World War II, with the demand for polyethylene as a component of war materiel greatly diminished, manufacturing plants in the United States and Britain had excess capacity to devote to civilian uses. The paths taken on opposite sides of the Atlantic Ocean were initially quite different. In the United States, with its active packaging industry, the use of polyethylene was expanded into film markets, while in Britain the emphasis was on molded items. Over time the paths converged as the benefits of using polyethylene in a variety of markets became apparent. With the change of product emphasis, the relative importance of the attributes of polyethylene changed. Some of its selling points were toughness, clarity, lightness, aesthetic appeal, and nontoxicity.

A. New Production Facilities Brought On-Line

In the years following World War II, Union Carbide and du Pont improved upon the production methods pioneered by ICI. Both of these companies made significant discoveries that led to great improvements in both the quantity and quality of the polyethylene produced. An important insight into the nature of the product was provided by Fox and Martin [7], who determined from infrared spectroscopy that polyethylene made by the high pressure process was branched to a significant extent. The branches principally comprised ethyl and butyl groups at a level of about one branch per 50 backbone carbon atoms. This insight led to investigation into the effect of branch content on the mechanical properties of polyethylene. Branching was found to affect the physical properties of polyethylene greatly (e.g., the fewer the branches, the higher the density and accompanying stiffness); rheological properties were also affected. From a commercial standpoint this was very important because it led to the predictable control of mechanical properties by the variation of polymerization conditions.

As a matter of course, improved control of the polymerization process led to the tailoring of specific grades of polyethylene to meet particular application needs. With the opening of new markets and the improvement of material properties, the demand for polyethylene increased, and production facilities were expanded to meet the growing need. The divergent courses taken by the polyethylene application markets in Britain and the United States led to the development of quite different grade slates in the two countries.

The number of companies manufacturing polyethylene in the United States increased after an antitrust judgment against du Pont and ICI forced the latter to license their patent to several American companies other than Union Carbide and du Pont. Companies in other countries soon entered the field. Within a decade of the end of the war, polyethylene plants were operating or being built in at least a dozen countries by more than a score of companies.

B. New Markets Open

In the ten years following World War II, the variety of products made from polyethylene expanded dramatically. During the war the limited supply of polyethylene had been used predominantly for small molded parts and extruded cable and wire insulation required by the electronics industry for the war effort. As the supply of polyethylene available for peacetime products became more plentiful, new markets were opened, supply and demand stimulating each other.

The market for film made from polyethylene was soon recognized in the United States, where its clarity, flexibility, toughness, and heat sealability made it desirable in the flourishing packaging field. Film was initially made by extrusion casting techniques similar to those developed for cellophane, which polyethylene largely displaced. A significant development in the film manufacturing industry came about with the design of equipment for producing polyethylene films by the bubble-blowing process. Film blowing was significantly faster than the casting process, and product attributes were more readily controlled. The new process quickly gained acceptance and spread rapidly to other countries. Apart from its widespread use in the packaging industry, polyethylene film entered such markets as agriculture and construction, where its resistance to permeation by water was valued. Polyethylene films were used in construction as moisture barriers under concrete slabs and foundations. In the field of agriculture, polyethylene films were used for covering greenhouses and—when made opaque by blending in carbon black—as a ground cover to inhibit the growth of weeds.

In Britain, early postwar emphasis was placed on the injection molding of household items. The size and complexity of such items increased as more sophisticated and larger injection molding machines were developed. The lightness and toughness of molded polyethylene articles enabled them to displace many products made from traditional materials such as metal or ceramics. Some early products that gained acceptance were washbowls, storage containers, and mixing bowls.

Polyethylene continued to be used as an insulator in the electrical field. Improvements in available grades enlarged its market share as it became usable for insulating cables for carrying ever higher loads.

The extrusion coating of polyethylene onto paper and cardboard opened up new packaging markets. Even in very thin layers, polyethylene is resistant to permeation by aqueous liquids. Thus, cardboard coated with polyethylene can be formed into packages to contain liquids such as milk and fruit juice. The polyethylene layer also forms the closure upon heat sealing. Extrusion coated cardboard cartons are much lighter than glass bottles and are less hazardous if dropped.

Another significant development in the packaging area was the development of the bottle-blowing process. By this process a tubular parison of molten polyethylene is inflated to fill a hollow mold, creating a thin-walled bottle. The resulting bottles are much lighter and tougher than their glass counterparts. The viscoelastic nature of molten polyethylene permits the molding of bottles and containers with a wide variety of profiles and degrees of complexity.

Polyethylene as a raw material for the extrusion of water piping was slowly accepted. Its resistance to chemical degradation was a major reason for its introduction into this area, but its tendency to deform slowly over time limited its use to thick-walled cold water applications.

Polyethylene's excellent resistance to chemical attack made it a desirable material for the construction of storage and conveyance installations for the handling of corrosive liquids. Methods were developed to construct tanks and conduits by welding molded slabs and rotomolded tubes of polyethylene.

In the years following World War II, polyethylene saw limited use as fiber. Methods were developed for spinning molten polyethylene into monofilaments ranging in diameter from 0.001 to 0.050 in. Fabrics woven from polyethylene feel waxy and are not comfortable worn against the skin, thus limiting their use in apparel. Twines and ropes that exhibited good flexibility could be woven from polyethylene, but they were not widely used, in part because of their relatively low tensile strength and their tendency to stretch under sustained loading.

C. Inherent Material Limitations Hinder Exploitation

Despite its overall balance of desirable properties, the polyethylene available in the decade following World War II exhibited a number of characteristics that limited its penetration into various markets. Several of the properties that hindered its progress in some areas were precisely those that made it desirable in others; others were more general in nature. The heart of the trouble lay in the restricted range of properties available from resins produced by the high pressure process.

Three of the principal hindrances to the enlargement of the polyethylene market were its low tensile strength, its flexibility, and its low softening temperature. Its lack of strength and rigidity kept polyethylene out of most structural applications. The low softening point restricted applications to those with service temperatures less than approximately 90°C, effectively excluding it from any markets that involved exposure to boiling water; sterilization, food processing, etc. and electrical uses where transient overloads could melt the insulation off conductors. All these limitations had the same origin, the high degree of short-chain branching that hindered the formation of crystallites in terms of size and perfection. Branches formed during polymerization are almost entirely excluded from crystallites when the molten material cools, leaving only the segments of linear backbone between them available to crystallize. The more branches, the shorter the available lengths of the chain that can crystallize, resulting in smaller crystallites and lower crystallinity. As the modulus of polyethylene crystals is approximately two orders of magnitude greater than that of the noncrystalline regions, the crystallinity level has a pronounced effect on the stiffness. Tensile strength is similarly affected by the degree of branching. The low softening temperature is the result of the melting of thin crystallites. It was readily appreciated that higher density materials, i.e., those with fewer branches, would outperform the available polyethylene resins in many key areas.

A potential problem associated with the long-term use of polyethylene is its tendency to "creep," that is, to deform gradually under sustained load. In extreme cases creep can lead to rupture. Creep is accelerated by higher temperatures. The problem is evident in applications such as high pressure tubing, in which the effect of a catastrophic rupture after an extended period of time can

Commercial Development of Polyethylene 33

be quite spectacular. Creep is a problem with all types of polyethylene, but it is especially prevalent in those with modest levels of crystallinity such as those made by the high pressure process. Structural applications were thus doubly barred to polyethylene, owing to both its initial lack of strength and its long-term dimensional instability.

Another impediment to the marketing of polyethylene was its susceptibility to cracking when placed in a hostile environment. It was found that many grades of polyethylene developed cracks when stressed in the presence of certain organic liquids. This property was variously termed solvent embrittlement, environmental cracking, or environmental stress cracking. In due course the term "environmental stress cracking" gained general acceptance because it accurately sums up the problem. The effect was more pronounced if the sample contained residual orientation from molding, had surface imperfections such as scratches or nicks, and was subject to multiaxial stresses. The range of organic liquids causing this effect was broad, the effect being most severe for various polar compounds such as alcohols, esters, ketones, and detergents. The molecular weight of polyethylene was found to play an important role; the higher the molecular weight, the less prone was the material to crack. Environmental stress cracking led to the premature failure of products in such diverse applications as food storage containers, pipes, and cable insulation.

The key to avoiding, or at least ameliorating, many of these defects lay in the ability to tailor polyethylene resins to usage requirements. A major portion of the research into the nature of polyethylene in the years following World War II was aimed at elucidating its molecular structure. From an early date it had been recognized that synthetic polymers were unlike regular chemical compounds in that they did not consist of a single molecular species. Polyethylene was found to be heterodisperse with respect to molecular weight and branching distribution. The distributions of chain lengths and branch concentrations are critical to the properties of the resin. Various anomalies in the character of polyethylene were found, indicating that the branch concentration was not identical for all molecular weights in a given resin [8–11]. Polyethylene with a broader molecular weight distribution was shown to have a higher concentration of long-chain branches than resins with a narrower molecular weight distribution, while the higher molecular weight species had a disproportionately high number of long-chain branches.

The tailoring of polyethylene resins could be achieved either during polymerization, by regulating reaction conditions to affect branching and molecular weight, or by post-reactor treatments such as cross-linking or blending. Research along both these lines was actively pursued, and great progress was made, especially in regard to control of the polymerization process. By 1955 control of polymerization conditions had improved to the point that polyethylene with a density of 0.94 g/cm^3 could be produced in a high pressure reactor. Molecular weights could also be tailored to match the needs of the fabrication equipment and the required properties of the product. Cross-linking of polyethylene permitted its use for short periods of time at temperatures above its softening point in applications where dimensional stability was not critical.

V. MAJOR INNOVATIONS WIDEN THE RANGE OF POLYETHYLENE

A. The Advent of High Density Polyethylene

Prior to the major discoveries of Karl Ziegler's research group in West Germany and the researchers of Phillips Petroleum in the United States, unbranched polyethylene had been produced in small quantities at a number of research facilities. In each case the significance of the high density polyethylene so produced went unappreciated, even when the material was studied with a view to commercialization [12]. That the significance of such discoveries should be overlooked is somewhat surprising given the acknowledged limitations imposed upon low density polyethylene by its high levels of branching.

1. Ziegler Polymerization

After World War II Karl Ziegler headed a research group at the Max Planck Institute in West Germany that was investigating the reactions of certain organometallic compounds, including triethyl aluminum. E. Holzkamp, a graduate student, found that ethylene could be dimerized to form butene in the presence of triethyl aluminum. This reaction was fortuitous, involving trace amounts of nickel from the stainless steel reaction vessel that combined with the triethyl aluminum to form a catalyst. The potential for polymerizing ethylene was recognized, and various transition metals were investigated with respect to their ability to form similar, but more effective, catalysts. Chromium complexes were found to catalyze the polymerization of ethylene to form a mixture of oligomers containing some high polymer. On 26 October 1953, H. Breil, another of Ziegler's graduate students, succeeded in producing significant quantities of polyethylene using a zirconium complex catalyst. The infrared vibrational spectrum of this material exhibited a very weak peak assigned to methyl groups, at 2962 cm^{-1}, which is prominent in high pressure polymerized polyethylene. The significance of this finding lies in the fact that methyl-terminated short alkyl branches are the principal source of this peak. Thus the essential linearity of the new product was demonstrated. H. Martin, a senior staff member with the group, succeeded in polymerizing polyethylene with a titanium complex at such modest temperatures and pressures that the polymerization could be performed in a glass reaction vessel.

The new form of polyethylene, with its negligible branching, displayed many properties that were superior to those of the highly branched resins previously available. Among the most significant improvements was a softening point elevated by approximately 30°C over those of products of the high pressure process. Other improvements lay in the fact that its stiffness and strength were also increased. With its higher degree of crystallinity and concomitant higher density it was named high density polyethylene (HDPE), the older type of polyethylene becoming known as low density polyethylene (LDPE).

The significance of Ziegler's discovery was recognized by the Nobel Prize Committee in 1963; they awarded a joint Nobel Prize for Chemistry to Ziegler and G. Natta for their respective work in the field of ethylene and propylene polymerization. Catalysts of the type pioneered by Ziegler and Natta are now known generically as Ziegler–Natta catalysts.

2. The Phillips Process

At about the same time that Ziegler's group was working on the polymerization of ethylene using transition metal organic complexes, researchers at the Phillips Petroleum company in the United States was investigating a similar reaction catalyzed by various supported transition metal oxides. Building upon wartime observations that reactors sometimes became plugged with a waxy solid when the goal was to produce butadiene from ethylene, P. Hogan and R. Banks investigated the fouling of a reactor packed with chromium salts and fed with propylene. Their initial interest lay in the synthesis of lubricating oils. Experiments with ethylene as the feedstock resulted in the production of a high molecular weight ethylene polymer. Experimentation along these lines was continued despite the fact that Phillips played no part in the contemporary polymer industry. Their product proved to be similar to the high density polyethylene produced by Ziegler's low pressure, low temperature polymerization process. The Phillips reaction took place in a hot solvent at the modest pressure (relative to the low density polyethylene process) of 500 psi, using a supported chromium oxide as the catalyst. Subsequently the Phillips high density polyethylene was found to have a slightly higher density than the Ziegler–Natta type of materials, indicating a greater degree of linearity [13].

3. The Standard Oil Process

Concurrent with the development of the Phillips process, Standard Oil of Indiana developed a similar ethylene polymerization process [14–17]. The basis of this process was the catalysis of ethylene to high density polyethylene using a supported molybdenum oxide catalyst under relatively modest conditions of temperature and pressure. The product has a range of densities similar to that available from the Phillips process. This system was not vigorously pursued and did not gain the acceptance of the Ziegler–Natta or Phillips processes.

B. Cross-Linking Methods Discovered

In 1948 M. Dole treated low density polyethylene with high energy radiation from the heavy water pile at the Argonne National Laboratories [18]. The resulting product had very different tensile properties from the starting material, its extension at break was severely reduced, and it maintained its dimensional stability at elevated temperatures. The change of physical properties was interpreted as being due to the formation of covalent carbon–carbon bonds linking the backbones of adjacent chains; i.e., the molecular chains had been bound together by cross-links to form a network. Similar work was carried out by A. Charlesby in the early 1950s [19].

In the 1950s G. Oster and his coworkers [20,21] discovered that polyethylene could be cross-linked by ultraviolet radiation if appropriate sensitizers were incorporated. As ultraviolet radiation does not penetrate deeply into polyethylene, the cross-linked portion forms a skin on thick parts. This reaction can also be used to incorporate chemically functional groups onto the otherwise inert polyethylene backbone.

In the early 1960s it was discovered that polyethylene could be cross-linked by the decomposition of various organic peroxides such as dicumyl peroxide. Organic peroxides decompose homolytically to form free radicals that cause cross-linking, decomposition rates increasing exponentially with temperature. It is possible to select a peroxide that has decomposition characteristics such that it can be blended with polyethylene in the melt at temperatures that do not result in cross-linking. When the temperature of the melt is subsequently increased, the peroxide decomposes to form free radicals that effect cross-linking.

The most recently developed commercial method of cross-linking polyethylenes involves the formation of alkoxysilyl bridges between adjacent chains. This process was developed in the late 1960s and early 1970s [22].

1. Heat Stability of Polyethylene Improved

A key attribute of cross-linked polyethylene is that it does not flow when heated above its crystalline melting temperature. Ordinarily, when polyethylene is raised to elevated temperatures, the crystallites that bind the material into its solid state melt, and the material becomes a viscous liquid. In the case of cross-linked polyethylene, when the crystallites melt, the cross-links remain intact, preserving the relative positions of the otherwise liquidlike chains. Thus, molten cross-linked polyethylene will soften and sag but not flow. This property is of major significance to the applicability of polyethylene in areas where the possibility of short-term high service temperatures exists.

The development of cross-linking led to the use of polyethylene in many markets from which it had previously been excluded. One such major application of cross-linked polyethylene is in the insulation of high voltage electrical cables where transitory overloads can generate sufficient heat to melt conventional polyethylene coverings.

VI. HIGH DENSITY POLYETHYLENE BECOMES A MAJOR PRODUCT

Ziegler patented his new form of polyethylene in 1953 [23]. Several chemical companies showed immediate interest in the new high density polyethylene, appreciating that its improved hardness, strength, and elevated softening temperature relative to low density polyethylene conferred upon it some very marketable attributes. Another perceived advantage of the low pressure polymerization process was that it would be less dangerous to operate than the high pressure process. Ziegler licensed only the chemical composition of his catalysts and the resulting product, leaving industry to work out production details. The diverse approaches of the various licensees to solving the technical details of commercial polymerization led to innovation and variation among the plants of the initial producers. Hoechst in West Germany became the first commercial producer in 1955. Hercules, who shared licensing and technical information with Hoechst, opened the first U.S. plant in 1957. The density of early grades of high density polyethylene made by Ziegler–Natta catalysts was in the region of 0.94 g/cm^3.

Phillips Petroleum started commercial production of their high density polyethylene in 1956. Their product had a density of approximately 0.96 g/cm^3, with an accompanying improvement of stiffness and strength relative to the Ziegler–Natta type of polyethylenes. Phillips licensed their invention to many companies; by the middle of the 1970s their process had come to dominate the worldwide high density polyethylene market.

Standard Oil sold a number of licenses to manufacture high density polyethylene using their process, the first commercial product being made in Japan in 1960. The Standard Oil invention was not marketed as aggressively as the other two methods of producing high density polyethylene and played a minor role in the development of the global polyethylene industry.

The companies already manufacturing low density polyethylene recognized the threat to their markets and took steps to modify their process to facilitate the production of resins of higher density. In 1956 ICI released a grade of low density polyethylene with a density that matched that of the early Ziegler–Natta high density polyethylene resins. Despite the inherent advantages of the low pressure polymerization process, the high pressure production facilities remained in use producing low density polyethylene.

A. New Markets Explored

The enhanced physical properties of high density polyethylene came at the expense of increased processing difficulties. The principal market originally identified were those of injection- and blow-molded articles and extruded film, sheet, and pipe. Initially the fabrication of usable articles from high density polyethylene resins was beset by problems. Converters accustomed to molding low density polyethylene were not equipped to handle the new polyethylene resins. A major problem was the shrinkage of molded articles upon cooling. High density polyethylene, being more crystalline than low density polyethylene, tends to shrink more upon solidification. When shrinkage was not uniform, due to uneven cooling or large changes in profile, the result was often warped products. The troubles with warping were particularly noticeable in large items, where the increased strength of high density polyethylene was likely to prove most advantageous. The early grades of high density polyethylene were also susceptible to environmental stress cracking, in part due to their rapid crystallization, which froze internal stresses into the final product. Despite its improved strength and modulus, high density polyethylene was found to be prone to creep, albeit to a lesser extent than low density polyethylene. Thus high density polyethylene did not find immediate acceptance in most markets. One limited application where it performed extremely well was in hula hoops. The demand for hula hoops consumed a substantial portion of the high density polyethylene initially produced in the US. The breathing space provided by the demand for high density polyethylene to meet the nonexacting requirements of hula hoops gave producers time to work out some of their production problems. It was found that the copolymerization of ethylene with small amounts of a second monomer reduced the product's density and made it easier to process.

With the combined capabilities of high and low pressure production facilities it became possible to produce polyethylene resins with densities in a range of 0.91–0.96 g/cm^3. This increased the range of products over those available prior to 1955 and permitted polyethylene to penetrate new markets and increase its utilization in existing ones. Research and development continued on both the high and low pressure polymerization processes with the goal of tailoring resins to meet the requirements of more specialized markets.

VII. POLYETHYLENE BECOMES MORE SPECIALIZED

From the 1940s onward, research had been directed toward the copolymerization of various monomers with ethylene. The commercialization of the Ziegler and Phillips processes provided an avenue whereby the properties of polyethylene could be modified by polymerizing ethylene with 1-alkene (α-olefin) comonomers. Efforts to incorporate comonomers were not restricted to the low pressure processes; various polar comonomers were added to the feedstock in the high pressure process to change the character of the product.

A. Copolymerization with α-Olefins

1. The Development of Linear Low Density Polyethylene

Commercial copolymerization of ethylene with a secondary monomer began in the late 1950s as a method of inhibiting the crystallization of high density polyethylene to reduce its shrinkage upon cooling. The initially low level of comonomer content was increased to yield a type of polyethylene with characteristic properties that is now known as linear low density polyethylene (LLDPE). The molecules that make up such resins have very few long-chain branches and thus may be considered to be "linear," while the short-chain branches of the comonomer hinder crystallization and reduce the density of the product to the range of conventional low density polyethylene. The comonomers used are typically α-olefins, principally 1-butene, 1-hexene, and 1-octene. The carbon atoms participating in the vinyl group of the comonomer are incorporated into the polymer backbone; the remaining carbon atoms form pendant groups referred to as short-chain branches. Propylene may also be used, but it is generally added in far higher proportions and the copolymer so produced has negligible crystallinity and is considered to be a rubber (and thus outside the scope of this work). The first commercial linear low density polyethylene was marketed in 1960 by du Pont. Union Carbide quickly followed suit with resins made by their innovative gas-phase polymerization process.

One of the chief attributes of linear low density polyethylene that made it highly attractive was its high film tear strength. This, coupled with its relatively high clarity compared to high density polyethylene, ensured its wide acceptance as a film grade material used for packaging; where it currently finds extensive use.

The incorporation of extremely high levels of comonomer can reduce the degree of crystallinity, and hence the resin density, to levels well below that of low density polyethylene. Such products are variously termed ultralow density or very low density polyethylene (ULDPE and VLDPE, respectively). These materials are flexible, clear, and elastomeric. They can be used neat or blended into other polymers as impact strength and clarity modifiers.

2. Comonomer Distribution Controlled in Linear Low Density Polyethylene

One of the problems of Ziegler–Natta type linear low density polyethylene lies in its nonuniform distribution of branches. The α-olefins that form the short-chain branches do not react with the catalyst at the same rate as ethylene and thus are incorporated into the polymer at some level other than the stoichiometry of the feedstock. Because of the variety of reaction sites on the surface of the catalyst, some molecules have a substantially higher comonomer content than others. In practice, comonomers tend to be more concentrated in the shorter molecules. Thus, Ziegler–Natta type linear low density polyethylene resins may be simplistically viewed as a blend of higher molecular weight, lightly branched polyethylene and lower molecular weight, more highly branched polyethylene. In addition to the broad composition distribution, the molecular weight distribution is also generally broad ($\overline{M}_w/\overline{M}_n \approx 3.5$–$4.5$). In recent years efforts have been made to limit both the compositional and molecular weight distributions in the belief that a family of resins with narrower distributions will prove to be more useful in some applications than those with broad distributions. This has been achieved by the use of improved catalysts, known as metallocenes, which have only one type of reaction site. Resins made with these new catalysts have the potential for being more readily tailored to the increasingly demanding requirements of the end user.

B. Copolymerization with Polar Comonomers

In the 1960s du Pont introduced copolymers comprising ethylene and polar comonomers. Such copolymers are produced exclusively by the high pressure polymerization process. Polar comonomers poison Ziegler–Natta and metal oxide catalysts, destroying their capacity to polymerize ethylene. Typical polar comonomers include vinyl acetate, acrylic acid, and methacrylic acid. The properties of such copolymers are governed both by the morphological considerations that control the properties of other types of polyethylene and by specific interactions between the polar comonomer units. The effect of the polar comonomers is somewhat obscured by the fact that high pressure polymerized polyethylene resins are inherently highly branched. Thus a fairly high comonomer incorporation level (5% by weight or more) is required to realize noticeable effects.

The specific interactions of the polar units in ethylene-methacrylic acid and ethylene-acrylic acid copolymers can be modified by neutralizing the acid function with an alkali to produce a family of materials known as ionomers. Thus, in addition to relatively weak van der Waals forces developed between polar units, aggregates of metal cations and anionic comonomers form. Such aggregates act as cross-links in the solid state while permitting viscous flow in the melt. Given a sufficiently high proportion of polar comonomer, with the appropriate neutralization level, ionomers exhibit elastomeric properties. The physical properties of such materials can be systematically varied over a wide range by controlling comonomer type and content, the degree of neutralization, and the nature of the cations introduced.

C. Manipulation of Molecular Weight Distribution

Early metal oxide and Ziegler–Natta type catalysts contained several types of active sites, producing high density polyethylene and linear low density polyethylene resins with relatively broad molecular weight distributions. The high and low molecular weight tails associated with such resins disproportionately affected the processing and physical strength of products made from them. Excessive quantities of low molecular weight material can exacerbate such problems as environmental stress cracking and low stress embrittlement or can leach out to contaminate foodstuffs stored in polyethylene packaging. Excessive amounts of very high molecular weight material can make processing more difficult and result in internal stresses being frozen into molded items. Over the years, increased knowledge of the molecular characteristics that resulted in processing and physical problems spurred research into more controlled polymerization. Cleaner feedstocks, closer monitoring and control of polymerization conditions, and improved catalysts reduced the breadth of the molecular weight distribution. When the molecular weight distribution is narrowed too far, however, the overall processability of polyethylene suffers owing to changes in the viscosity characteristics of resins. This may be counteracted by operating two or more polymerization reactors in series or parallel, each run under different reaction conditions. The product of such reactors is effectively a polymer blend. The net result of research over the last 40 years has been polyethylene resins that are increasingly tailored to meet the requirements of the processor and the end user.

BIBLIOGRAPHY

Renfrew A and Morgan P, Polythene, New York: Interscience.
McMillan FM. The Chain Straighteners. London: Macmillan, 1979.
Roy. Inst. Chem. The Discovery of Polythene, Lecture series, 1964, No. 1.

REFERENCES

1. H von Pechmann. Berichte 31:2643, 1898.
2. E Bamberger, F Tschirner. Chem Ber 33:955, 1900.
3. MEP Friedrich, CS Marvel. J Am Chem Soc 52:376 1930.
4. WH Carothers, JW Hill, JE Kirby, RA Jacobson, J Am Chem Soc 52:5279, 1930.
5. H Koch G Ibing Brennstoff-Chem 16:141, 1935.
6. EW Fawcett, RO Gibson, MW Perrin, JG Patton, EG Williams, B Patent 471,590, Sep. 6, 1937 (1937).
7. JJ Fox, AE Martin. Trans Faraday Soc 36:897, 1940.
8. FW Billmeyer. J Am Chem Soc 75:6, 1953.

9. FW Billmeyer. J Am Chem Soc 75:118, 1953.
10. V Desreux MC Spiegels. Bull Soc Chim Belg 59:476, 1950.
11. F Bebbington, E Hunter, RB Richards. XIIth Int Congr Pure Appl Chem, New York, September 1951.
12. FM McMillan, The Chain Straighteners. London: McMillan, 1979.
13. JP Hogan, RL Banks. US Patent 2,825,721 (1958).
14. E Field, M Feller. US Patent 2,726,231 (1955).
15. EF Peters, A Zletz BL Evering. Ind Eng Chem 49:1879, 1957.
16. M Feller, E Field. Ind Eng Chem 49:1883, 1957.
17. M Feller, E Field. Ind Eng Chem 51:155, 1959.
18. M Dole. J Macromol Sci Chem A15:1403, 1981.
19. A Charlesby. Proc Roy Soc (Lond) A215:187, 1952.
20. G Oster. J Polym Sic Lett Ed 22:185, 1956.
21. G Oster, GK Oster H Moroson. J Polym Sci 34:671, 1959.
22. MG Scott. US Patent 3,646,155, 1972.
23. K. Ziegler, Ger Patent 878,560 1953.

3
Production Processes

I. HIGH PRESSURE POLYMERIZATION

A. Free Radical Chemical Processes

The polyethylene products known as low density polyethylene (LDPE) resins are produced exclusively by high pressure free radical polymerization. The chemistry involved in their production is deceptively simple, requiring little more than an appropriate source of free radicals and conditions of high temperature and pressure. The free radicals initiate the polymerization process when the monomers have been forced into close proximity by high pressure. Termination of chain growth occurs when the free radical on a growing chain is transferred to another chain or is quenched by another radical. In practice, numerous competing side reactions occur that result in branching and premature chain termination. The nature of the product is controlled by the initiator concentration, temperature, pressure, availability of vinyl comonomers, and the presence of chain transfer agents.

1. Initiation

High pressure polymerization of ethylene is initiated by the decomposition of various molecules to produce free radicals. A radical species then abstracts a hydrogen atom from an ethylene monomer to form an incipient polymer chain. Various initiators are used, the most common types being oxygen, organic peroxides and various azo compounds. The role of oxygen is not clearly understood, but it probably involves the formation of organic peroxides in situ [1].

Oxygen as initiator $RH + O_2 \rightarrow ROOH$

where:
R = an alkyl group (presumably ethyl).

Peroxides and azo compounds decompose in the reaction vessel under appropriate conditions to form two or more species, each of which bears an unpaired electron. Examples of these reactions include:

Initiator	Reaction
Hydroperoxide	$ROOH \rightarrow RO\bullet + HO\bullet$
Dialkyl peroxide	$ROOR' \rightarrow RO\bullet + R'O\bullet$
Tertiary perester	$RCOOCOCR'_3 \rightarrow RCO\bullet + R'_3COCO\bullet$
Azodialkyl	$RNNR' \rightarrow RN\bullet + R'N\bullet$

R and R' = an alkyl or aryl group or various other organic moieties.

After decomposition of the initiator the process continues with the free radical species attaching itself to an ethylene molecule, the unpaired electron relocating to the opposite end of the monomer:

$$R\bullet + CH_2{=}CH_2 \rightarrow R{-}CH_2{-}CH_2\bullet$$

2. Chain Propagation

Growth of the polyethylene chain proceeds when the free radical on the end of a growing chain reacts with an ethylene molecule brought into close proximity by the force of the high pressure. The incoming ethylene attaches to the end of the chain via a carbon–carbon covalent bond, and an unpaired electron is transferred to the new chain end.

$$\sim\sim\sim\sim CH_2{-}CH_2\bullet + CH_2{=}CH_2 \rightarrow \sim\sim\sim\sim CH_2{-}CH_2{-}CH_2{-}CH_2\bullet$$

3. Comonomer Incorporation

Various comonomers containing a vinyl group can be incorporated into the growing chain. The most frequently used comonomers are vinyl acetate and methacrylic acid. Their incorporation follows the same scheme as the addition of ethylene to the end of the growing chain. Due to their polar nature—which stabilizes intermediate transition states during addition to the chain end, thus lowering the energy of activation—such comonomers are incorporated in preference to ethylene.

4. Chain Branching

Chain branching occurs when the terminal radical responsible for chain growth abstracts a hydrogen atom from a preexisting polyethylene chain. The result is the termination of growth at its original site and continued propagation at a new one. When radical transfer occurs intramolecularly it results in short chain

Production Processes

branching (SCB); intermolecular transfer gives rise to long-chain branching (LCB). The frequency and type of chain branching are controlled largely by the polymerization conditions. Hence the ultimate properties of the resin can be controlled to some extent by altering the reaction conditions. As a general rule, higher temperatures promote branching.

Inter- or intramolecular transfer of a radical results in the new growth of a polymer chain from somewhere along the length of a pre-existing chain.

The probability of intermolecular hydrogen abstraction from a given molecule, which leads to long-chain branching, is proportional to the length of the molecule. Thus, long-chain branching is more prevalent at higher molecular weights.

Short-chain branching occurs when the growing end of a chain turns back on itself, allowing the abstraction of a hydrogen atom only a few bonds away from the chain terminus; this process is known as "backbiting." Chain growth continues from the location of the new radical, leaving the original chain end as a short branch. As a result of the approximately tetrahedral arrangement of the bonds linking carbon atoms to neighboring atoms, ethyl and butyl branching is prevalent [2]. More complex branches are formed to a lesser extent. Short chain branches tend to form small clusters on the main chain separated by linear runs of polyethylene. Some examples of backbiting reactions are shown in Figure 1.

5. Chain Transfer

Chain transfer is the process by which the growth of a polyethylene chain is terminated in such a way that the free radical associated with it transfers to another molecule on which further chain growth occurs, i.e., the number of free radicals and growing chains remains constant. The molecule to which the free radical is transferred can be either ethylene or a deliberately added chain transfer agent (CTA) (also known as a telogen) such as a solvent molecule. The net effect of adding a chain transfer agent is to reduce the average molecular weight of the resin. As a general rule, chain transfer is controlled by altering reaction conditions rather than by the addition of chain transfer agents.

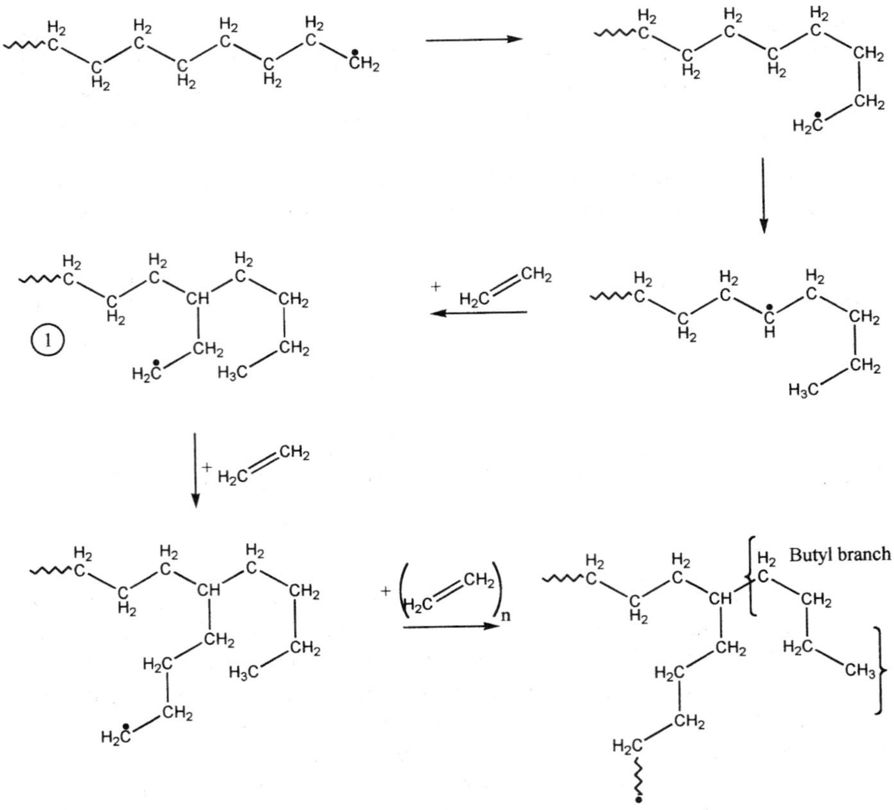

(a)

Figure 1 Examples of backbiting reactions occurring during high pressure polymerization. (a) Formation of butyl branch; (b) Formation of 2-ethylhexyl branch; (c) Formation of paired ethyl branches.

Examples of chain transfer include

$$\sim\sim\sim\sim CH_2-CH_2\bullet + CH_2=CH_2 \rightarrow \sim\sim\sim\sim CH=CH_2 + CH_3-CH_2\bullet$$

$$\sim\sim\sim\sim CH_2-CH_2\bullet + CH_2=CH_2 \rightarrow \sim\sim\sim\sim CH_2-CH_3 + CH_2=CH\bullet$$

and

$$\sim\sim\sim\sim CH_2-CH_2\bullet + RH \rightarrow \sim\sim\sim\sim CH_2-CH_3 + R\bullet$$

(b)

(c)

where
R = an alkyl or aryl group or some other organic moiety.

In each case the newly formed radical species is capable of initiating chain growth.

6. Termination

Complete termination of chain growth is brought about when two radicals, at least one of which is at an active chain end, meet and quench each other. The quenching radical can be another growing chain end, an initiator fragment, or an ethylene radical. Various impurity molecules can prematurely terminate the growth of a chain, so great care is taken to ensure that all reactants are extremely pure.

When the unpaired electrons that make up radicals meet, they generally combine to form a covalent bond.

$$\sim\sim\sim\sim CH_2-CH_2\bullet + CH=CH_2\bullet \rightarrow \sim\sim\sim\sim CH_2-CH_2-CH=CH_2$$
$$\sim\sim\sim\sim CH_2-CH_2\bullet + RO\bullet \rightarrow \sim\sim\sim\sim CH_2-CH_2-OR$$

When two growing chain ends meet, the result may be chain coupling to form a single polymer molecule or disproportionation to leave the chains as separate molecules.

$$\sim\sim\sim\sim CH_2-CH_2\bullet + \bullet CH_2-CH_2\sim\sim\sim\sim$$
$$\rightarrow \sim\sim\sim\sim CH_2-CH_2-CH_2-CH_2\sim\sim\sim\sim$$
$$\sim\sim\sim\sim CH_2-CH_2\bullet + \bullet CH_2-CH_2\sim\sim\sim\sim$$
$$\rightarrow \sim\sim\sim\sim CH=CH_2 + CH_3-CH_2\sim\sim\sim\sim$$

B. High Pressure Production Facilities

Since the first production of low density polyethylene in a continuous pilot plant in 1937, there has been an extraordinary divergence of manufacturing processes. Despite the diversity of plants in use, they all share certain characteristics. Figure 2 outlines the key components common to high pressure polymerization facilities. This process scheme and all subsequent ones are, of necessity, greatly simplified. A full description of the many technical difficulties that must be overcome in producing polyethylene is beyond the scope of this work.

Fresh ethylene **a** also known as make-up ethylene because it forms only about 10–20% of the reactor feed is fed into a primary compressor **1**. The fresh ethylene is joined by recycled feedstock **b** and **c**. The primary compressor elevates the ethylene pressure to approximately 1500–4000 psi, and it is then transferred **d** into the secondary compressor **2**, which boosts the pressure to approximately 15,000–22,500 psi. The pressurized ethylene **e** is fed into the reactor **3**.

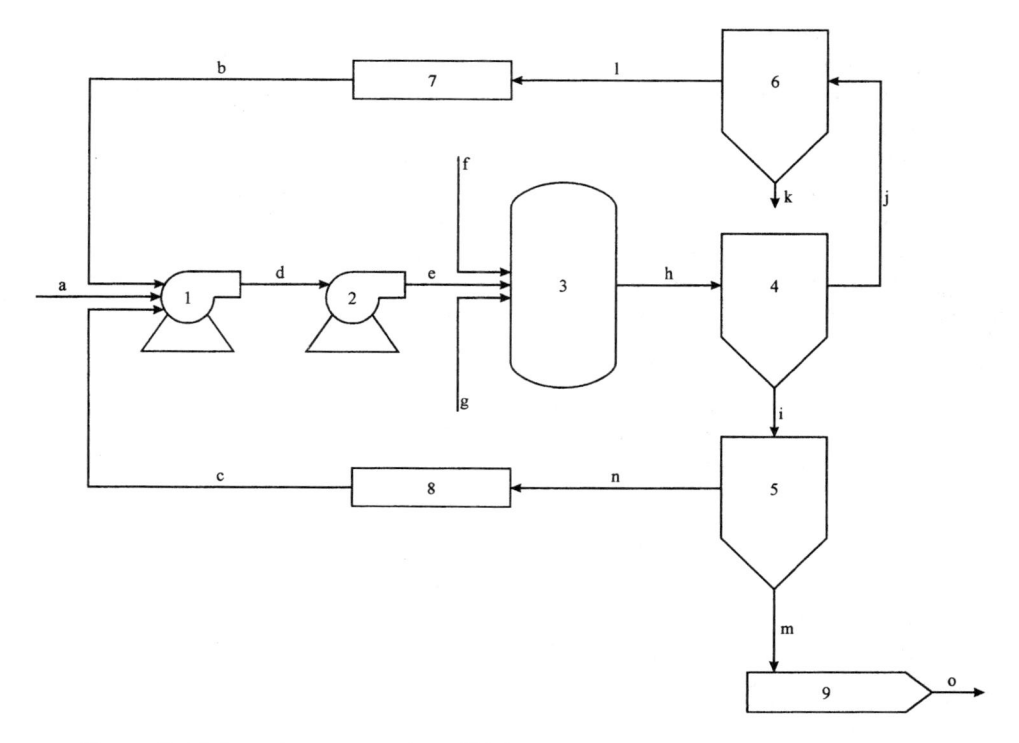

Figure 2 Schematic representation of high pressure polymerization of ethylene. 1, Primary compressor; 2, secondary compressor; 3, reactor; 4, high pressure separator; 5, low pressure separator; 6, low pressure separator; 7,8, coolers; 9, extruder. a, Fresh ethylene; b,c, recycled ethylene; d, intermediate pressure ethylene; e, high pressure ethylene; f, catalyst; g, chain transfer agent; h, ethylene, oils, waxes, and polyethylene; i, ethylene and polyethylene; j, ethylene, oils, and waxes; k, oils and waxes; l, ethylene recycle; m, polyethylene; n, ethylene recycle; o, LDPE pellets.

The initiator **f** and chain transfer agent **g** can be metered into the ethylene stream as it enters the reactor or at various points within it. From the reactor the product stream **h** containing a mixture of unreacted ethylene, oils, waxes, and polyethylene proceeds to a two stage separation process. The product stream is initially let down into a high pressure separator **4** wherein the polyethylene precipitates and is drained off with some ethylene **i** to a low pressure separator **5**. The low molecular weight oils and waxes remain in solution in the bulk of the ethylene, and this stream **j** is let down into a separate low pressure separator **6**. Here the ethylene is stripped from the oils and waxes, which are discharged in waste stream **k**. The ethylene for recycle **l** proceeds to a cooler **7**, from which it is piped to the primary compressor to join the make-up feed. In the low pressure separator **5** the ethylene is flashed off and the polyethylene **m** is removed. The ethylene stream **n** is recycled via a cooler **8** to the start of the process. Many variants of the separation process exist, with different arrangements of separators that can recycle unreacted ethylene to either or both of the compressors. The polyethylene product is fed into an extruder **9**, where it is homogenized and blended with additives, principally antioxidants. The product is extruded as thin strands that are chopped into pellets to form low density polyethylene resin **o** that is ready for packaging and shipping. Often the product is transferred directly into railway hopper cars for transportation, but it may be stored in silos prior to loading into sacks, boxes, trucks, etc.

High pressure polymerization plants are more costly to build, operate, and maintain than low or medium pressure plants. Despite these drawbacks, new low density plants are still occasionally being commissioned. It is unlikely that many new low density polyethylene plants will be commissioned in the future.

C. High Pressure Reaction Conditions

Commercial high pressure polymerization of ethylene is relatively inefficient with less than 20% of the feedstock being consumed on each pass through the reactor. The reason for this is primarily a matter of heat transfer. The polymerization of ethylene is extremely exothermic, producing approximately 800 calper gram of polyethylene.

$$nC_2H_4 \rightarrow (C_2H_4)_n, \quad + 22 \text{ kcal/mol}$$

This heat must be removed to maintain stable reaction conditions. If a temperature of approximately 300°C is exceeded, ethylene and polyethylene decompose rapidly, yielding more heat and excess gaseous products.

$$C_2H_4 \rightarrow C + CH_4, \quad + 30 \text{ kcal/mole}$$
$$C_2H_4 \rightarrow 2C + 2H_2, \quad + 11 \text{ kcal/mol}$$
$$\frac{1}{n}(C_2H_4)_n \rightarrow C + CH_4, \quad + 8 \text{ kcal/mol}$$

The elevated temperature and pressure increase the rate of the decomposition, and unless they are quickly brought under control the reaction runs away. All reactors are fitted with rupture disks designed to burst when a set overpressure is reached, thus venting the reaction vessel to the atmosphere in the event of a runaway reaction. The liberated gases form a highly explosive mixture with air and can ignite spontaneously if sufficiently hot, in what is euphemistically known as "aerial decomposition." The likelihood of an aerial decomposition can be reduced by pumping large amounts of water into the vent line. The problem of

Production Processes

runaway reactions was encountered very early in the development of high pressure polymerization when experimental reactors were damaged or burst by overpressure.

The molecular structure of low density polyethylene is principally governed by the reaction conditions used in its production. To optimize the yield and properties of the final resin it is necessary to balance the various reactions involved with initiation, propagation, branching, chain transfer, and termination. The principal control variables are the reaction temperature and pressure. The type of initiator employed is of importance only with respect to its decomposition rate and overall concentration. The concentration and efficiency of chain transfer agents are secondary variables, which are not always employed.

Reactor pressures from 7500 to 50,000 psi have been reported, but a normal working range of 15,000–22,500 psi is more typical. Reaction temperatures are normally to be found between 180 and 200°C but can lie within a range of 100–300°C.

The molecular weight of low density polyethylene tends to increase as the reaction pressure is increased. Elevated pressure forces a greater number of ethylene monomers into proximity with the growing chain end, thus promoting chain growth. When the pressure is maintained in the range of 1500–3000 psi, the resulting products are predominantly oils and waxes with molecular weights in the range of 100–500. An increase in the pressure up to 7,500 psi increases the average molecular weight to approximately 2000. The pressure inside a reactor can be deliberately varied to systematically alter the reaction conditions. This yields a resin that has a broader molecular weight distribution than would otherwise be produced. Pressure fluctuations take the form of pulses with various profiles of increasing and decreasing pressure.

The level of branching in a low density polyethylene resin rises as the polymerization temperature increases. Higher temperatures promote the random motion of the growing chains, increasing the probability that they will adopt configurations conducive to backbiting. The increased branching level reduces the degree of crystallinity, resulting in a resin of lower density.

The type of initiator used is dictated largely by the reaction conditions. Initiators with a decomposition half-life of a few seconds at the reaction temperature are normally chosen. In general, organic peroxides decompose at lower temperatures than those required for oxygen initiation, thus permitting lower reaction temperatures and consequently a decreased level of branching. Azo initiators are typically used at the lowest reaction temperatures, remaining viable at temperatures below 100°C. As the reaction temperature is increased, the rate of chain propagation rises, reducing the demand for initiator. If the reaction temperature is sufficiently high, ethylene monomers spontaneously decompose to form radicals that initiate polymerization. This is generally undesirable because the temperatures involved are so high that the likelihood of runaway reactions is increased.

The product of such autoinitiated polymerization has extremely low density. In general the concentration of added initiator is less than 200 ppm. Thus the resulting resins, are not greatly contaminated by initiator residues and normally require no purification prior to use. Certain initiator residues can impart an off taste or smell to resins, making them undesirable in food packaging applications.

Telogens help control the average molecular weight and molecular weight distribution of low density polyethylene by transferring the radical from a growing chain end to another molecule to start a new chain. The net result of the addition of a telogen is to reduce the average molecular weight of the product. Low molecular weight alkanes, such as butane, are commonly used as telogens. The use of telogens is more common in tubular reactors than in autoclaves.

Reactors can take one of two forms: either an autoclave, with a height-to-diameter ratio in the region of 5–20, or a tubular reactor, with a length-to-diameter ratio from a few hundred up to tens of thousands. An autoclave reactor is typically 10–15 ft in diameter, whereas a tubular reactor might be only 1 in. in diameter but may reach 2000 ft in length. As may well be imagined, these two divergent reactor geometries pose uniquely different chemical engineering problems requiring disparate control conditions.

Tubular and autoclave reactors with their disparate profiles require different methods of temperature control. The ethylene entering an autoclave reactor is precooled, so that it can absorb some of the heat generated by the polymerization reaction already in progress. As the temperature of the incoming stream rises, the initiator decomposes. The surface-to-volume ratio of autoclave reactors is so low that external cooling has little effect. Autoclave reactors are stirred vigorously to reduce the likelihood of localized hot spots. Typical average residence times of ethylene within an autoclave are in the region of 3–5 min. As unreacted ethylene exits the autoclave with the polyethylene product it carries away excess heat. In a tubular reactor the incoming ethylene is preheated to decompose the initiator, thereby starting polymerization. Once the reaction is under way, the excess heat is removed by external cooling, which is effective given the narrow diameter of the tube. The residence time of reactants within a tubular reactor is typically 20–60 sec.

The difference between the essential lack of mixing in the tubular reactor and the high levels of mixing in the autoclave presents distinct opportunities for the control of reaction conditions and hence the molecular structure of the products. One of the main differences lies in the fact that tubular reactors tend to produce a less homogeneous resin than autoclaves. With no large-scale mixing in a tubular reactor, the relative concentrations of initiator, ethylene, and telogen perforce vary along its length. The telogen is not noticeably consumed in the reaction; therefore its concentration relative to the ethylene increases along the length of the reactor. Thus the molecules made in the early part of the tube will typically have higher molecular weights than those made in the later part. In tubular reactors fresh initiator can be injected at various points. The product from a tubular reactor might thus mimic a blend of two or more components made in separate autoclave reactors.

The necessity to preheat the ethylene before it reaches a tubular reactor can lead to fouling problems. In cases where the initiator decomposes prematurely, or when the monomer itself autoinitiates polymerization, polyethylene can be produced in the preheater. Fouling occurs principally as a coating on the walls of the reactor inlet piping, where it can reduce flow. This unwelcome material can be high molecular weight or cross-linked polyethylene. When the coating sloughs from the walls it can cause blockages downstream in the reactor and separation system. If high molecular weight or cross-linked polyethylene makes it into the final product it can cause processing problems for the converter. A common problem associated with fouling is the occurrence of inhomogeneities ("gels") seen in films, sheets, and thin-walled parts.

A single reactor is by no means the only possible configuration employed; patents disclose configurations including multiple reactors in series and parallel, with one or both types within the same production line. Multiple reactors permit the production of resins with a wider variety of properties than that available from a single reactor.

Overall, the molecular structure of low density polyethylene resins is controlled largely by the reaction conditions, in contrast to high density polyethylene (HDPE) resins, whose structure is influenced by a combination of the catalyst and the reaction conditions. The precise control of operating conditions is crucial to producing a desirable homogeneous resin. Sensors abound throughout the system, their outputs being analyzed by one or more computers that adjust the various flow rates, initiator concentration, heat exchanger settings, etc. to optimize productivity.

Typical products of high pressure polymerization processes have melt indices (a measure of viscosity in the molten state) in the range 0.2–50 and densities falling between 0.90 and 0.94 g/cm^3. The effect of variations in the molecular character of low density polyethylene resins on properties is discussed in Chapter 5.

II. ZIEGLER–NATTA TYPE CATALYZED POLYMERIZATION

Ziegler–Natta type catalysis is one of two methods used commercially to produce high density polyethylene, the other being metal oxide catalysis. Ziegler–Natta catalysis is very flexible; the variety of catalyst systems that fall into this family is immense. In addition to ethylene, many other alkenes may also be polymerized, to produce either homopolymers when reacted in isolation or copolymers when the feedstock is a mixture of alkenes. Ziegler–Natta type polymerization takes place under conditions of relatively low temperature and pressure, well below those used in free radical polymerization but overlapping to some extent with the medium temperature and pressure conditions required by metal oxide catalysis.

A. Ziegler-Natta Catalysis

The variety of Ziegler–Natta catalysts is immense; numerous books and thousands of papers and patents have been written on the subject. It is beyond the scope of this work to go into the detailed chemical nature of these catalysts and their properties. What follows is therefore a much abbreviated outline of this extensive field. Readers wishing to learn more about this subject are directed to the bibliography at the end of this chapter.

Ziegler–Natta catalysts consist of a complex of a base metal alkyl or halide with a transition metal salt. Base metals from groups I–III of the periodic table may be used in combination with transition metals from groups IV–VIII. Naturally, some of these combinations are preferred for one type of polymerization or another. A classic example of a Ziegler–Natta catalyst suitable for the polymerization of ethylene to high density polyethylene is the complex of triethyl aluminum ($AlEt_3$) with titanium tetrachloride ($TiCl_4$). The proposed reaction mechanism is shown in Figure 3.

The active center is thought to comprise a titanium atom coordinated with four chlorine atoms and an alkyl group in an octahedral configuration with an empty site. An incoming ethylene molecule coordinates with the titanium at the vacant site, thereafter inserting between the metal and the alkyl. A new vacancy is thus generated at the apical position. Repetitive addition of ethylene molecules generates a polyethylene chain.

Ziegler–Natta catalysts used in solution polymerization are generally soluble, i.e., homogeneous catalysis, while those used in gas-phase reactors are supported on materials such as silica, i.e., heterogeneous catalysis. From the products of Ziegler–Natta catalysis it has been determined that the catalysts carry a mixture of active sites, which, owing to their slightly different chemical environments, polymerize the feedstock differently. Sites that tend to produce longer chains are less likely to incorporate alkyl comonomers, whereas shorter chains typically contain more comonomer. Polar comonomers deactivate Ziegler–Natta catalysts.

B. Low Pressure Production Facilities

When Ziegler patented the low pressure polymerization of ethylene to produce high density polyethylene, he disclosed only the chemistry involved, not the process. Accordingly, licensees of his patent had to develop the plant required to

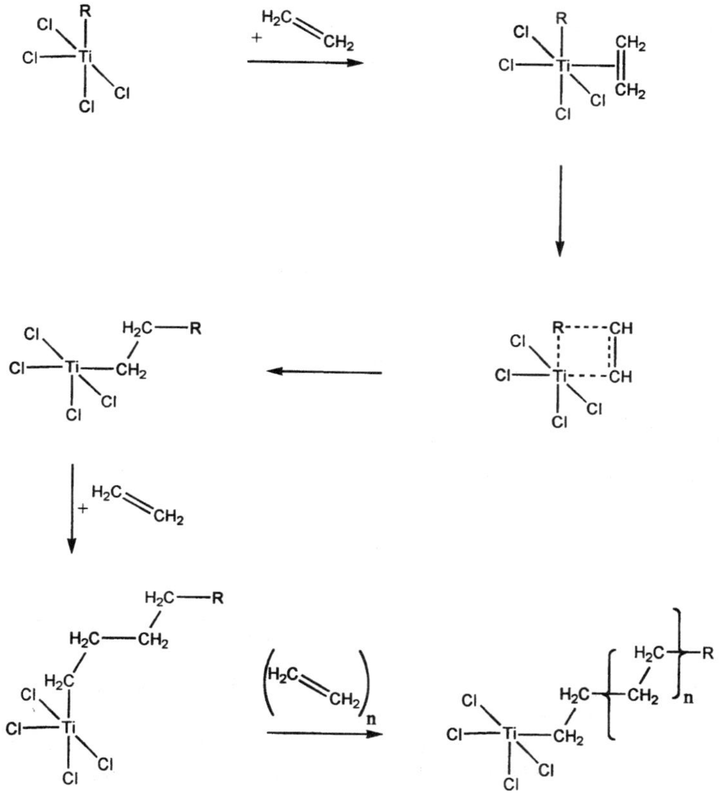

Production Processes

Figure 3 Ziegler–Natta catalyzed polymerization of ethylene.

commercialize the procedure. The result has been a wide variety of equipment incorporating the ingenuity of many technical groups. Low pressure polyethylene production facilities are somewhat more diverse than their high pressure production counterparts. Figure 4 outlines some of the key components common to many, but by no means all, low pressure polymerization facilities.

Ethylene feedstock **a** is fed into a compressor **1** that compresses it to the required polymerization pressure. The pressurized ethylene **b** is fed into a jacketed reactor **2**, where it is mixed with catalyst and cocatalyst **c** and a solvent **d** from tanks **3**, **4**, and **5**, respectively. Heat released by the polymerization reaction is removed by external cooling through the jacket and vaporization of solvent **e**, which is cooled and liquefied in a condenser **6**, then returned to the reactor **f**. From the reactor the product stream **g**, consisting of polyethylene, solvent, and

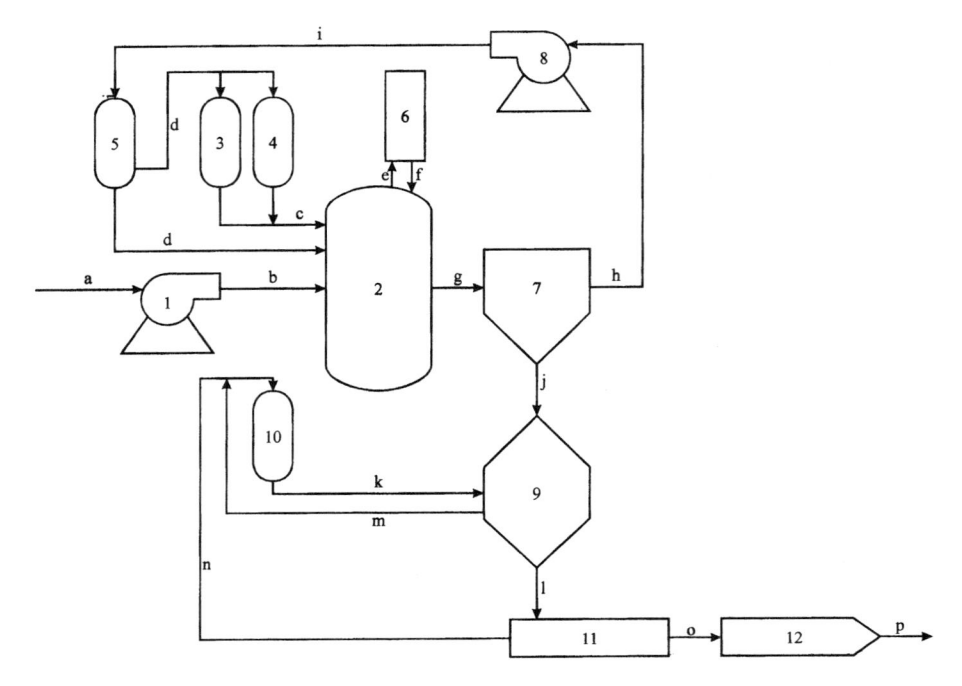

Figure 4 Schematic representation of low pressure polmerization of ethylene. 1, Compressor; 2, reactor; 3, catalyst reservoir; 4, cocatalyst reservoir; 5, solvent reservoir; 6, condenser; 7, separator; 8, compressor; 9, de-ashing unit; 10, solvent reservoir; 11, dryer; 12, extruder. a, Ethylene; b, pressurized ethylene; c, catalyst–cocatalyst mixture; d, solvent; e, solvent vapor; f, condensed solvent; g, polyethylene, catalyst, and solvent; h, solvent vapor; i, recycled solvent; j, polyethylene and catalyst; k, de-ashing solvent; l, wet polyethylene; m, recycled de-ashing solvent; n, recycled de-ashing solvent; o, raw polyethylene; p, polyethylene pellets.

catalyst, is fed into a separator **7**. Solvent is flashed off **h** to a compressor **8** for recycling **i**. Polyethylene and catalyst **j** are fed into a de-ashing unit **9**, where the catalyst is deactivated and dissolved with a de-ashing solvent **k** from tank **10**. ("Ash" is the common name for catalyst or catalyst residue entrapped in the product.) Wet polyethylene **l** is fed from the de-ashing unit into a dryer **11**, and the excess solvent is recycled **m**. Solvent residues are stripped from the polyethylene in the dryer and recycled **n**. The dry polyethylene powder **o**, consisting of granules approximately 500–1,000 μm in diameter, is transferred to an extruder **12**, where it is homogenized and blended with additives consisting primarily of antioxidants. The product is extruded as thin strands, which are chopped into pellets to form high density polyethylene resin **p**, which is then ready for packaging and shipping.

Variations on the above scheme abound. It is possible to dispense with the solvent entirely and perform the reaction in the gaseous state. Such a process is known as gas-phase polymerization, the reaction taking place within a fluidized bed in which the catalyst is supported on an inert substrate. De-ashing is not always desired or required. The product is not necessarily pelletized; sometimes the granular product is simply dry blended to incorporate additives and homogenize it. As in the case of high pressure polymerization, the low pressure process is not limited to a single reactor; two or more can be used in series or in parallel to produce a resin with the desired molecular characteristics.

C. Low Pressure Reaction Conditions

The low pressure polymerization of ethylene is substantially more efficient than the high pressure process in terms of the percentage of monomer converted to polymer on each pass through the reactor. In the case of the solution or slurry processes, wherein the ethylene is dissolved in, or diluted with, an inert solvent, the conversion may approach 100%. The difference lies in the heat transfer mechanisms of the two systems. In the case of high pressure polymerization, surplus monomer acts as a heat sink, necessitating a large excess to meet this need. In the case of low pressure, solvent-diluted, ethylene polymerization, the inert solvent acts as a heat sink, being boiled off to a condenser. The solvent is cooled and liquefied in the condenser prior to its return to the reactor. In addition to removing heat, solvent helps prevent fouling by flushing the product out of the reactor. In gas-phase polymerization the conversion of gaseous ethylene to solid polyethylene generates heat that is offset to a large extent by an accompanying reduction in pressure as gas is converted to solid. The balance of the heat is removed by external cooling coils and heat transfer to the cool incoming ethylene gas.

The molecular characteristics of the resins produced by Ziegler–Natta processes are controlled by the nature of the catalyst, the presence of chain transfer, agents, and the reaction conditions. A major factor governing the selection of control variables is an economic one. Resin attributes must be offset against marketability and production costs. As is the case with all commercial polyethylene production, it is necessary to balance the various reactions of initiation, propagation, chain transfer, and termination. The principal variations in the product are its average molecular weight and molecular weight distribution. Plants displaying a wide variety of designs are currently being operated in a diversity of manners with many different catalysts. Certain plant configurations and operating conditions give rise to products that have specific individual characteristics, such as small amounts of long chain branching, the inclusion of pendant methyl groups, and other defects of the linear structure. The effects of molecular characteristics on the properties of polyethylene are discussed in Chapter 5.

Ziegler–Natta catalysts are unlike those used in most commercial reactions in that they are incorporated within the raw product. During polymerization the catalyst molecules are surrounded with, and become encapsulated by, the polyethylene particles that they have generated. The entrapped catalyst is removed from the reactor along with the product. The presence of catalyst residues in the product can lead to problems of chemical degradation, corrosion of processing equipment, discoloration, and contamination of sensitive materials (such as food or medicines) that come in contact with finished items. To avoid such problems, the raw polyethylene is often subjected to a de-ashing step to deactivate and remove the catalyst residue. The solvent used for de-ashing is ordinarily an aliphatic alcohol, a common choice being butanol. In noncritical applications or when the catalyst activity is extremely high—and therefore its potential for contamination negligible—the de-ashing step is eliminated. If the de-ashing step is to be bypassed, catalyst activities in excess of 100,000 g of product per gram of transition metal are generally required. In addition to contamination of the resin with catalyst residue, there is another compelling reason to use highly active catalysts: cost. Ziegler–Natta catalysts are expensive to produce and require special handling and transportation. The goal is to use as little catalyst as possible, thereby holding the cost to less than 5¢ per pound of product. Highly active catalysts reduce the cost of production not only directly but also indirectly, as the step of de-ashing is no longer required.

Low pressure polyethylene production facilities are designed to operate in one of three modes: with the reaction mixture in a gas phase, in a liquid phase, or as a slurry. If they are operated under the conditions of gas or slurry polymerization, the reaction temperature is maintained well below the melting temperature of the product, typically 30–100°C. This prevents melting or softening of the product, which could lead to agglomeration of the polyethylene granules into clumps that could foul the reactor. Solution reactors, in which the feedstock and product are dissolved in an inert solvent, can be operated at higher temperatures, typically 100–200°C. To a large extent the molecular weight of the product increases as the reaction temperature decreases. This is due to a decreased rate of chain transfer relative to chain propagation, the chain transfer reactions being the principal molecular weight limiter. The reaction temperature is selected in conjunction with the operating pressure, the two factors being interdependent to some extent. Operating pressures can vary from as low as atmospheric up to 300 psi. Pressures in the upper part of this range are normally associated with the slurry or gas-phase processes. Typical residence times for the ethylene and catalyst in the reactor are on the order of 1–4 hr, with extremes being 0.5–10 hr.

Catalysts used for slurry or gas-phase polymerization are normally supported on an inert substrate, commonly silica or alumina. These types of catalysts are referred to as "heterogeneous" as they reside in a separate phase from the ethylene. In gas-phase reactors the heterogeneous catalyst is supported in a fluidized bed to ensure that it forms an intimate mix with the monomer. The use of a fluidized bed also lessens the chance of the reactor being fouled by densely packed polyethylene granules. Catalysts used in solution reactors are generally not supported and are referred to as "homogeneous" because they are dissolved in the solvent with the ethylene. Supported catalysts frequently have a higher activity than nonsupported ones. Products made using heterogeneous catalysts are less likely to require de-ashing than those made with homogeneous ones. The support medium can cause some problems during subsequent conversion of the resin. The usual support media remain trapped in the polymer and are present when the resin is converted to end product. Silica and alumina are somewhat abrasive and can accelerate wear in extruders and other processing equipment.

Ziegler–Natta catalyst systems invariably contain a variety of active sites due to differences in physical or chemical structure. Each type of site behaves differently, having a unique activity level or a propensity to cause branching, chain transfer, or rearrangement of molecular chains. Disparate active sites will react to changes in reaction conditions in different ways. Accordingly, each site will produce polyethylene molecules with a different distribution of chain lengths, branching, saturation, etc. It is therefore desirable to use a catalyst that has a limited range of active sites, the activity of which can be more closely controlled by reaction conditions. The newly developed metallocene catalyst systems meet this need, each catalyst containing only one type of active site. The purity and reproducibility of a catalyst system is of paramount importance to manufacturers. Batches are often checked independently in a pilot reactor before being used in commercial production.

In gas- and slurry-phase polymerization the catalyst particles act as templates on which polyethylene granules grow. Thus the shape of the catalyst and support determines the shape of the product granules. This is especially important if the material is not pelletized prior to shipping and use. The size and shape of a resin's granules affect its packing, mixing, and transfer characteristics. The factors controlling bulk flow properties of polymer particles are quite complex, and unsuitable attributes can lead to poor processability. Problems for the end user may include poor dispersion of pigment, plugging of pneumatic transfer lines, or fouling of extruder feed mechanisms.

The molecular weight of polyethylene produced with a specific catalyst under a given set of reaction conditions can be controlled to some extent by the presence or absence of a chain transfer agent, invariably hydrogen. The addition of hydrogen limits the average molecular weight of chains by terminating the growth of one chain and initiating a new one.

Low pressure polyethylene production lines are not limited to a single reactor. Two or more reactors may be arranged in series or parallel to generate a more

III. METALLOCENE POLYMERIZATION

Metallocene polymerization is used to produce a distinctive range of ethylene–α-olefin copolymers that are less polydisperse than those available from Ziegler–Natta catalyst systems. This is achieved because each catalyst contains only one type of active site, all of them polymerizing the available monomers in an identical fashion. The fact that each catalyst consists of one type of active site has earned metallocenes the name of *single-site catalysts* (SSC) or *uniform-site catalysts*, the former being more widely used. The net result is a uniform polymeric product that has a most probable molecular weight distribution ($\overline{M}_w/\overline{M}_n \approx 2.0$) and homogeneous comonomer incorporation. Metallocene polymerization takes place under mild reaction conditions similar to those used in Ziegler–Natta production facilities.

A. Metallocene Catalysis

Metallocene catalysts make up a family that includes many hundreds of variations. In their most general form, metallocene catalysts comprise a metal atom from group IV of the periodic table (titanium, zirconium, or hafnium) attached to two substituted cyclopentadienyl ligands and two alkyl, halide, or other ligands with a methylalumoxane $(-\text{MeAlO}-)_n$ cocatalyst commonly known as MAO. The cyclopentadienyl rings may be linked ("bridged") by a silicon or carbon atom to which are attached hydrogen atoms, alkyl groups, or other substituents. The cyclopentadienyl rings may be part of a larger indenyl ring structure. Examples of metallocene catalysts are shown in Figure 5. When one of the cyclopentadienyl rings is replaced by a heteroatom, such as nitrogen, attached to the bridging atom, the molecule is sometimes referred to as a "constrained geometry catalyst," an example of which is shown in Figure 5d. In order to obtain high catalytic activity, a large excess of methylalumoxane must be used. Molar ratios of aluminum to the group IV metal atom typically range from 50 to 1000. Methylalumoxane is expensive and remains as a contaminant in the resin, so high catalyst activity is crucial from the standpoints of economy and product purity. Other cocatalysts such as fluorinated organoboron compounds are also used. The coordination polymerization that occurs at the active site of a metallocene catalyst is similar to the reaction that occurs at the active site of a Ziegler–Natta catalyst, as illustrated in Figure 3. The molecular characteristics of metallocene resins are controlled by the structure of the catalyst, the monomer/comonomer feedstock

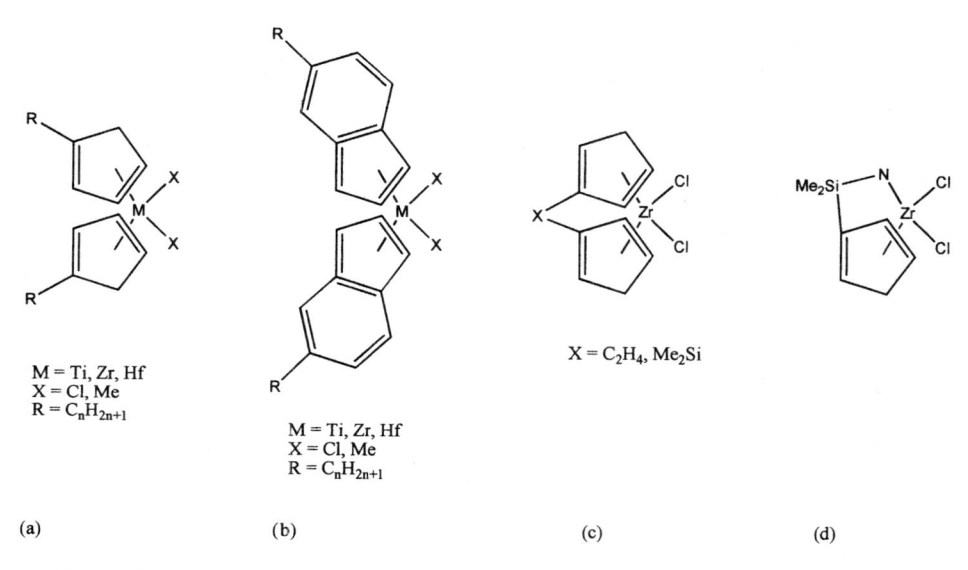

Figure 5 Structure of gneric metallocene catalyst and examples of specific catalyst molecules. (a) Generic metallocene structure; (b) generic metallocene with indenyl substituents; (c) bridged metallocene; (d) "constrained geometry catalyst."

ratio, and polymerization conditions. Reviews of the development of metallocene catalysts can be found in papers by Horton [3] and Kaminsky [4].

IV. METAL OXIDE CATALYZED POLYMERIZATION

The metal oxide catalyzed polymerization of ethylene takes place under conditions of medium pressure and temperature. It is practiced according to two methods, the Phillips process and the Standard Oil process (also known as the Indiana process); the former is based on chromium oxide catalysis, whereas the latter uses molybdenum oxide. The Phillips process dominates the field of metal oxide catalysis. Chromium oxide catalysis is the most widely used method for the production of high density polyethylene, accounting for a little more than half the worldwide output.

A. Chromium Oxide Catalysis

The first step in preparing Phillips-type catalysts is the impregnation of a support of highly porous silica or aluminosilicate of low alumina content with an aqueous solution of chromic acid or chromium trioxide. After drying, the catalyst is activated by heating to 500–700°C in an oxidative environment; this creates surface silyl chromate species, which are precursors to the active site [5].

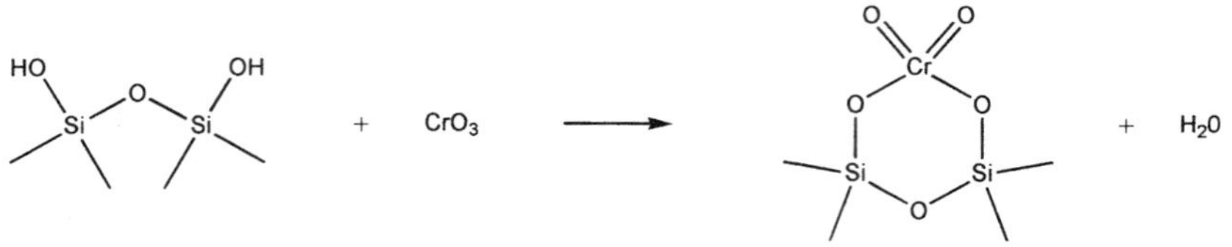

Reduction to a lower valence (active) state takes place in ethylene at high temperature and may occur in the reactor [6].

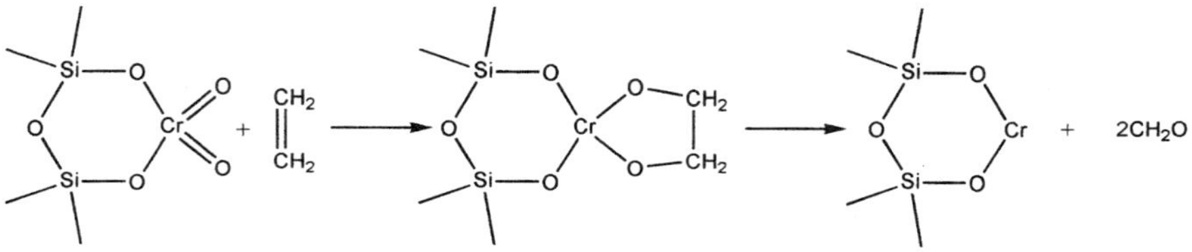

The active site is thought to comprise a chromium–carbon bond that complexes an incoming ethylene molecule, which then proceeds to insert between the chromium and the carbon [5].

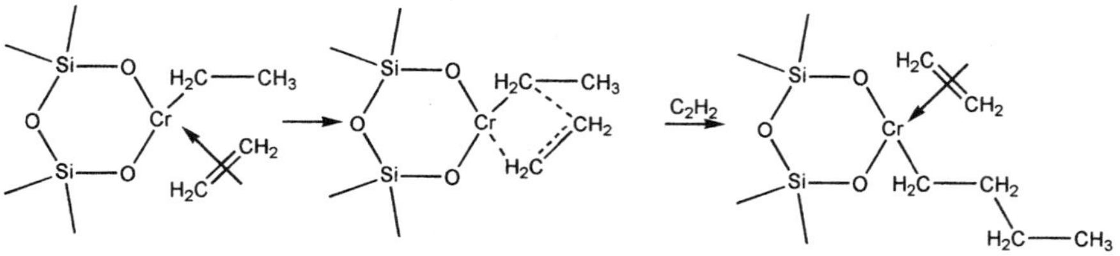

The simplified reaction scheme shown above is just one of many possible mechanisms. The exact oxidation state of the active chromium is a matter of some debate; every valence state from Cr(II) to Cr(VI) has been proposed [7]. Variations on the basic mechanism abound, with numerous supports and additives being reported. One of the more successful variations includes a chromium titanium complex [8].

The nature of the support plays a key role in sustaining catalyst activity over prolonged periods. Friable catalyst supports fragment as the polymer particles grow, exposing new precursor sites that are reduced to the active form by ethylene in the reactor. Such supports expedite the growth of polymer at many active centers.

B. Medium Temperature and Pressure Production Facilities

There are two commercial polyethylene production processes catalyzed by metal oxides that employ conditions of temperature and pressure intermediate between those used in the Ziegler–Natta and high pressure processes. Both the Phillips and Standard Oil processes were fully developed prior to licensing, but since their introduction the technical ingenuity of various production teams has led to a divergence of operating procedures and equipment. As in the case of the Ziegler–Natta process, the reaction may be carried out in the gas phase, in a slurry, or in solution. A schematic diagram illustrating the key components of commercial plants is presented in Figure 6.

Details of the equipment used vary with the polymerization conditions.

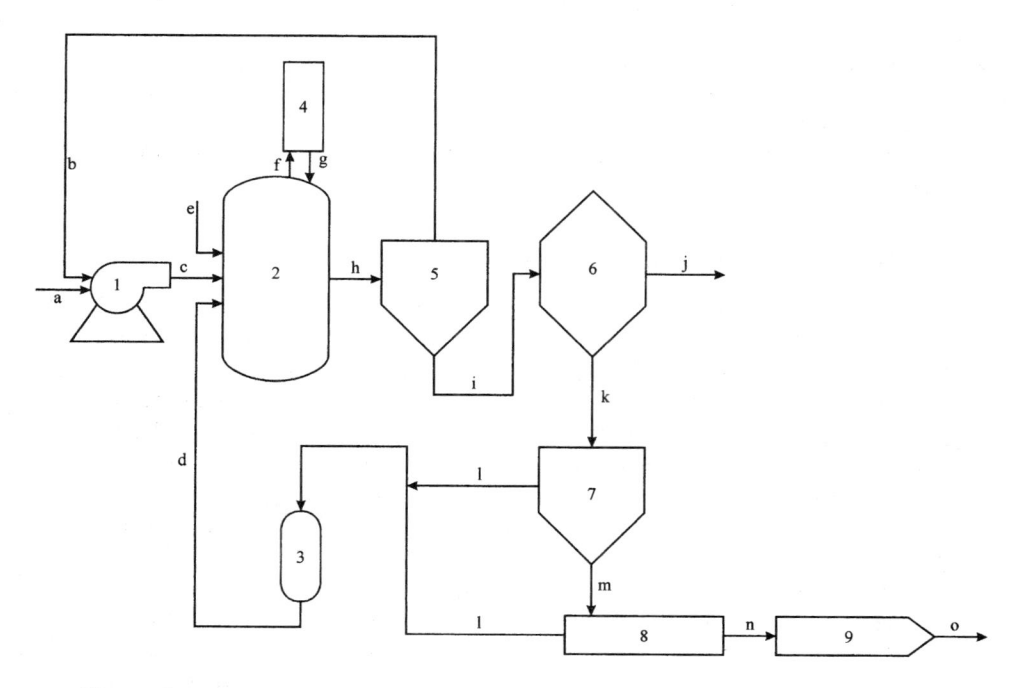

Figure 6 Schematic representation of Phillips polymerization process. 1, compressor; 2, reactor; 3, solvent reservoir; 4, condenser; 5, flash drum; 6, filter/centrifuge; 7, separator; 8, dryer; 9, extruder. a, Fresh ethylene; b, recycled ethylene; c, pressurized ethylene; d, solvent; e, catalyst; f, solvent vapor; g, condensed solvent; h, polyethylene, ethylene, solvent, and catalyst; i, polyethylene, solvent, and catalyst; j, spent catalyst; k, polyethylene and solvent; l, recycled solvent; m, damp polyethylene; n, raw polyethylene granules; o, polyethylene pellets.

Fresh (make up) ethylene **a** and recycled ethylene **b** enter a compressor **1**. The pressurized monomer **c** is fed to a reactor **2**, where it is diluted by solvent **d** from tank **3** and catalyst **e** is injected. The reactor is a stirred vessel that is equipped with an external cooling jacket and a condenser **4**. Heat released by polymerization is removed by the external cooling and the vaporization of solvent **f**, which is cooled and liquefied before being returned to the reactor **g**. The raw product **h** comprising polymer, monomer, catalyst, and solvent is fed to a flash drum **5**. The unreacted ethylene is evaporated off at low pressure and returned for recycling **b** after appropriate purification. Polyethylene, catalyst, and solvent **i** are transferred to a filtration or centrifuge unit **6** where the spent catalyst is removed **j**. The polyethylene and remaining solvent **k** is passed to a separator **7**, where the solvent is removed for recycling **l**. Damp polyethylene **m** is conveyed to a dryer **8** for removal of the residual solvent. Dry granules **n** are transferred to an extruder **9**, where they are homogenized and blended with additives, principally antioxidants.

Some extruders are equipped to handle damp polyethylene, drying the raw product concurrently with extrusion. The product is extruded as thin strands that are chopped into pellets to form high density polyethylene resin **o**, which is then ready for packaging and shipping in a manner common to other polyethylene resins.

C. Medium Pressure Reaction Conditions

Medium pressure and temperature polyethylene production facilities have much in common with Ziegler–Natta type polymerization plants. The reaction can be carried out in the gas phase, in a slurry, or in solution. There is a fair degree of overlap between the ranges of temperature and pressure employed in the two processes, but on the whole the Phillips and Indiana processes are run slightly hotter and under greater pressure. Removal of excess heat produced by polymerization is also accomplished in similar ways. Conversion of ethylene to polyethylene can reach 98% in some solvent-diluted processes. Unless otherwise stated, the following description of operating conditions relates primarily to the Phillips process, which dominates this type of production.

The principal reaction variables are the operating temperature and pressure and the catalyst type and concentration. The controllable resin attributes are principally the average molecular weight and the molecular weight distribution. To a lesser extent, the incorporation of a small degree of long-chain branching is effected by chain transfer reactions. Comonomers can also be incorporated, but this occurs on a less frequent basis than in the Ziegler–Natta processes. Monitoring and control of reaction conditions are performed by computer.

The selection of reaction conditions depends on the mode of operation in which the plant is run and the choice of solvent (if used). Overall reaction temperatures are generally in the range of 30–200°C. Solution polymerization, requiring

Production Processes

dissolution of the product, takes place in the upper part of this range, typical reaction temperatures being 120–170°C. The selection of an operating temperature is limited by the solution temperature of polyethylene in the solvent; poorer solvents require higher temperatures. Some common solvents are n-alkanes such as heptane and hexane and various saturated alicyclic solvents such as cyclohexane. In the slurry process it is not necessary that the solvent dissolve the product; thus, lower reaction temperatures are used, typically in the range of 30–100°C. In solvent-diluted processes the liquid holds 0.1–30% by volume of the monomer, typically 2–15% in the slurry process and 25–30% in the solution process. The use of a moving or fluidized bed in the slurry process permits loadings of feedstock in the solvent greater than could otherwise be tolerated. Gas-phase reactors are always run at temperatures below the melting point of the product, in the range from 30°C to approximately 100°C. This prevents the polyethylene granules from fusing into lumps that could foul the reactor. Gas-phase reactions take place in a fluidized bed consisting of reaction product granules into which the catalyst is injected. With no inert solvent to help absorb the heat of polymerization, it is very important to avoid hot spots in the bed to prevent fouling. The molecular weight of the product is dependent upon the reaction temperature. All else being equal, an increase in reaction temperature will generate a product with a lower molecular weight; at the same time, the conversion rate of monomer to polymer will increase.

Operating pressures varying from 25 to 800 psi have been reported. Solvent-diluted polymerization generally takes place under pressures of 300–700 psi, while gas-phase polymerization normally takes place at pressures of less than 100 psi. Additional cooling of the reaction mixture can be accomplished by precompression of the monomer to a higher pressure than that required for polymerization. When the ethylene expands into the reactor it cools the reactants. The Indiana process is generally run in liquid hydrocarbon solvents at temperatures in excess of 100°C and at a pressure of approximately 1000 psi.

Catalysts in medium pressure polyethylene production processes are metal oxides, or some complex thereof, supported on a refractory oxide such as alumina or silica or a mixture of the two. Catalysis is always heterogeneous, the catalyst remaining solid under reaction conditions. The Phillips process uses chromium based catalysts, and the Indiana process uses molybdenum-based ones. Both types of catalysts can also be used to polymerize other 1-alkenes. The complexity of catalysts has increased since their introduction in 1957. Present day catalysts can consist of metal oxides combined with various metal hydrides or organic complexes. Compounds incorporating metals other than chromium can be used as cocatalysts. The complexity is such that there is some degree of overlap between metal oxide and Ziegler–Natta catalysts. In addition to acting as a diluent for the monomer, solvents can also play a role as a cocatalyst, changing the valence state of the metal. The use of chain transfer reagents is little mentioned in the patent literature.

Catalysts are introduced into the reactor by a carrier stream of solvent or monomer. The productivity of the process is controlled by the reactivity of the catalyst and the rate at which it is injected. Metal oxide catalysts are extremely sensitive to poisoning by a variety of impurities. The effect of catalyst deactivation is to lower the productivity and to decrease the average molecular weight of the product. Catalyst poisons include air and water, both of which must be rigorously excluded from the solvent and monomer streams. The greater the volume of solvent used, the greater is the potential for catalyst poisoning. This is an important factor in the move away from solvent-diluted processes toward a greater use of gas-phase polymerization.

Heterogeneous catalyst residues can be removed from the product by filtration or centrifugation. This step can be bypassed if the catalyst is sufficiently active that its presence in the final product is of no consequence. Problems associated with an excess of catalyst in the product can include a green off-color caused by chromium salts and abrasive wear of processing equipment by alumina or silica. Spent catalyst recovered in the de-ashing step can be processed to reactivate it and then returned to the reactor.

BIBLIOGRAPHY

Boor J. Ziegler–Natta Catalysts and Polymerizations. New York: Academic Press, 1979.
Kaminsky W, Sinn H, eds. Transition Metals and Organometallics as Catalysts for Olefin Polymerization. Berlin: Springer-Verlag, 1987.
Masters C. Homogeneous Transition-Metal Catalysis. London: Chapman and Hall, 1981.
Parshall GW, Ittel SD. Homogeneous Catalysis: The Applications and Chemistry of Catalysis by Soluble Transition Metal Complexes. New York: Wiley, 1992.
Quirk RP, ed. Transition Metal Catalyzed Polymerizations: Alkenes and Dienes Part B. Harwood Academic, New York, 1983.

REFERENCES

1. P Ehrlich, RN Pittilo. J Polym Sci 43:389, 1960.
2. WL Mattice, FC Stehling. Macromol 14:1479, 1981.
3. AD Horton. Trends Polym Sci 2:158, 1994.
4. W Kaminsky. Macromol Chem Phys 197:3907, 1996.
5. JP Hogan. J Polym Sci Part A-1 8:2637, 1970.
6. TJ Pullukat, RE Hoff, M Shida. J Polym Sci Polym Chem Ed 18:2857, 1980.
7. J Boor, Ziegler–Natta Catalysts and Polymerizations. New York: Academic Press, 1979, pp 280–284.
8. TJ Pullukat, M Shida, RE Hoff. In: RP Quirk, ed. Transition Metal Catalyzed Polymerizations: Alkenes and Dienes. New York: Harwood Academic, 1983, pp 697–712.

4
Morphology and Crystallization of Polyethylene

I. SEMICRYSTALLINE MORPHOLOGY

Semicrystalline polymers are those that consist of two or more solid phases, in at least one of which molecular chain segments are organized into a regular three-dimensional array, and in one or more other phases chains are disordered. The noncrystalline phases form a continuous matrix in which the crystalline regions are embedded. Most polyolefins are semicrystalline; their specific morphology is governed by molecular characteristics and preparation conditions. Polyethylene is no exception to this rule; it is all but impossible to prepare a solid specimen of polyethylene that is not semicrystalline. All commercial polyethylene products are semicrystalline. The physical properties exhibited by polyethylene products are governed by the relative proportions of the crystalline and noncrystalline phases and their size, shape, orientation, connectivity, etc. with respect to one another.

A. Three-Phase Morphology

Solid polyethylene consists of a three-phase morphology as shown schematically in Figure 1. Submicroscopic crystals, called crystallites, are surrounded by a noncrystalline phase comprising a partially ordered layer adjacent to the crystallites and disordered material in the intervening spaces.

The ordered phase of semicrystalline polyethylene consists of crystallites in which molecular chain segments are packed in regular arrays. The thickness of crystallites in molded high density polyethylene samples is commonly in the range of 80–200 Å with lateral dimensions of up to several micrometers. Low density and linear low density polyethylene samples typically have somewhat thinner crystallites with smaller lateral dimensions. The noncrystalline regions separating crystallites can vary from approximately 50 Å to 300 Å. Linearly

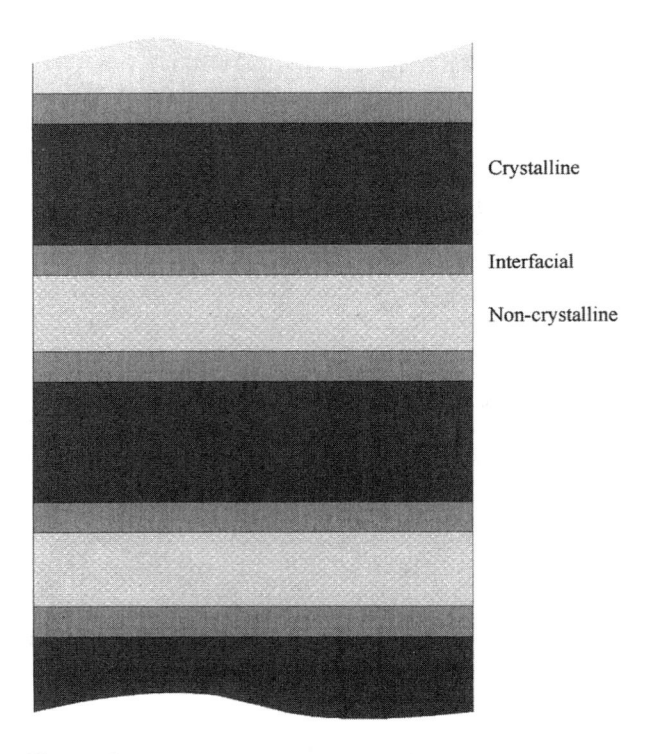

Figure 1 Schematic representation of the three phases present in solid polyethylene.

extended segments of polyethylene molecules traverse the thickness of the crystallites approximately perpendicular to their lateral dimensions. Allowing for some degree of chain tilt, the length of an extended molecular segment is slightly greater than the thickness of the crystallite it inhabits. The extended length of a typical polyethylene molecule may be 10,000 Å or more, which is many times the thickness of the crystallites. In such a sample it is clear that segments from any given molecule must traverse the thickness of one or more ordered and disordered regions many times.

Disordered molecular segments in the noncrystalline regions are normally continuous with those in the crystallites. These segments comprise the three types shown in Figure 2. Noncrystalline segments can traverse the intercrystalline zone to connect to an adjacent crystallite, they can double back to attach themselves to the crystallite from which they originated, or they can terminate in a chain end. These three configurations are known respectively as ''tie chains,'' ''loops,''

Morphology and Crystallization

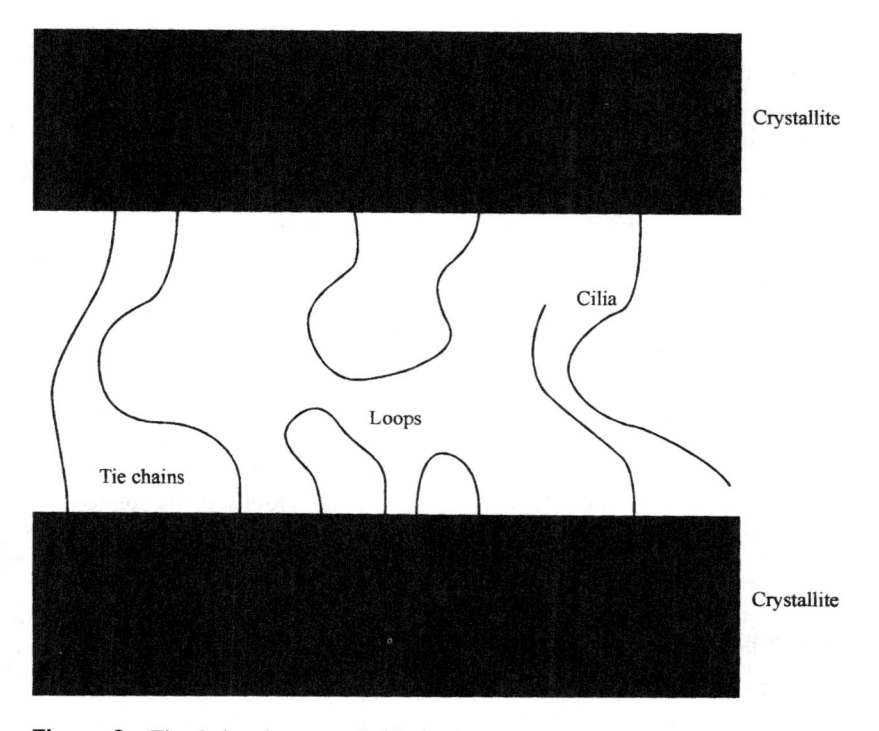

Figure 2 Tie chains, loops, and cilia in the noncrystalline phases of polyethylene.

and "cilia." The relative proportions and nature of chains following each of these trajectories is the subject of ongoing investigation. It is widely accepted that the ratio of chains returning to the same crystallite versus those spanning the disordered regions is a function of the molecular weight, branching level, and crystallization conditions pertaining to the sample. The degree of connectivity between neighboring crystallites plays a major role in determining the physical properties of a sample.

At the boundary between disordered regions and crystallite surfaces exists a third phase made up of chain segments that exhibit varying degrees of order as they traverse it. This third phase is termed the interfacial region, interface, or partially ordered region. The character of the interfacial region is very important because it serves to link the two primary phases. Without the interfacial layer to connect the disordered and crystalline regions, polyethylene would be a weak material. The nature of this third phase is the subject of much discussion. Attempts to measure or define it invariably provide somewhat contradictory information [1–3]. Defining the character of the interface is problematic because it consists of chain segments displaying a range of conformations. Those chain segments adjacent to the crystallite surface would be expected to be substantially more ordered than those in the proximity of the disordered region. The boundary between the disordered and partially ordered regions is inevitably indistinct and should not be considered a sharp line of demarcation.

The term "amorphous" is widely used to describe the noncrystalline regions as a whole. This can be misleading, because the noncrystalline phase encompasses both disordered and partially ordered regions, which are not necessarily isotropic. Even the disordered chain segments are not in truly random configurations. Constraints placed upon the noncrystalline regions by the growth of crystallites during solidification reduce the number of degrees of freedom available to even the most disordered of regions. In the case of commercially fabricated items, orientation is frozen into the noncrystalline regions when the crystallites form. To avoid ambiguity, regions outside crystallite boundaries are referred to generically in this volume as noncrystalline regions, being subdivided into disordered and interfacial (partially ordered) regions.

B. The Importance of Semicrystallinity

The concept of semicrystallinity is important because polyethylene can be considered to be a composite of crystalline and noncrystalline regions. Polyethylene that consisted solely of crystalline matrices would be a friable material, and a totally amorphous sample would be a highly viscous fluid. In practice, of course, polyethylene is a tough, resilient material. The arrangement of the three phases with respect to each other, their relative proportions, and their degree of connectivity determine the properties of a polyethylene sample. Neither pure crystalline nor pure amorphous polyethylene samples are available, so the properties of each phase must be extrapolated from those of partially crystalline samples.

Given the estimated properties of each phase and assuming a model of connectivity via the interface, it is possible to explain the mechanical behavior of polyethylene samples. To carry out this type of analysis it is desirable to have an accurate knowledge of the relative proportions of each of the three phases. In practice the single term "degree of crystallinity" is frequently used to characterize the semicrystalline nature of polyethylene samples. Quantification of the three phases of polyethylene can be made experimentally by several methods [1–3], the degree of crystallinity by many more [4–6]. The experimental determination of the relative amounts of each of the phases is addressed in Chapter 6. Some of the most commonly used and most important descriptors of a polyethylene sample, such as density and stiffness, are closely related to its crystallin-

Morphology and Crystallization

ity level. Many of the physical properties of polyethylene can be inferred from a knowledge of its degree of crystallinity.

II. POLYETHYLENE CRYSTAL UNIT CELLS

In crystallographic terms the *unit cell* is the smallest entity that contains all the information required to construct a complete crystal. Unit cells consist of parallelepipeds, such as cuboids and rhombohedrons, containing a small number of atoms. The lengths of the sides can be identical or different, as can the angles made by the intersection of any two faces. The lengths of axes are designated a, b, and c, and the angles between faces, α, β, and γ, as illustrated in Figure 3. A complete crystal can be constructed from a unit cell by translating it repeatedly along each of its axes a distance equal to the length of that axis. The result of this process is shown in Figure 4.

The unit cells of most nonpolymeric compounds conatain an integral number of complete molecules. In contrast, polymeric unit cells contain short segments from one or more molecular chains. By convention, the c axis of a polymeric unit cell is designated as being parallel with the chain axis of its molecular segments.

Polyethylene exhibits three types of unit cells—orthorhombic, monoclinic, and hexagonal—all of which are relatively simple compared to other polyolefins and to polymers in general. The orthorhombic unit cell is by far the most common; for all practical purposes it may be considered to be the only one present in commercial samples.

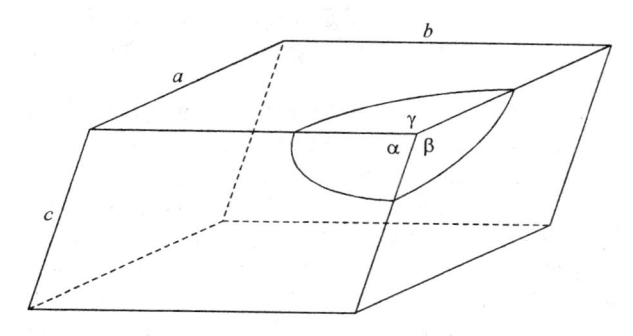

Figure 3 Generic unit cell.

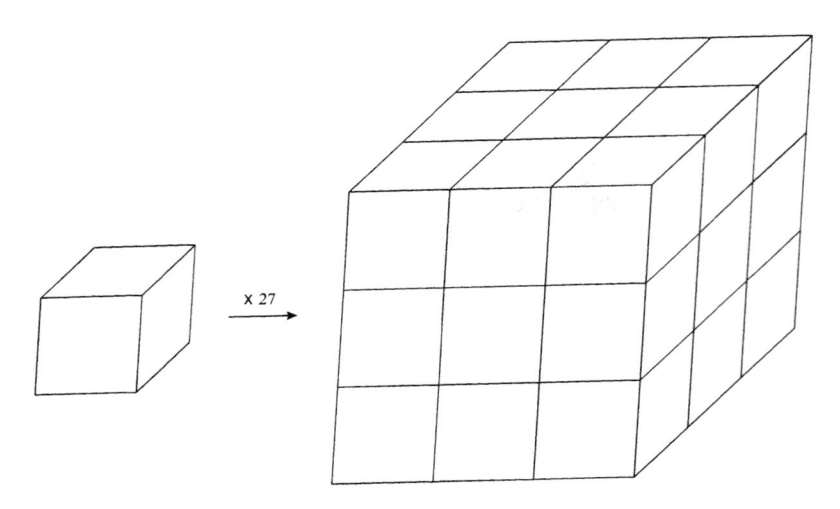

Figure 4 Crystal matrix built up by translation of a unit cell.

A. Orthorhombic Unit Cell

The orthorhombic unit cell is a cuboid, each of its axes having a different length while the angles made by adjoining faces are all 90°. Each unit cell contains a complete ethylene unit from one chain segment and parts of four others from surrounding chain segments, for a total of two per unit cell. The orthorhombic unit cell is variously illustrated in Figure 5.

The dimensions of the a, b, and c axes of an unperturbed polyethylene unit cell are reported to be 7.417, 4.945, and 2.547 Å respectively [7–9]. These values were measured for high density polyethylene at room temperature. The density of a unit cell with these dimensions is 1.00 g/cm^3. This value is widely accepted and is commonly used in the calculation of the degree of crystallinity from sample density.

The dimensions of the orthorhombic unit cell are not constant. Low density polyethylene and linear low density polyethylene have larger a and b axis dimensions than high density polyethylene, while the length of the c axis remains essentially constant [10–13]. Experiments involving X-ray diffraction [14], ^{13}C NMR spectroscopy [15], and the chemical digestion of the noncrystalline regions of linear low density polyethylene samples [16] indicate that branches larger than the methyl group are largely excluded from the crystalline lattice. There is some indication that a small percentage of ethyl branches can be incorporated into the crystalline region of rapidly quenched samples [17]. Thermodynamic calculations support the hypothesis that branches larger than a methyl group cannot be accom-

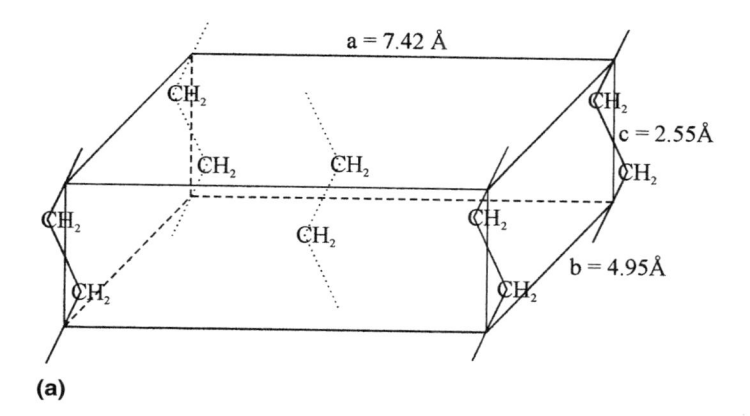

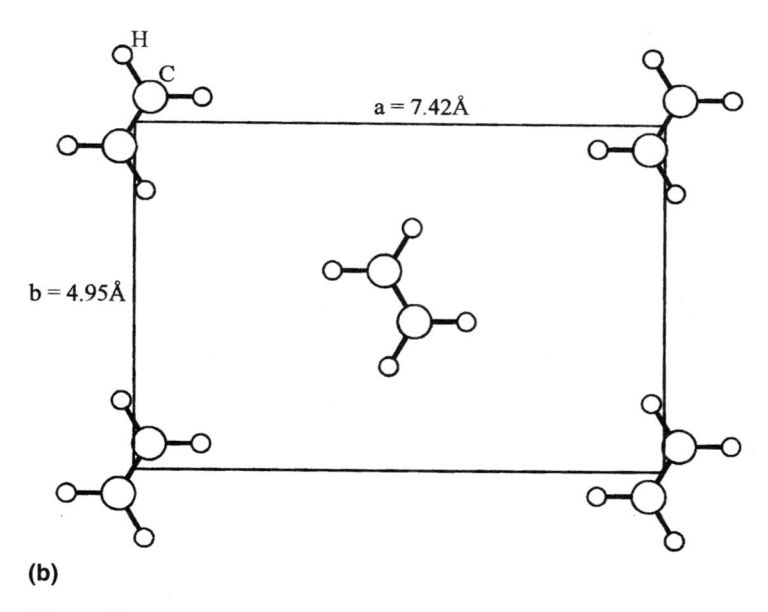

Figure 5 Polyethylene orthorhombic crystal habit. (a) Orthogonal view; (b) view along the c axis; (c) space-filling representation viewed along the c axis.

(c)

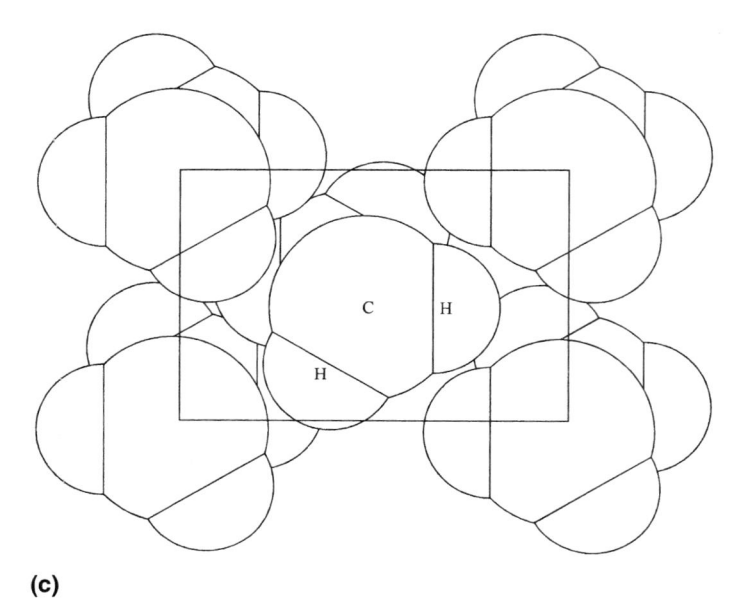

Figure 5 Continued

modated by the crystal lattice without its disruption [18]. However, it is clear that the extent of expansion of the unit cell is related to the comonomer content of the sample. A possible explanation of the unit cell expansion involves the concentration of short-chain branches in the interfacial regions due to their exclusion from crystallites. The high concentration of branches causes overcrowding of the interface, resulting in the underlying crystallite expanding slightly to relieve the steric interference.

B. Monoclinic Unit Cell

The monoclinic crystal form of polyethylene (also referred to as the triclinic form) is a metastable phase formed under conditions of elongation [18,19]. It may be present to a small extent in commercial samples that have undergone cold working after initial molding. Temperatures in excess of 60–70°C cause it to revert to the orthorhombic form [20]. The monoclinic phase is sometimes present in nascent granules of ultrahigh molecular weight polyethylene due to

Morphology and Crystallization

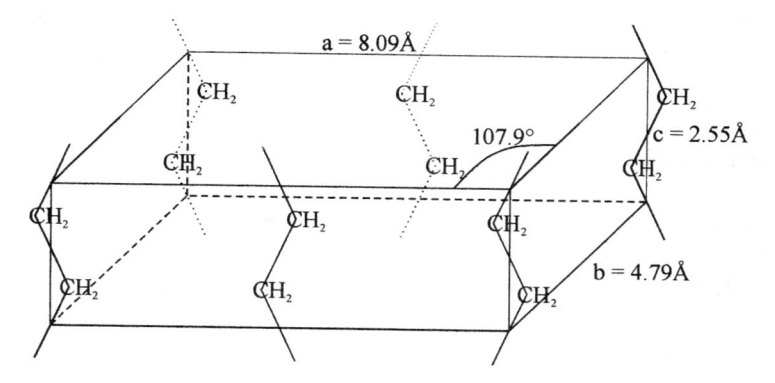

Figure 6 Polyethylene monoclinic unit cell.

expansion during the polymerization process [21,22]. The configuration and dimensions of the monoclinic unit cell are shown in Figure 6.

C. Hexagonal Unit Cell

The hexagonal crystal form of polyethylene is a laboratory curiosity produced by crystallization at extremely high pressures [23,24]. It is not produced under any conditions currently pertaining to commercial processes. The hexagonal phase is also sometimes known as the "rotator" phase, because individual chain

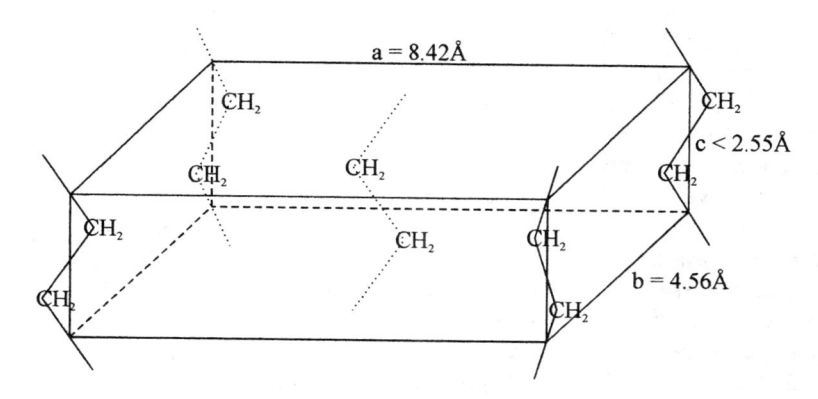

Figure 7 Polyethylene hexagonal unit cell.

stems are rotated at random phase angles with respect to their neighbors. The dimensions and configuration of the hexagonal unit cell are shown in Figure 7.

III. CRYSTALLITES

Polymer crystallites are microscopic crystals that are embedded in a matrix of noncrystalline material. Our understanding of the nature of crystallites has evolved over the years since polyethylene was discovered. An improved knowledge of the crystallization process and the use of modern analytical techniques have given us a clear understanding of these entities.

A. Fringed Micelles

Shortly after the discovery of polyethylene its semicrystalline nature was demonstrated by wide-angle X-ray diffraction [25]. Diffraction patterns collected from solid samples exhibited both the amorphous scattering haloes characteristic of disordered materials and the discrete rings peculiar to microcrystalline samples. From the angular dispersion of the crystalline rings, the average dimensions of the crystalline phases were calculated to be on the order of 100–200 Å The calculated size of the crystalline domains was considerably smaller than that usually associated with nonpolymeric crystals. The term "crystallite" was applied to these small crystalline regions embedded in a disordered matrix. Given that the low density polyethylene available at that time was known to have molecular lengths at least an order of magnitude greater than the crystallite thickness, it was necessary to posit a morphology wherein the two known dimensions were reconciled. Thus was born the "fringed micelle" model, a representation of which is shown in Figure 8.

In the fringed micelle model, crystallites are envisaged as small bundles ("micelles") of parallel extended linear chain segments disposed randomly in a matrix of disordered chains. Unlike the crystals of nonpolymeric materials that consist of an integral number of identical molecules, it was recognized that the crystallites in polyethylene contain chain portions from many molecules. The "fringed" designation came from the manner in which noncrystalline chain segments were assumed to be attached to the crystalline stems via partially aligned noncrystalline chain segments. The fringe was required to splay out to accommodate the reduced density in the noncrystalline region relative to that of the crystalline regions. The lateral dimensions of micelles were thought to be limited by the packing of the fringes. The addition of crystalline stems to a growing micelle results in an increasingly large deviation between the orientation of the crystalline stems and the adjoining segments in the fringe. This causes increased steric hin-

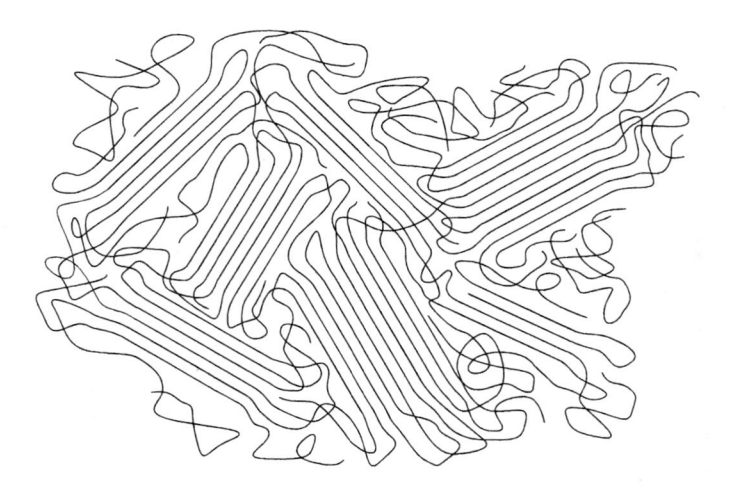

Figure 8 Fringed micelle model.

drance in the fringe, which eventually prevents the addition of further chain segments to the micelle. This effect is illustrated in Figure 8.

B. Lamellae

The fringed micelle model prevailed for more than a decade until the discovery, in the mid-1950s, that various polymers could crystallize from solution to form lozenge-shaped crystals with lateral dimensions several orders of magnitude greater than their thickness [26–29].

1. Polyethylene Crystals Grown from Solution

When polyethylene crystallizes from very dilute solution at elevated temperatures, it forms lamellar crystals with thicknesses of the order of 100 Å—similar to the dimensions previously calculated for melt-crystallized samples—and lateral dimensions varying from a few micrometers up to more than 100 μm. These structures were termed "single crystals" because their electron diffraction scattering patterns were consistent with that of a single extended crystal. It has since been observed that polyethylene crystals grown from dilute solution have a distinct layer of noncrystalline material at their basal faces [30]. Figure 9 shows an electron micrograph of solution crystals of high density polyethylene grown from a dilute solution in xylene.

Electron diffraction also revealed that the c axis of the unit cell was almost

Figure 9 Electron micrograph of polyethylene crystals grown from dilute solution from xylene. (From Ref. 95.)

perpendicular to the lateral dimensions of such crystals [28]. As the molecular length of polyethylene was known to be many times greater than the crystal thickness, some form of molecular folding had to be invoked. The existence of such folding was never in doubt, but the particulars of its nature were the subject of much discussion. The surface of solution crystals was observed to be microscopically smooth, leading to the suggestion that the crystals were composed of crystalline stems linked to their neighbors by a series of regular tight folds [31–33]. This model required that consecutive chain segments from a single molecule be laid down successively on the growing face of the crystal in a mode known as (tight fold) adjacent reentry. An opposing view held that many molecules contributed chain segments to each layer of the growing face. The resulting basal surface was thought to consist of an irregular array of loops connecting crystalline stems in the manner of a contemporary telephone switchboard [34]. These two models are shown schematically in Figure 10.

Between the two extremes, various compromise models were proposed. Two such models are illustrated in Figure 11.

Although of much academic interest, solution crystals have no commercial significance.

2. Lamellae Crystallized from the Melt

Transmission electron microscopy of the replicas generated from fracture surfaces of polyethylene samples broken at liquid nitrogen temperatures, in the late

Morphology and Crystallization

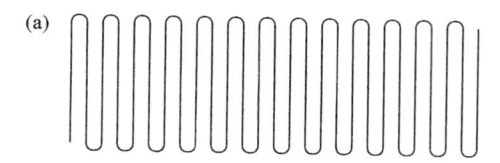

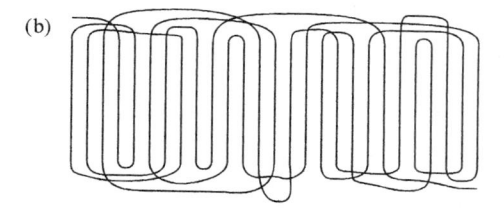

Figure 10 (a) Adjacent reentry with tight folds. (b) Switchboard model.

1950s, revealed that lamellar crystals were present in melt-crystallized specimens [35–37]. The newly observed structures were found to have thicknesses similar to those of solution crystals, with a similar aspect ratio. With the discovery of lamellae, the fringed micelle model inevitably lost favor.

In the last four decades the technology for imaging the semicrystalline morphology of polyethylene has improved dramatically. With the advent of permanganic acid etching of fractured and cut surfaces [38] and chlorosulfonic acid staining of microtomed sections [39], it became possible to image lamellar morphologies directly in the electron microscope. Recent developments using ruthenium tetroxide vapor staining techniques [40,41] and atomic force microscopy permit the routine study of the arrangement of lamellae in polyethylene. Structures with dimensions as small as a few tens of angstroms can be clearly distinguished when correctly oriented with regard to the viewing axis. Transmission electron micrographs that show stacked lamellar morphologies in very low molecular weight and low molecular weight fractions of high density polyethylene are shown in Figures 12 and 13, respectively. Examples of electron micrographs of linear low density polyethylene and low density polyethylene are shown in Figures 14 and 15, respectively. In general, the higher the degree of crystallinity of a sample, the better the overall organization of lamellae within it. Some highly branched samples with a very low degree of crystallinity (less than approximately 15% by volume) exhibit highly fragmented lamellae somewhat akin to fringed micelles.

The interpretation of polyethylene morphology is not an exact science. All techniques for morphological characterization rely on averaging the properties

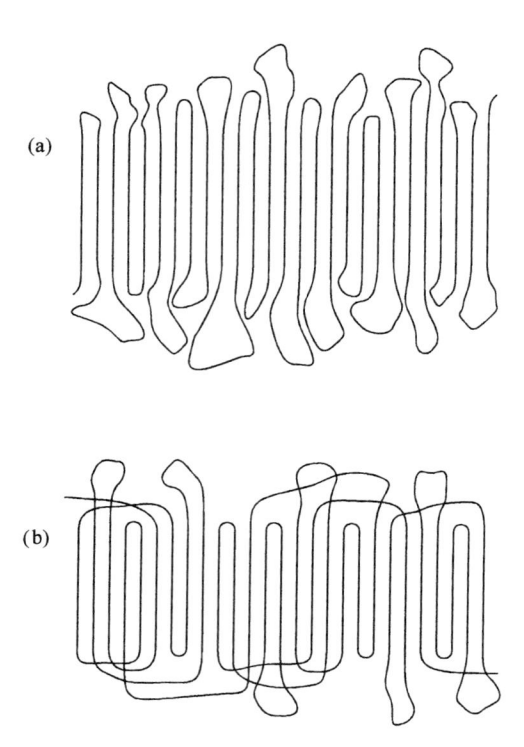

Figure 11 (a) Loose loops with adjacent reentry. (From Ref. 96.) (b) Composite model. (From Ref. 97.)

of a group of atoms; the sample size may vary, but there is never an exact determination of the trajectories of molecules. Great care must be exercised when analyzing data from electron microscopy and other techniques that provide information on morphology. Individual electron and atomic force micrographs are rarely representative of the material as a whole. Many photomicrographs must be examined in order to build up an overall picture of the morphology of a sample. Staining techniques reveal structures at varying levels of contrast depending on a multitude of conditions, some of which are outside the control of the operator. Dimensions of crystallite structures derived from electron microscopy should be compared with data from other techniques, such as small-angle X-ray diffraction and longitudinal acoustic mode Raman spectroscopy, before conclusions are drawn.

The terms "lamella" and "crystallite" are often used interchangeably when referring to polyethylene morphology. Strictly speaking, a lamella is a specific type of crystallite, which in turn is a microscopic crystal. As the major growth habit of semicrystalline polyethylene is lamellar, the two terms are used interchangeably in this work unless a clear distinction is made.

Morphology and Crystallization

Figure 12 Regimented arrays of lamellae in a very low molecular weight ($M_w \approx 11{,}000$) fraction of high density polyethylene quench-crystallized from the melt. (From Ref. 98.)

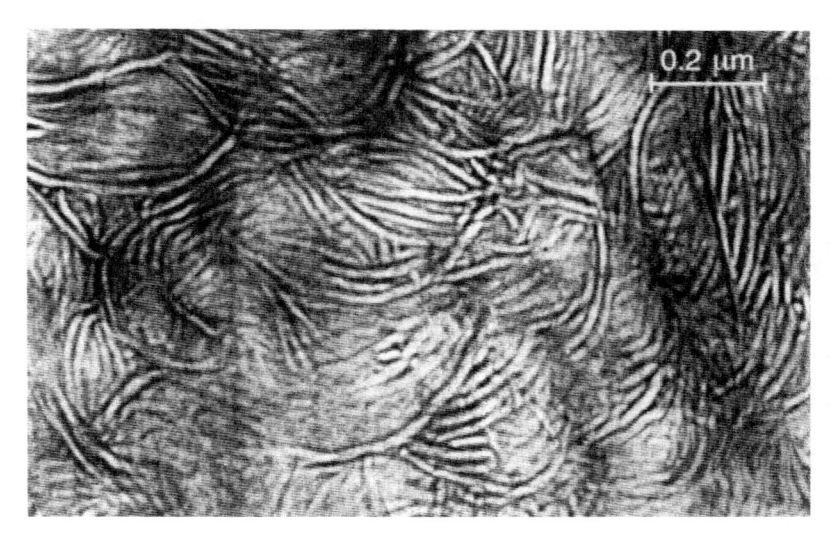

Figure 13 Partially ordered lamellar arrays in a low molecular weight ($M_w \approx 46{,}000$) fraction of high density polyethylene quench-crystallized from the melt. (From Ref. 98.)

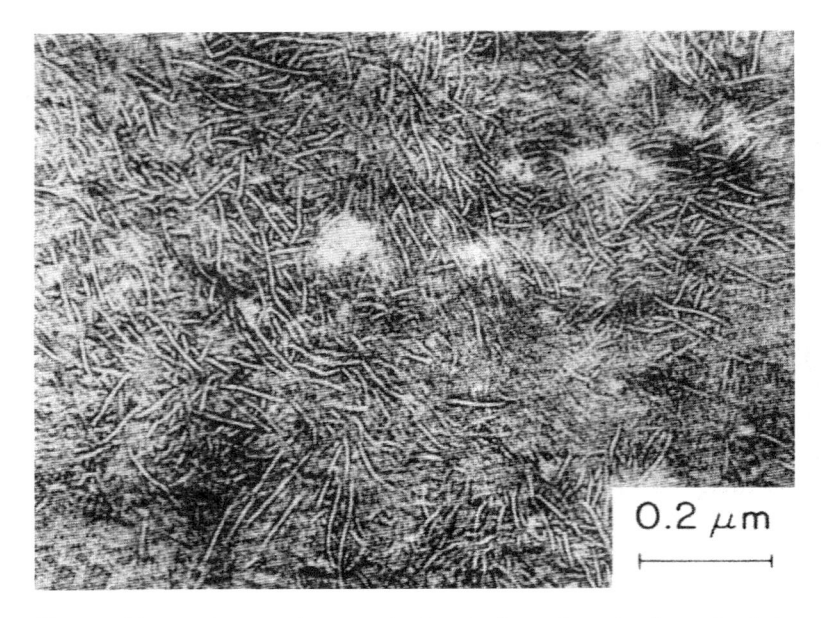

Figure 14 Partial lamellar organization in linear low density polyethylene ($M_w \approx$ 100,000; density ≈ 0.920 g/cm^3) crystallized from the melt. (From Ref. 41.)

Figure 15 Disorganized lamellae in low density polyethylene ($M_w \approx 450,000$; density ≈ 0.918 g/cm^3) crystallized from the melt. (From Ref. 41.)

IV. CRYSTALLIZATION MECHANISMS

Polyethylene crystallizes from the molten state or solution when prevailing conditions make the crystalline state more stable than the disordered one. The processes by which polyethylene crystallizes reflect the properties of the disordered state from which the ordered phase condenses. Thus, for instance, levels of chain entanglement, molecular dimensions, and viscosity all play important roles. The factors affecting the structure of the disordered state are both intrinsic to the molecules and extrinsic to the surrounding conditions. The principal molecular factors are the molecular weight, molecular weight distribution, and concentration, type, and distribution of branches. External factors include temperature, pressure, shear, concentration of solution, and polymer–solvent interactions.

The driving force behind crystallization is thermodynamic. The system strives to achieve the lowest possible free energy state, but the process is impeded by factors that affect its rate, such as viscosity, chain entanglements, and noncrystallizable entities such as cross-links and branchpoints. As with all thermodynamic processes occurring at constant pressure, the direction of change is governed by the free energy of the competing states according to

$$\Delta G = \Delta H - T \Delta S$$

where

ΔG = change of Gibbs free energy
ΔH = change of enthalpy
T = absolute temperature
ΔS = change of entropy

In the case of a polyethylene melt transforming to a semicrystalline solid, ΔG will be negative, i.e., the change will be favorable, when the enthalpy released upon crystallization exceeds the loss of entropy multiplied by the absolute temperature. The lower the temperature for a given system, the greater will be the driving force of crystallization. From a kinetic point of view, the increased viscosity associated with decreasing temperature slows the crystallization process. The precise mechanism of crystallization is determined by the balance of thermodynamic and kinetic factors. The kinetics of crystallization is addressed in Section V of this chapter. As the process of crystallization is physical in nature, it is, of course, reversible when the prevailing conditions no longer favor the crystalline state.

The lowest energy configuration of an isolated polyethylene chain is the linearly extended all-trans form, which is also known as the "planar zigzag." This configuration is shown in Figure 16. The arrangement by which pairs of hydrogen atoms attached to carbon atoms alternate from side to side along the chain provides for minimum steric interference.

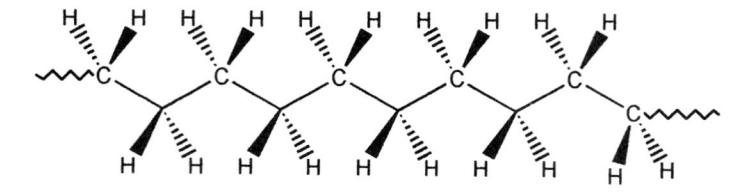

Figure 16 Segment of polyethylene molecule in an all-trans configuration.

The physical mechanisms of the crystallization process control the resulting morphology of the solid state and hence the properties of the product. Despite the fact that polyethylene has been studied intensively for over 60 years, crystallization mechanisms are incompletely understood. On a macroscopic scale, polyethylene morphologies can be accurately described, but the trajectories of the individual molecules, which are controlled by crystallization mechanisms and which ultimately determine properties, are known only to a first approximation. The study of polyethylene crystallization and morphology involves the deduction of molecular mechanisms from crystallization kinetics and solid-state bulk properties.

The formation of polyethylene semicrystalline morphologies involves two distinct processes: crystallite initiation and crystallite development. The former is usually referred to as "nucleation" or, more specifically, "primary nucleation." Crystallite development proceeds by the addition of chain stems to the surfaces of a nucleus or growing crystallite and by "secondary nucleation," whereby new lamellae are presumed to initiate from the surfaces of existing ones. It is by crystal growth and secondary nucleation that the semicrystalline morphology permeates the complete body of a sample. Primary nucleation, crystal growth, and secondary nucleation can all occur simultaneously. The term "ternary nucleation" is sometimes used to describe the manner by which molecular segments are laid down on the growing face of a crystallite. (It should be noted that there is a lack of consistency in the scientific literature regarding the definition of secondary and ternary nucleation. When reading accounts of research, care must be taken to ensure that the precise meaning of the term "nucleation" is understood.)

A. Nucleation

1. Primary Nucleation

Primary nucleation is the process by which the formation of new, independent crystallites is initiated. It takes place when a bundle of adjacent chain stems adopt parallel, linearly extended configurations and pack together to form an assembly

Morphology and Crystallization 85

of unit cells that exceeds a critical size. The structure so formed is the nucleus of a crystallite upon which further growth takes place. The formation of a nucleus can occur spontaneously due to statistical variation within the disordered phase, in which case it is termed "homogeneous nucleation," or alternatively it can take place on a preexisting surface, in which case it is termed "heterogeneous nucleation." If the preexisting surface is the core of a polyethylene crystallite that was not completely destroyed by dissolution or heating, the process is termed "self-nucleation." During polyethylene crystallization, heterogeneous nucleation is by far the most prevalent process. In the following sections concerning nucleation, only melt crystallization is discussed, as this is most pertinent to commercial processes.

a. Homogeneous Nucleation. Homogeneous nucleation is the process whereby nuclei form spontaneously in a polyethylene melt as it cools. When a polyethylene melt is cooled below the equilibrium melting temperature of polyethylene crystals, the conversion to a semicrystalline state becomes thermodynamically favorable. In the absence of heterogeneous nuclei, crystallization will not occur until the melt is supercooled by 50°C or more [42]. The reason for this supercooling is the energy barrier that homogeneous nuclei must overcome to reach stability. This principle is illustrated in Figure 17. Homogeneous nuclei form when statistical variation within a polymer melt results in the formation of

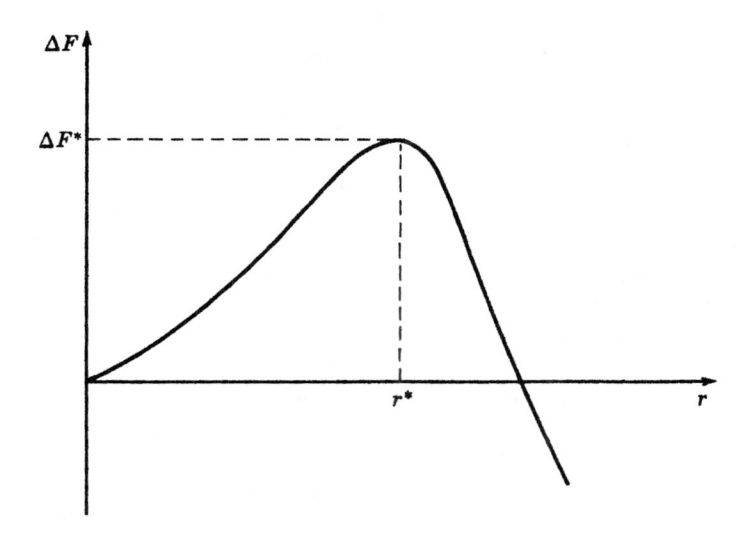

Figure 17 Schematic representation of the free energy barrier to homogeneous nuclei formation in terms of surface-to-volume ratio for spherical nuclei. (From Ref. 99.)

ordered assemblies of chain segments larger than a critical size. The critical size is a function of the surface area of the incipient nuclei. When embryonic nuclei form, a new surface is produced, the area of which depends upon the dimensions of the structure. The formation of an interface between the solid and liquid states is costly from an energetic standpoint; if it is not overcome by an accompanying reduction in enthalpy, the embryonic nucleus will be unstable. The surface-to-volume ratio of the incipient nucleus is important; more compact nuclei will be favored over extended ones [43].

The energy barrier decreases as the tempreature falls, thus lowering the critical nucleus size. It follows that the rate of formation of stable nuclei will rise as the critical size is reduced, the net result being an increasing nucleation rate as the tempreature falls.

As the formation of homogeneous nuclei is a statistically random occurrence, it can be facilitated by systematic perturbation of the molten state. The easiest method of introducing anisotropy into the melt is by orientation. Orientation preferentially aligns molecular segments, thereby reducing the entropic barrier that must be overcome to form a stable nucleus. Even a small perturbation of the melt is effective in initiating crystallization; however, the greater the orientation, the greater will be the effect. Orientation is invariably introduced during the commercial fabrication of polyethylene items. Elevated pressure is also efficacious in raising the primary nucleation rate, but the effect is small in view of the extremely high pressures that must be applied. Thus pressure plays a negligible role in promoting nucleation during commercial processing.

b. Heterogeneous Nucleation. Heterogeneous nucleation is the initiation of crystallite growth by foreign bodies within the molten phase. These inclusions can take many forms; virtually any contaminant that is solid at the crystallization temperature, such as catalyst residues or dust particles, can act as a nucleating agent. The mechanism is the same in all cases; a group of extended polyethylene chains deposits on the surface of preexisting solid, thereby reducing the free energy of the system. Thus, foreign solids act as preferential nucleation sites because they lower the critical size of polyethylene nuclei relative to homogeneous nucleation. Crystalline contaminants with crystallographic spacings matching those of the polyethylene unit cell are particularly effective nucleating agents. Favored nucleation sites include crystal grain boundaries, cracks, discontinuities, and cavities. Heterogeneous nucleating agents are rarely deliberately added to polyethylene, a sufficient concentration occurring naturally to effectively promote crystallization. It has been convincingly shown that heterogeneous nuclei are the primary initiators of crystallization in polyethylene [42]. A review of the mechanisms by which heterogeneous nucleation is believed to occur can be found in Ref. 42.

Morphology and Crystallization

c. Self-Nucleation. Self-nucleation occurs when molten polyethylene contains small seed crystals that were not destroyed during the melting process. These preexisting polyethylene crystalline entities act as nuclei upon which crystallites can develop. Self-nucleation is uncommon during the fabrication of polyethylene products, because the conditions inherent in commercial processes are too severe to permit preexisting polyethylene crystallites to persist.

2. Secondary Nucleation

When polyethylene crystallizes it does so in a way that results in the replacement of a disordered melt with a semicrystalline morphology in which lamellae are aligned with their neighbors to a greater or lesser extent. The degree of alignment is related to the degree of crystallinity of the sample; the higher the level of crystallinity, the greater will be the organization of lamellae. The reason for the ordering of lamellae is that the vast majority of them are not initiated independently by primary nucleation events but rather are initiated at the surface of preexisting lamellae. Thus secondary nucleation is responsible for the orderly filling of a solid sample with semicrystalline structures.

The effects of secondary nucleation can be seen clearly in the electron microscope for both solution- and melt-crystallized samples. Solution-grown crystals exhibit such features as terracing of progressively smaller crystallites stacked upon a larger one and spiraling stacks of lamellae overgrowing a crystal surface. Sectioning of melt-crystallized samples often reveals Y-shaped lamellae with a secondary lamella branching at a shallow angle from a primary lamella (such a structure is visible in the center of Figure 13). An example of secondary lamellar growth occurring from solution is shown in Figure 18. The process by which secondary nucleation takes place is not well defined. It is postulated that cracks, faults, or other discontinuities on the surface of crystallites serve as nucleation sites.

B. Crystal Growth Mechanisms

1. Primary Crystallization

The structural details of the arrangement of polyethylene chains within crystallites are clearly understood. The dimensions of the unit cell can be unequivocally determined for any sample and change little with variations in molecular or processing variables. The same cannot be said of the details of crystallite surfaces. As the configurations of the individual chains comprising the lamellar surface cannot be observed directly, our understanding comes from deductions based upon a variety of analytical techniques that reveal bulk morphology or average

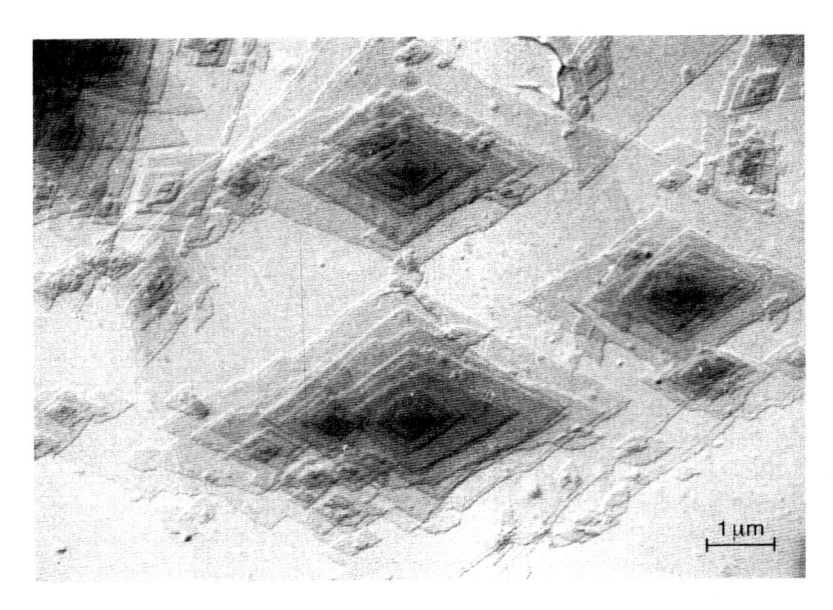

Figure 18 Overgrowth of spiraled terraces on solution-grown crystals. (From Ref. 100.)

chain segment properties. Such deductions are strongly biased by underlying assumptions regarding the mechanisms of crystallization.

The chain trajectories postulated for the basal planes of solution-grown crystals greatly influenced subsequent discussions regarding the nature of melt-crystallized polyethylene. It was assumed by many that because the thicknesses of solution crystals and melt-crystallized lamellae were similar, the mechanisms leading to their formation and hence their surface structures would be analogous. For almost two decades a model embodying adjacent reentry with tight folds for melt-crystallized lamellae, based on the structure postulated for solution-grown crystals, was embraced by the majority of investigators [44]. A small body of researchers refuted this supposition, advancing models in which a substantial proportion of the chains comprising lamellar surfaces adopted configurations other than tight folds, such as nonadjacent reentry, adjacent reentry with loose loops, and the traverse of the intercrystalline regions to form tie chains [45–47]. Some of the most notable models suggested were based upon concepts proposed for solution crystal morphology. Various models are illustrated schematically in Figure 19.

At the extremes of the range of models hypothesized are adjacent reentry with tight folds [44] and the model in which all chains that leave a crystallite enter a partially ordered region, from which they can either span the interlamellar

Morphology and Crystallization

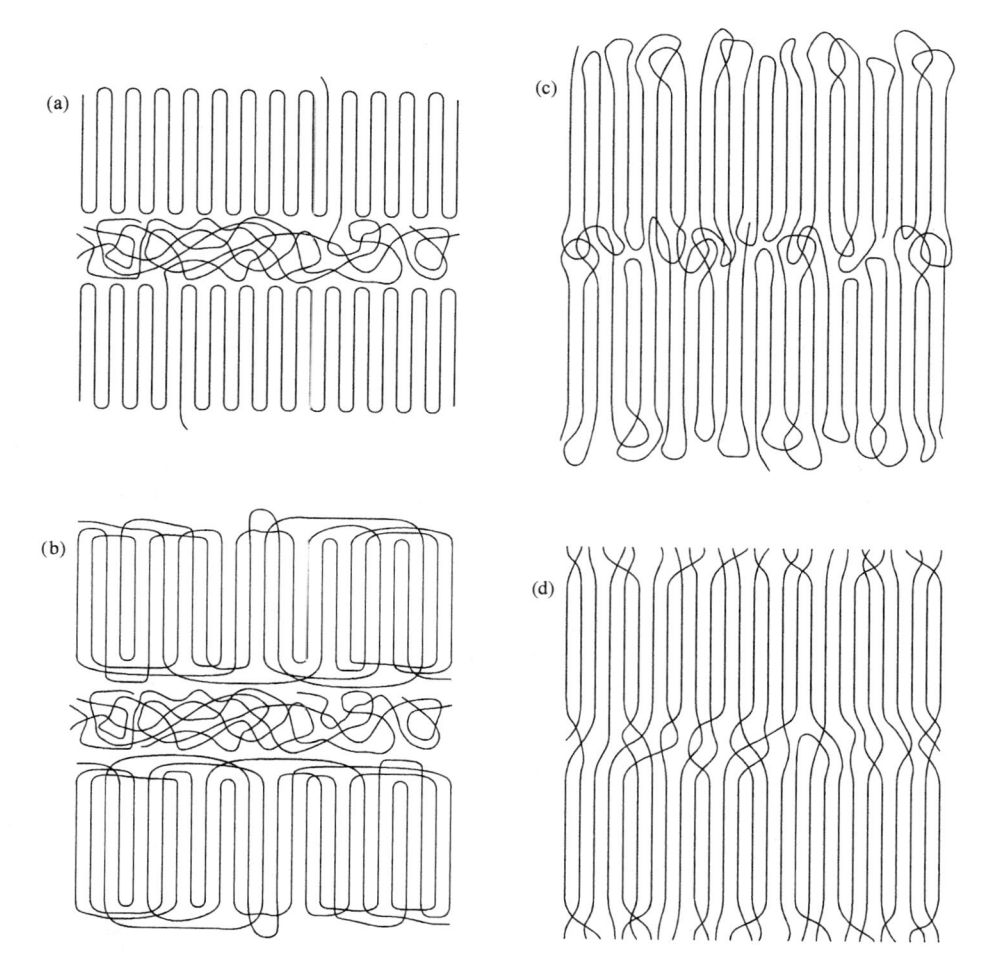

Figure 19 (a) Tight folds with adjacent reentry. (From Ref. 33.) (b) Switchboard model. (From Ref. 34.) (c) Loose loops with adjacent reentry. (From Ref. 96.) (d) Departure of chains from immediate environs of crystallite. (From Ref. 45.)

zone or return to the lamella from which they originated [45]. The latter is often referred to as the "Flory model." It is important to appreciate that these two extremes require diverse modes of crystallization and result in very different interfaces between ordered and disordered domains. The key difference between the two extremes lies in the number and type of connections between lamellae. The adjacent reentry model requires that chain stems be laid down continuously from a single molecule in a series of hairpin bends until its length is exhausted. The basal surfaces of the lamellae, which consist primarily of tight folds, are thus relatively smooth, with very few chains making the transition between the crystalline and disordered zones. In this model there is no partially ordered layer between crystalline and noncrystalline regions. At the other extreme, the Flory model requires that a single molecule be incorporated into a number of different lamellae, connecting them by a series of tie chains spanning the noncrystalline regions. Loose loops emanating from lamellar surfaces can intertwine with those from neighboring lamellae to form additional physical links between adjacent crystallites. According to this model the lamellae are intimately connected by a network of tie chains and entangled loops. The Flory model requires an interface between the crystalline and disordered regions. Naturally, a rigorous understanding of the properties of polyethylene requires that the correct structural model be adopted. For instance, it may well be imagined that the response of the postulated morphologies to applied stress would be very different. A crucial difference between the adjacent reentry and Flory models lies in the mechanism of crystallization.

For a molecule to adopt a regular series of hairpin bends it would have to crystallize independently of its neighbors. Only one molecule could be laid down on a growing crystal face at a time. This was thought to occur by a process of ''reeling in,'' whereby a chain was drawn out of the surrounding melt and laid down on the growing face of a crystallite in a series of tight folds. This is illustrated schematically in Figure 20. This process involves the longitudinal translation of chain segments along a path defined by adjacent molecules, the driving force being entropy. When a chain segment is deposited on a growing crystal face, the molten portion of the chain to which it is directly attached will be deprived of some of its degrees of freedom. This local reduction in entropy will be compensated by a general translation of the molten portion of the molecule toward the crystallite, which will tend to increase the freedom of movement of those chain segments closest to the crystallite. The trajectory followed by the molecule is determined by interactions with neighboring chains—such as entanglements—that permit longitudinal slippage but restrict latitudinal motion. When sufficient freedom of movement has been generated in the segments adjacent to the growing face, crystallization will proceed. In practice, the process would be continuous, an entropy gradient being generated along the length of the molten chain.

This mode of crystallization requires large-scale motion of individual polymer chains involving very rapid chain translation with respect to the growth of the crystallite. A single polyethylene molecule is assumed to be deposited on a crystal in its entirety before another one can begin the process. Given the high crystallization rate of polyethylene, the molecular translation of chains in the molten regions would have to be extraordinarily swift. This process has been modeled mathematically [48,49].

Morphology and Crystallization

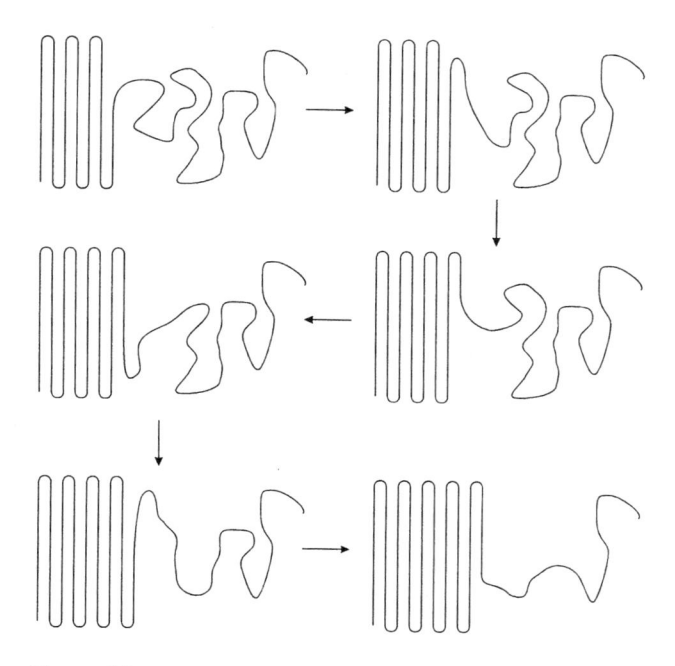

Figure 20 Formation of tight fold adjacent reentry by reeling in of polymer chains from the melt.

The process required to produce lamellar structures that correspond to the Flory model is illustrated schematically in Figure 21. Large-scale motion of polyethylene molecules during crystallization is not required. Molten chains crystallize essentially in situ, the only cooperative motion required being that of relatively short lengths of chain, in the region of 100–300 Å, necessary for parallel alignment of chain segments. Longitudinal chain translation is required for molecular segments to adopt favorable alignment, but this takes place on a more local scale than that required to form a series of adjacent reentry loops. Many different molecules contribute chains to a single layer of the growing crystal face.

According to Flory's model, few chains participate in adjacent reentrant tight folds. The majority of chains that leave the crystalline matrix become an integral part of an interfacial zone. In the simplest version of Flory's model, in which every crystalline chain stem is attached to a chain segment in the interfacial region, there is a density anomaly at the interface [45,50]. Because the chains in the interface are partially randomized, they make angles of less than 90° with the surface of the crystallite. When an imaginary plane is drawn parallel to the crystallite surface, the minimum cross-sectional area will be presented by chains

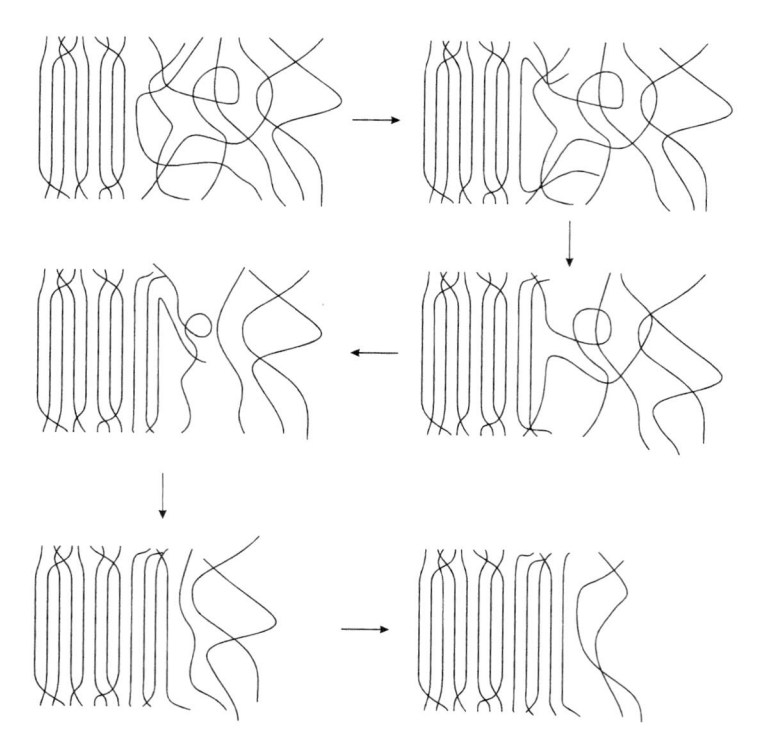

Figure 21 Formation of Flory-type lamellar structures.

perpendicular to the surface. Thus the total area of chains in a section cut parallel to the crystallite surface will be greater in the interface than in the crystal. As the density of the sample is directly proportional to the cross-sectional area of chains in a section, this model predicts an interface with a density greater than that of the crystal. It is not possible to envisage any packing of disordered chains that could have a density greater than that of the crystal. A certain degree of steric hindrance may be relieved if the chains in the crystallite are not normal to the lamellar surface, thus decreasing the density of chains at the lamellar surface [51,52]. Further steric hindrance is relieved when a certain proportion of the chains leaving a crystal return to the same crystal after a brief foray into the interfacial region.

The two opposing models share a common assumption that individual chain segments span the complete thickness of the lamellae. In neither case is it postulated that chain folding occurs within the body of crystallites. In both models the deposition of chain segments on the growing surface requires that disordered chains in the proximity of the growth face be displaced to make room for growth of the crystallite. If the disordered segments closest to the growing face are not to be incorporated into the crystal structure, due to branchpoints or the like, they must diffuse away if they are not to hinder lamellar growth.

Various experimental data cast doubt upon the validity of the adjacent reentry model. Two of the most compelling experiments are those involving the study of deuterated and cross-linked polyethylene. Small-angle neutron scattering experiments conducted on blends of regular and deuterated polyethylene revealed that the radius of gyration of a molecule—a measure of its overall dimensions—does not change significantly as the sample passes from the molten to the solid form, i.e., no large-scale motion of chains takes place [53–56]. Second, cross-linked polyethylene forms lamellae despite the fact that large-scale movement of chains is prevented by cross-links [57]. These observations confirm that large-scale motion of chains is not required for the formation of polyethylene lamellae.

Several alternative models were proposed that attempted to reconcile tight fold adjacent reentry with the observed scattering phenomena; these included the central core model [58] and the variable cluster model [59], shown schematically in Figures 22 and 23, respectively. These two models are based upon the premise

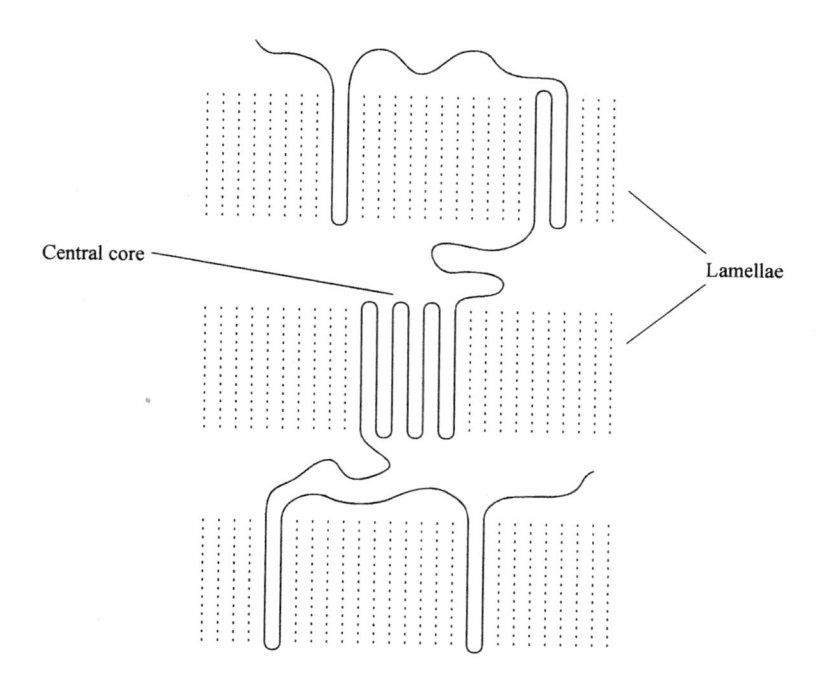

Figure 22 Central core model. (From Ref. 58.)

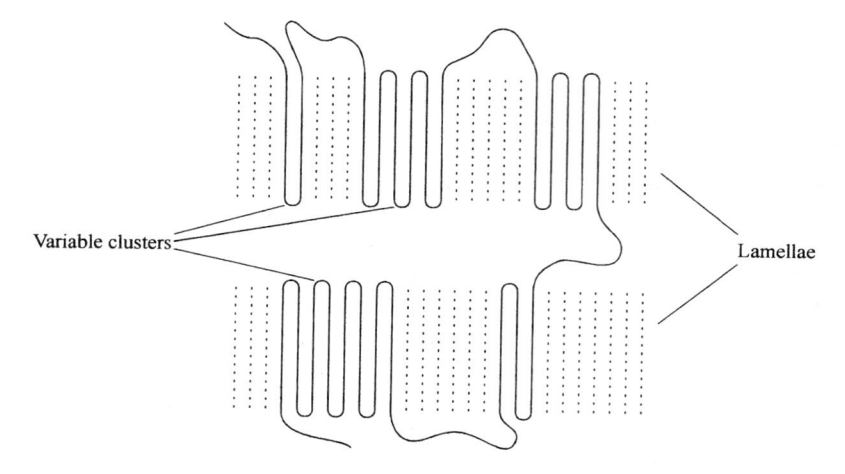

Figure 23 Variable cluster model. (From Ref. 59.)

that molecules in the solid should have a radius of gyration similar to those in the melt and that adjacent reentry occurs with a probability of ≥ 0.65 [60,61]. These models also satisfy the requirement that there be no density anomaly in the interface.

In 1979 a conference was held with the principal goal of airing the differences between the diverse morphological models. The reader's attention is directed to the proceedings of this conference for a full exposition of the various models [62]. Since that time the concept of widespread adjacent reentry with tight folds has largely fallen out of favor. The present-day consensus is that solid polyethylene consists of three phases as outlined in Section I of this chapter. It is accepted that an undetermined portion of chain stems in crystallites are connected to their nearest neighbors by folds of varying degrees of tightness but that the majority of stems exhibit some degree of connectivity with the disordered regions. Figure 24 represents the three-phase structure of polyethylene as it is presently understood.

Although the nature of the crystalline and disordered phases is now well understood, the precise nature of the interface has yet to be established. Efforts continue in this field using various experimental and theroretical tools.

   a. Regimes of Crystallization. Three modes of crystallization are postulated to describe the lateral growth of lamellae. These are termed regimes 1, 2, and 3. These regimes describe the manner in which chain stems are laid down upon the lamellar growth faces.

Morphology and Crystallization

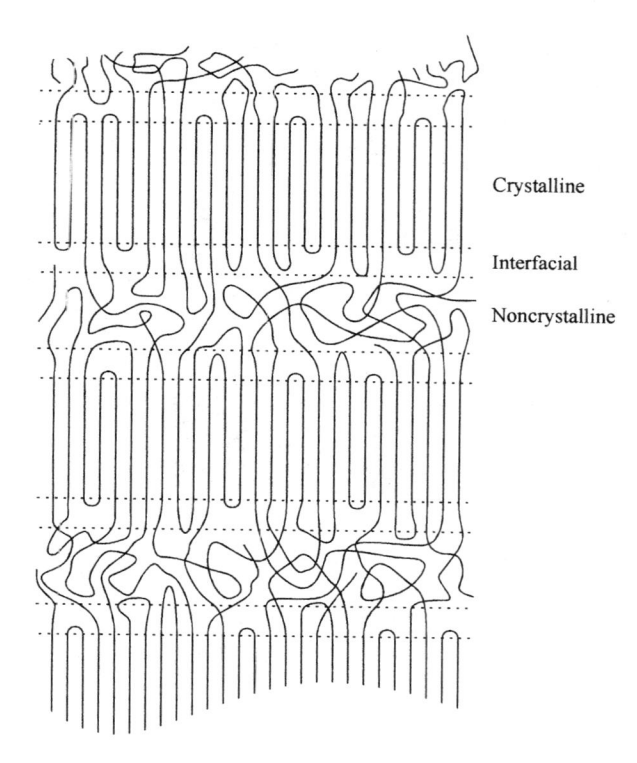

Figure 24 Representation of the three-phase morphology of polyethylene.

Regime 1. In regime 1 an extended chain segment is laid down on the unblemished growth face of a crystallite, followed by crystallization of segments in sequentially adjacent positions until the surface is completely covered. The process subsequently occurs repeatedly, advancing the growing face of the crystallite through the disordered phase. This regime is illustrated schematically in Figure 25.

Regime 2. In regime 2, several chain segments are laid down independently on a growth face, followed by subsequent additions of chains in positions adjacent to the initial segments. Addition takes place until the growth face is completely covered. Further growth takes place by repetition of this process. This regime is illustrated schematically in Figure 26.

Regime 3. Regime 3 is more complicated than either of the other two regimes, with crystal growth occurring in more than one crystallographic layer simultaneously. According to this regime, nucleation events on the growing sur-

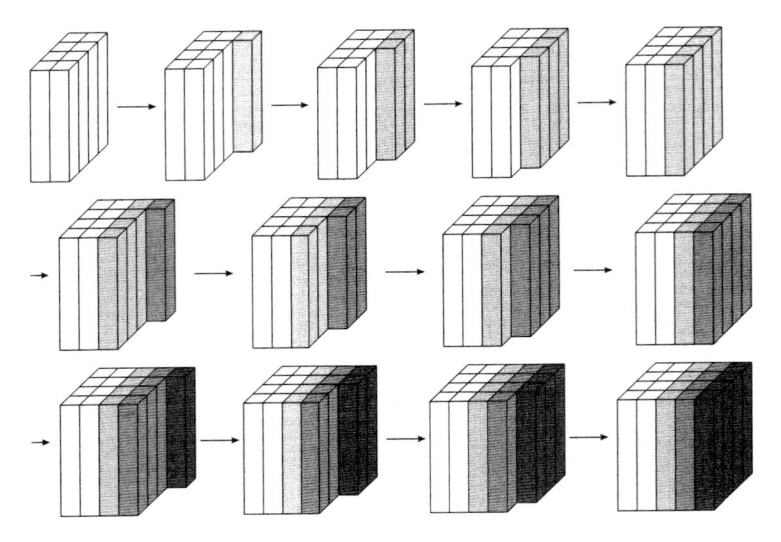

Figure 25 Lateral growth of lamella according to regime 1.

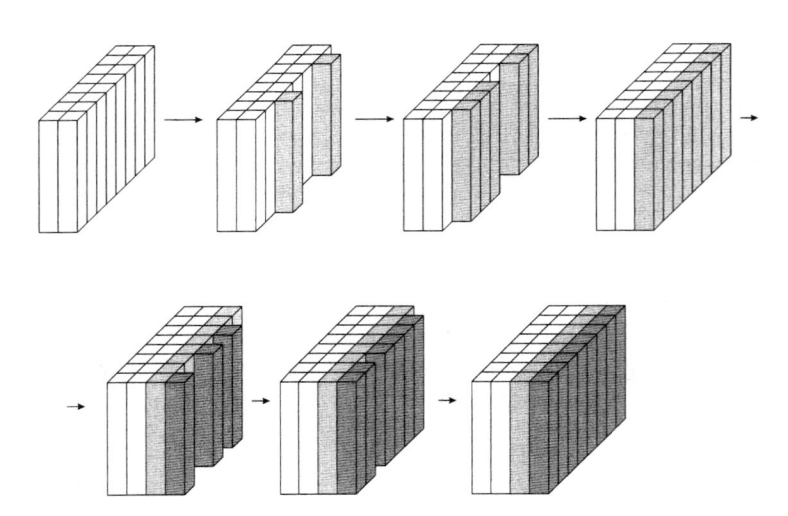

Figure 26 Lateral growth of lamella according to regime 2.

Morphology and Crystallization

face of a crystallite occur profusely, with many isolated chain segments being laid down simultaneously or in very rapid succession. Before sequential addition of further molecular segments has time to entirely cover the growth surface, additional nucleation events take place on the partially completed layer; thus crystallization can occur simultaneously at several layers. This regime is illustrated schematically in Figure 27.

Comparison of Crystallization Regimes. The primary cause of the differences between the three regimes involves the relative rates of initial chain segment deposition versus that of sequential addition. If the initial deposition is slow relative to sequential addition, then regime 1 will be favored, changing to regime 2 and then to regime 3 as the deposition rate increases. In commercial processes, regimes 2 and 3 dominate. The adjacent reentry model of crystal growth requires that regime 1 be followed. Growth according to Flory's model can take place by any of the three regimes. Each of the three regimes has been identified by crystal growth kinetics, giving further credence to the Flory model as the basis for polyethylene crystallization.

There is some evidence to suggest that disordered chain sequences in the melt do not pass directly to the crystal phase upon crystallization at high tempera-

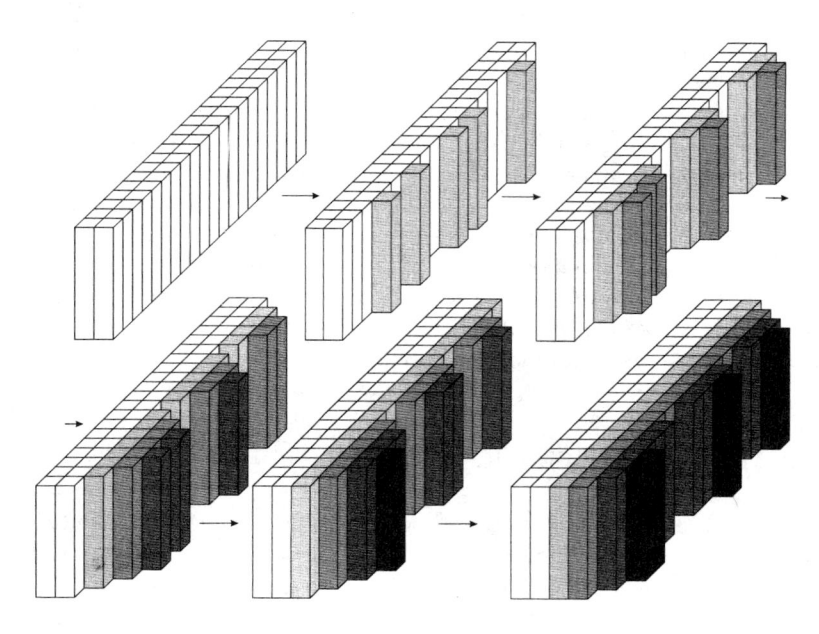

Figure 27 Lateral growth of lamella according to regime 3.

ture [69]. It is possible that lamellae initially form as liquid crystalline smectic structures in which extended chain sequences are approximately parallel. Subsequently, adjacent sequences pack together to form a true crystal phase. It is postulated that prior to the packing process molecular sequences can slide back and forth to exclude branches or relieve steric interference within the interface.

3. Secondary Crystallization

The crystallization of polyethylene does not always cease when a sample is cooled to room temperature. Secondary crystallization may continue at room temperature, albeit at a much slower rate. Various samples undergo physical changes at ambient temperature over a period of time ranging from a few hours to several weeks after molding. The physical property changes reflect an observed gradual increase in degree of crystallinity. This effect is most noticeable in samples that initially exhibit a modest degree of crystallinity. Thus, ultrahigh molecular weight polyethylene has been observed to slowly increase in density at room temperature after being rapidly cooled from the melt [70]. Low density polyethylene resins undergo similar density changes, and their tensile yield characteristics change significantly [71]. Increased density can be explained in two ways: (1) an increase of the crystalline fraction or (2) improved packing within the noncrystalline regions that results in increased density of the noncrystalline regions.

The mechanisms involved in secondary crystallization are unclear, but it appears that residual stresses imposed during primary crystallization can be relieved by local molecular motion [72]. The changes associated with secondary crystallization are modest in comparison with those of primary crystallization. The degree of crystallinity is unlikely to increase by more than 2–3%, even after prolonged conditioning at room temperature. Samples crystallized slowly at higher temperatures are unlikely to exhibit secondary crystallization, because their chain segments initially have the opportunity to adopt thermodynamically stable conformations. Several modes of secondary crystallization have been proposed, any or all of which may take place to some extent: thickening of preexisting crystallites is possible; lamellae may anneal to relieve crystal defects; or thin, poorly ordered crystallites may form in interlamellar zones [73].

V. CRYSTALLIZATION KINETICS

Crystallization kinetics is the outward and measurable manifestation of the crystallization process. It is not of itself important to the solid-state properties of polyethylene; however, crystallization kinetics reflects the route by which polyethylene solidifies from the disordered state. Thus, the study of crystallization kinetics contributes to an understanding of the mechanisms of crystallization and the trajectories of the molecules that make up the solid state. The crystallization kinetics of a sample is governed by two processes: the rate of nucleation and the rate of crystal growth. A complete understanding of the crystallization mechanisms of polyethylene will eventually lead to their control via tailoring of molecular properties and crystallization conditions and hence to the control of solid-state properties. Currently our ability to predict the physical properties of a polyethylene resin based on its molecular structure and crystallization mechanism is rudimentary.

The conditions under which polyethylene crystallizes influence the mechanisms by which the process takes place. Therefore, controlling the rate of crystallization regulates the properties of the product within limits imposed by its molecular character. For example, two polyethylene samples with very different molecular characteristics may be made to behave similarly in the solid-state by appropriate control of their crystallization rates. Thus, rapidly quenched high density polyethylene with a molecular weight of 500,000 and a broad molecular weight distribution exhibits tensile characteristics similar to those of a linear low density polyethylene with a molecular weight of 100,000 and a narrow molecular weight distribution that is slowly cooled from the melt [74].

In commercial processes the conditions under which crystallization takes place are constantly changing. The crystallization of fabricated items occurs as the sample is being cooled, never attaining a steady state until ambient temperature is reached. Many fabricated items start to crystallize while still flowing or undergoing deformational or relaxational processes. Conditions change not only with time but also with location within the sample. Polyethylene acts as an effective thermal insulator; thus conditions in the core of a thick sample may be very different from those in the region in contact with the walls of a mold. In addition, branched samples tend to fractionate upon cooling, the less branched molecules crystallizing first, leaving those with a higher concentration of branches in the melt.

Although rarely explicitly stated, crystallization kinetics is an important factor in commercial fabrication processes. The rate at which a polyethylene resin crystallizes determines its suitability to a given processing technique or end use. A grade of polyethylene that crystallizes rapidly may be suitable for injection molding processes, whereas a grade that crystallizes at a slower rate would be more desirable for film blowing. Naturally, other factors, such as the physical requirements of the end product, viscosity, and overall production costs, must also be taken into account. Most polyethylene suppliers offer a range of resins intended for different fabrication processes. In each case the melt properties, which strongly influence crystallization, are carefully tailored to suit the specific requirements of the conversion process.

The data presented in the following discussion largely refer to the crystallization kinetics of quiescent melts under isothermal conditions—a very different situation from that pertaining to the dynamic conditions found in commercial conversion processes. In an industrial context, quantitative considerations regarding the crystallization kinetics of polyethylene are often ignored, being subordinate to the more pragmatic question, Can a particular resin be processed economically to form the desired article using the equipment available?

Albeit somewhat different in detail, the principles governing static melt-crystallization also apply to the crystallization of oriented melts. Therefore, in the following discussion emphasis is placed on qualitative principles rather than quantitative considerations. A full exposition of the quantitative aspects of polymer crystallization kinetics can be found in the works of Mandelkern and Wunderlich listed in the bibliography at the end of this chapter.

A. Factors Affecting Crystallization Kinetics

Many of the factors that determine the rate of crystallization of polyethylene also influence the morphology of a sample and its overall degree of crystallinity. Thus, for example, high melt viscosity that hinders the motion of molten chains during the crystallization process and hence decreases the crystallization rate also inhibits the formation of thick lamellae. Similar parallels may be drawn concerning molecular characteristics such as molecular weight distribution, long-chain branching, and the length of short-chain branches. Independent variables exist that can be used to influence the rate of crystallization within the limits imposed by the molecular characteristics of the resin, the two principal ones being temperature and orientation. Slow crystallization may be the result of either thermodynamic or kinetic factors. At high temperatures close to the equilibrium melting temperature, the crystallization rate will be low because there is a small thermodynamic driving force. Conversely, crystallization occurs slowly at low temperatures, because molecular factors such as high molecular weight and branching inhibit the movement of chains. The combination of the molecular characteristics of a polyethylene resin and the conditions under which it solidifies determines the attributes of the solid product.

The kinetics of crystallization can be followed by a number of techniques, none of which directly measures the growth of individual crystallites or follows individual molecules as they are incorporated into the semicrystalline morphology. Currently it is only possible to measure bulk properties and draw conclusions regarding the molecular processes involved. Two of the most common methods of studying the crystallization process are differential scanning calorimetry, in which heat flow is measured as a function of time and temperature, and hot stage optical microscopy, in which the progress of growth fronts of spherulites is observed.

1. Effect of Temperature and Orientation

The overall rate of crystallization of any polymer is controlled by competing kinetic and thermodynamic factors. The viscosity of the melt increases as the

Morphology and Crystallization

temperature decreases, thus inhibiting the diffusion of molecular segments into favorable positions from which to precipitate on to the growing faces of crystallites. On the other hand, a reduction in temperature lowers the energy barrier to the formation of stable nuclei and also increases the change of Gibbs free energy associated with crystallization. Thus kinetic factors tend to slow crystallization as the temperature falls, but energetic factors favor the process. Accordingly, the highest rate of crystallization for any given polymer typically occurs at an absolute temperature of approximately the equilibrium melting temperature multiplied by 0.8. In the case of linear polyethylene, the maximum rate would be expected to occur at approximately 60°C. However, polyethylene crystallizes so quickly that even during rapid quenching from the melt the process is essentially complete before the temperature corresponding to the maximum rate can be reached. The very high crystallization rate of polyethylene, especially that of unbranched resins, compared to that of other polymers is due to a combination of its high enthalpy of crystallization and the great flexibility of the polyethylene chain.

The temperature at which any lamella will be stable is a function of its thickness and the interfacial free energy associated with its basal planes. The greater the thickness, the higher the temperature at which it can exist. Consequently, thicker lamellae form at elevated temperatures. In addition, elevated crystallization temperatures generally lead to higher overall levels of crystallinity. Higher levels of crystallinity increase the modulus of samples at the expense of a longer crystallization time. This process can be taken to the extreme; at very high temperatures (in excess of 128°C), high density polyethylene may take several weeks to crystallize, forming highly crystalline specimens that are stiff but brittle.

The addition of linear chain stems to crystallite growth faces requires that adjacent noncrystalline sequences migrate out of the way to accommodate them. This process is facilitated by free volume within the melt and rapid chain motion. The higher the temperature, the greater will be the free volume and the faster the various vibrations, rotations, and assorted motions of chain segments will take place. This translates to a lower melt viscosity and decreased hindrance to chain movement. This promotes the probability that a linear sequence will adopt a conformation favorable for deposition on the growing face of a crystallite. Conversely, because the temperature is higher, the stable thickness of a crystal will be increased, requiring that longer extended chain sequences be deposited on the growth face.

The effects of isothermal crystallization at various temperatures on the ultimate degree of crystallinity and crystallization rate of high density polyethylene are illustrated in Figures 28 and 29.

Most commercial crystallization takes place under conditions of considerable shear or orientation, which tends to enhance the crystallization rate by reducing the associated change of entropy. However, deformation need not be severe

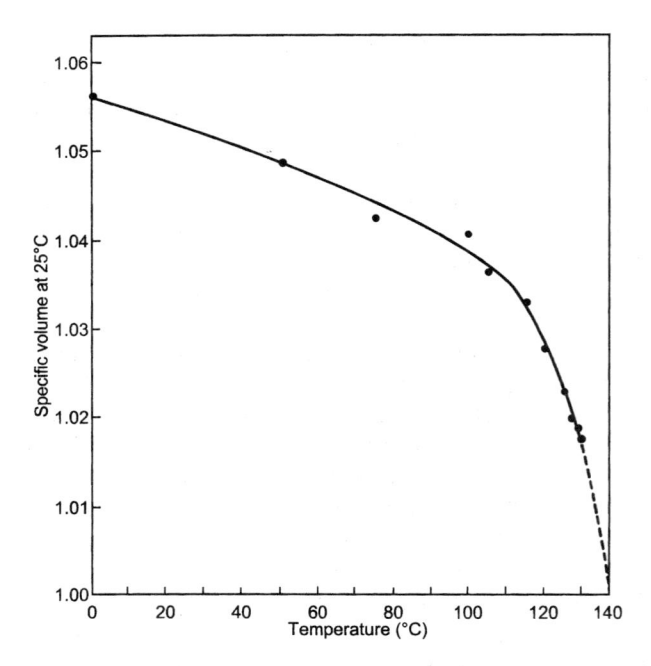

Figure 28 Specific volume as a function of crystallization temperature for a high density polyethylene resin. (From Ref. 101.)

to enhance crystallization. Polymer melts consist of statistically random assemblies of chains in which small overall changes may result in relatively large local variations that can initiate crystallization. This area is little studied, although the principle is widely accepted. Localized orientation of molecules reduces the entropy loss associated with the formation of nuclei, lowering the energy barrier and thus promoting nucleus formation. The effect may be readily observed on the hot stage of an optical microscope. A quiescent molten sample of high density polyethylene cooled to 127°C under a glass cover slip will remain uncrystallized for several minutes. However, a slight movement of the cover slip, subjecting the sample to shear, will instantly induce crystallization.

2. Effect of Molecular Characteristics

The crystallization of polyethylene is a cooperative process. It requires the concerted motion of chain segments from various molecules to effect longitudinal translation of linear sequences into positions favorable for crystallization. Molecular migration involves the slippage of chain segments past one another; hence

Morphology and Crystallization

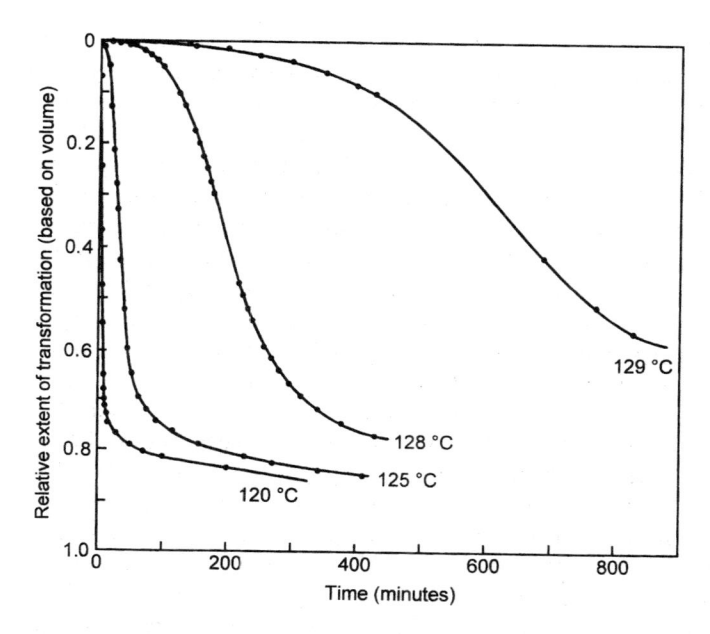

Figure 29 Extent of crystallization as a function of time for high density polyethylene. (From Ref. 102.)

the characteristics of the molecules are important to the movement. The fact that polyethylene resins do not consist of a discrete species of molecules but rather contain a distribution of many chain lengths and with varying compositions complicates the crystallization process.

Unlike most crystalline materials, the molecules of polyethylene are not incorporated into crystallites in their entirety. The presence of noncrystallizable entities, such as branches and entanglements, divides a polyethylene molecule into linear sequences that may or may not be of sufficient length to crystallize. The shorter the sequences, the less likely they are to form stable crystallites at a given temperature. The linear sequences between branches are not identical; rather, there is a statistical distribution of lengths, determined by the concentration and distribution of branches along the backbone. The distribution of linear sequences may be highly skewed or even bimodal, as it is not necessary that all chains in a polyethylene resin contain the same concentration of branches. Ziegler–Natta type linear low density polyethylene resins invariably contain high molecular weight species with long linear runs and lower molecular weight chains with shorter crystallizable sequences [75]. Only linear segments with sufficient freedom of movement to migrate into configurations parallel to growing lamellar surfaces may crystallize. Not all polyethylene resins are crystallizable; some highly branched materials, such as high comonomer content ethylene-vinyl acetate copolymers or ethylene-propylene rubbers, contain negligible levels of crystallinity.

During the crystallization of branched polyethylene, linear chain sequences are incorporated into crystallites, excluding the branches into the noncrystalline regions. Thus, as the sample cools, the noncrystallizable sequences are concentrated in the remaining melt. The branches are specifically concentrated in the interfacial region purely as a matter of exclusion from the lamellae [76]. Thicker crystallites are more thermodynamically stable than thinner ones, and there is a tendency for the longest linear sequences to crystallize first as a sample cools, leaving the shorter ones to crystallize subsequently. At lower temperatures, thinner lamellae are formed and the crystallization rate is reduced. Thinner lamellae form as a consequence of having shorter crystallizable linear sequences available; these crystallites are thermodynamically stable only at lower temperatures. The crystallization rate decreases owing to both thermodynamic and kinetic considerations. The shorter linear sequences are capable of forming only thin lamellae, which does not result in as great a decrease in Gibbs free energy as the formation of thicker ones, and thus crystallization is less thermodynamically favorable. The remaining melt has an increased viscosity because it is richer in branches and has lower thermal energy, each of which inhibits the mobility of chains. The free movement of chains is hindered by bulky side groups for a number of reasons: Ionic species may form permanent or transient micelles, the mass of the branch will damp the kinetic movements of the chain in its vicinity, and the bulk of the branch makes it more likely to meet resistance when slipping between other chains during longitudinal movement. The rise in viscosity increases the time required for crystallizable sequences to precipitate on crystallite growth faces. If noncrystallizable species lie at the growth face of a crystallite, they must diffuse away before further lamellar development can take place. If diffusion occurs at a slower rate than crystal growth can take place, the rate of growth will be inhibited. This is a commonly observed phenomenon. The initial growth of spherulitic radii is linear as a function of time, indicating that the growth is an interface-controlled process. As crystallization progresses, the rate of spherulitic growth decreases to a dependence upon the square root of time, indicating that growth is a diffusion-controlled process governed by the rate at which noncrystallizable sequences can diffuse away from the growth front [77]. Some highly branched species may be incapable of crystallizing under the prevailing conditions. Such species are found entirely in the interlamellar regions or at the interfaces between spherulites. The effect of increasing branch content and temperature on the isothermal crystallization rate of various hydrogenated polybutadienes is illustrated in Figure 30 (hydrogenated polybutadiene is chemically analogous to ethyl-branched linear low density polyethylene).

Morphology and Crystallization

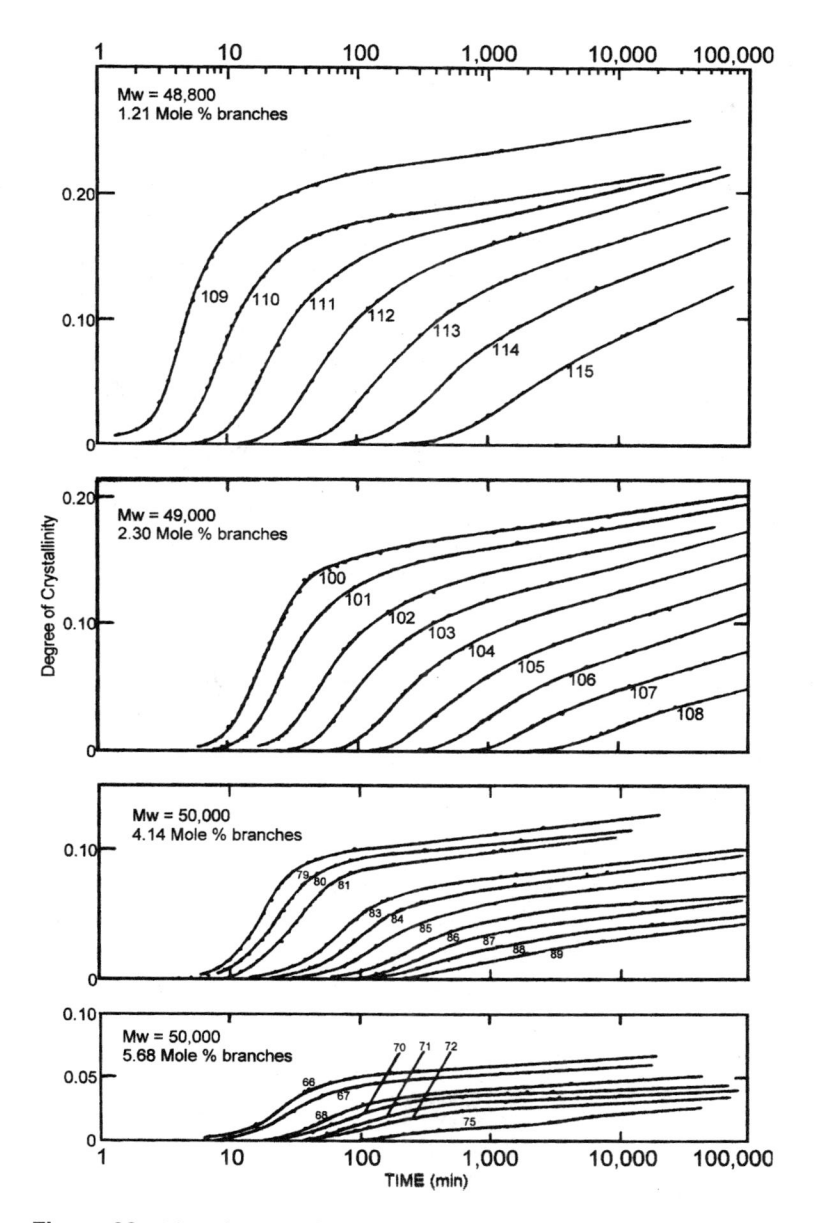

Figure 30 Plot of extent of crystallization for hydrogenated polybutadienes as a function of time, crystallization temperature °C, and branch content. (From Ref. 103.)

In some cases it is possible to develop two distinct populations of lamellae that have different thicknesses within the same sample. The causes of such distributions vary but can generally be attributed to nonuniformity of the molecular nature of the sample—broad molecular weight or wide composition distribution—or to nonuniform crystallization conditions [78]. An extreme case involves the solidification of blends of linear and branched polyethylene; the linear molecules crystallize first to form a network of thick lamellae between which thinner lamellae of the less stable low density polyethylene form at lower temperatures.

High molecular weight species will be entangled at more points along their length than shorter ones, which hinders the overall translation of such molecules and increases melt viscosity. The exact nature of entanglements is a matter of speculation, but they undoubtedly include a wide range of structures from true knots to transient steric effects. From a practical viewpoint, an entanglement may be considered to be any entity that significantly impedes the motion of the chain segments of which it is composed at a given instant in time. At present there is no method available for analyzing the nature of entanglements, although their existence is not in doubt. The higher the average molecular weight of a polyethylene resin, the greater its viscosity and the lower its crystallization rate. The distri-

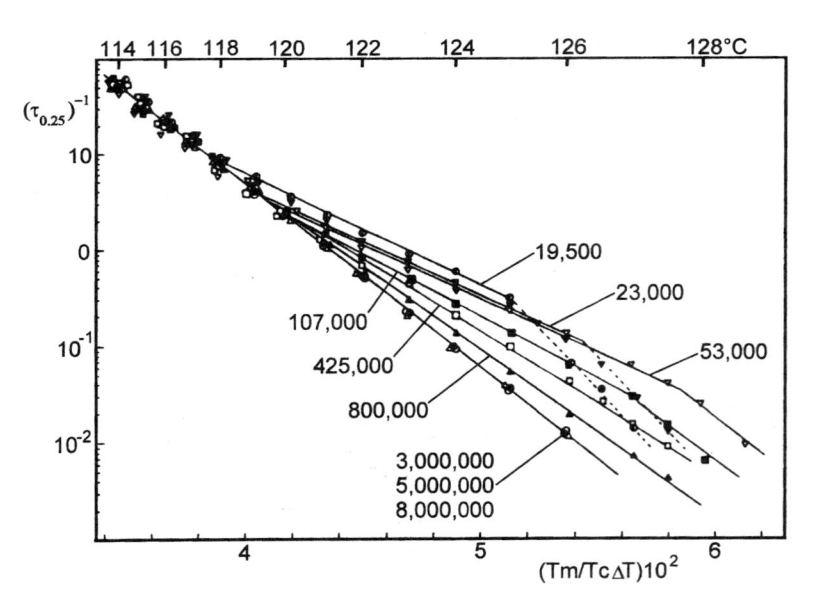

Figure 31 Plot of time to develop 25% of the total crystallinity for a series of linear polyethylene resins of different molecular weights as a function of isothermal crystallization temperature. (From Ref. 104.)

Morphology and Crystallization

bution of chain lengths in a sample is also important. Very short chains, although not of a crystallizable length, may be highly mobile, effectively increasing the free volume in their vicinity and facilitating the movement of longer crystallizable sequences. The effect of molecular weight on the crystallization rate of linear polyethylene samples as a function of isothermal crystallization temperature is illustrated in Figure 31.

The effect of branching on the crystallization of polyethylene is greater than the effect of increasing molecular weight. A fiftyfold increase in the weight-average molecular weight of linear polyethylene, from 60,000 to 3,000,000 results in a decrease in degree of crystallinity from 78% to 52%, as determined by density [79], whereas the addition of 1.9 ethyl branches per 100 carbon atoms to the backbone of a polyethylene with a molecular weight average of 104,000 can reduce it to less than 50% crystallinity [80].

VI. INTERLAMELLAR CONNECTIONS

Connections between neighboring crystallites come in two forms: covalent and ionic. Covalent links are present in virtually all polyethylene specimens, whereas ionic links require the presence of a polar comonomer. Covalent connections can be subdivided into two categories: direct tie chains and entangled loose loops emanating from adjacent lamellae. These two types of connections are illustrated in Figure 32. Ionic links are created when anionic groups attached to polyethylene chains form small clusters with metal cations, or by hydrogen bonding between polar species attached to adjacent chains. The nature of ionic links is addressed in Chapter 7.

Interlamellar connections cannot be directly imaged; therefore, their existence and properties have to be inferred from material properties. For the sake of simplicity, no distinction is drawn here between the effects of ionic and covalent links. Unless otherwise noted, the term "tie chain" is used generically to encompass all types of intercrystallite connections.

Interlamellar connections are crucial to the mechanical properties of polyethylene because they transmit forces between crystallites. Tie chains determine or influence a variety of mechanical properties, such as ductility, toughness, and modulus. Without the benefit of interlamellar links, polyethylene would be a brittle material with little physical strength.

The nature of tie chains can be described using four parameters, one chemical and three physical: (1) the strength of the polyethylene backbone and ionic linkages (if present); (2) the concentration and distribution of tie chains; (3) the distribution of the degrees of freedom of motion, i.e., how taut the links are; and (4) the angle that the links make with respect to an applied force and the lamellar surfaces. The latter three physical parameters depend on the molecular nature of

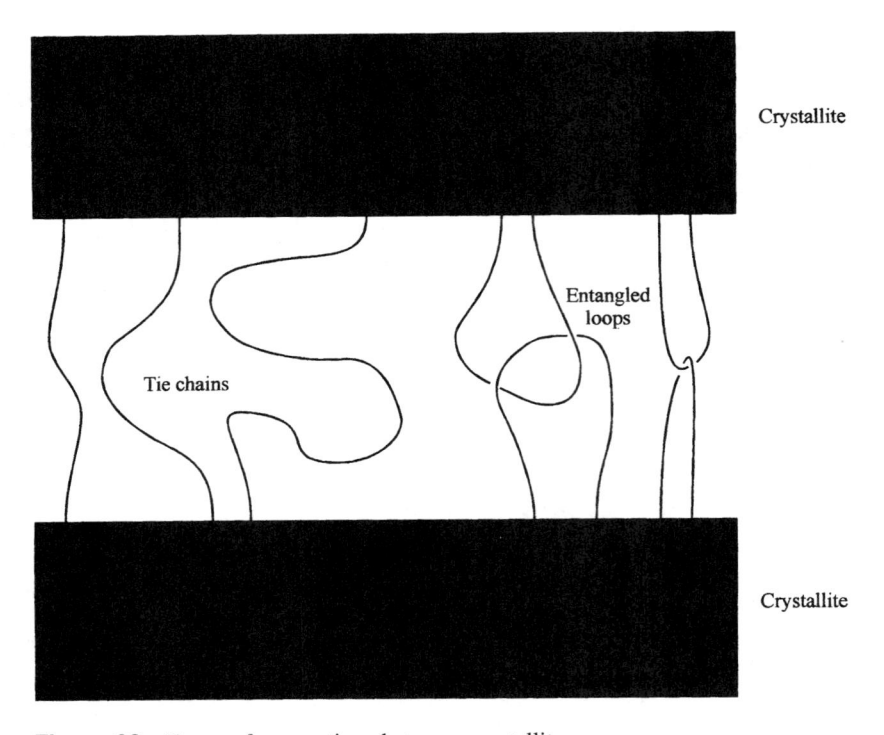

Figure 32 Types of connections between crystallites.

a specimen and the method by which it is prepared. The physical parameters are not independently variable; any alteration of specimen preparation or molecular character will affect all three.

The physical properties of tie chains have not been experimentally quantified. Attempts have been made to estimate tie chain concentration [63,64,81,82], but the results should not be considered absolute. Based on information obtained from morphological studies, small-angle neutron scattering, and computer modeling, some generalized qualitative predictions can be made regarding tie chains [83–86]. The wider an interlamellar region, the fewer will be the tie chains that span it. The smaller the radius of gyration of molecules in the melt, the smaller will be the number of tie chains in a solid crystallized from the melt. Tie chains spanning wide interlamellar regions are less likely to be taut than those spanning narrow regions. Rapidly quenched samples will have more tie chains than those that are crystallized slowly. Specimens crystallized from solution will have fewer tie chains than those crystallized from the melt.

The higher the concentration of tie chains, the greater will be the connectivity between neighboring lamellae. This results in localized cooperative effects under applied load. For instance, tension will tend to disrupt crystallites, but an individual lamella cannot deform without an accompanying deformation of those intimately linked to it. Thus, highly crystalline specimens with a large number of tie chains spanning relatively narrow interlamellar regions will deform cooperatively across a given cross section, i.e., by necking. A sample with poorly developed crystallinity, and therefore fewer interlamellar links, is more likely to deform homogeneously.

Only tie chains that are taut can transmit stress. The higher the concentration of taut connections, the greater the load that can be carried. A corollary to this is that if there are an insufficient number of tie chains to transmit an applied stress, the tie chains will fail, primarily by breakage or by pulling out of one of the crystallites that they link. If the taut links between a pair of lamellae break, adjacent tie chains will experience an increased load. The heightened stress level will increase the likelihood that the newly stressed tie chains will in turn break. The net effect of this is a domino-like breaking of tie chains, resulting in brittle failure across the width of a specimen. This occurs when the strength of lamellae exceeds the load-bearing capabilities of interlamellar connections. The topics of polyethylene specimen deformation and failure are addressed fully in Chapter 8.

VII. CRYSTALLIZATION PRODUCTS

Polyethylene can solidify to form a surprisingly wide variety of semicrystalline morphologies. The aim of this section is to provide a brief overview of some of the most common and significant structures developed. For a more complete survey of polyethylene morphologies, the reader's attention is directed to the works of Woodward and Bassett listed in the bibliography.

A. Crystallization from Solution

When polyethylene crystallizes from solution it can adopt a wide variety of habits depending upon the crystallization conditions and the molecular nature of the resin. The key variables controlling crystallization are specific interactions between the polymer and the solvent, concentration, temperature, and molecular elongation. As a general rule, the lower the concentration, the higher the temperature, and the lower the average molecular weight of the resin, the more regular will be the crystals that form. Some of the most notable products of solution crystallization are described below. Polyethylene products derived from solution are of little commercial interest except for gel-spun fibers.

1. Single Crystals

The term "single crystal" is applied loosely to a wide variety of crystalline morphologies derived from quiescent dilute polymer solutions (typically less than 1% by weight). (From a crystallographic point of view, the term "polyethylene single crystal" is a misnomer. Polyethylene crystals grown from solution are coated with a noncrystalline overlayer composed of loops.) The classic polyethylene single crystal is a lozenge-shaped lamella with lateral dimension many times greater than its thickness. There are a vast number of variations on this basic structure. Electron micrographs of some of the most common types of single crystals are shown in Figures 9, 18, and 33. For a survey of the types of single crystals, the reader is directed to the works of Bassett, Geil, and Woodward listed in the bibliography.

a. Lamellar Thickness. The thickness of polyethylene single crystals varies widely depending upon the conditions under which they were grown and the molecular nature of the resin. The thickness of single crystals of linear polyethylene is controlled primarily by the polymer–solvent interaction parameter

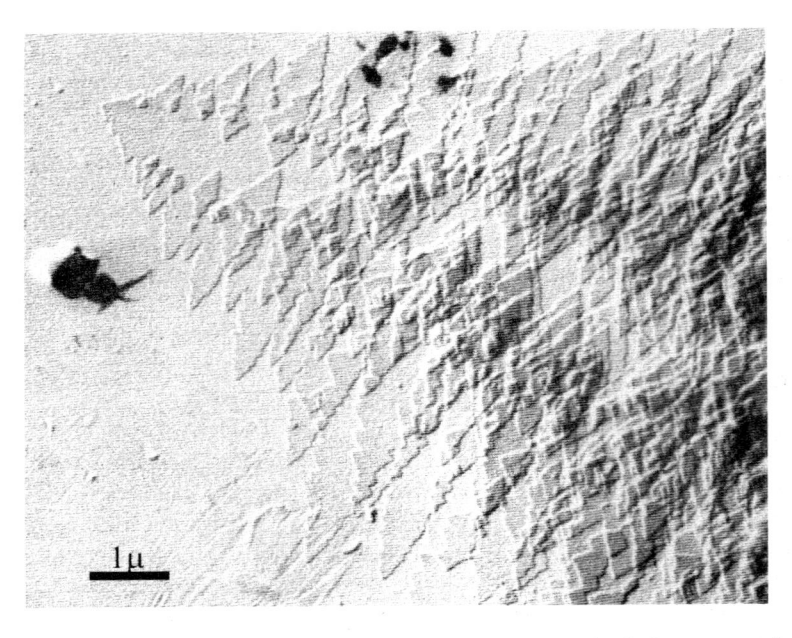

Figure 33 Dendritic single crystals of high density polyethylene grown from dilute solution in xylene. (From Ref. 105.)

Morphology and Crystallization 111

and the crystallization temperature; the molecular weight plays a secondary role [87]. The structure of short-chain branched polyethylene single crystals is primarily dependent upon the concentration and distribution of branches, which limits the maximum thickness, and secondarily upon the polymer–solvent interactions and the crystallization temperature. The effect of crystallization temperature on the thickness of linear polyethylene single crystals is shown in Figure 34, and the effect of branch concentration in hydrogenated polybutadiene crystals is illustrated in Figure 35.

2. Loosely Connected Lamellae

When polyethylene is crystallized from relatively concentrated solutions (2% or more by weight), mats of crystals are formed that are composed of individual lamellae loosely connected to their neighbors by tie chains. An example can be prepared by allowing a 2% solution of ultrahigh molecular weight polyethylene in xylene to evaporate to dryness at room temperature. The result is a flexible film that can be drawn by a factor of more than 10 before it breaks.

3. Shish Kebab Structures

When a concentrated solution of high molecular weight polyethylene is subjected to high shear at temperatures just above its quiescent crystallization temperature, fibers of polyethylene precipitate. When dried and viewed under the electron microscope these fibers are found to have a "shish kebab" structure, as shown

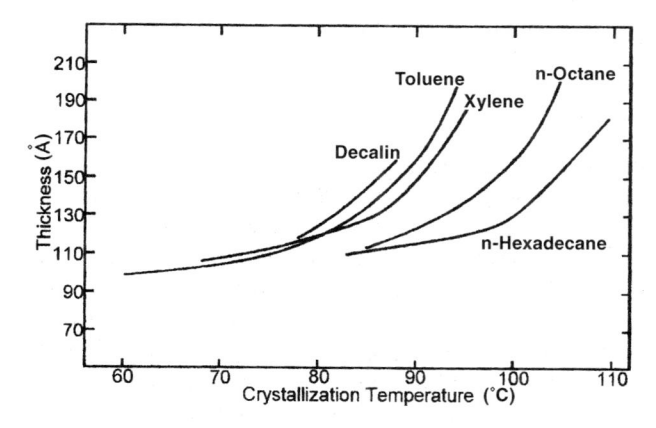

Figure 34 Effect of crystallization temperature on the thickness of single crystals of high density polyethylene grown from dilute solution in various solvents. (From Ref. 106.)

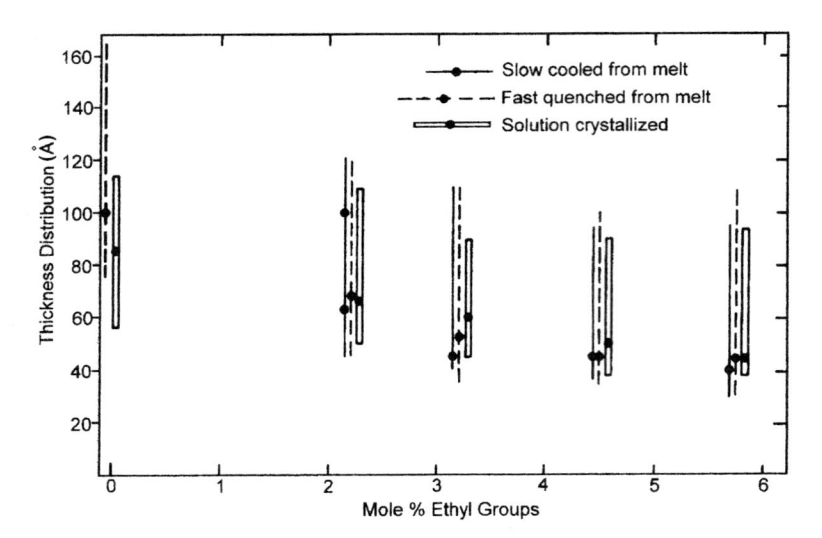

Figure 35 Effect of co-unit content on the thickness of crystals of hydrogenated polybutadiene grown from dilute solution in xylene and from the melt. (From Ref. 107.)

in Figure 36. The spine of the fiber consists of high molecular weight chains that are almost fully extended, upon which are overgrowths of lamella-like disks, centered upon the fiber, with their c axes parallel with the long axis of the fiber. The arrangement of molecules in a shish kebab structure is illustrated schematically in Figure 37. The thickness of the crystallites is similar to that of solution crystals grown from quiescent solutions. Conditions suitable for the formation of shish kebabs can be found in rapidly stirred solutions [88].

4. Gel-Spun Fibers

Highly oriented fibers can be spun from concentrated solutions (gels) of ultrahigh molecular weight polyethylene at high temperature [89]. Strands of viscous gel are extruded downward from small holes in a horizontal plate at the top of a high tower; simultaneously they are drawn rapidly downward and the solvent is stripped off. The resulting fibers are taken up on a drum at the foot of the tower. Details of this process are given in Chapter 8. The fibers so produced consist of essentially parallel molecules that are highly extended. The degree of crystallinity of such fibers is very high, as there are few entanglements and no branches to prevent the well-aligned molecules from crystallizing. The extremely high orientation endows the fibers with high elastic modulus, approaching the theoretical limit estimated for perfectly aligned crystalline polyethylene.

Morphology and Crystallization

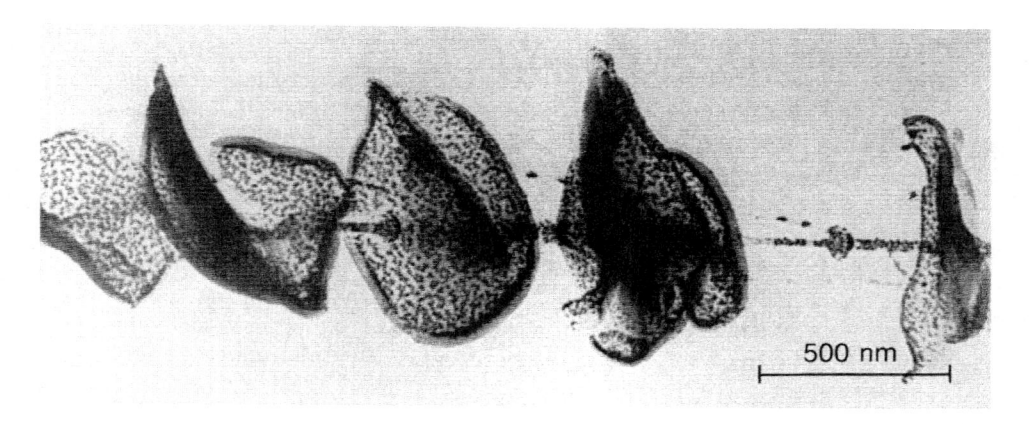

Figure 36 Electron micrograph of "shish kebab" structure grown from a stirred dilute solution of ultrahigh molecular weight polyethylene in xylene. (From Ref. 108.)

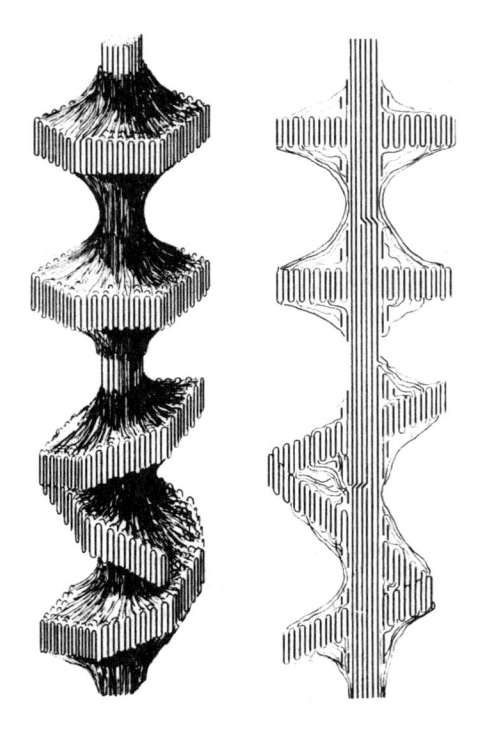

Figure 37 Representation of arrangement of molecular chains in shish kebab structure. (From Ref. 88.)

B. Crystallization from the Molten State

When polyethylene crystallizes from the molten state, the crystallites that make up the rigid portion of the semicrystalline structure are organized with respect to each other at various levels. On a local scale lamellae tend to align themselves parallel with their neighbors, while on a larger scale bundles of lamellae can form stacks, sheaves, and spherulites. Local organization can vary from isolated, curved, and disjointed lamellae in branched polyethylene to well-regimented stacks of lamellae in highly crystalline linear samples. As a rule of thumb, the higher the degree of crystallinity, the greater the organization of the lamellae. Large-scale organization of crystallites can exist on a scale of many tens of micrometers and is frequently referred to as supermolecular structure.

1. Quiescent Crystallization

a. Local Arrangement of Lamellae. There is a general tendency for lamellae to align themselves parallel with their nearest neighbors. The formation of parallel lamellae is a natural consequence of the outgrowth of one lamella from another; such lamellae are bound to share common crystallographic axes. This tendency toward alignment increases as the separation between lamellae decreases. The higher the degree of crystallinity, the greater the local alignment of lamellae. Examples of small-scale organization of lamellae are shown in Figures 12 and 13.

Lamellar Thickness. The factors that control lamellar thickness are discussed in Section V. Within the limits imposed by molecular considerations, the thicknesses of lamellae are determined by the temperature at which they form. Increasing the molecular weight or branch content of a sample will restrict the configurations available for crystallization and hence will limit the maximum lamellar thickness attainable. Figures 38, 39, and 40, respectively, illustrate the effects of temperature, molecular weight, and branch content on the thickness of lamellae crystallized from the melt.

b. Large-Scale Arrangement of Lamellae—Spherulites and Sheaves. The most commonly recognized (but not universal) example of supermolecular organization is the spherulite, which consists of bundles of lamellae radiating from a common nucleus. When grown in isolation, spherulites are approximately spherical. In practice, a multitude of spherulites nucleate and grow concurrently until the growing surfaces of neighboring spherulites impinge upon each other. Growth continues into the remaining molten regions until the whole volume is pervaded. When all growth has ceased, the resulting spherulites are irregular polyhedrons. Spherulites exist in a wide range of sizes and relative degrees of

Morphology and Crystallization

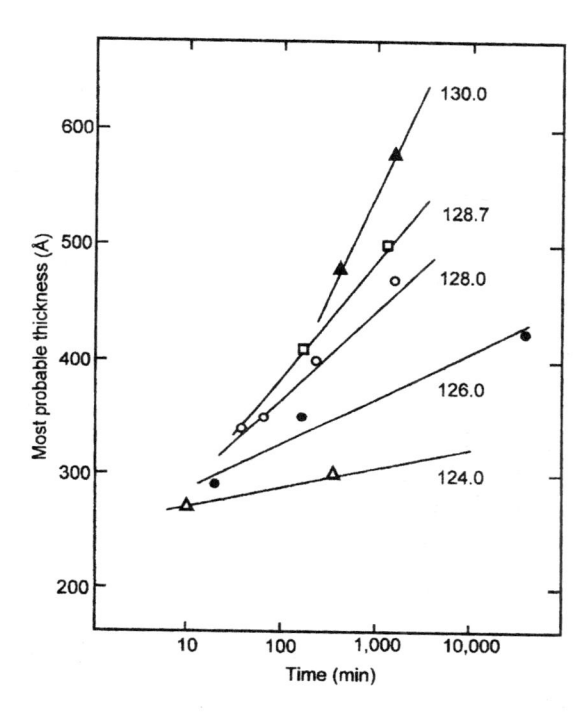

Figure 38 Effect of crystallization temperature on the thickness of lamellae of a narrow molecular weight fraction ($M_w \approx 70{,}000$) of high density polyethylene crystallized from the melt. (From Ref. 109.)

internal organization. The perfection of the supermolecular organization is roughly dependent upon the molecular nature of the resin. To a first approximation, the lower the molecular weight and the lower the branch content, the higher the degree of organization of crystallites within spherulites. As each spherulite is centered on a single primary nucleus, the average size is inversely proportional to the density of nucleation.

If the concentration of nucleation events is extremely dense, there will be insufficient room for the spherulites to mature. Under such circumstances a multitude of lamellar bundles, termed "sheaves," are produced. Toward either end of the bundle the lamellae splay out from each other to form a waisted structure. Sheaves are so named because of their outward similarity to wheat sheaves (albeit on a very different scale). Each sheaf is equivalent to the core of a spherulite. The middle of each sheaf consists of a small bundle of parallel lamellae.

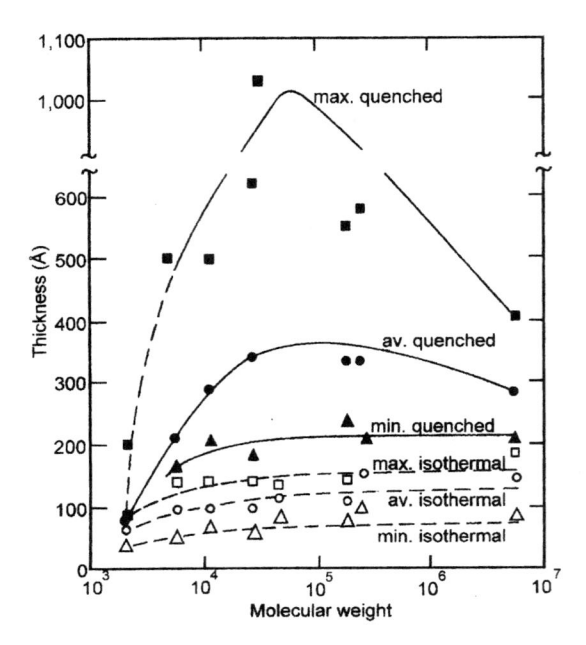

Figure 39 Effect of average molecular weight on the thickness of lamellae of narrow molecular weight fractions of high density polyethylene crystallized from the melt. (From Ref. 98.)

Spherulites in thin films are visible when viewed between crossed polars under an optical microscope. They appear as "Maltese crosses" due to the alignment of chains within lamellae that radiate from the nucleus. An optical micrograph of spherulites grown in a thin film of high density polyethylene is shown in Figure 41.

When radial lamellae spiral as they grow outward from the nucleus, the resulting spherulites appear to be ringed when viewed between crossed polars. This effect is shown in Figure 42.

The degree of ordering within polyethylene spherulites can be determined by examining the scattering pattern produced when a collimated beam of light is shone through a thin film mounted between crossed polars [90–92]. The scattering pattern takes the form of four-leaf clover, the lobes angled at 45° to the polarization planes. A typical scattering pattern is shown in Figure 43. The degree of ordering of the spherulite is reflected in the shape, size, and intensity of the scattering pattern. The clearer and better defined the lobes, the better is the organization of the lamellae within the spherulites. The degree of perfection of spheru-

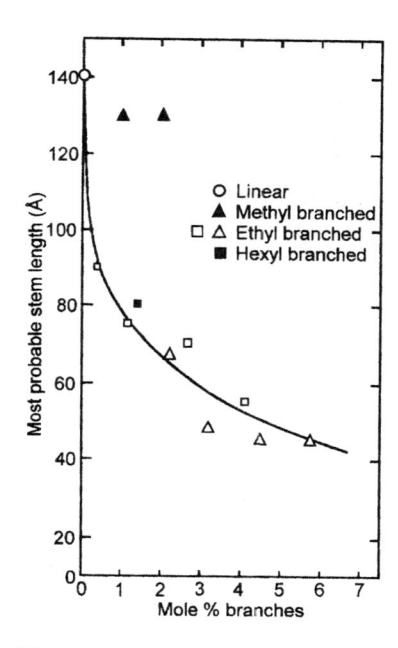

Figure 40 Effect of co-unit content on the thickness of lamellae of linear low density polyethylene crystallized from the melt. (From Ref. 110.)

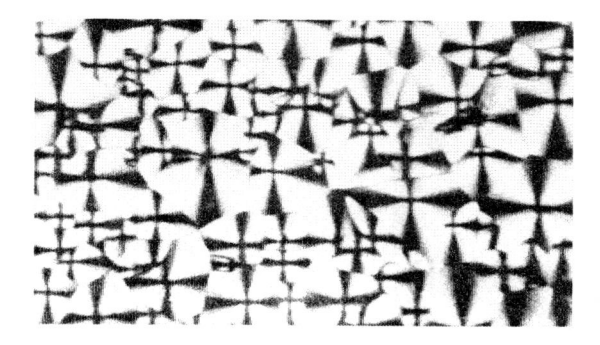

Figure 41 Spherulites in a thin film of high density polyethylene viewed between crossed polars under an optical microscope. (From Ref. 111.)

Figure 42 Ringed spherulites in a thin film of high density polyethylene viewed between crossed polars under an optical microscope. (From Ref. 112.)

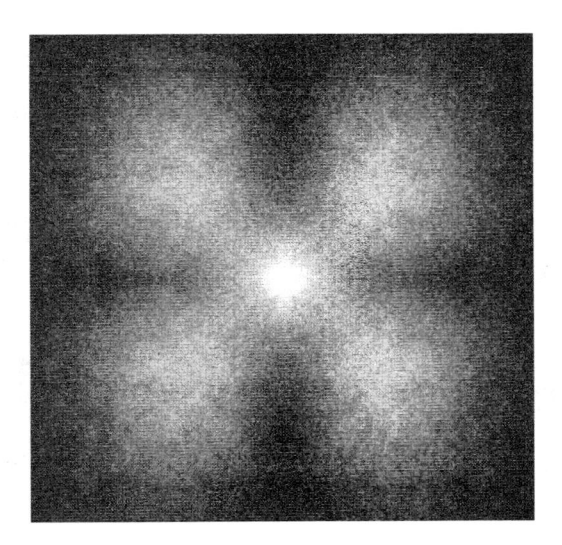

Figure 43 Small-angle laser light scattering pattern generated from a film of high density polyethylene between crossed polars.

lites increases with decreasing molecular weight for linear polyethylene [93] and with decreasing co-unit content for branched samples [94].

2. Crystallization from Oriented Melts

In most commercial processes crystallization occurs from anisotropic melts. The degree of orientation strongly influences the supermolecular structure of the solid state. In most molded items, not all regions experience the same degree of orientation; thus different parts of the product display different supermolecular morphologies.

When orientation is low, an essentially spherulitic supermolecular morphology is frequently adopted, but as orientation increases, different structures are formed. Due to the statistically random nature of molecules in the solid state, certain sequences are liable to become aligned to a greater extent than others during the orientation process. The extent of alignment and the number of chain sequences participating are functions of the degree of orientation, the molecular weight distribution, and the branching characteristics of a sample. As the overall alignment of the molecules increases, the orientation of the crystalline c axes in the lamellae with respect to the deforming force tends to improve. Thus an increase in melt orientation results in a higher proportion of the lamellae with lateral dimensions perpendicular to the deforming force.

At very high levels of orientation, a small proportion of the polyethylene molecules—those with the highest molecular weight—become fully extended over some or all of their length. These fully extended segments act as nuclei upon which crystallization subsequently takes place. The net result is a stack of lamellae with their c axes aligned almost perfectly with the macroscopic deformation. This supermolecular morphology has much in common with the shish kebab structures described above. When such shish kebabs form from the molten state, they are sometimes referred to as ''cylindrites.''

BIBLIOGRAPHY

Bassett DC. Principles of Polymer Morphology. London: Cambridge Univ Press, 1981.
Geil PH. Polymer Single Crystals. New York: Interscience, 1963.
Mandelkern L. Crystallization of Polymers. New York: McGraw-Hill, 1964.
Mandelkern L. Polym Eng Sci 7:232, 1967.
Mandelkern L. J Phys Chem 75: 3909, 1971.
Mandelkern L. J Macromol Sci Phys A15(16):1211, 1981.
Mandelkern L. Acc Chem Res 23:380, 1990.
Woodward AE. Atlas of Polymer Morphology. Munich: Hanser, 1989.
Wunderlich B. Macromolecular Physics. Vol. 2. New York: Academic Press, 1976.
Condensed State of Macromolecules. Faraday Disc Chem Soc, Vol 68, 1979.

REFERENCES

1. G R Strobl, W Hagedorn. J Polym Sci Polym Phys Ed 16:1181, 1978.
2. R Kitamaru, F Horrii, K Marayama. Macromolecular, 19:636, 1986.
3. H Hagemann, RG Snyder, AJ Peacock, L Mandelkern. Macromolecular 22:3600, 1989.
4. R Chiang, P J Flory. J Am Chem Soc 83:2857, 1961.
5. P J Flory, A Vrij, J Am Chem Soc 85:3548, 1963.
6. W O Statton. J Polym Sci Part C 18:33, 1967.
7. M I Bank, S Krimm. J Appl Phys 39:4951, 1968.
8. S Kavesh, J M Schultz. J Polym Sci Polym Chem Ed 8:243, 1970.
9. R Kitamaru, L Mandelkern. J Polym Sci A-2 8:2079, 1970.
10. P R Swan. J Polym Sci 56:403, 1962.
11. P R Swan. J Polym Sci 56:409, 1962.
12. M Heink, K-D Häberle, W Wilke. Colloid Polym Sci 269:675, 1991.
13. E M Antipov, S D Artamonova, I V Samusenko, Z Pelzbauer. J Macromol Sci-Phys B30:245, 1991.
14. C G Vonk, H Reynaers. Polym Commun 31:190, 1990.
15. F Laupretre, L Monnerie, L Barthelemy, JP Vairon, A Sauzeau, D Roussel. Polym Bull 15:159, 1986.
16. C France, P J Hendra, W F Maddams, H A Willis. Polymers 28:710, 1987.
17. S Hosoda, H Nomura, Y Gotoh, H Kihara. Polymers 31:1999, 1990.
18. K Shirayama, S I Kita, H Watabe. Makromol Chem 151:97, 1972.
19. T Seto, T Hara, K Tanaka. Jpn J Appl Phys 7:31, 1968.
20. P J Hendra, M A Taylor, H A Willis. Polymers 26:1501, 1985.
21. S Ottani, B E Wagner, R S Porter. Polym Commun 31:370, 1990.
22. D Hofmann, E Schulz, D Fanter, H Fuhrmann, D Bilda. J Appl Polym Sci 42:863, 1991.
23. D C Bassett, S Block, G J Piermarini. J Appl Phys 45:4146, 1974.
24. M Yasuniwa, R Enoshita, T Takemura. Jpn J Appl Phys 15:1421, 1976.
25. C W Bunn. Trans Faraday Soc 35:482, 1939.
26. R Jaccodine. Nature 176:305, 1955.
27. E W Fischer. Z Naturforsch 12a:753, 1957.
28. A Keller. Phil Mag 2:1171, 1957.
29. P H Till. J Polym Sci 24:301, 1957.
30. R K Sharma, L Mandelkern, Macromolecules 2:266, 1969.
31. A Keller. Makromol Chem 34:1, 1959.
32. P H Lindenmeyer. Science 147:1256, 1962.
33. D C Bassett, F C Frank, A Keller. Phil Mag 8:1753, 1963.
34. J B Jackson, P J Flory, R Chiang. Trans Faraday Soc 59:1906, 1963.
35. E W Fischer. Z Naturforsch 12a:753, 1957.
36. R Eppe, E W Fischer, H A Stuart. J Polym Sci 34:721, 1959.
37. C Sella, J Trillat, Compt Rend 248:410, 1959.
38. R H Olley, A M Hodge, D C Bassett. J Polym Sci Polym Phys ED 17:627, 1979.
39. G Kanig. Prog Colloid Polym Sci 57:176, 1975.

40. J S Trent, J O Scheinbeim, P R Couchman. Macromolecules 16:589, 1983.
41. H Sano, T Usami, H Nakagawa. Polymers 27:1497, 1986.
42. RL Cormia, FP Price, D Turnbull. J Chem Phys 37:1333, 1962.
43. L Mandelkern. Crystallization of Polymers. New York: McGraw-Hill, 1964, pp. 241ff.
44. P H Geil. J Polym Sci 47:65, 1960.
45. P J Flory. J Am Chem Soc 84:2857, 1962.
46. L Mandelkern. Polym Eng Sci 7:232, 1967.
47. L Mandelkern. J Polym Sci Polym Symp 50:457, 1975.
48. P G de Gennes. J Chem Phys 55:572, 1971.
49. J D Hoffman. Polymers 23:656, 1982.
50. F C Frank. Faraday Disc Chem Soc 68:7, 1979.
51. G M Stack, L Mandelkern, I G Voigt-Martin. Macromolecules 17:321, 1984.
52. J Martinez-Salazar, P S Barham, A Keller. J Polym Sci Polym Phys Ed 22:1085, 1984.
53. J Schelten, D G H Ballard, G D Wignall, G Longman, W Schmatz. Polymers 17:756, 1976.
54. D Y Yoon, P J Flory. Polymers 18:509, 1977.
55. D Y Yoon. J Appl Cryst 11:531, 1978.
56. D Y Yoon, P J Flory. Faraday Disc Chem Soc 68:288, 1979.
57. A J Peacock, H A Willis, P J Hendra. Polymers 28:705, 1987.
58. J D Hoffman, C M Guttman, E A DiMarzio. Faraday Disc Chem Soc 68:177, 1979.
59. C M Guttman, E A DiMarzio, J D Hoffman. Polymers 22:597, 1981.
60. J Schelten, D G H Ballard, G D Wignall, G Longman, W Schmatz. Polymers 17:756, 1976.
61. D M Sadler, A Keller. Polymers 17:37, 1976.
62. Condensed State of Macromolecules. Discussion of the Faraday Society. Vol 68. 1979.
63. A Lustiger, N Ishikawa. J Polym Sci Polym Phys Ed 29:1047, 1991.
64. J T Yeh, J Runt. J Polym Sci Polym Phys Ed 29:371, 1991.
65. P J Flory, D Y Yoon, K A Dill. Macromolecules 17:862, 1984.
66. S C Mathur, WL Mattice. Macromolecules 21:1354, 1988.
67. S C Mathur, K Rodrigues, W L Mattice. Macromolecules 22:2781, 1989.
68. R C Lacher, J L Bryant, L N Howard. J Chem Phys 85:6147, 1986.
69. G Kanig. Colloids Polym Sci 269:1118, 1991.
70. K W McLaughlin. Personal communication.
71. A J Peacock. Unpublished observation.
72. K Schmidt-Rohr, H W Spiess. Macromolecules 24:5288, 1991.
73. D W Noid, B G Sumpter, B Wunderlich. Polymers 31:1254, 1990.
74. R Popli, L Mandelkern. J Polym Sci Polym Phys Ed 25:441, 1987.
75. R Alamo, R Domszy, L Mandelkern. J Phys Chem 88:6587, 1984.
76. K Rodrigues, S C Mathur, W L Mattice. Macromolecules 23:2484, 1990.
77. L Mandelkern, Crystallization of Polymers. New York: McGraw-Hill, 1964, pp 225ff.
78. N Alberola. J Mater Sci 26:1856, 1991.
79. L Mandelkern, A J Peacock, Stud Phys Theor Chem 54:201, 1988.

80. A J Peacock, L Mandelkern. J Polym Sci Polym Phys Ed 28:1917, 1990.
81. N Brown, I M Ward. J Mater Sci 18:1405, 1983.
82. Y Huang, N Brown. J Polym Sci Polym Phys Ed 29:129, 1991.
83. E W Fischer, K Hahn, J Kugler, U Struth, R Born, M Stamm. J Polym Sci Polym Phys Ed 22:1491, 1984.
84. E W Fischer. Polym J 17:307, 1985.
85. C M Guttman, A A DiMarzio, J D Hoffman. Polymers 22:1466, 1981.
86. R C Lacher. Macromolecules 19:2639, 1986.
87. L Mandelkern, A J Peacock. Polym Bull 16:529, 1986.
88. A J Pennings. J Polym Sci Polym Symp 59:55, 1977.
89. P Smith, P J Lemstra. J Mater Sci 15:505, 1980.
90. R S Stein, M B Rhodes. J Appl Phys 31:1873, 1960.
91. L Mandelkern. Disc Faraday Soc 68:310, 1979.
92. J Maxfield, L Mandelkern. Macromolecules 10:1141, 1977.
93. L Mandelkern, M Glotin, R A Benson. Macromolecules 14:22, 1981.
94. D L Wilfong, G W Knight. J Polym Sci Polym Phys Ed 28:861, 1990.
95. S J Spells, S J Organ, A Keller, G Zerbi. Polymers 28:697, 1987.
96. E W Fischer, R Lorenz. Kolloid-Z Z Polym 189:197, 1963.
97. T Kawai, Makromol Chem 90:288, 1966.
98. I G Voigt-Martin, L Mandelkern. J Polym Sci Polym Phys Ed 22:1901, 1984.
99. L Mandelkern. Crystallization of Polymers. New York: McGraw-Hill, 1964, p 242.
100. A Keller. Morphology of crystalline polymers. In: R H Doremus, B W Roberts, D Turnbull, eds. Growth and Perfection of Crystals. Proc Int Conf Crystal Growth, Cooperstown. New York: Wiley, 1958, pp 499–528.
101. L Mandelkern, A S Posner, A F Diori, D E Roberts. J Appl Phys 32:1509, 1961.
102. R H Doremus, B W Roberts, D Turnbull, eds. Growth and Perfection of Crystals. New York: Wiley, 1958.
103. R G Alamo, L Mandelkern. Macromolecules 24:6480, 1991.
104. J G Fatou, C Marco, L Mandelkern. Polymers 31:1685, 1990.
105. P H Geil, D H Reneker. J Polym Sci 51:569, 1961.
106. L Mandelkern. J Phys Chem 75:3909, 1971.
107. R C Domszy, R Alamo, P J M Mathieu, L Mandelkern. J Polym Sci Polym Phys Ed 22:1727, 1984.
108. M J Hill, P J Barham, A Keller. Colloid Polym Sci 258:218, 1980.
109. G M Stack, L Mandelkern, I G Voigt-Martin. Polym Bull 8:421, 1982.
110. R Alamo, R Domszy, L Mandelkern. J Phys Chem 88:6857, 1984.
111. R S Stein. In: R W Lenz, R S Stein, eds. Structure and Properties of Polymer Films. Polym Sci Technol, Vol 1. New York: Plenum, 1973, p. 10.
112. G Chiu, R G Alamo, L Mandelkern. J Polym Sci Polym Phys Ed 28:1207, 1990.

5
Properties of Polyethylene

I. INTRODUCTION

The principal value of polyethylene lies in its desirable balance of physical properties in the solid state and its chemical inertness. These qualities in combination with its low cost and ready processability make it the material of choice for a wide variety of uses. The balance of physical properties can be controlled by judicious selection of the resin and processing parameters, thereby producing items useful in an extremely broad range of applications. In this chapter the macroscopic properties of polyethylene, with particular emphasis on commercially relevant attributes, are discussed with respect to the key molecular and morphological variables that influence them.

The physical properties of solid polyethylene are determined by its semicrystalline nature. The factors that control semicrystalline morphology were discussed at length in Chapter 4, and the principles introduced there are now applied to explain the physical attributes of the solid state. When dealing with polyethylene it should always be kept in mind that many of its most important properties are attributable to a combination of the characteristics of its crystalline and noncrystalline components and the connections linking them. On the positive side, polyethylene is a tough flexible material that is chemically inert and has a high electrical resistance. Against this must be balanced its dimensionally instability under prolonged load and its relatively low softening temperature. Polyethylene is thus very useful in short-term or non-stress-critical applications such as food wrapping, storage containers, and piping but ineffective as an engineering resin or where high temperature stability is required, such as in structural components or underhood automotive applications.

The chemical inertness of polyethylene and its excellent electrical resistance stem from the covalent nature of its carbon–carbon and carbon–hydrogen bonds. From an electronic polarity standpoint, the two primary types of bonds in polyethylene are well matched with little dipole moment. The result is that polyethylene molecules are largely resistant to chemical attack and little affected by electrical fields. Ionomers and other polar copolymers, such as ethylene-vinyl acetate and ethylene-methacrylic acid, are a special case, and their properties are discussed separately in Chapter 7.

The molecular characteristics of a polyethylene resin control its melt rheological properties. These characteristics include the distribution of molecular lengths and the number and type of branches (if any). Except in the case of polar copolymers, there is very little interaction between adjacent polyethylene chains in the melt. The combination of limited chain interaction and a flexible backbone of carbon–carbon bonds results in polymer melts that are highly mobile on a local scale.

II. PHYSICAL PROPERTIES OF SOLID POLYETHYLENE

A. Density

"Density" is one of the descriptors most commonly used when discussing polyethylene resins. This is primarily because many of the physical properties of a polyethylene sample can be predicted to a fair approximation based solely upon its density. When measured under controlled conditions the density of different resins can be used as one factor to help predict their relative properties. The relationship between certain mechanical properties and the density of a sample arises from the semicrystalline nature of polyethylene. The higher the proportion of crystalline phases, the higher the density. The relationship between the ordered and disordered regions in a polyethylene sample controls its material properties, and it is this relationship, via the degree of crystallinity, that density probes. Thus, knowledge of the density of the sample reveals something of its semicrystalline morphology.

The factors that govern the density of a polyethylene sample are those that influence its degree of crystallinity. Thus density is a function of molecular weight characteristics, branch content, and preparation conditions. When all other factors remain constant, the density of a specimen will increase as the branch content, molecular weight, or rate of crystallization decrease or the degree of orientation increases. Of these factors, branch content is the most influential, followed jointly by molecular weight and degree of orientation and lastly by the rate of crystallization. Thus the samples that have the lowest densities are those that are highly branched, regardless of other factors. Conversely, the most dense are unbranched resins with a low molecular weight that have been crystallized slowly or are highly oriented. Figure 1 illustrates the effect of increasing branch content on the density of a series of linear low density polyethylene (LLDPE) samples having weight-average molecular weights ranging from 65,000 to 130,000. Also included for comparison is a high density polyethylene (HDPE)

Properties of Polyethylene

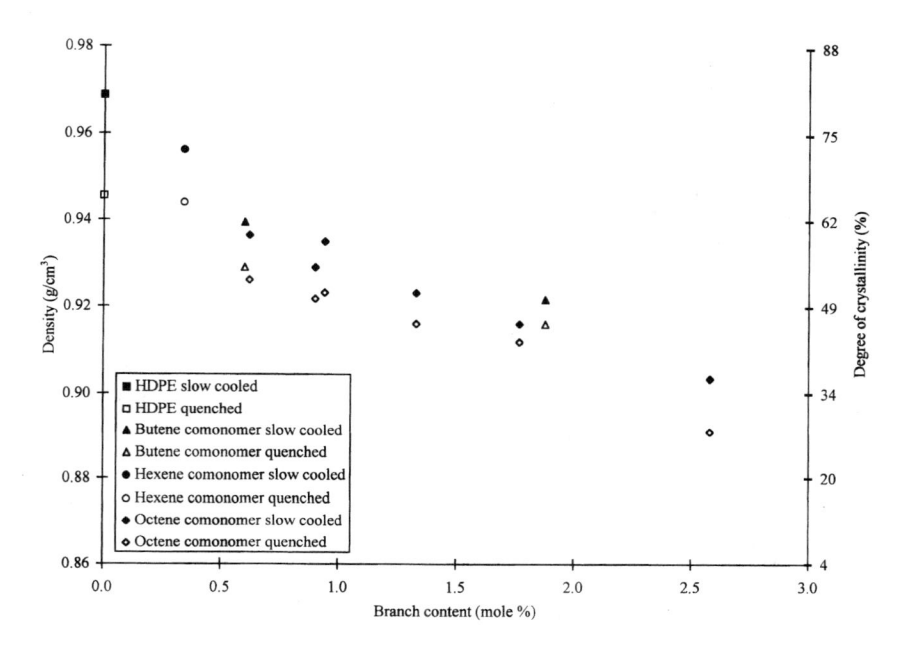

Figure 1 Plot of density as a function of branch content for compression-molded high density and linear low density polyethylene resins. (From Refs. 2 and 103.)

resin having a weight-average molecular weight of approximately 61,000. For reference, the degree of crystallinity calculated from density is plotted on the right-hand axis. Each material was compression molded and then crystallized under one of two regimes, by quench cooling or by slow cooling in air between thick aluminum plates. Two relationships are readily apparent from this figure: Density drops as the branch content increases (particularly at low branch contents), and, to a lesser extent, an increased cooling rate reduces the density; the effect being more pronounced at lower co-unit contents (i.e., higher densities). The densities of linear low density polyethylene resins are essentially independent of the length of the branch. The degree of crystallinity falls by just over 50% in going from the slow-cooled high density polyethylene sample to its linear low density polyethylene counterpart that has 2.6 branches per 100 atoms in the backbone. As the comonomer content is increased further, the density continues to fall, but at a reduced rate.

Figure 2 illustrates the effect of increasing molecular weight on the density of high density polyethylene samples (note that molecular weight is plotted on a logarithmic scale). Crystallization procedures were similar to those of Figure 1. Density falls as molecular weight rises, with the crystallization rate playing a

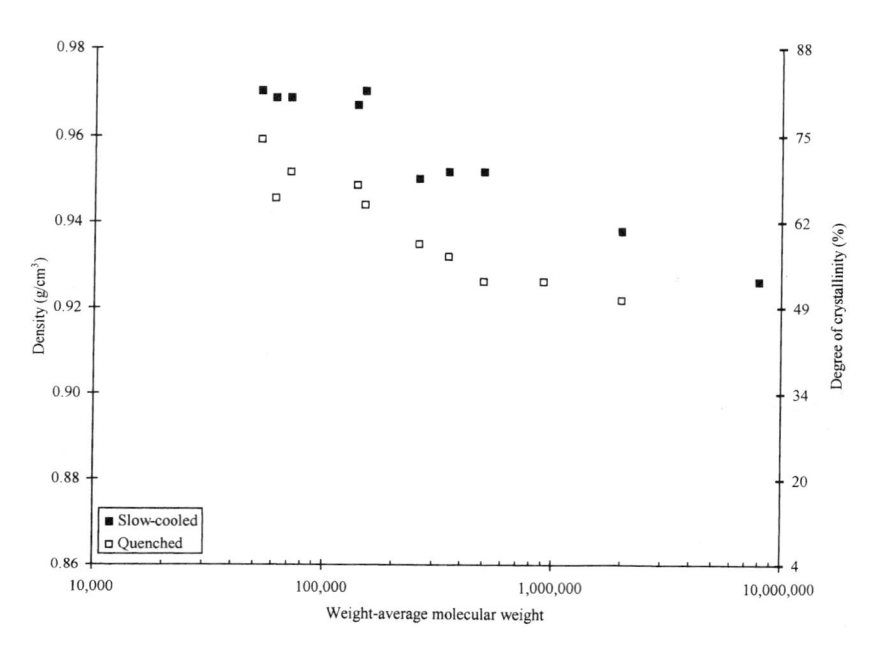

Figure 2 Plot of density as a function of molecular weight for compression-molded linear polyethylene resins. (From Ref. 103.)

secondary role. A tenfold increase in molecular weight results in an approximately 15% reduction of the degree of crystallinity.

Due to difficulties associated with accurately determining the branch distribution and molecular weight of low density polyethylene (LDPE), too few data exist to plot the effects of these characteristics on their densities. It would be expected that the density of low density polyethylene samples would follow relationships qualitatively similar to those of linear low density polyethylene. The relationship between density and degree of orientation is clouded because the accurate determination of orientation is complicated and highly drawn samples often contain voids that make the accurate determination of density difficult.

The inclusion of nonolefinic comonomers such as vinyl acetate, methacrylic acid, and norbornene, or chemical modification with such elements as chlorine and oxygen, destroys the simple relationship linking density and degree of crystallinity. Invariably, polyethylene resins that contain elements other than carbon and hydrogen have an elevated density in comparison to homopolymers or olefin branched materials with similar degrees of crystallinity.

Local density within a specimen does not necessarily reflect its bulk density. Differences in cooling rate and shear effects can result in a distribution of densities within a sample. The most common manifestation of this is in thick injection-molded parts. The skin cools quickly against the chilled mold surface, which inhibits the development of crystallinity, while the core cools more slowly, allowing it to develop a higher degree of crystallinity.

Within broad limits, the densities of the different types of polyethylene fall within the ranges indicated in Figure 10 of Chapter 1. These ranges are subject to some flexibility depending on polymerization and crystallization conditions; it would be possible to find extreme examples of each type of polyethylene outside the ranges quoted.

B. Mechanical Properties

The mechanical properties of a polyethylene specimen can be loosely defined as those attributes that involve the physical rearrangement of its component molecules or distortion of its initial morphology in response to an applied force. On a macroscopic scale, the exercise of a mechanical property results in a dimensional change to the sample. Such physical rearrangements of a sample's morphology occur when it is subjected to external stresses, which may take the form of tension, compression, shear, torque, or combinations thereof. To a large extent the mechanical properties of polyethylene prescribe its realm of application, defining material performance under the influence of external forces. The topic of mechanical properties spans a broad range, covering a multitude of attributes including elastic modulus, impact resistance, hardness, and creep.

In this chapter no attempt is made to list the mechanical properties of all the polyethylene resins available. It is more important to understand the basic relationships that govern such properties. The nature of a specimen's response to applied stress can be correlated with its morphological and molecular characteristics; it is these relationships that are emphasized. The mechanical properties of a specimen are controlled by its processing history within the limits imposed by its molecular characteristics. The nature of the molecular mechanisms involved in the physical deformation of polyethylene is discussed in Chapter 8.

The typical mode of polyethylene deformation is one of yielding and necking followed by strain hardening. Localized yielding is especially noticeable in samples with higher degrees of crystallinity, in which necks form that may have a cross-sectional area of less than one-tenth of that of the original specimen. A prime example of this can be seen in high density polyethylene films such as those used for grocery bags. When stretched perpendicular to its principal orientation direction, the film yields in discrete regions and thins down preferentially to form one or more necks. As elongation continues, the necks grow and merge, encompassing the complete sample. The final stage, strain hardening, occurs when the necked region draws homogeneously prior to break.

The mechanical propterties of polyethylene may be divided into two broad categories: (1) low strain properties such as yield stress and initial modulus and (2) high strain properties, typified by ultimate tensile strength and draw ratio at break. To a first approximation, the low strain properties are controlled by a sample's morphological features, and the high strain properties by its molecular characteristics.

1. Tensile Properties

Tensile properties of polymers are measured on instruments that record the force required to elongate a sample as a function of applied elongation. The details of tensile testing equipment are discussed in Chapter 6. Various conventions exist for the representation of the deformational characteristics of polymer samples. The applied load may be plotted as stress, i.e., force per cross-sectional area of the specimen. This can be somewhat misleading, because the cross-sectional area of a sample is not constant with either location or time. It is common to plot the load as "engineering stress," that is, force per unit area based upon the original cross section of the specimen. Load can be plotted against relative or percent strain of the sample compared to the original gauge length. Relative strain—commonly known as the draw ratio—is the ratio of the deformed sample length to its original length. Percent strain is the increase in length of the sample, multiplied by 100, divided by its original length.

$$\text{Draw ratio} = \frac{\text{deformed length}}{\text{original length}}$$

$$\text{Percent strain} = \frac{\text{increase in sample length}}{\text{original length} \times 100}$$

In the case of polyethylene, which generally deforms by necking, the use of strain as the abscissa, instead of elongation, is rarely appropriate, because different portions of the sample experience different levels of strain concurrently. Plots showing the deformational characteristics of polyethylene are often referred to as stress–strain plots, but this is commonly a misnomer, except in highly specialized cases. In the following discussion the tensile characteristics of polyethylene are illustrated with figures according to the force versus elongation convention.

A schematic force versus elongation curve illustrating the major tensile phenomena of polyethylene is shown in Figure 3. The progression of shapes of a specimen is plotted along the top of the figure. Most tensile samples start off as a "dogbone" (or "dumbbell"), the enlarged ends of which ("tabs") are gripped by the jaws of the tensile tester. The central portion, with parallel sides, is called the gauge region, and its length is termed the gauge length. Initially the

Properties of Polyethylene 129

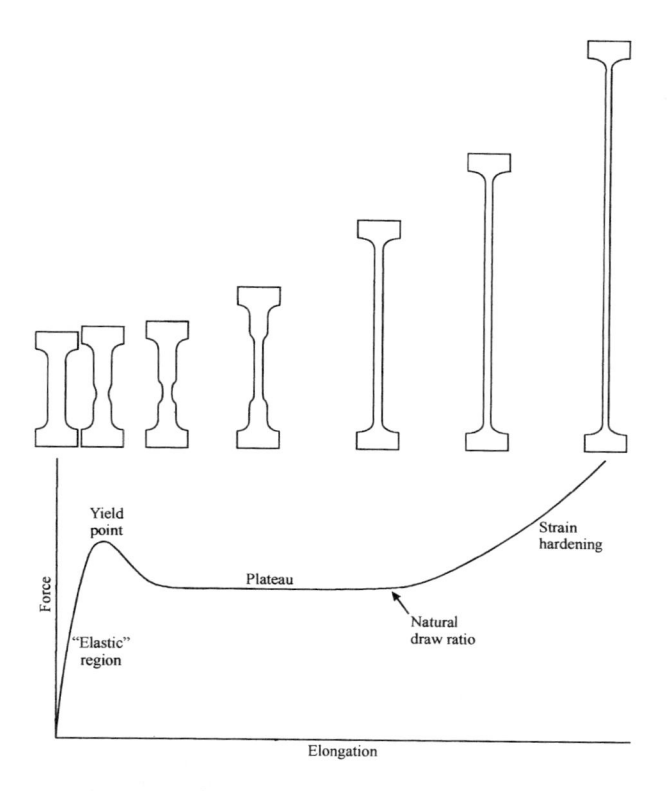

Figure 3 Generalized force versus elongation curve for polyethylene illustrating principal tensile phenomena.

gauge region elongates homogeneously until it reaches a point at which one cross-sectional slice yields independently of the rest of the specimen. The onset of heterogeneous elongation corresponds to the yield point. As elongation continues, the incipient neck becomes better established until it forms a sharply defined region. Upon further elongation the neck propagates, growing to encompass the entire gauge length. The force required for neck propagation is essentially invariant, resulting in a "plateau" in the force versus elongation curve. The strain at the end of the plateau is termed the "natural draw ratio." Subsequent deformation, termed "strain hardening," is homogeneous, with the necked region elongating uniformly until the sample breaks.

The precise shape of a force versus elongation curve is determined by the initial morphology and molecular characteristics of the sample. Figures 4–7 depict the shapes of a series of experimentally derived force versus elongation

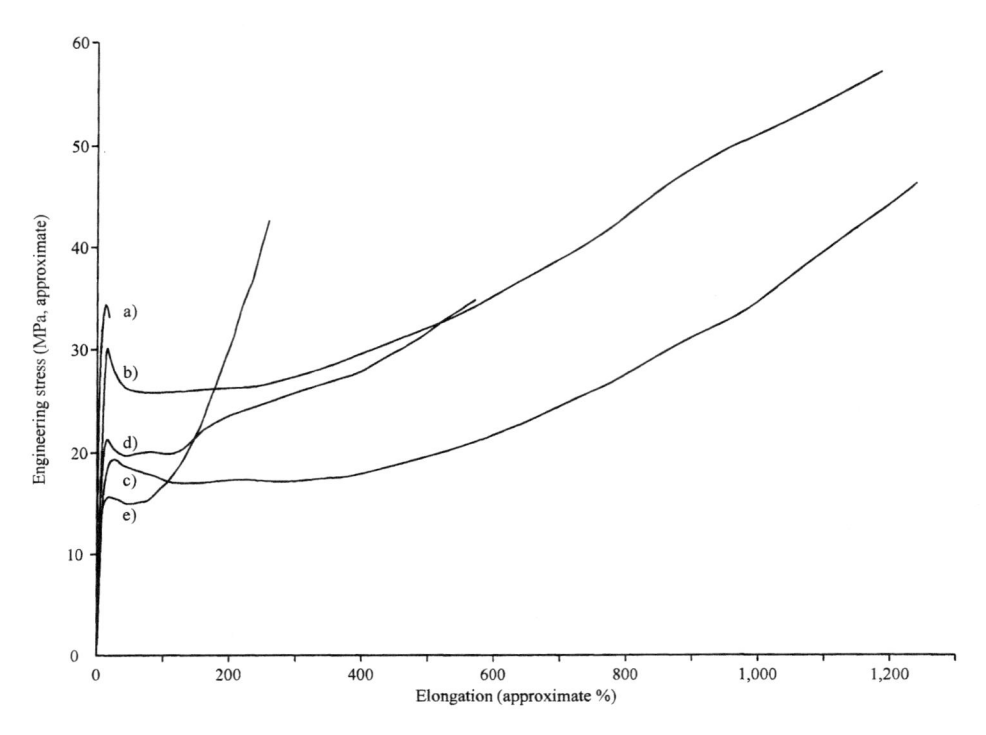

Figure 4 Force versus elongation curves for (a) high density polyethylene of modest molecular weight ($\overline{M}_w \sim 150{,}000$) crystallized isothermally at 128.5°C; (b) high density polyethylene of modest molecular weight ($\overline{M}_w \sim 150{,}000$) crystallized slowly; (c) high density polyethylene of modest molecular weight ($\overline{M}_w \sim 150{,}000$) quench cooled; (d) high density polyethylene of medium molecular weight ($\overline{M}_w \sim 500{,}000$) quench cooled; (e) ultrahigh molecular weight (linear) polyethylene ($\overline{M}_w \sim 3{,}000{,}000$) quench cooled.

curves corresponding to a variety of isotropic polyethylene samples. It is readily apparent that there is no universal shape of force versus elongation curve that applies to all polyethylene resins.

Unless otherwise stated, the data discussed in the following sections refer to isotropic samples drawn at room temperature at elongation rates in the range of 0.5–4.0 in./min.

a. Elastic Modulus. When a polyethylene sample is subjected to external stress there is an initial deformation prior to yield that is homogeneous and is largely recoverable when the stress is removed. This initial region of elasticity can vary from 1% to 2% for highly crystalline samples up to 50% or more in

Properties of Polyethylene

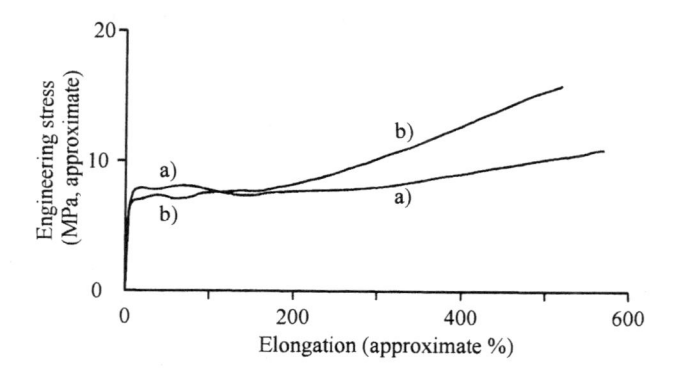

Figure 5 Force versus elongation curve for (a) low density polyethylene quench cooled; (b) ethylene-vinyl acetate copolymer with high comonomer content (~28 wt%) quench cooled.

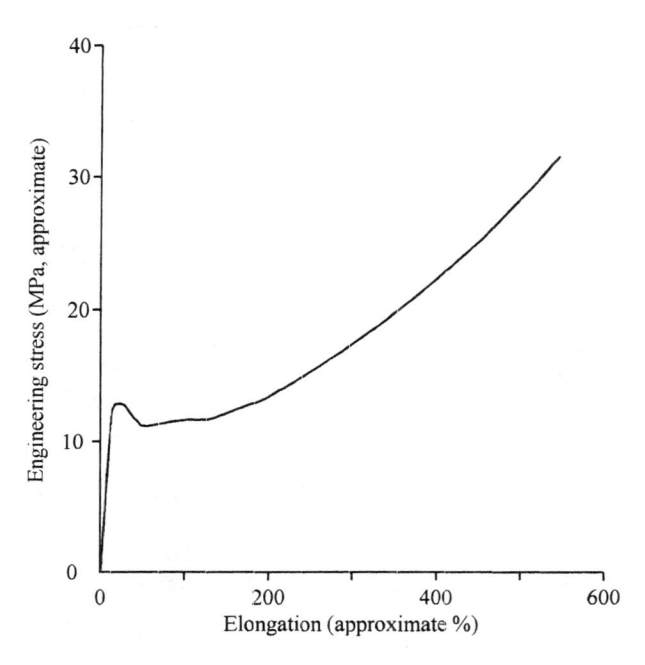

Figure 6 Force versus elongation curve for linear low density polyethylene of modest molecular weight ($\overline{M}_w$ ~ 100,000) and low comonomer content (~1 mol%) quench cooled.

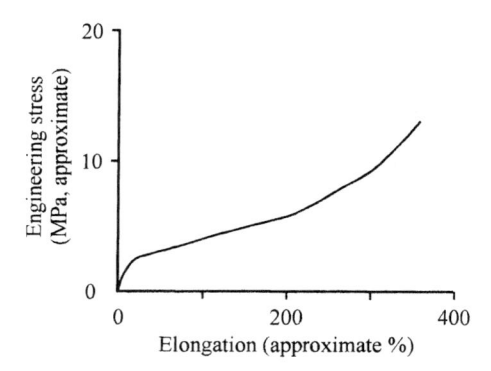

Figure 7 Force versus elongation curve for very low density polyethylene of modest molecular weight ($\overline{M}_w \sim 100{,}000$) and low comonomer content (~ 6 mol%) quench cooled.

high co-unit copolymers and ionomers. The relationship between force and elongation is not Hookean, even at very low deformations. Thus the elastic constant of this region, which is the stress required to deform the sample by a given strain, decreases as a function of elongation. The elastic constant is variously referred to as the "initial modulus," "tensile modulus," "Young's modulus," "elastic modulus," or simply the "modulus" of the sample. The elastic modulus of a sample is a measure of its rigidity; the higher the modulus, the stiffer the sample. The value of elastic modulus is normally derived from the initial slope of the force versus elongation plot. The two most commonly used units are pounds per square inch (psi) and meganewtons per square meter (MN/m^2) [also known as megapascals (MPa)]; (1 psi $\approx$ 0.0069 MN/m^2; 1 $MN/m^2 \approx$ 145 psi). The determination of elastic modulus is addressed in Chapter 6.

For the majority of isotropic samples, the elastic modulus increases approximately linearly with the degree of crystallinity. Except in the case of certain very high co-unit copolymers and ionomers, the data for linear and branched polyethylenes follows the same approximate relationship. This is illustrated in Figure 8, in which the elastic moduli for several different types of polyethylene are plotted as a function of degree of crystallinity.

Experiments performed on high density polyethylene elongated in the solid state to varying draw ratios reveal a strong dependence of elastic modulus on molecular orientation; this relationship is illustrated in Figure 9. By gel or solution spinning polyethylene, the degree of orientation can be increased further, and the elastic modulus is increased accordingly [1]. At the highest degrees of orientation, the elastic modulus of the resulting fibers approaches that calculated for perfectly aligned polyethylene. Table 1 lists the elastic modulus range of the various types of polyethylene and ultradrawn high molecular weight polyethylene

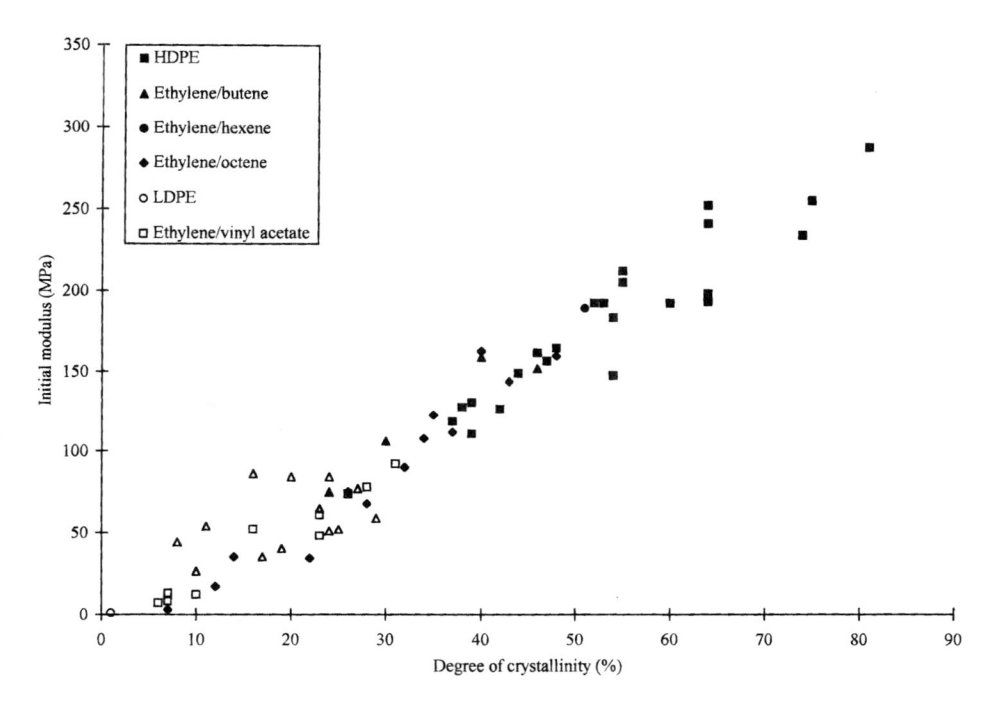

Figure 8 Plot of initial modulus as a function of degree of crystallinity (from Raman spectroscopy) for various isotropic polyethylene samples. (From Ref. 2.)

fibers; for comparison, the table also lists the moduli of a selection of semicrystalline polyolefins, engineering polymers, and nonpolymeric materials.

b. Yield Phenomena. Yielding occurs in a polyethylene specimen when it ceases to deform homogeneously and starts to deform heterogeneously. Up to the yield point, deformation is principally elastic, whereas afterwards the sample takes on a permanent set. Examination of Figures 4–7 reveals that the nature of the yield point varies greatly with the type of polyethylene examined and the conditions under which it was crystallized. In samples with degrees of crystallinity greater than approximately 40%, the yield point corresponds to the first maximum in the force versus elongation curve. Samples having lower degrees of crystallinity do not exhibit a clearly defined maximum at the yield point. In such cases the value of the yield force may be estimated by extrapolating the curve from before and after the first inflection and taking the intersection of the lines as an imaginary yield point. In linear low density and low density polyethylene samples, two distinct maxima may occur in close succession. In other cases an inflection may be followed by a diffuse maximum, as illustrated in Figure 5. The

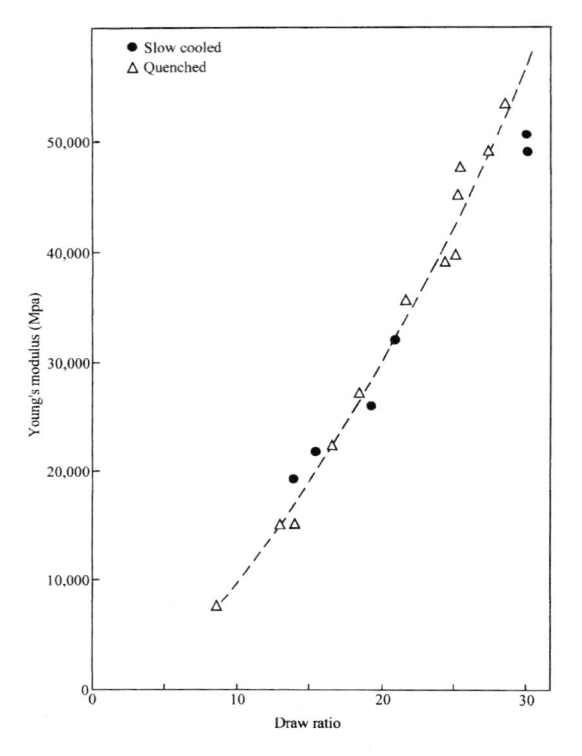

Figure 9 Plot of Young's modulus of high density polyethylene as a function of orientation. (From Ref. 104.)

mechanisms associated with multiple yield peaks are the subject of speculation but may correspond to the yielding of bimodal distributions of lamellar populations [2–4]. At very low levels of crystallinity there may be no distinguishable yield point, as illustrated in Figure 7. Such materials range from highly viscous fluids to elastomers as their molecular weight increases from a few tens of thousands up to several hundred thousand.

The sharpness of the yield peak exhibited during force versus elongation measurements reflects the distinctness of the neck observed visually. Highly crystalline samples, which exhibit distinct yield peaks, initially deform in a localized region to create a neck. The neck is highly oriented, having a much smaller cross-sectional area than the undeformed regions that coexist in series with it. Samples with very low levels of crystallinity exhibit neither localized necking nor a distinct yield peak. Between the extremes lies a continuum of peak distinction and neck definition.

Properties of Polyethylene

Table 1 Elastic Modulus of Various Types of Polyethylene and Selected Polymers

Material	Elastic modulus (10^3 psi)
Very low density polyethylene	<38
Polyethylene ionomer	<60
Ethylene-vinyl acetate copolymer	7–29
Low density polyethylene	25–50
Linear low density polyethylene	38–130
High density polyethylene	155–200
Ultradrawn polyethylene fibers	>29,000
Polytetrafluoroethylene	58–80
Nylon 6	100–464
Acrylonitrile-butadiene-styrene (ABS)	130–420
Polypropylene (isotactic)	165–225
Nylon 6,6	230–550
Poly(methyl methacrylate)	325–450
Polystyrene ("crystal")	330–485
Polycarbonate	345
Poly(vinyl chloride) (unplasticized)	350–600
Acetal (polyoxymethylene)	400–520
Poly(ethylene terephthalate)	400–600
Lead	2000
Glass	8700–14,500
Carbon steel	30,000

Source: Ref. 97.

The tensile yield stress (also known as the tensile yield strength) is the force at which the sample yields, divided by its cross-sectional area. In practice, the actual cross section of the sample at yield is rarely measured, the area of the undeformed specimen being used instead. This is a reasonable approximation providing that the elongation at yield is only a few percent. For isotropic samples the yield stress at room temperature is closely correlated with the degree of crystallinity and thus with sample density. Yield stress as a function of degree of crystallinity (calculated from density) for various samples is plotted in Figure 10. This plot contains data covering a wide range of density, incorporating all the major types of polyethylene. The most striking observation regarding these data is that all samples, regardless of type or source, fall within the same envelope. On close inspection it is observed that high density polyethylene samples of relatively low crystallinity tend to have a higher yield stress than branched samples of similar crystallinity. Maximum yield stress ($\sim$33 MN/m^2) is obtained

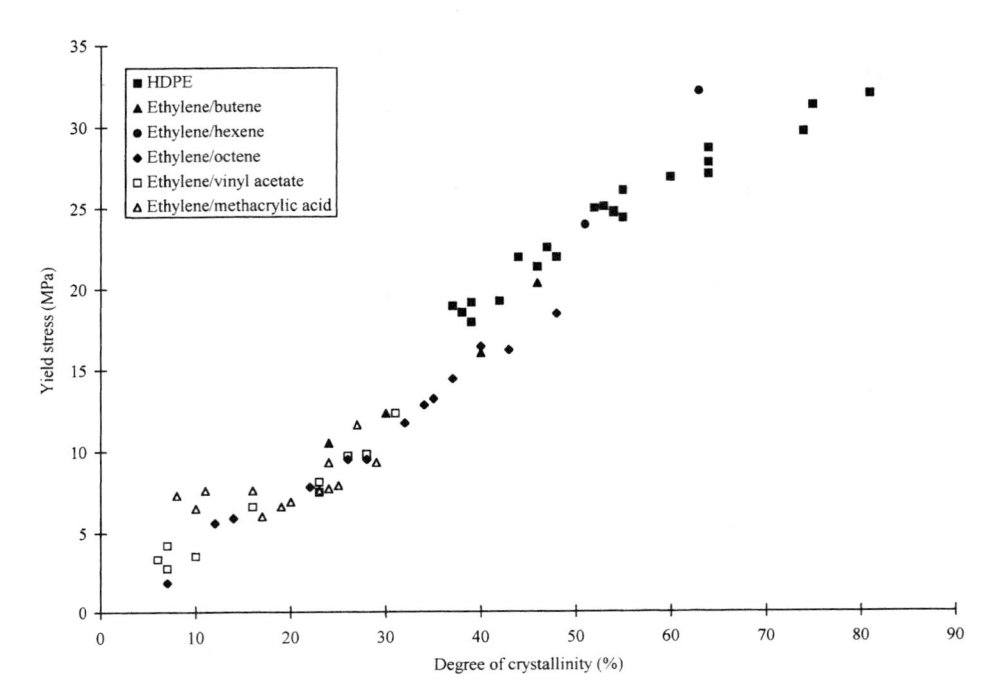

Figure 10 Plot of yield stress as a function of degree of crystallinity (from density) for various isotropic polyethylene samples. (Data from Refs. 105 and 106.)

for samples with degrees of crystallinity of approximately 80%. Above this crystallinity level, samples tend to be brittle. It is of interest to note that the break stress of brittle samples is approximately constant, corresponding roughly to the maximum yield stress attainable in ductile samples.

The yield stress of isotropic samples is closely correlated with their initial modulus. Figure 11 illustrates the relationship between the yield stress and the initial modulus of a series of linear and branched polyethylene samples. There is some indication that linear and branched samples follow separate, but similar, relationships.

Table 2 lists representative yield stress values of the various types of polyethylene; for comparison, the table also provides the yield stresses of a selection of other polymers.

The yield stress of a specimen is of great interest from a practical point of view. In many cases it represents the maximum permissible load that a sample can withstand while still performing its assigned role. Once a sample has yielded,

Properties of Polyethylene

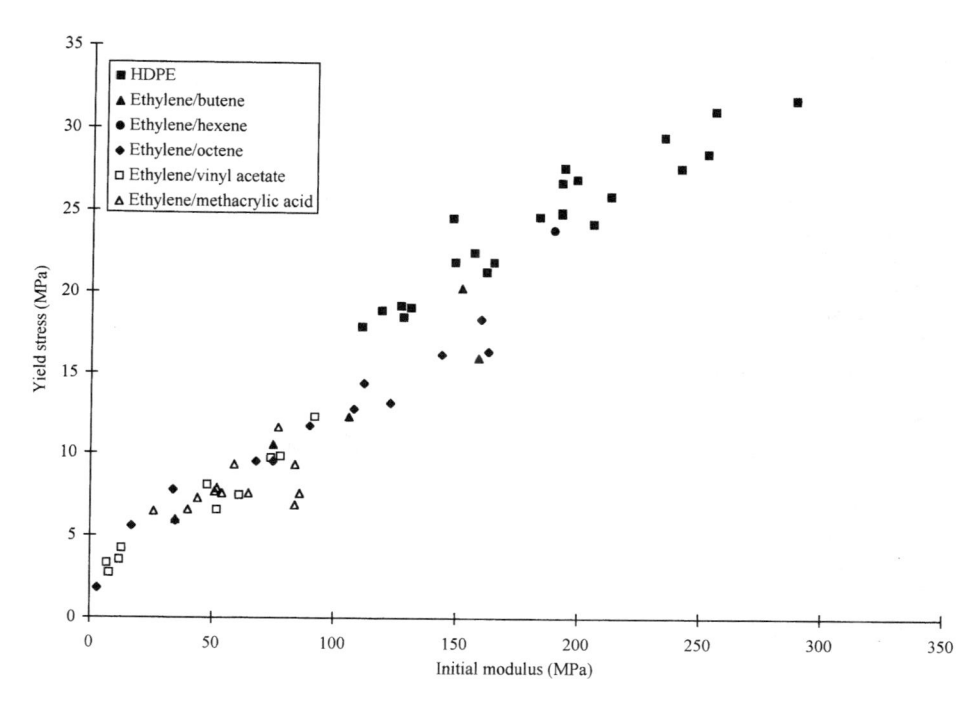

Figure 11 Plot of yield stress as a function of initial modulus for various isotropic polyethylene samples.

its dimensions are irrevocably changed and it may no longer meet the requirements for continued service. In cases where there is a distinct yield maximum in the force versus elongation curve, the force required to propagate a neck along the length of a sample is lower than the yield stress. Once such a sample has yielded, it will continue to elongate unless the applied load is removed.

The elongation at yield of a sample is the strain corresponding to the yield point. It is routinely quoted in terms of percent strain relative to the undeformed sample length, but it is perfectly valid to express it as a draw ratio. Qualitatively, the elongation at yield decreases as the yield stress and elastic modulus of a sample increases. Its value can range from 1% to 2% for highly crystalline samples to more than 50% for samples with very low degrees of crystallinity. The value of the elongation at yield is much less important than the yield stress, because in typical applications specimens have to withstand applied stress rather than applied strain.

Table 2 Yield Stress of Various Types of Polyethylene and Selected Polymers

Polymer	Yield stress (psi)
Very low density polyethylene	<1,100
Polyethylene ionomer	<1,600
Linear low density polyethylene	1,100–4,200
Ethylene-vinyl acetate copolymer	1,200–1,600
Low density polyethylene	1,300–2,800
High density polyethylene	2,600–4,800
Acrylonitrile-butadiene-styrene (ABS)	4,300–6,400
Polypropylene (isotactic)	4,500–5,400
Poly(vinyl chloride) (unplasticized)	5,900–6,500
Polystyrene ("crystal")	6,400–8,200
Nylon 6,6	6,500–12,000
Nylon 6	7,400–13,100
Poly(methyl methacrylate)	7,800–10,600
Poly(ethylene terephthalate)	8,600
Polycarbonate	9,000
Acetal (polyoxymethylene)	9,500–12,000

Source: Ref. 97.

c. *Draw Ratio at Break.* The term "draw ratio at break" refers to the strain of the sample at the point of tensile failure. Other terms corresponding to the same elongation include "ultimate draw ratio," "elongation at break," and "ultimate tensile strain." It may be expressed as a draw ratio—the ultimate sample length divided by its original length—or as percent elongation with respect to the original length. Hereafter the term "draw ratio" is used exclusively. Ideally the draw ratio at break would be determined from the ultimate extension of a portion of the sample relative to its length prior to deformation. In practice, this is rarely done; its practical measurement is addressed in Chapter 6.

The draw ratio at break of a polyethylene sample is a function of its molecular nature and its initial orientation. The molecular characteristics that facilitate drawing are similar to those that promote the development of high degrees of crystallinity. Features that hinder the slippage of chains past one another during crystallization also inhibit the drawing process. The two principal inhibitors to chain movement are entanglements and branch points. Thus, high molecular weight linear polyethylene resins and branched samples have lower draw ratios at break than low molecular weight unbranched samples. The initial morphology of an isotropic sample has little effect on its ultimate draw ratio, except when

Properties of Polyethylene

extreme crystallization conditions have been employed. Examples of preparation conditions severe enough to change the draw ratio at break include slow crystallization of low molecular weight samples, which generate abnormally high degrees of crystallinity, and crystallization from solution. Very highly crystalline samples tend to be brittle, while solution-crystallized mats of ultrahigh molecular weight polyethylene draw to a much greater extent than corresponding melt-crystallized samples. Invariably, specimens retract somewhat after failure. Thus, the draw ratio measured after break will always be somewhat less than the actual draw ratio at break. This effect is most noticeable for highly branched samples in which the retraction can amount to one-half of the total elongation at the point of failure. In high density polyethylene samples the retraction is typically one-tenth of the total elongation. Linear low density samples may retract as much as one-fourth of their total elongation. When oriented samples are drawn parallel with their orientation direction they draw to a lesser extent than their isotropic counterparts. The greater the initial orientation, the lower will be the draw ratio at break.

Figure 12 illustrates the effect of molecular weight on the draw ratio at

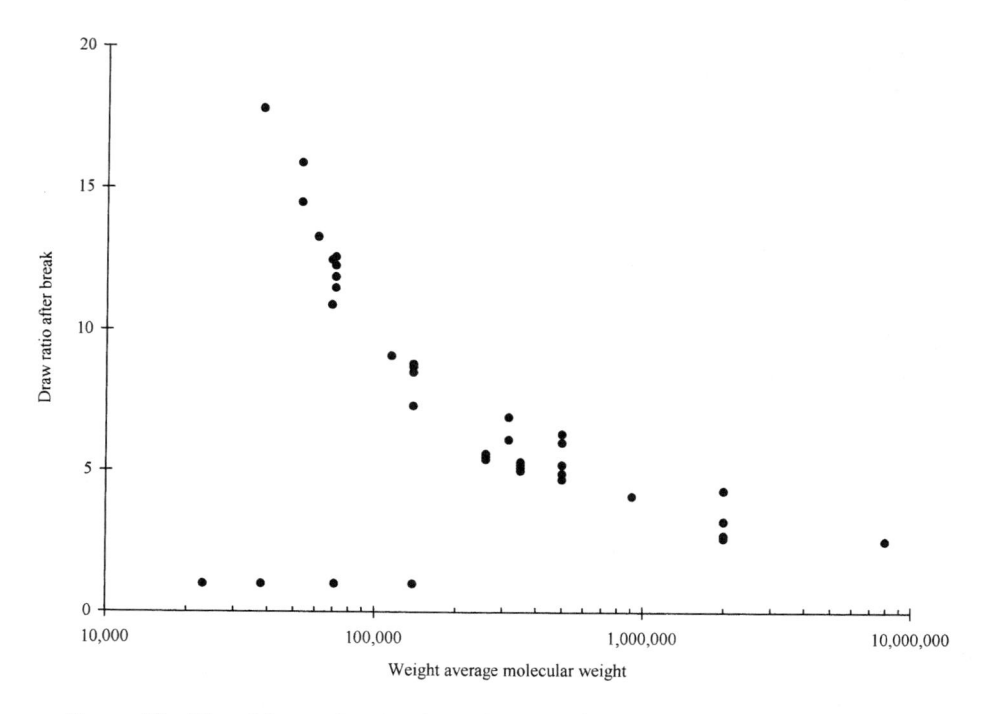

Figure 12 Plot of draw ratio at break as a function of molecular weight for compression-molded linear polyethylene resins. (Data from Refs. 103 and 105.)

break of a number of isotropic high density and ultrahigh molecular weight linear polyethylene resins. Samples of each material were crystallized from the melt under widely differing conditions to generate a wide range of morphologies and degrees of crystallinity. The data reveal two separate relationships, one for samples that fail in a ductile manner and the other for brittle specimens. The brittle specimens, which are all highly crystalline, exhibit negligible drawing prior to failure (although microscopic examination of the fracture surfaces reveals localized deformation). The ductile specimens all fall close to a single line, revealing a monotonic decrease of draw ratio at break with increasing molecular weight, dropping from 15 or more for a resin with $\overline{M}_w \approx 50,000$ to the vicinity of 3 for ultrahigh molecular weight polyethylene. The draw ratio at break of the ductile samples is largely unaffected by their degree of crystallinity or morphology.

Figure 13 shows the effect of molecular weight on the draw ratio after break for a series of short-chain branched polyethylene resins. The solid line indicates the relationship for linear polyethylene resins from Figure 12. Branched samples invariably have lower extensibility at break than linear samples of the

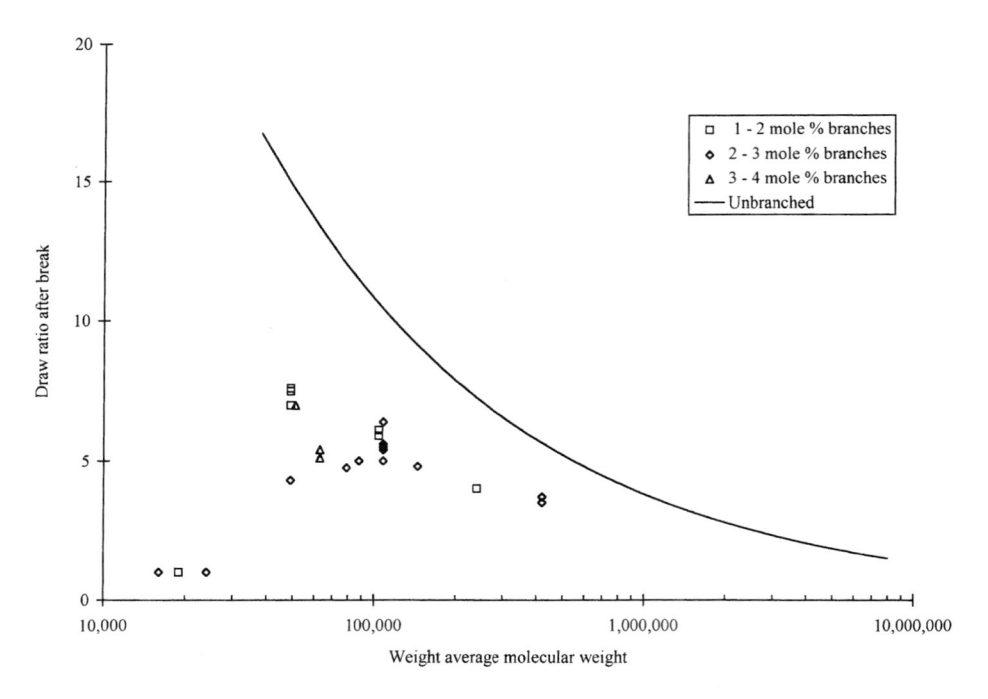

Figure 13 Plot of draw ratio after break as a function of molecular weight for compression-molded branched polyethylene resins. (From Ref. 106.)

Properties of Polyethylene

same molecular weight. As a whole, the data show a large amount of scatter, but they can be resolved into several families as a function of branch content. For ductile samples at a given molecular weight, the draw ratio at break falls as their comonomer content increases. Similarly, for a given comonomer content, the draw ratio at break of ductile samples falls as the molecular weight increases. The molecular weight corresponding to the transition between brittle and ductile behavior increases as the comonomer content increases.

d. Ultimate Tensile Stress. The ultimate tensile stress—also known as the "tensile strength," "ultimate tensile strength," "breaking stress," or "stress at break"—of a sample is the force required to break it divided by its cross-sectional area. In absolute terms the cross-sectional area of the sample at the point of break should be used to calculate the breaking stress, in which case it may also be referred to as the "true ultimate tensile stress." More often the original cross-sectional area is used. Care must be exercised when comparing values that a consistent method is used. In this work the term "engineering breaking stress" is used when the calculation involves the undeformed cross-sectional area; when the actual cross-sectional area at break is used, the term "true ultimate tensile stress" will be applied. The true ultimate tensile stress exceeds the engineering breaking stress by a factor close to the draw ratio at break.

The true ultimate tensile stress of a sample cannot be directly correlated with the morphological features of the undeformed specimen. It depends largely upon the draw ratio at break of the sample, insofar as it results in a reduction of the cross-sectional area. Thus for high density samples the true ultimate tensile stress is approximately inversely related to their molecular weight, as shown in Figure 14. The data exhibit a fair degree of scatter at any given molecular weight; obviously some factors other than molecular ones must play secondary roles. For two high density polyethylene specimens with very different molecular weights but identical initial dimensions, it is normal for the one with the lower molecular weight to have a higher true ultimate tensile stress because of its higher elongation at break, even though the higher molecular weight sample may require a greater force to break it.

The values of true ultimate tensile strength of branched polyethylene samples are generally lower than those of high density samples, largely because of the higher draw ratios obtainable for the high density specimens. The relationship between true ultimate tensile strength and molecular characteristics of short-chain branched samples is shown in Figure 15. Above a critical molecular weight, the true ultimate strength falls as a function of increasing molecular weight and branch content.

e. Temperature Effects. The tensile properties of polyethylene samples are strongly influenced by temperature, especially between room temperature and their melting ranges. Elastic modulus and yield stress fall monotonically with

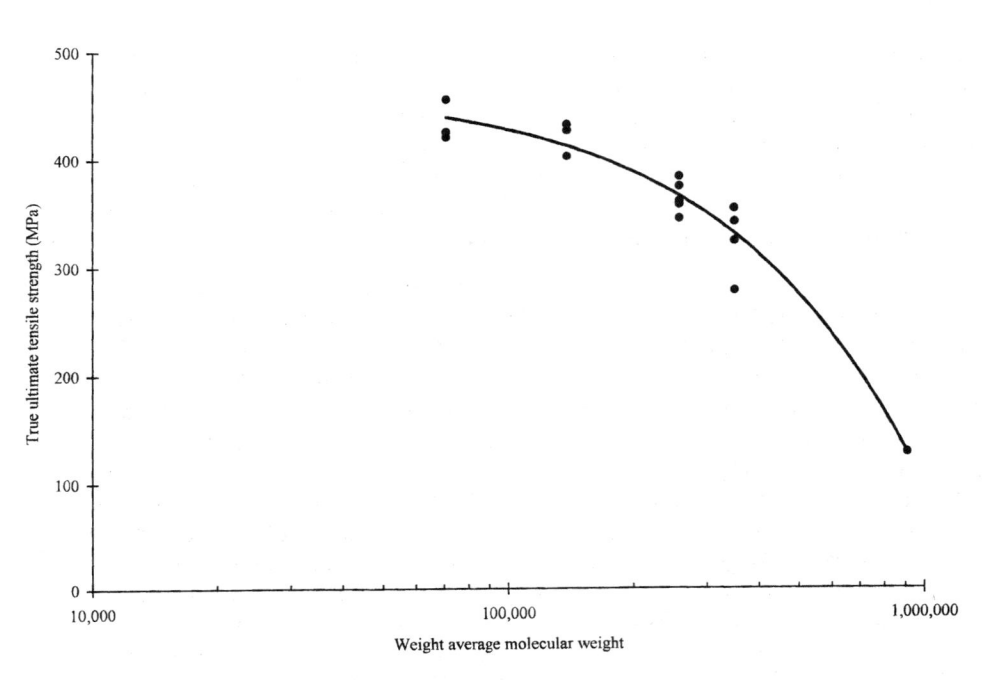

Figure 14 Plot of true ultimate tensile stress as a function of molecular weight of linear polyethylene samples. (Data from Refs. 103 and 105.)

increasing temperature between the glass and melting transitions. Draw ratio at break and true ultimate tensile strength show little variation between the glass transition and room temperature. The draw ratio at break of lower molecular weight linear polyethylene samples rise rapidly to a maximum before falling as the sample undergoes the transition from the solid to the melt; all other samples show a more modest dependence on temperature. The response of samples to temperature is influenced greatly by their molecular characteristics; branched samples fall into a separate category from linear samples, the properties of which are affected greatly by molecular weight.

The relationship between yield stress and deformation temperature for various branched and linear polyethylene samples is shown in Figure 16. The data fall into two discrete categories corresponding to linear and branched materials. Regardless of the nature or concentration of branches, the branched polyethylene samples follow a similar pattern. The yield stress drops rapidly from about the glass transition temperature to 0°C, then more slowly, falling to zero at approximately the peak melting temperature of the sample. The linear samples display a steadier decline of yield stress as a function of temperature over the same range.

Properties of Polyethylene

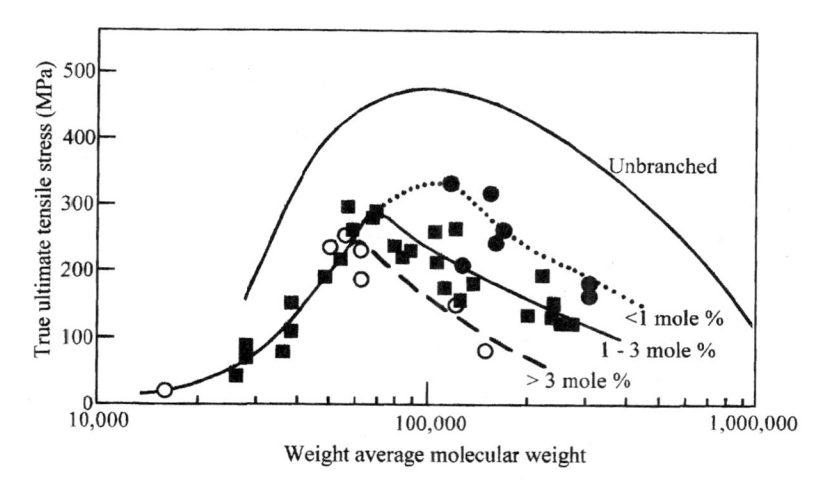

Figure 15 Plot of true ultimate tensile stress as a function of branch content for various ethylene copolymers. (From Ref. 106.)

Even at the approximate peak melting temperature of the linear samples, they all exhibit a small, but measurable, yield stress. At each temperature there is an approximate correlation between the values of yield stress for the linear samples and their degree of crystallinity; this further correlates with their relative molecular weights. The relationship of elastic modulus to drawing temperature follows a pattern similar to that of the yield stress, with the notable exception that the moduli of branched samples are significantly greater than those of the linear materials at $-100°C$.

Below the glass transition temperature, all polyethylene samples fail in a brittle manner; thus they all have a draw ratio at break of approximately 1.0 below about $-100°C$. The relationship between the draw ratio at break and the deformation temperature is shown in Figure 17. Once again the branched and the linear samples fall into distinct categories. The branched samples show a tight grouping, regardless of the nature or concentration of the branch, gradually diverging as the temperature increases. The increase of draw ratio at break as a function of drawing temperature is fairly modest, reaching a broad maximum of 6–8 at around 50–60°C. Linear samples show much more variation than branched ones. From $-100°C$ to $0°C$, linear polyethylene samples exhibit a modest increase in draw ratio at break with little differentiation between samples. As the temperature increases further, the draw ratio at break rises more rapidly, the greatest increase occurring for the specimens of lowest molecular weight. The draw ratio at break for specimens with an average molecular weight of 1 million or

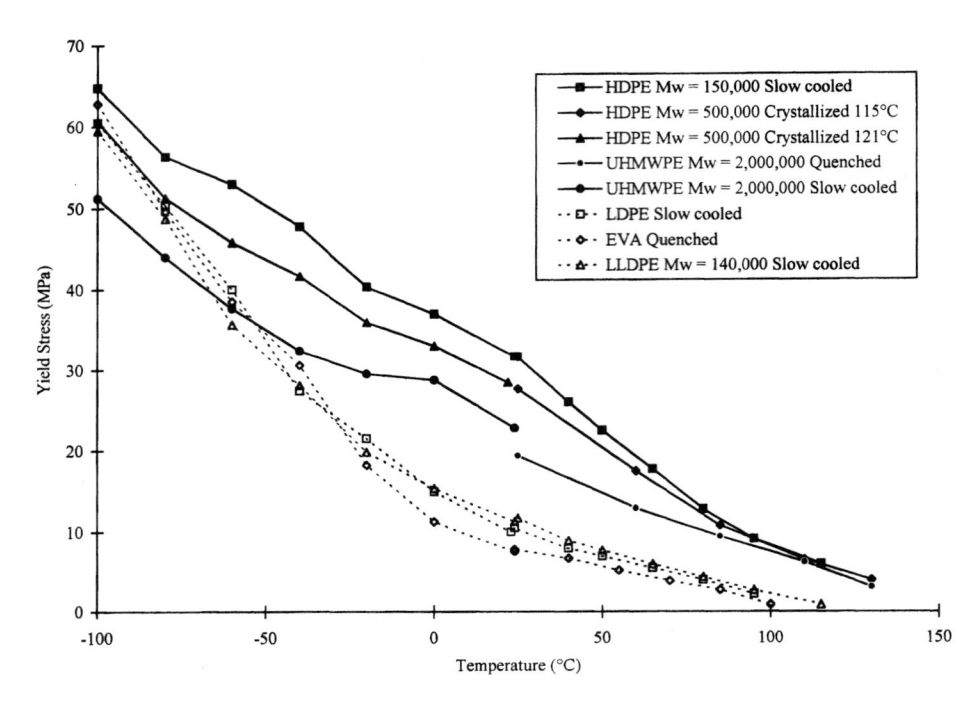

Figure 16 Plot of yield stress as a function of draw temperature for various polyethylene samples. (From Ref. 107.)

more continues to increase up to their peak melting temperature. Samples with an average molecular weight below 1 million reach a maximum below their peak melting temperature.

The relationship between true ultimate tensile stress and deformation temperature for various branched and linear polyethylene samples is shown in Figure 18. Below approximately 0°C the true ultimate tensile strength of any given sample is little affected by temperature. Above 0°C the values for branched samples fall off monotonically to zero at approximately their peak melting temperature. The values for the linear samples are strongly influenced by draw ratio at break. The two samples that exhibit a maximum in draw ratio also exhibit a maximum in true ultimate tensile stress. The higher molecular weight samples show either little variation with drawing temperature or a slight decrease.

f. Elongation Rate Effects. The rate at which specimens are deformed greatly affects their response to the applied stress. The effect of increasing deformation rate on the low strain portions of the force versus elongation curve is

Properties of Polyethylene

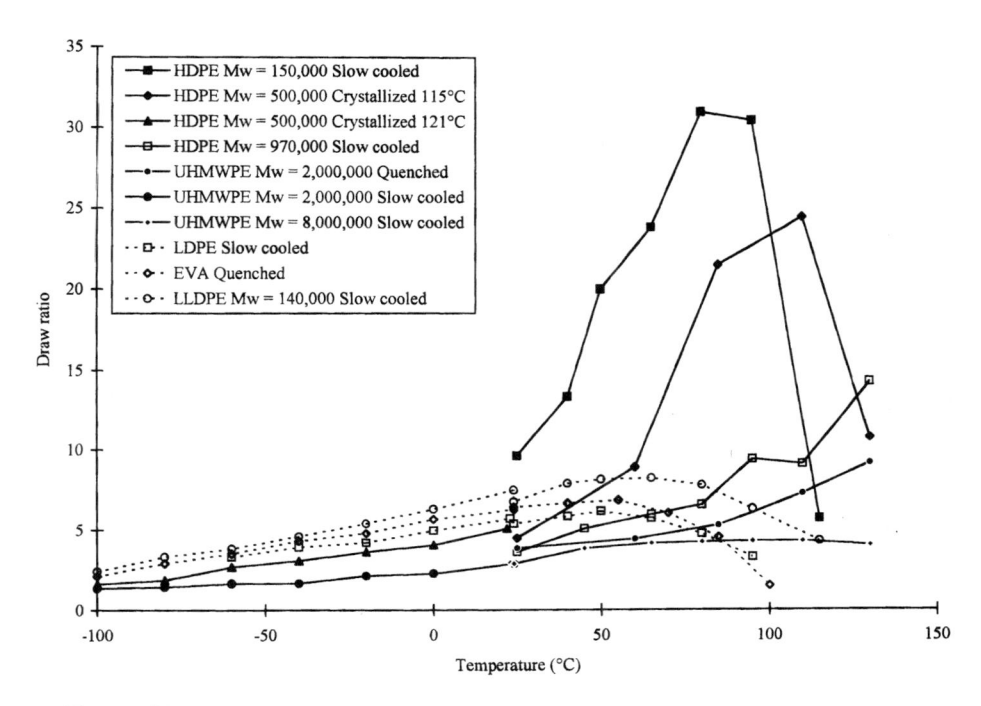

Figure 17 Plot of draw ratio at break as a function of draw temperature for various polyethylene samples. (From Ref. 107.)

similar to the effect of increasing a sample's degree of crystallinity or decreasing the drawing temperature. Faster draw rates give rise to increased elastic moduli, higher yield stresses, lower yield elongations, sharper yield peaks in the force versus elongation curve, and a better defined neck as observed visually. When taken to extremes, a rapid application of strain can convert an otherwise ductile sample into a brittle one. At the other end of the scale, decreasing the deformation rate results in a less well defined neck. The effects of changing the draw rate on the sample's response can be understood in terms of relaxation phenomena, the mechanisms of which are addressed in Chapter 8 with regard to the orientation of polyethylene.

4. Compressive Properties

The compressive modulus of isotropic polyethylene, like its tensile modulus, is low in comparison with those of many other semicrystalline polymers, especially the so-called engineering resins. However, unlike the tensile modulus, the com-

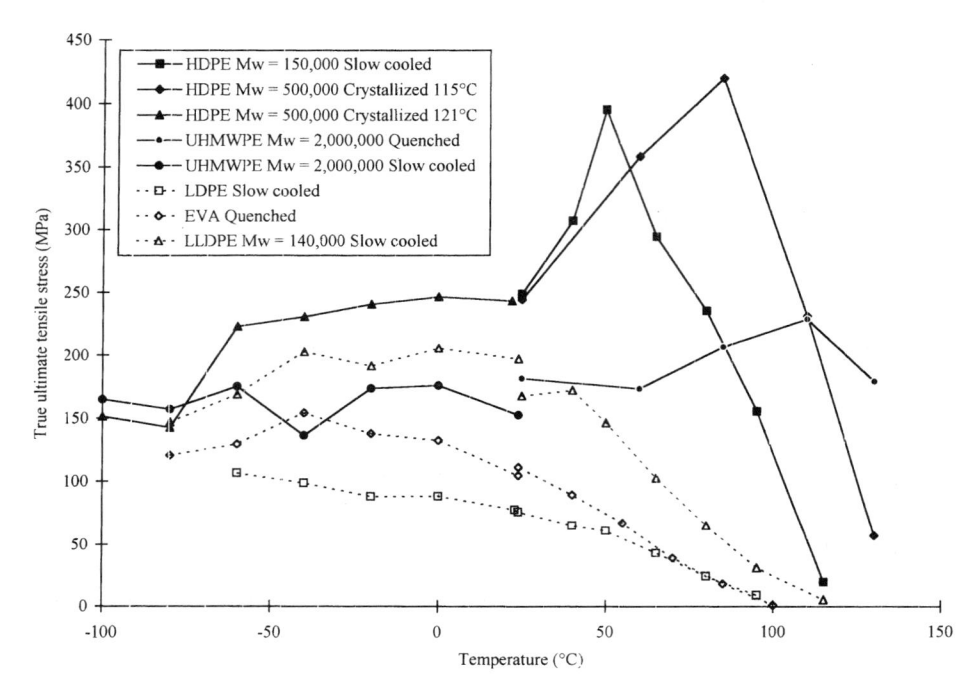

Figure 18 Plot of true ultimate tensile stress as a function of draw temperature for various polyethylene samples. (From Ref. 107.)

pressive modulus cannot be improved by orientation. Because of its low compressive modulus, combined with its ductility and high creep, polyethylene is not used in situations that call for high compressive strength. Accordingly, the bulk compressive modulus of solid polyethylene is not routinely measured, and there are few data relating it directly to structural parameters. Insofar as compressive deformation involves the local rearrangement of a sample's initial morphology, the compressive modulus follows the same trends as the elastic modulus with respect to a sample's degree of crystallinity.

The local compressive strength of polyethylene is of more interest commercially than its bulk compressive modulus. Local compressive strength is normally referred to as "hardness," "microhardness" or "microindentation hardness" (MH). Microhardness is important in terms of the retention of a good surface finish on molded articles.

Microhardness can be determined from the dimensions of an indentation left by a stylus having a known profile applied to a specimen with a known force. The resulting value is quoted in terms of force per unit area, typically in pounds

Properties of Polyethylene

per square inch (psi) or megapascals (MPa). Hardness can also be determined from the penetration depth of a needle forced into a sample by a spring. Two of the most common such hardness testing devices are the Shore and the Rockwell. The units of measurement of these tests are dimensionless. The deformation caused by the indentor or needle involves rearrangement of the initial morphology and hence depends on structural parameters similar to those involved in the short-range tensile deformation of polyethylene. The microhardness of a sample is thus strongly correlated with its tensile yield stress and elastic modulus and hence its degree of crystallinity. For a wide range of polyethylene samples, microhardness can be linearly related to degree of crystallinity [5]. This relationship is illustrated in Figure 19. Microhardness increases when a sample is annealed, rising with increased crystallinity and lamellar thickness [6]. When the microhardness of various polymers is compared, their ranking is similar to that of the elastic modulus. The microhardness values of various polyethylene samples and selected other polymers are given in Table 3.

3. Flexural Modulus

The only flexural property of any practical significance to the use of polyethylene is its modulus. The flexural modulus of polyethylene influences its behavior as a packaging material when used as sheet or film and in blown containers, especially

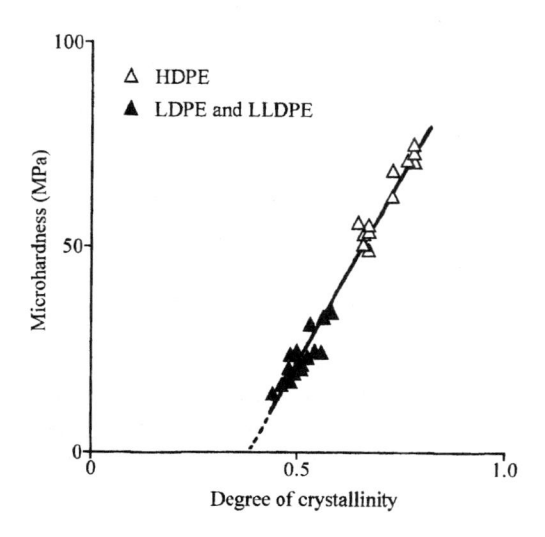

Figure 19 Plot of microhardness as a function of degree of crystallinity (from density) for high density polyethylene, low density polyethylene, and linear low density polyethylene. (From Ref. 5.)

Table 3 Microhardness of Various Types of Polyethylene and Selected Polymers

Polymer	Microhardness	
	Shore D	Ball indentation (MPa)
Ethylene-vinyl acetate copolymer	17–45	
Polyethylene ionomer	25–66	
Low density polyethylene	44–50	13.5
Linear low density polyethylene	55–66	
High density polyethylene	66–73	53.5
Polytetrafluoroethylene	50–65	41
Poly(vinyl chloride) (unplasticized)	66–85	115
Nylon 6	72	62.5
Polypropylene (isotactic)	74	72.5
Nylon 6,6	75	72.5
Polystyrene ("crystal")	78	110
Poly(ethylene terephthalate)		120
Poly(methyl methacrylate)		172

Source: Refs. 60 and 98.

bottles. Due to its relatively high flexibility, polyethylene is rarely used in load-bearing applications. The absolute values of flexural modulus are of similar magnitude to the tensile modulus, being controlled by the same morphological characteristics that control tensile and compressive properties.

4. Rupture Properties

The rupture properties of polyethylene discussed in this section are those involving the formation of fresh surfaces under the influence of abruptly applied tensile or flexural stresses. This section deals principally with impact and tearing failure arising from the application of stresses of the order of magnitude of the yield stress. Low stress brittle failure and rupture after cold drawing, respectively, are discussed in the sections on long-term mechanical properties and tensile properties. Two principal types of polyethylene rupture exist, these being crack propagation through thick specimens such as the wall of a pipe, and the tearing or puncture of thin specimens such as sheet and film. The rupture of polyethylene involves two processes that absorb energy: inelastic deformation and the formation of new surface area. The greater the resistance to rupture exhibited by a sample, the greater is its perceived toughness. In the brief discussion that follows, no distinction is made between impact and tear resistance, as similar factors are active in both. A complete discussion and thorough analysis of the rupture mechanisms of polyethylene are outside the scope of this work; readers who wish to become better acquainted with this complex field are directed to the works of Williams, Ward, and Kausch listed in the bibliography.

Polyethylene samples display a wide range of rupture properties depending upon the molecular nature of the resin, the conditions under which specimens are prepared, and the regimen under which they were tested. Thus a low molecular weight high density polyethylene ($\overline{M}_w \approx 50{,}000$) may be brittle or ductile, depending on whether it is quenched from the melt or slowly cooled, whereas low density polyethylene is ductile under all but the most extreme conditions (sub-glass transition temperatures and very high strain rates). Linear low density polyethylene is the toughest type of polyethylene, requiring the greatest input of energy to rupture it. In common with low stress brittle failure, the preexistence of stress concentrators such as notches, voids, or inclusions is extremely important; rupture invariably originates at such discontinuities. Sample configuration is also crucial in determining the nature of rupture; it controls the relative amounts of energy required to deform the sample and create new surface area. Thicker and wider samples have a greater propensity than thin ones to fail in a brittle manner; with a larger proportion of the energy going into crack propagation.

The factors controlling the rupture of polymeric items (especially the semicrystalline ones) are so complex that tear and impact resistance are generally quoted in simplistic terms, ignoring many of the variables that affect toughness. Typically the fracture strengths of polymers are quoted in terms of the energy required to break a sample of standard dimensions under standardized testing conditions by the impact of a falling weight or swinging pendulum. Values of tear strength are quoted in terms of the energy required to tear a film using a standard test configuration. The values so obtained are relevant only in terms of comparison with other materials tested under identical conditions. Different specimen configurations and testing conditions can deliver highly discrepant results. Standard testing configurations generally bear little resemblance to actual usage. This must always be taken into consideration when reviewing impact and tear resistance data for polymers, especially when different materials are being compared. The results of such impact tests cannot be used for engineering calculations with any degree of certainty.

The energy expended during the rupture process is distributed in various ways. Initially, when the sample is deformed to a small extent, energy is absorbed by elastic deformation. As the deformation becomes larger and stresses increase, yielding occurs, absorbing more energy, principally in the region in which a crack will develop. Once a critical deformation condition (dependent upon the material properties and the sample configuration) is reached, a crack is initiated, invariably originating at a defect, such as an inclusion or notch, and either occurring adventitiously or created deliberately. When the crack is initiated, some of the stored elastic energy is released, being used to break molecular chains, create and break fibrils, disentangle molecules, and create fresh surface area. As deformation continues, the propagation of the crack through the specimen absorbs more energy. When the sample ruptures completely, the remaining stored elastic energy is released. Thus the overall input of energy goes into deforming the sample and creating new surface area. Permanent deformation occurs primarily in the vicinity of the crack tip as it proceeds through the sample; this region of deformation is sometimes referred to as the "outer plastic zone." The overall mechanism follows a similar course to that of low stress brittle failure, but there is less time for molecular disentanglement and hence more breakage of chains.

The overall toughness of a sample is determined by the two energy-absorbing processes. In general, linear low density polyethylene is the toughest of the different classes of polyethylene, exhibiting excellent impact and puncture resistance. Low density polyethylene is so readily deformed that the inelastic deformational component of its fracture resistance is very low, even though crack propagation is relatively difficult. Thus for a given density, linear low density polyethylene has greater toughness than low density polyethylene. In contrast to low density polyethylene, high density polyethylene requires much more energy to deform it. However, high density polyethylene is relatively notch-sensitive, so it does not deform to any great extent before crack initiation and therefore the crack propagates readily, thus giving it the lowest overall toughness of the different types of polyethylene. Comparisons of the three different types of polyethylene are difficult because the mechanisms involved are sensitive to sample configuration. Changes in sample dimensions result in different amounts of energy being absorbed by deformation and crack propagation. In order to evaluate the crack initiation and crack propagation energies separately it is necessary to use the so-called R method [7] or employ an instrumented impact tester [8]. As a rule of thumb, a direct measurement of the crack initiation energy and of that required for crack propagation can be made only if the deformation zone around the notch tip is smaller than the dimensions of the sample. In the case of low density polyethylene, the deformation zone is so large that a sample would have to be many inches thick and wide for such measurements to be made, whereas a high density polyethylene sample need be only about 1/2 in. thick. From a theoretical standpoint the limiting factor is the size of the sample necessary for plane strain conditions to exist [7].

Many of the factors that control the fracture resistance of polyethylene to high speed deformation are similar to those that influence the process of low stress brittle failure. Crack propagation proceeds most rapidly in samples that have the fewest tie chains linking adjacent crystallites; this is a function of molecular weight, branching characteristics, and preparation conditions [7,8]. The lower the molecular weight of a sample, the lower is its propensity to form an extended region of energy-dissipating fibrils in advance of the crack tip [8]. In

Properties of Polyethylene 151

general, rupture occurs most readily in samples of high density polyethylene with low molecular weight that have been cooled slowly, i.e., those samples with the highest degrees of crystallinity, which are also those that are the stiffest. The length of short-chain branches plays a role in determining the puncture resistance of linear low density polyethylene; thus at a given density and molecular weight, ethylene-octene copolymers are somewhat tougher than ethylene-hexene resins, which in turn are significantly tougher than ethylene-butene samples [9].

Testing conditions also play a large part in determining impact and tear resistance. The lower the testing temperature, the lower will be the measured toughness of the sample. At temperatures from subambient to about 60°C, linear low density polyethylene exhibits greater toughness than high density polyethylene; this advantage is lost at higher temperatures due to its lower crystalline melting temperature [8]. The higher the strain rate, the lower will be the fracture resistance in thick samples. The existence of a preexisting notch in a specimen greatly affects its impact strength or tear resistance; samples that are notched inevitably show lower rupture resistance. The fact that high density polyethylene is more notch-sensitive than linear low density polyethylene, which in turn is more notch-sensitive than low density polyethylene, should always be taken into account when comparing impact and tear resistance.

Sample preparation plays an important part in determining rupture resistance. The most important factors affecting commercial samples are the degree and direction of orientation. Failure occurs most readily in planes parallel to the chain orientation direction and is a function of the degree of anisotropy. Notch sensitivity is also a function of orientation within the sample; notches that occur parallel to orientation will have a greater effect than those occurring in the transverse direction.

Given the large disparities between the mechanisms of failure exhibited by the different classes of polyethylene (and incidentally many other semicrystalline polymers), it is very difficult to make quantitative comparisons between the fracture toughness and tear strengths of polyethylenes and other polymers. There is no comprehensive comparison of the crack initiation and crack propagation energies of the different types of polyethylene available in the literature. In part this is because the acquisition of such data is quite time-consuming and also because such data are relevant only to the stringent conditions required by rigorous testing. Table 4 lists the Izod impact strengths of polyethylene and various other polymers. These numbers should be considered comparative only; even the relative ranking can change depending upon notching, temperature, strain rate, etc.

5. Long-term Mechanical Properties

The morphology of polyethylene, and hence the shape of items manufactured from it, is not stable when it is subjected to prolonged stress. This is so even

Table 4 Izod Impact Strength of Various Types of Polyethylene and Selected Polymers

Polymer	Izod impact strength (ft-lb/in. of notch)
Linear low density polyethylene	0.35– No break
High density polyethylene	0.4–4
Polyethylene ionomer	7– No break
Low density polyethylene	No break
Very low density polyethylene	No break
Ethylene-vinyl acetate copolymer	No break
Ultrahigh molecular weight polyethylene	No break
Poly(methyl methacrylate)	0.2–0.4
Poly(ethylene terephthalate)	0.25–0.7
Polystyrene ("crystal")	0.35–0.45
Polypropylene (isotactic)	0.4–1.4
Poly(vinyl chloride) (unplasticized)	0.4–2.2
Nylon 6,6	0.55–2.1
Nylon 6	0.6–3
Polystyrene ("high impact")	0.95–7
Acetal (polyoxymethylene)	1.1–2.3
Acrylonitrile-butadiene-styrene (ABS)	1.5–12
Polytetrafluoroethylene	3
Polycarbonate	12–18

Source: Ref. 97.

when the applied stress is much lower than that required to induce instantaneous yielding. Morphological instability can manifest itself as creep, stress relaxation, crazing, brittle failure, and environmental stress cracking. Each of these manifestations is deleterious to a greater or lesser extent. All of these long-term instabilities involve the gradual rearrangement of molecules either on a local basis, as in the case of brittle failure and stress cracking, or throughout a large portion of the sample, as in creep and stress relaxation.

    a. Creep. The term "creep" is used to describe the gradual deformation of a sample under prolonged loading, which may be either constant or intermittent. The applied forces required to induce creep are lower than those required to permanently deform the material in the short term, i.e., lower than the instantaneous yield stress. Such forces can take the form of tension, compression, torsion, shear, or any combination thereof. Creep occurs on a macroscopic scale, resulting in the deformation of large portions of specimens. It occurs most often in samples that do not contain abrupt surface discontinuities such as notches or deep

Properties of Polyethylene

scratches or inhomogeneities within the bulk such as grit or voids, which act as stress concentrators. When stress concentrators are present, low stress brittle failure is more common.

The strain that a sample exhibits after a given time is the sum of three components: the instantaneous elastic strain upon loading (proportional to the elastic modulus), the delayed elastic strain, and the Newtonian component, which accounts for the viscous flow of the material. In contrast to the mechanical properties described earlier, creep takes place on a time scale of hours to years rather than seconds to minutes. Due to the extended periods of time over which creep takes place, data are normally discussed in terms of time on a logarithmic scale. The nature of creep deformation depends on the testing conditions, but in all cases the strain response to the applied load is nonlinear with respect to time. Figure 20 depicts the general response of polyethylene to applied load on a log time basis. The response of strain to load progresses in three distinct phases. In the first phase the strain increases gradually, with the specimen deforming homogeneously. After a critical time (t_c) the deformation increases rapidly as the specimen yields and develops a neck. The neck rapidly propagates through the length of the specimen until it encompasses the whole gauge region. Thereafter the strain remains essentially constant, negligible deformation occurring even after extended periods of time. Prior to the onset of necking, much of the strain

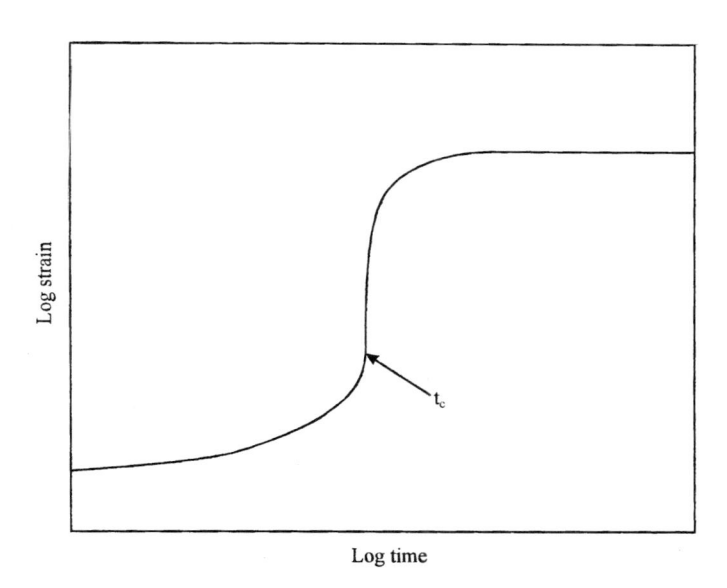

Figure 20 Illustration of the general form of strain as a function of log time for a polyethylene sample under an applied load.

is recoverable if the load is removed. Once a sample has yielded, the deformation is largely irreversible. An increase in load or temperature will cause the strain versus time curve to move toward higher strains and shorter times, i.e., up and to the left. The general nature of the curve remains constant when the curve is shifted, but its precise shape changes; this phenomenon is illustrated in Figure 21 for varying loads on medium density polyethylene.

Creep is a relaxation phenomenon involving the gradual release of local stress in a sample by short-range molecular rearrangement. The higher the applied stress or the greater the freedom of molecular motion, the faster the relaxation will take place. Accordingly, creep is more pronounced at greater loadings and higher temperatures or in samples with a large amorphous fraction, i.e., a low degree of crystallinity. It is especially prevalent in samples above their glass transition temperature, which is the case in virtually all applications of polyethylene. Invariably, creep is an undesirable phenomenon because it results in the distortion of the original shape of an article. Creep is especially disadvantageous when the precise shape of an item is critical to its continued performance. It can manifest itself as sag in load-bearing items, such as fuel tanks or packing crates, or as a change of profile in parts such as the teeth on gear wheels. Polyethylene fares poorly in the realm of creep, even in relation to other inexpensive thermoplastics such as polypropylene and poly(vinyl chloride). In comparison to engineering thermoplastics, such as nylon, polyoxymethylene, and poly(ether ether

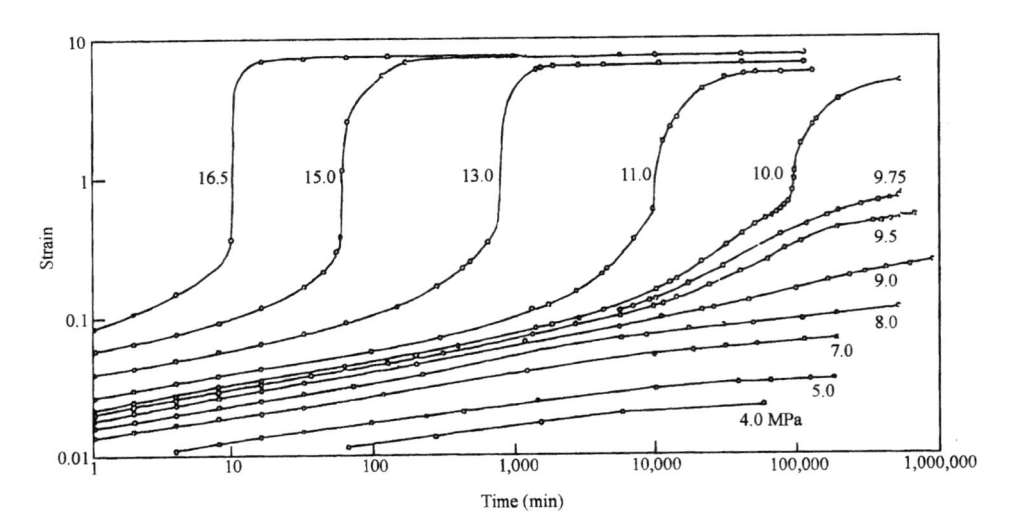

Figure 21 Strain as a function of time for medium density polyethylene subjected to various levels of stress. (From Ref. 12.)

Properties of Polyethylene

ketone), polyethylene is particularly deficient. The poor creep performance of polyethylene excludes it from a wide range of applications for which it would otherwise be suitable.

The response of a sample to prolonged loading is quantified in terms of the instantaneous creep compliance, $J(t)$, which is the strain divided by the stress at a given time. The creep compliance of polyethylene is nonlinear with respect to time.

$$J(t) = \frac{e(t)}{\sigma}$$

where

$J(t)$ = creep compliance at time t
$e(t)$ = strain at time t
σ = stress

The low-strain creep response of polyethylene over an extended period of time–up to several years—can (under favorable conditions) be predicted from short-term tests by applying the principle of time–temperature superposition. According to this principle the response of a sample to a given load as a function of time may be modeled using a series of elevated temperatures [10]. This principle is illustrated in Figure 22. When creep compliance is plotted against log time for a given load at low strain levels, the shapes of the curves generated are similar. They can be overlain be translating them horizontally and vertically by appropriate distances. The master curve so generated can be consulted to predict the effect of loading at times up to two orders of magnitude greater than the original testing period. In practice this principle works very well provided that the stresses and times involved do not exceed those required to initiate necking.

If the criterion for failure is taken as the onset of necking, a composite time–temperature superposition curve as a function of stress can be generated to predict extended use failure times [11] as shown in Figure 23.

The acquisition of a complete set of creep data applicable over 50 years or more requires that a large number of samples be tested over a range of loads at various temperatures for extended periods of time. Although the equipment required for such testing is relatively simple, a large battery of instruments is required if the data are to be obtained within a reasonable length of time (less than a year or so). Such testing on a scientifically rigorous basis is very expensive from the points of view of time, space, and capital investment. Given the manifestly poor creep performance of polyethylene, it is not surprising that research laboratories choose not to investigate such properties intensively, their resources being better employed on more creep-resistant materials. Accordingly there are

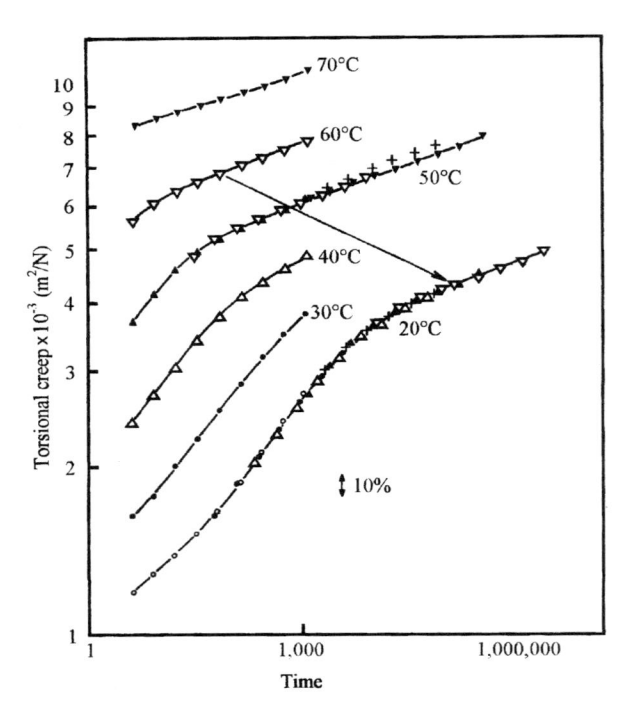

Figure 22 Composite curve of creep compliance as a function of time for small strain torsion of low density polyethylene. (From Ref. 10.)

few data available to relate creep properties to molecular and morphological characteristics on a quantitative basis.

On a qualitative basis the factors that affect creep can be rationalized in terms of the effect of morphological and molecular characteristics on the potential for stress relaxation within a sample. Accordingly, high crystallinity samples, with thick lamellae and small interlamellar distances, are less susceptible to creep than low crystallinity specimens. Hence an increase in density will result in a decrease of creep. Oriented samples are less prone to creep than isotropic ones, the improvement correlating with the increase in orientation. Susceptibility to creep is also related to a sample's processing and conditioning history. Different sections of a molded item may exhibit different creep responses depending upon their shear history [12]. The elapsed time between molding a sample and testing its creep response is important, because subtle changes in morphology can take place after molding. Exposure of samples to the elements can also result in morphological and molecular changes that affect creep—generally adversely. On a

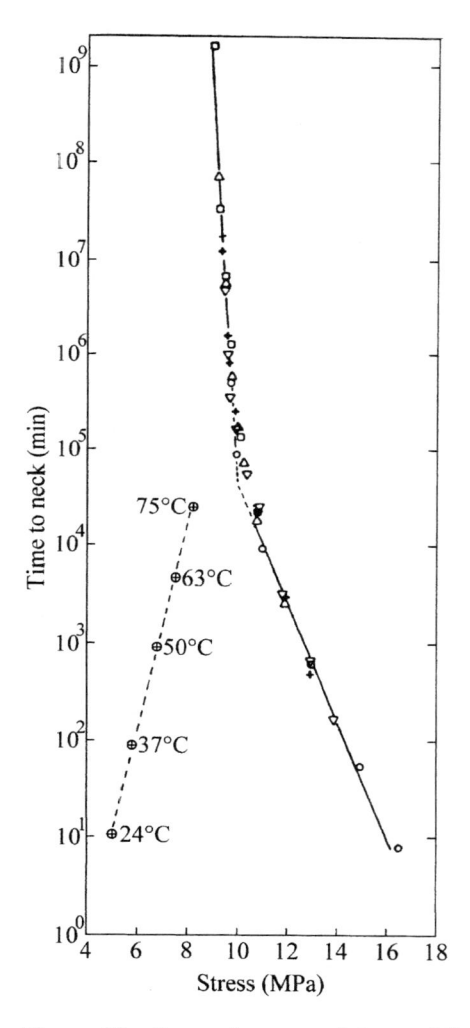

Figure 23 Composite curve of time to failure (necking) as a function of tensile stress for linear low density polyethylene. Circles containing crosses indicate the amount of shift required at each temperature to superpose the data. (From Ref. 11.)

molecular basis creep is a function of average molecular weight, molecular weight distribution, and branch content. A polyethylene resin with $\overline{M}_W > 200{,}000$, a broad molecular weight distribution, and 8–10 short-chain branches per 1000 carbon atoms in the backbone is highly resistant to creep when molded into pipes for gas and water transport [13].

  b. Stress Relaxation.   Stress relaxation is the phenomenon by which the stress required to maintain a constant strain in a sample decreases as a function of time. It proceeds incrementally by the cooperative rearrangement of molecular segments on adjacent chains. Its mechanism is thus closely related to that involved in creep; hence the same factors that control creep also control stress relaxation. The general characteristics of stress relaxation are illustrated in Figure 24. As strain is increased to its plateau value, stress rises rapidly. Maximum stress is experienced at the point at which maximum strain is achieved. Thereafter the stress level falls, the rate gradually decreasing as a function of time. Given sufficient time, the stress can approach zero.

  Stress relaxation is measured in terms of the stress relaxation modulus, which is the stress divided by the strain at a given time. The principle of time–temperature superposition applies to stress relaxation, thus composite master

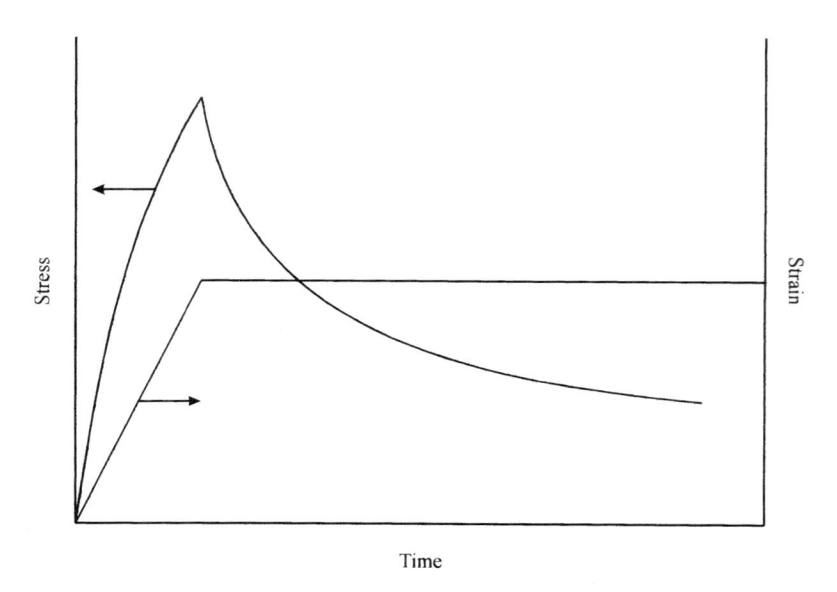

Figure 24 Plot of stress as a function of time after deformation, illustrating stress relaxation.

Properties of Polyethylene

curves can be generated to predict stress over extended periods of time. An example of four such master curves is shown in Figure 25.

In common with creep, polyethylene shows a high propensity toward stress relaxation. Failures attributable to stress relaxation include leakage from compression fittings and joints formed by forcing flexible polyethylene tubing over a rigid pipe. Given its susceptibility to stress relaxation, polyethylene is rarely used in applications in which it is likely to occur. There are therefore few data available that relate stress relaxation to the morphological or molecular nature of polyethylene.

c. Low Stress Brittle Failure. In addition to creep, polyethylene can fail in a brittle manner when subjected to sustained low levels of loads. This phenomenon is variously known as "low stress brittle failure," "creep rupture," "creep crack growth," "long-term static fatigue failure," or "long-term brittle failure." This mode of failure is distinguished from ductile failure (creep) in that the deformation takes place on a microscopic basis, i.e., the strain occurs over a thin cross section, ultimately leading to complete penetration of the sample. Low stress brittle failure is initiated at surface or bulk inhomogeneities that act as stress concentrators. The most common types of stress concentrators are scratches, notches, or incisions on the surface, but they can also take the form of voids or

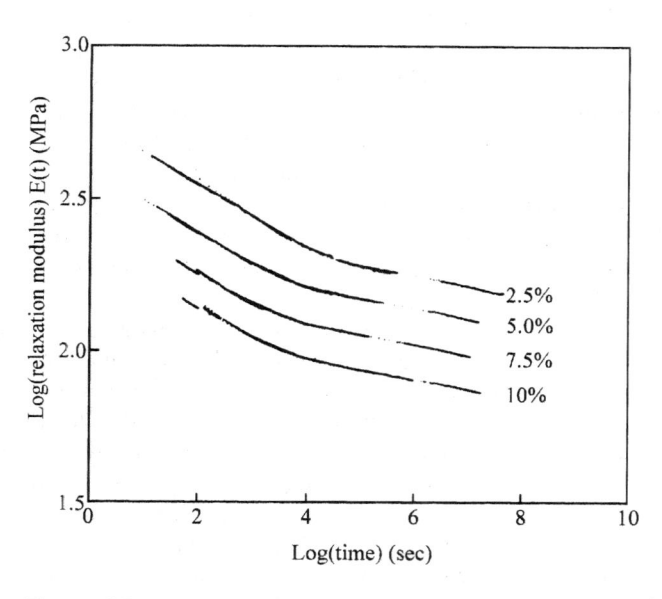

Figure 25 Composite plots of log relaxation modulus as a function of log time for medium density polyethylene at various strains at 23°C. (From Ref. 108.)

inclusions within the bulk. Joints formed by welding may be another point of weakness [14]. Such discontinuities drastically reduce the period of time for which a sample can sustain a given load. In practice, great care is taken to avoid such irregularities due to their highly deleterious effects. The stress required to cause the failure of a notched sample in a given period of time is significantly lower than that required to cause an unnotched sample to yield under identical conditions. The stresses involved are generally less than one-half of the short-term yield stress.

For slow crack growth to occur, the stress must be sufficiently low that the creep rate is lower than the rate of molecular disentanglement at the crack tip [15]. This translates to a critical stress above which the sample fails in a ductile mode and below which it fails in a brittle manner. A plot of time to failure (sample penetration) as a function of applied stress is shown in Figure 26. The mode of failure can be predicted by observing the initial material response to load; if a craze immediately develops at the tip of the notch, the sample will ultimately fail in a brittle manner.

The growth of a crack by low stress brittle failure follows a characteristic sequence, illustrated schematically in Figure 27. The separation of the lips of the crack at the surface of the sample (crack opening displacement) can be used to follow the process of crack growth, as shown in Figure 28. Upon initial loading of the sample, the opening of a preexisting notch is increased. A craze, consisting of a small, highly strained, microvoided fibrillar region, forms at the tip of the notch. Fracture of the sample is initiated when the fibrils rupture and a new craze forms behind it, further into the sample. This process is repeated at irregular intervals, the crack extending into the sample by fits and starts. After a period of time the rate of crack growth accelerates and the crack advances rapidly until the remaining cross-sectional area is insufficient to sustain the applied load, i.e., the stress in the remaining ligament exceeds the short-term yield stress. At this point the ligament fails in a ductile manner. During the process of crack growth, strain is localized at the tip of the crack while the bulk of the sample retains its original form.

Much effort has been expended to determine the factors that control low stress brittle crack growth because it is a major cause of failure in long-term applications of polyethylene, the most important of which are gas and water distribution pipes and geomembrane liners for waste disposal sites—landfill, toxic and nuclear. Pipes generally fail at inclusions or scratches introduced during installation. Geomembranes fail at sites of physical damage such as gouges, sharp folds, and abrasion; at seams due to inappropriate welding conditions; and at stress points caused by thermal contraction. The most important factor that controls brittle failure is the preexistence of a stress concentrator; without this, failure will inevitably be ductile. The various sources of stress concentration are reviewed by Peggs and Carlson [16]. Increased stress and temperature enhance the

Properties of Polyethylene 161

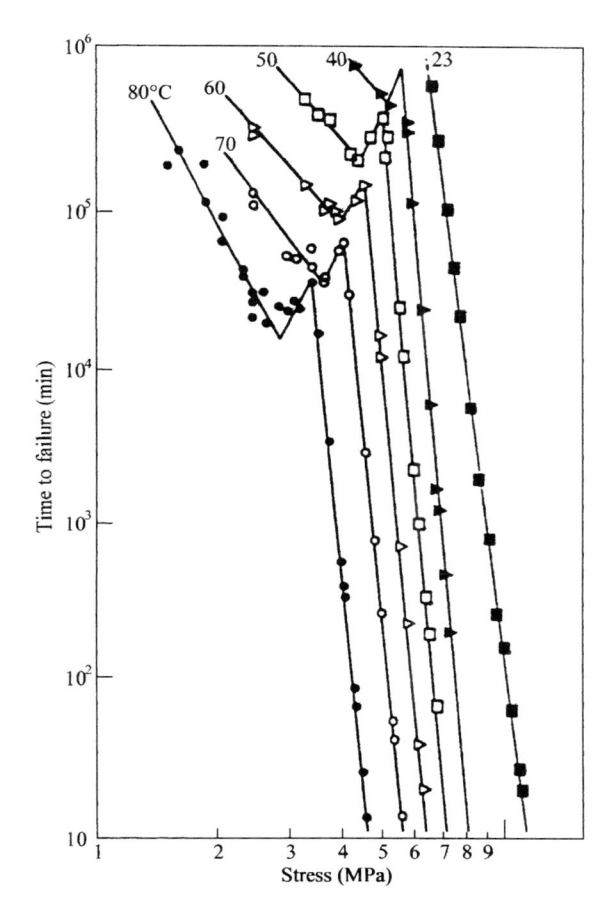

Figure 26 Plot of time to sample penetration as a function of tensile stress at various temperatures for notched linear low density polyethylene. (From Ref. 23.)

crack growth rate by increasing the rate of molecular disentanglement in the fibrils [17]. Figure 28 illustrates the effect of increased stress on the time to penetration of linear low density polyethylene. Deeper and sharper notches enhance stress concentration at the notch tip and hence increase the rate of crack growth. The effect of notch depth on the rate of crack opening is illustrated in Figure 29. Processing factors are also important but have been insufficiently quantified to permit generalizations to be made regarding their effects.

In addition to the physical factors that influence crack growth, the morphological and molecular nature of the polyethylene resin are very important. It has

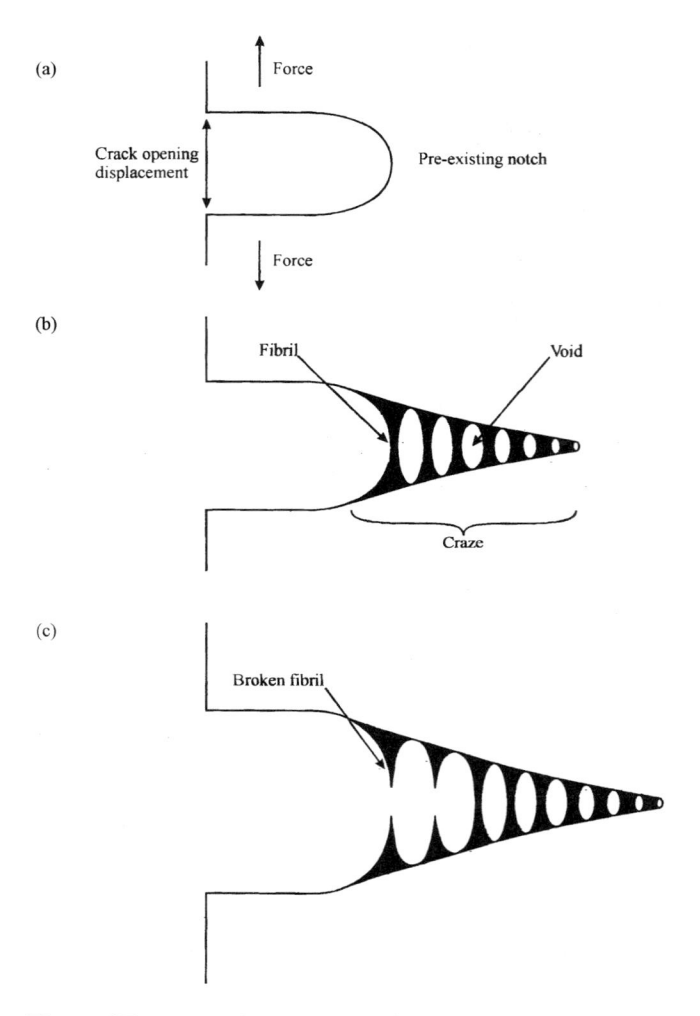

Figure 27 Schematic illustration of the course of low stress brittle failure. (a) Preexisting notch subjected to opening force; (b) formation of craze; (c) fracture begins as fibrils break.

been postulated that the rate of crack growth is inversely related to the concentration of tie chains that presumably hamper the disentanglement of fibrils [17,18]. The postulated relationship of crack growth to tie chain concentration should be used only as a guide; many other factors must also be considered, including the configuration of tie chains, the ratio of tie chain concentration to lamellar thick-

Properties of Polyethylene 163

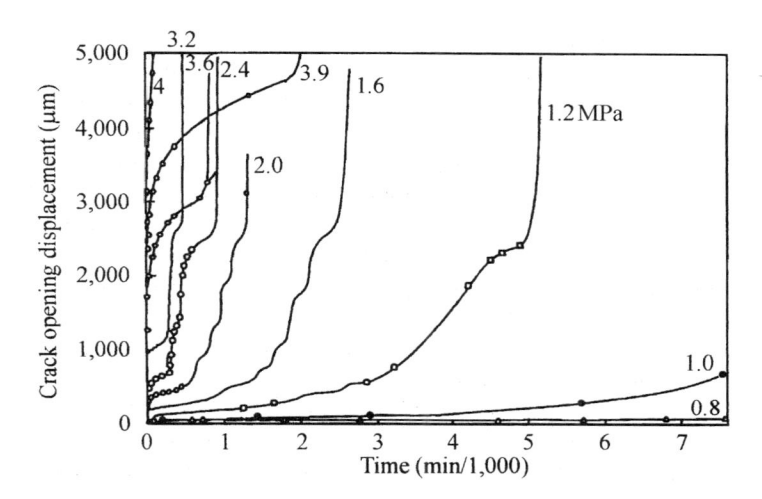

Figure 28 Plot of crack opening as a function of time for linear low density polyethylene at 80°C for various stresses. (From Ref. 109.)

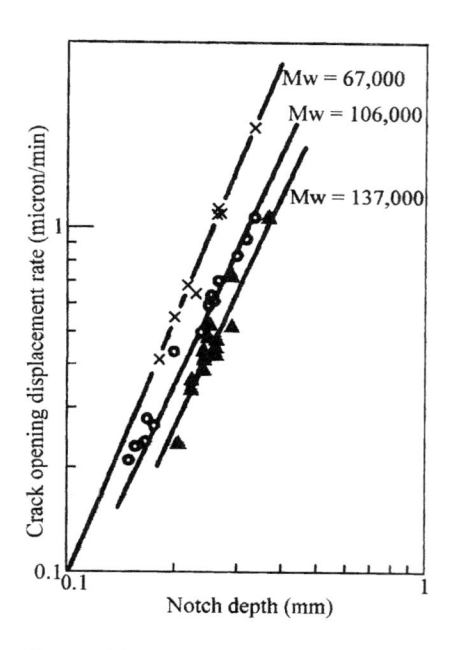

Figure 29 Plot of rate of crack opening as a function of notch depth for linear low density polyethylene at 80°C for various stresses. (From Ref. 17.)

ness, and the degree of crystallinity. The morphological and molecular factors that influence tie chains are discussed in Chapter 4. Thicker lamellae and a higher degree of crystallinity enhance the likelihood of brittle break [17], but the time to crack initiation increases proportionally to lamellar thickness [19]. Conversely, the thicker the noncrystalline regions, the shorter will be the crack initiation time. Annealing samples of linear low density polyethylene at various temperatures up to the crystalline melting peak results in a decrease in the rate of slow crack growth concomitant with an increase in the overall degree of crystallinity and average lamellar thickness [20]. The morphological deformation mechanisms that are active in low stress brittle failure are addressed in Chapter 8. Some of the molecular characteristics that have been related to the crack growth rate include average molecular weight and short branch content, an increase in either of which slows the growth of cracks [21]. The effect of increasing branch content on the time to failure is shown in Figure 30. Gas and water distribution pipes may be made from either medium density or high density polyethylene. Due to a combination of various factors, neither type of resin stands out as being demonstrably superior with regard to its tendency to low stress brittle failure.

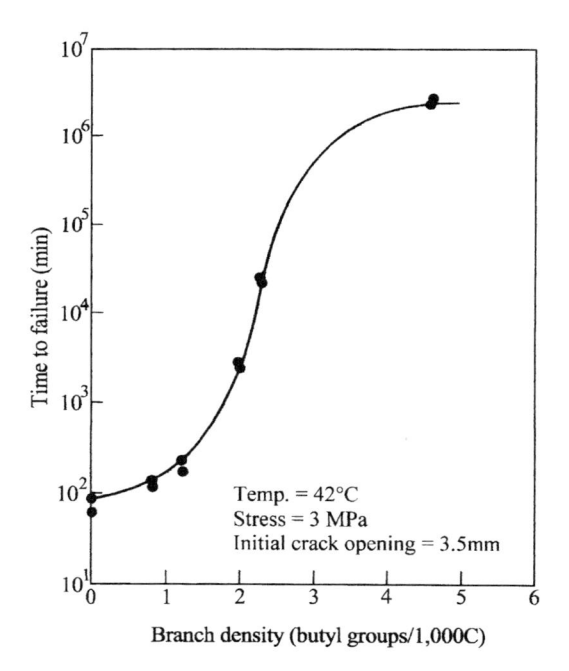

Figure 30 Plot of time to failure as a function of branch density in polyethylene. (From Ref. 21.)

Properties of Polyethylene 165

The penetration of a polyethylene material in use can have drastic and far-reaching consequences. It is required that the service life of gas and water distribution pipes exceed 50 years; for geomembranes, indefinite lifetimes may be required. Obviously, it is not feasible to test samples for such long periods of time prior to use, so it is crucial that usable lifetime predictions should be available from short-term tests. The importance of such predictions is increasing; 80% or more of new gas pipe is made from polyethylene [11], which accounts for approximately 15% of all gas pipe presently in service [13]. Recently it was demonstrated that the mechanisms of low stress brittle failure can be duplicated at short times when the sample is subjected to high pressure during tensile testing [22]. If this method can be generalized, it may provide a method of screening polyethylene resins for use in long-term, low-load applications. For the time being, given the extreme sensitivity of low stress brittle failure to notch configuration, it is very hard to predict the service life of pipes and geomembranes in which scratches and other defects have been introduced adventitiously during installation. The best that can currently be achieved is to take care to avoid the formation of stress concentrators and to predict service life based upon creep data where the time–temperature superposition principle is applicable.

Fatigue cracking is brittle failure induced by intermittent low stresses; it follows the same general trends as low stress brittle failure. Fatigue cracking takes place at a faster rate than low stress brittle failure on a basis of total time under load. In addition to the factors active for low stress brittle failure, fatigue cracking is controlled by the frequency of loading, relaxation time, and the waveform of the applied stress [23]. Square waves are more damaging than sine waves, which in turn are more damaging than triangular waves. Figure 31 illustrates the effect of loading frequency on the crack opening displacement for linear low density polyethylene sample. The number of cycles to failure decreases as the stress and temperature increase, as shown in Figure 32.

d. Environmental Stress Cracking. Environmental stress cracking (ESC) may be defined as the brittle failure of a stressed sample in the presence of a sensitizing agent (usually a liquid), failure occurring at shorter times than when stress is the only factor. The term "environmental stress cracking" is broadly applied to failure in the presence of two categories of liquid: solvents capable of swelling polyethylene and nonsolvents that are surfactants. Gases and some viscous solids may act as stress cracking agents, but they are generally less active than liquids. The agents that cause environmental stress cracking do so on a purely physical basis; no chemical reactions occur. Some authors make a distinction between solvent stress cracking, which involves localized swelling of the substrate by solvents prior to failure, and environmental stress cracking, which is brought about by nonsolvents [24]. Naturally there exist subtle differences between the mechanisms of failure caused by solvents and those caused

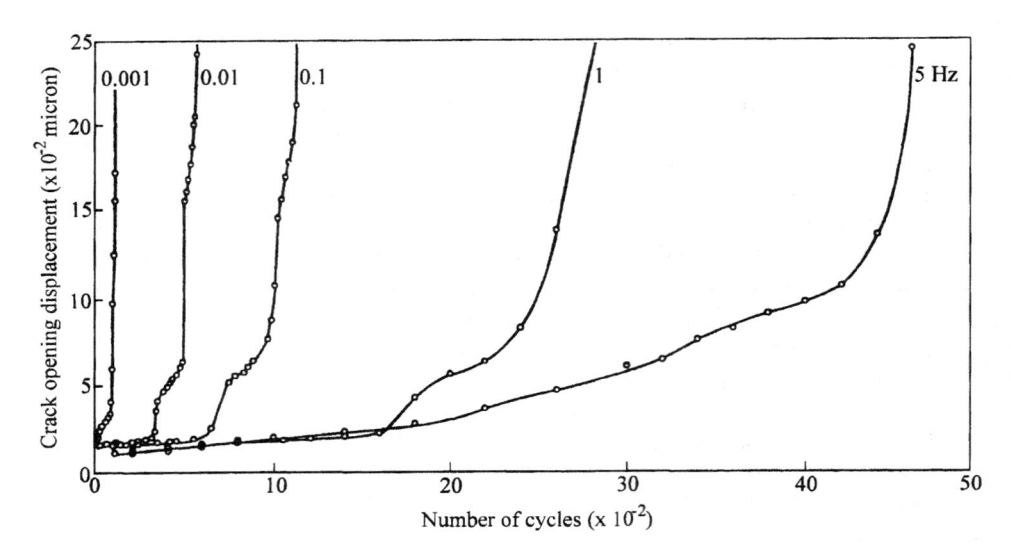

Figure 31 Plot of crack opening displacement as a function of number of cycles at various frequencies for a linear low density sample with 4.5 butyl branches per 1000 carbon atoms in the backbone. (From Ref. 23.)

by nonsolvents, but the factors important to the suppression of the phenomena are similar. Therefore, in the discussion that follows no distinction is made between the two types of failure, the term "environmental stress cracking" being used to cover both. The ability of a resin to withstand such processes is known as "environmental stress cracking resistance," commonly shortened to ESCR. The specific mechanisms of failure are addressed in Chapter 8.

Environmental stress cracking proceeds in a manner similar to low stress brittle failure but at a faster rate. Although not a prerequisite for the onset of cracking, a notch, a scratch, or some other type of stress concentrator accelerates the process. Left unchecked, environmental stress cracking will result in material failure by penetration. The stress required to drive the process may be either external, such as that induced by deformation, or internal, such as residual stresses incorporated during molding. A subtle distinction between the effects of surfactant- and solvent-related environmental stress cracking is that cracking in the presence of a surfactant requires polyaxial stress, whereas solvent stress cracking may be initiated by either uniaxial or polyaxial stresses. In practice, stresses are invariably polyaxial—except in the case of fibers under tension—so this distinction is largely moot. The fracture surface of an environmental stress crack looks similar to a low stress brittle failure surface. The majority of the

Properties of Polyethylene

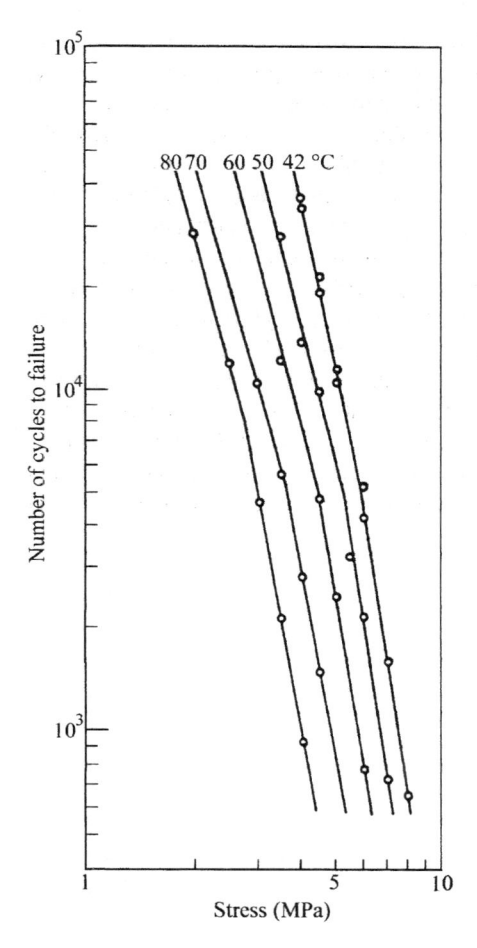

Figure 32 Plot of cycles to failure as a function of stress at various temperatures for a linear low density sample with 4.5 butyl branches per 1000 carbon atoms in the backbone. (From Ref. 23.)

material remains essentially undeformed, while that in the immediate vicinity of the fracture plane is highly deformed on a microscopic scale. Like low stress brittle failure, environmental stress cracking may take up to several years to completely penetrate the sample. In highly oriented specimens, cracks occur only parallel with the orientation axis. Examples of environmental stress cracking include the penetration of electrical cable insulation when it is bent in the presence of a sensitizing agent; such as grease inadvertently applied during installation, and the rupture of bottles containing detergents when stacked in boxes such that undue stress is exerted on those at the bottom of the pile. In years gone by, environmental stress cracking was a major problem associated with polyethylene, especially when high density polyethylene resins were first introduced. Such failures helped earn polyethylene a poor reputation. Since the factors that influence environmental stress cracking became well known, molecular tailoring of resins and improved molding practices have greatly reduced its occurrence.

Environmental stress cracking is fostered by high stresses and elevated temperatures. Stress cracking agents come in many varieties, ranging from aqueous solutions of surfactants to pure solvents and from simple hydrocarbons to silicone oils. The effectiveness of various sensitizing agents is shown in Table 5. The potency of a sensitizing agent is related to its ability to wet the polymer surface. The larger the surface area a stress cracking agent can cover, the more likely it is to encounter an area of stress concentration where it will be most effective. In general, the lower the viscosity and surface tension of a sensitizing agent, the more effective it will be; high viscosity stress cracking agents take longer to

Table 5 Effectiveness of Various Solvents as Environmental Stress Cracking Agents

Solvent	Time to failure (hr at 2000 psi)
Hexane	0.3[a]
Benzene	0.77[a]
Toluene	0.85[a]
Butyl acetate	1.0[a]
Xylene	1.1[a]
n-Propanol	1.7
n-Amyl alcohol	2.5
Dodecyl alcohol	3.4
Acetic acid	3.7
Isopropanol	6.5
Acetone	10.4[a]
Ethanol	13.2
Tricresyl phosphate	14.6
Diethylene glycol	28.5
Methanol	50.0
Water	55.0[b]

[a] Stretches.
[b] Cold draw.
Source: Ref. 99.

diffuse into samples than low viscosity ones and hence require more time to reach areas of stress concentration. The smaller the difference between the solubility parameter of the stress cracking agent and the polymer, the more effective the agent will be [25].

It should be emphasized that the presence of a stress cracking agent alone is not sufficient to initiate environmental stress cracking; applied stress—internal or external—must also be present. A critical minimum stress level is required for environmental stress cracking to take place on a finite time scale. In the presence of an aggressive sensitizing agent, the critical stress may be less than one-tenth of the short-term yield stress. Another key external factor influencing environmental stress cracking is the design of molded parts. Designs that incorporate highly stressed regions, such as thin sections with sharp angles or thick sections with different crystallization rates, tend to fail at these locations. Thus, proper mold design can reduce the tendency to undergo environmental stress cracking. The reduction of molded-in stresses also reduces the tendency to undergo low stress brittle failure. Samples that contain crazes are more prone to environmental stress cracking than those that do not. The high surface area of a craze permits ready access of stress cracking agents to the highly stressed molecules in that region.

Apart from the external influences that cause environmental stress cracking, the principal intrinsic factor is the molecular weight distribution of the sample. The greater the fraction of very low molecular weight chains in a sample, the greater will be its propensity to undergo environmental stress cracking. When a polyethylene resin that is susceptible to environmental stress cracking is treated with chloroform, which preferentially extracts the lowest molecular weight species, its susceptibility to environmental stress cracking is reduced and may even be eradicated. Conversely, adding a low molecular weight component to an otherwise resistant polyethylene resin promotes environmental stress cracking. On a general basis, the susceptibility of polyethylene resins to environmental stress cracking can be correlated with their melt indices; time to failure decreases as the melt index rises. For a homologous series of polyethylene resins made with the same catalyst system there is a fairly sharp cutoff of susceptibility to environmental stress cracking as the molecular weight increases. The disposition to undergo environmental stress cracking has been tentatively linked to a postulated low concentration of tie chains between crystallites [16]. This is reinforced by the observation that a resin slowly crystallized to develop high crystallinity will be more apt to experience environmental stress cracking than a specimen of the same resin that has been quench-cooled. In practice, the level of crystallinity in a specimen is determined by the need for short molding times and is not an independently controllable factor. Thus, molecular tailoring is far more important in reducing environmental stress cracking than morphological manipulation. With current polymerization processes it is possible to produce resins of sufficiently narrow molecular weight distribution that the low molecular weight chains that induce environmental stress cracking are present to a negligible degree.

C. Thermal Properties

Semicrystalline polymers in general differ from most crystalline solids in that they display a melting range rather than a discrete melting point. Polyethylene exhibits a range of melting phenomena that can occur at temperatures from as low as room temperature up to 140°C. The melting range is a consequence of the inevitable distribution of lamellar thicknesses in the solid state. In addition, solid polyethylene exhibits several secondary transitions due to localized molecular motions in the crystalline, disordered, or interfacial regions. The thermal characteristics of polyethylene, especially its relatively low melting and softening temperatures, are some of the primary elements that define its realm of applications.

1. Melting Range

Semicrystalline polymers do not exhibit melting points in the classic sense, i.e., as a sharply defined transition from the solid to the liquid state occurring at a discrete temperature. Thus, polyethylene undergoes a transition from the semicrystalline to the molten state that takes place over a temperature range that can span from less than 10°C up to 70°C. As it passes through this transition the semicrystalline morphology gradually takes on more of the characteristics of the amorphous state at the expense of the crystalline regions. The melting range is broad because it consists of a series of overlapping melting points that correspond to the melting of lamellae of various thicknesses. Thicker lamellae have higher melting points. A dispersion of lamellar thicknesses is a natural consequence of entanglements and chain branching that divide chain backbones into a series of discrete crystallizable sequences with a distribution of lengths. Further broadening of the distribution of crystallite sizes, and hence the melting range, occurs when crystallization occurs over a range of temperatures as the sample cools. The narrowest melting ranges are exhibited by low molecular weight linear polyethylene specimens that have been melt- or solution-crystallized at isothermal temperatures [26]. The broadest melting ranges occur in branched samples crystallized during rapid cooling [27]. The melting range of polyethylene can extend down as far as room temperature when the distribution of lamellar thicknesses includes a very thin component.

The melting characteristics of polymers are commonly investigated by means of differential scanning calorimetry (DSC), the principles of which are discussed in Chapter 6. DSC provides a trace, called a thermogram, that consists of the instantaneous heat capacity of a specimen plotted as a function of temperature. The greater the volume of crystallites that melt at a given temperature, the higher the sample's instantaneous heat capacity. Figure 33 shows schematic thermograms representing a variety of commercial polyethylene resins. As can be readily seen, the thermal characteristics of polyethylene samples can differ widely. The thermogram of slow-cooled high density polyethylene shown in Figure 33a exhibits a relatively narrow melting peak with only a small tail on the low temperature side. When the same material is quench-cooled it exhibits quite different thermal characteristics, as shown in Figure 33b, wherein the peak temperature is shifted several degrees lower and a prominent low temperature tail is present. The thermogram of quench-cooled linear low density polyethylene, shown in Figure 33c, continues the trend with an even more pronounced low temperature tail and a peak temperature approximately 10°C lower than that of the comparable high density polyethylene sample. In Figure 33d, the thermogram of quench-cooled low density polyethylene, the peak maximum is significantly broadened and is much less intense, occurring approximately 10°C below that of linear low density polyethylene; the dominant low temperature tail stretches down almost to room temperature.

There is an approximately inverse relationship between the position of the peak maximum and the overall breadth of the melting peak. Samples with lower molecular weights, lower levels of branching, and slower crystallization rates tend to have narrower melting distributions and elevated peak melting temperatures. The normalized area under the peak—which is a measure of the degree of crystallinity—can be approximately correlated with the temperature of the peak maximum and the sharpness of the melting range. These effects may be traced to the distribution of lamellar thickness, the formation of which is discussed in Chapter 4. From a practical standpoint it is often desirable to characterize the thermal characteristics of polyethylene resins by a single temperature value. For simplicity's sake the value quoted is generally that of the peak taken from the DSC thermogram. This value is often erroneously termed the melting point; it should more strictly be referred to as the "peak melting temperature." As illustrated in Figures 33a and 33b, the position of the peak depends on the conditions under which a sample is prepared. In addition, the peak melting temperature also depends on the manner in which the DSC experiment is performed (the factors affecting its position are discussed in Chapter 6). Given the inconstancy of the peak position, one should not place too great an emphasis on its value. Table 6 lists the effects of various molecular and morphological characteristics on peak melting temperature. Figure 34 shows the typical peak melting temperatures for the various classes of polyethylene. It should be emphasized that the data in Figure 34 are representative only; certain samples may exhibit characteristics outside the ranges quoted.

Polyethylene melts at temperatures that are relatively low in comparison with those of other commercial semicrystalline polyolefins and much lower than

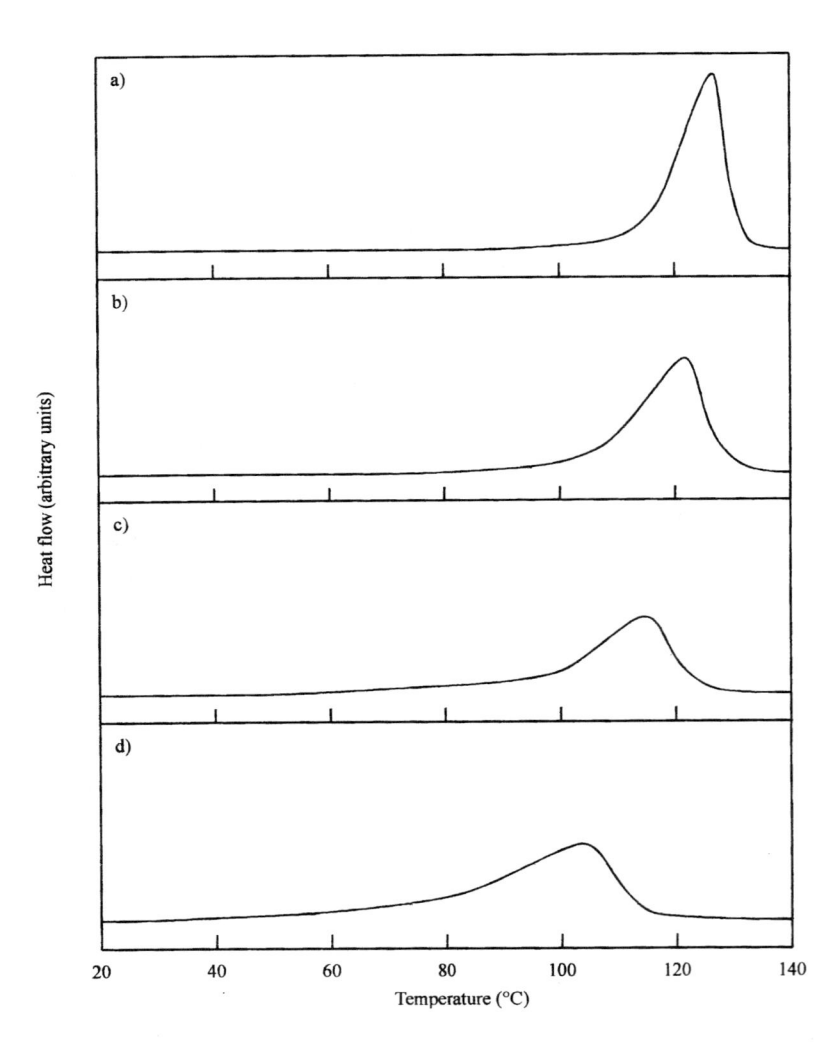

Figure 33 Schematic thermograms representing various polyethylene samples. (a) Slow-cooled high density polyethylene of moderate molecular weight; (b) quench-cooled high density polyethylene of moderate molecular weight; (c) quench-cooled linear low density polyethylene of moderate molecular weight; (d) quench-cooled low density polyethylene.

Properties of Polyethylene

Table 6 Effect of Molecular, Processing, and Morphological Characteristics on Peak Melting Temperature of Polyethylene

Variable	Effect on melting temperature	Notes
Increased branch content	Decrease	Very high branch contents reduce melting temperature to just above room temperature.
Increased molecular weight	Decrease	Drop of ~5°C for linear polyethylene increasing from 50,000 to 10,000,000.
Decreased density/crystallinity	Decrease	Branch content has greater effect than molecular weight.
Increased cooling rate	Decrease	Greatest effect on linear polyethylene.
Increased orientation	Increase	Greatest effect on high molecular weight linear polyethylene.

Source: Ref. 97.

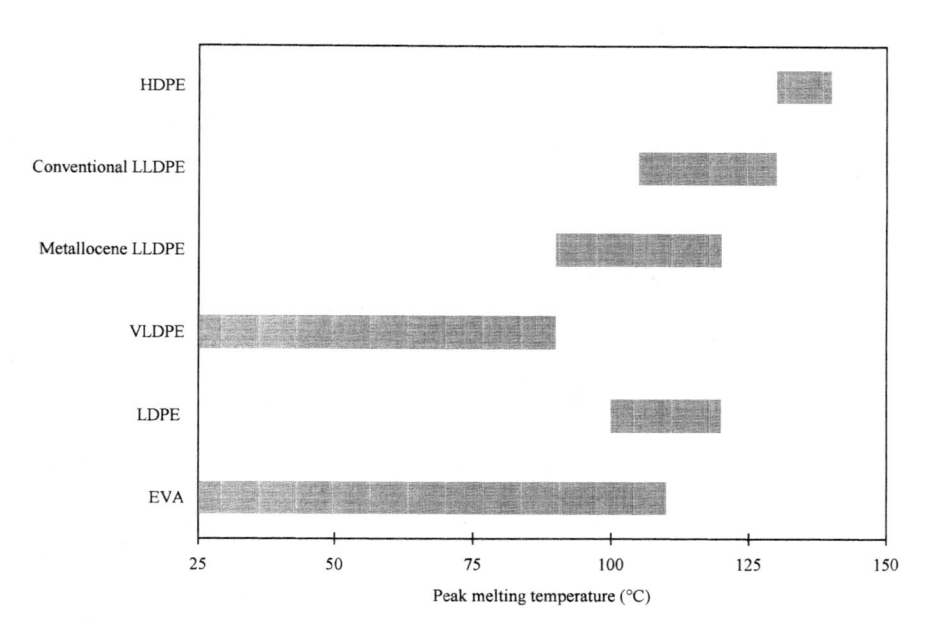

Figure 34 Typical peak melting temperatures for various classes of polyethylene.

those of the engineering thermoplastics (approximate maximum service temperature is addressed in the following section). A beneficial side effect of this is that polyethylene can normally be processed at lower temperatures than most other thermoplastics. The low melting temperature of polyethylene relative to other commercial polymers can be explained on a thermodynamic basis:

$$\Delta G = \Delta H_f - (T\Delta S + \Delta \zeta)$$

where

$\Delta G =$ change of Gibbs free energy
$\Delta H_f =$ heat of fusion
$T =$ absolute temperature
$\Delta S =$ change of entropy
$\Delta \zeta =$ change of interfacial free energy

When the Gibbs free energy of a system is reduced by a phase change, i.e., the change of Gibbs free energy (ΔG) associated with the transition is negative, the transformation is thermodynamically favored. In the case of the transition from the crystalline to the disordered state, the change of Gibbs free energy is the heat of fusion that must be introduced to disrupt the crystal lattice, minus the entropy increase associated with the conversion of the crystalline matrix to the disordered state multiplied by the absolute temperature plus the free energy released by the destruction of the order/disorder interface. As the temperature is increased, the sum of the entropic and interfacial terms eventually surpasses the heat of fusion, whereupon ΔG becomes negative. At the temperature at which this happens, the crystallites become thermodynamically unstable and ultimately melt. In the case of polyethylene, the increase of entropy upon changing from the crystalline to the disordered state is high relative to the heat of fusion, largely due to the flexibility of the molecular backbone. This results in a lower melting range for polyethylene in comparison with most other thermoplastics. Thinner crystallites melt at lower temperatures than thicker ones because they have a higher surface-to-volume ratio and hence a relatively greater contribution from the interfacial free energy term.

The theoretical melting point of polyethylene crystals of infinite thickness has been variously estimated to fall between 138°C and 146°C [28], a reasonable value lying toward the upper end of this range. In practice, the theoretical melting point of polyethylene is never attained, because all samples are polydisperse in molecular weight, are frequently branched, and are cooled too rapidly for infinitely thick lamellae to develop. The theory of polymer melting has been thoroughly addressed by Mandelkern [29], to which the reader's attention is directed for a full exposition of the melting phenomena of polyethylene.

The melting range of polyethylene can be shifted to higher temperatures by an increase in pressure [30]. This can be tracked by following the density of

Properties of Polyethylene

a sample as a function of temperature and pressure. Figure 35 illustrates this effect for high density polyethylene. The melting range of a sample is represented by the region in which the specific volume rises rapidly before leveling off after the peak melting temperature. It is readily seen that very high pressures must be exerted on a sample to influence its melting range significantly. In practice such pressures are rarely encountered.

2. Heat Distortion Temperature

The heat distortion temperature (HDT; also known as the heat deflection temperature) of a polymeric sample is the temperature at which it begins to show appreciable deformation under load in the short term. The HDT of a polymer may be used as a guide to its maximum service temperature. It is related to the elastic modulus of the crystalline and disordered regions, their relative proportions, and their structural relationship to each other. In the case of polyethylene, samples become more deformable as the temperature rises, primarily for three reasons: (1) The disordered regions become more flexible due to increased thermal motion; (2) the proportion of relatively rigid crystalline regions decreases as thinner crystallites melt; and (3) the translation of chain segments through crystallites

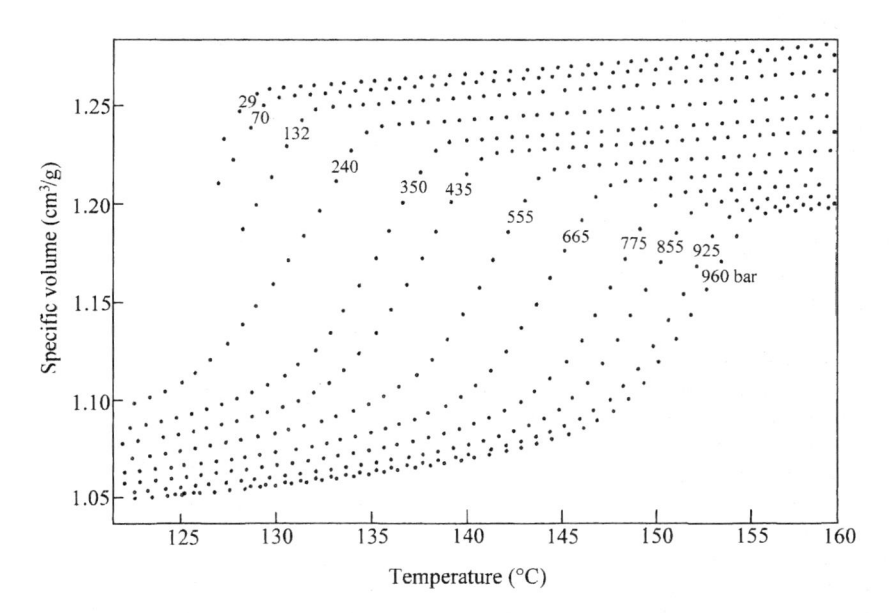

Figure 35 Plot of specific volume as a function of temperature for high density polyethylene at various pressures. (From Ref. 30.)

becomes easier. The heat distortion temperature increases with the degree of crystallinity and lamellar thickness. The heat distortion temperatures of a variety of polyethylene samples are shown in Table 7 together with those of a selection of semicrystalline polyolefins, engineering plastics. The heat distortion temperature of polyethylene is low relative to that of other polymers for two principal reasons: (1) The material is inherently less rigid than most other thermoplastics, so it starts off at a disadvantage, and (2) melting begins at a relatively low temperature. The determination of heat distortion temperature is addressed in Chapter 6.

Under conditions of applied stress and elevated temperature, thin crystallites are disrupted. Some of these can be melted by elevated temperature alone, while others require the combination of stress and elevated temperature to melt them. Thin lamellae are more susceptible to disruption by imposed stress because they are inherently less stable than thicker ones. As the stiffness of a polyethylene sample is directly related to its degree of crystallinity, any reduction in the volume of the crystalline phase due to melting results in increased flexibility. The linear chain sequences that comprise a crystallite are not fixed permanently in place. Under applied stress, which is transmitted by taut tie chains, individual linear

Table 7 Heat Distortion Temperature of Various Types of Polyethylene and Selected Polymers

Polymer	Heat distortion temperature (°C)	
	At 66 psi	At 264 psi
Low density polyethylene	40–44	
Linear low density polyethylene	40–80	
Polyethylene ionomer	45–52	34–38
Ultrahigh molecular weight polyethylene	68–82	43–49
High density polyethylene	82–91	
Poly(ethylene terephthalate)	21–66	75
Poly(vinyl chloride) (unplasticized)	60–77	57–82
Polytetrafluoroethylene	71–121	46
Polystyrene ("crystal")	76–94	68–96
Acrylonitrile-butadiene-styrene (ABS)	77–113	77–104
Polypropylene (isotactic)	107–121	49–60
Polycarbonate	138–142	121–132
Acetal (polyoxymethylene)	162–172	123–136
Poly(methyl methacrylate)	165–225	155–212
Nylon 6	175–191	68–85
Nylon 6,6	218–246	70–100

Source: Ref. 97.

sequences can slip through the crystalline matrices as long as they are not encumbered by large branches that prevent them from doing so. Such rearrangements are possible because the energy barrier associated with chain slippage is low and the crystallite volume remains unchanged, i.e., there is no net change of crystal enthalpy. The energy barrier to translating a polyethylene chain through the crystal lattice by the length of a monomer unit is low in comparison with other crystalline polymers because there are no bulky side groups to deform the lattice and no polar forces—such as the hydrogen bonds found in crystals of nylons—to overcome. Slippage of this sort relieves the stress on the tie chains that transmit the force in the sample, allowing the sample to flex slightly. Elevated temperatures increase local motion within the polymer; this increases the probability of molecular rearrangements that permit flexing of the specimen.

3. Heat of Fusion

The heat (enthalpy) of fusion (ΔH_f) of a sample is a measure of the amount of heat that must be introduced to convert its crystalline fraction to the disordered state. It is thus uniquely dependent upon the degree of crystallinity of the sample and the theoretical heat of fusion of a 100% crystalline sample. The heat of fusion of 100% crystalline polyethylene has been calculated to be 69 cal/g [31]. In the case of commercial polyethylene samples, heats of fusion range from essentially zero up to values approaching the theoretical maximum.

The factors controlling the heats of fusion are those that control the ordering of chains in the semicrystalline state. Disruptions to the linearity of the molecular structure reduce the potential for ordering, while crystallization conditions determine the degree of ordering realized. Effective discontinuities in the linearity of molecules may be permanent or transient. Permanent interruptions take the form of branches or cross-links that cannot be incorporated into the ordered regions. Entanglements form the basis of transient disruptions and depend on the molecular weight of the material. The factors that determine the actual degree of ordering realized, and hence the heat of fusion, are principally the rate of crystallization and the degree of orientation. The slower the crystallization process or the higher the degree of orientation, the greater will be the heat of fusion. Slow crystallization rates allow molecules to adopt configurations that are favorable to the formation of crystallites. Higher degrees of orientation force the molecules into alignment, facilitating the formation of crystallites. In practice, orientation has to be extreme for it to play a major role in determining the heat of fusion. Such extremes are usually found only in highly drawn or melt- or gel-spun fibers. The effect of orientation becomes more pronounced as the molecular weight of the sample increases.

The heat of fusion of a sample is principally of interest from the perspective of indicating the underlying degree of crystallinity. Its practical measurement by

Table 8 Heats of Fusion of Various Types of Polyethylene

Polymer	Heat of fusion (cal/g)	Degree of crystallinity (%)
Very low density polyethylene	<15	<22
Ethylene-vinyl acetate copolymer	7–35	10–50
Polyethylene ionomer	14–31	20–45
Linear low density polyethylene	15–38	22–55
Low density polyethylene	21–37	30–54
High density polyethylene	38–53	55–77
Ultradrawn polyethylene fibers	>62	>90

DSC is addressed in Chapter 6. Table 8 lists typical heats of fusion encountered for the various classes of polyethylene.

4. Heat Conduction

Polyethylene, in common with other nonpolar materials, has no free electrons that can readily conduct thermal energy. Therefore it conducts heat only by the transmission of vibrational or rotational energy from one chain segment to another, either inter- or intramolecularly. The transmission of thermal energy is more efficient in crystallites, where chain sequences are in closer proximity, than in disordered regions. Thus high density polyethylene is a better conductor of heat than low density polyethylene. Table 9 lists the heat conductivity of various polyethylene samples, selected polyolefins and engineering plastics, and some common nonpolymeric materials.

5. Heat Capacity

The heat capacity of a material is the amount of heat that must be introduced into a given amount of sample to raise its temperature by a given increment. The specific heat capacity of a material can be quoted for constant pressure or constant volume and is given the abbreviation C_p or C_v, respectively. Various units are used for specific heat capacity, typically being some combination of joules or calories per mole or grams per degree Celsius. In the case of polymeric systems, the mole unit pertains to the monomer rather than the polymer as a whole.

The value of the heat capacity of a polymer varies as a function of temperature. The heat capacity of polyethylene rises rapidly from a minimal value at absolute zero, leveling off somewhat as the temperature increases further. The value of the specific heat capacity of a polymer can be calculated quite accurately as a function of temperature based upon a knowledge of the atomic vibrations

Properties of Polyethylene

Table 9 Heat Conductivity of Various Types of Polyethylene, Selected Polymers, and Some Common Nonpolymeric Materials

Material	Thermal conductivity [cal·cm/sec/cm^2 · °C) $\times$ 10^{-4}]
Polyethylene ionomer	5.7–6.6
Low density polyethylene	8
Linear low density polyethylene	8–10
High density polyethylene	11–12
Polypropylene (isotactic)	2.8
Polystyrene ("crystal")	3.0
Poly(ethylene terephthalate)	3.3–3.6
Poly(methyl methacrylate)	4.0–6.0
Poly(vinyl chloride) (unplasticized)	3.5–5.0
Polycarbonate	4.7
Acetal (polyoxymethylene)	5.5
Nylon 6	5.8
Nylon 6,6	5.8
Polytetrafluoroethylene	6.0
n-Hexane	3.3
Glass	13–32
Water	15
Copper	9512

Source: Refs. 97 and 100.

of the monomer [32]. The calculated specific heat capacity at constant volume as a function of temperature for a series of polyolefins, including polyethylene, is shown in Figure 36.

On a molar basis, the heat capacity of polyethylene falls well below that of other polymers. The reasons for this lie in the simplicity of the polyethylene molecule. The six atoms that make up each monomer unit have a small number of vibrational modes compared to other monomers, and thus they have fewer degrees of freedom to absorb heat. For a given input of heat, less energy will be absorbed by vibrations in polyethylene and more by segmental motion than in other polymers. Thus the temperature of polyethylene will increase more as molecules become thermally agitated. On a mass basis, the differences are not as pronounced because there are fewer moles of monomer per gram for other polymers than there are for polyethylene.

From a processing point of view, low heat capacity is desirable because it means that less heat is required to raise the temperature of a polymer to an appropriate forming temperature. To some extent the relatively large heat of fusion of polyethylene offsets its low heat capacity, but it still requires significantly less

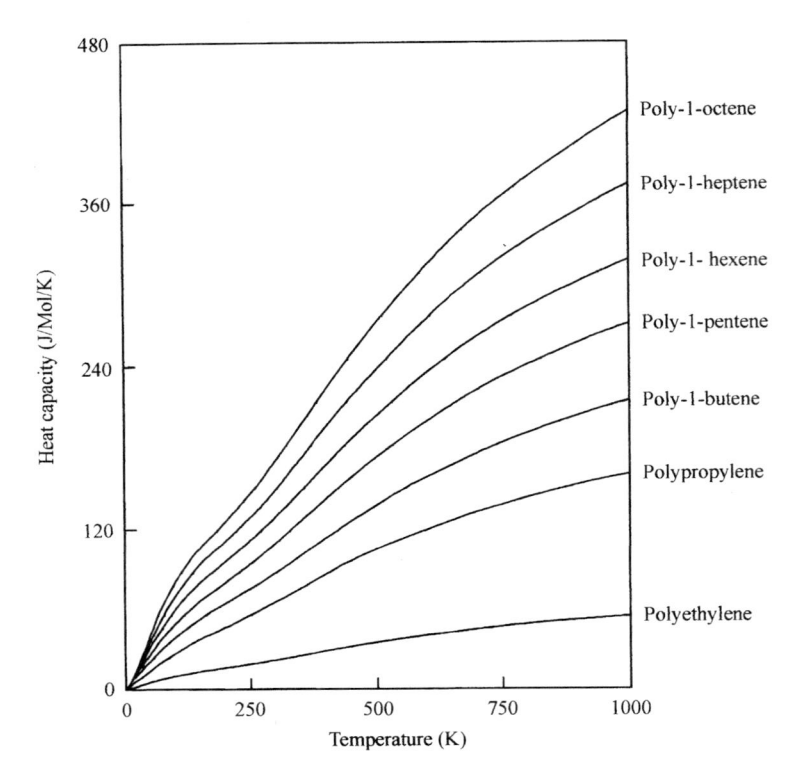

Figure 36 Calculated specific heat capacity of various polyolefins at constant volume as a function of temperature. (From Ref. 32.)

energy to raise polyethylene to its processing temperature (which is low to begin with) in comparison to other polymers. For example, it requires approximately 50% more energy per gram to raise isotactic polypropylene, with a degree of crystallinity of 60%, from room temperature to a processing temperature of 220°C than it does to raise polyethylene with a similar degree of crystallinity to a processing temperature of 180°C.

6. Thermal Expansion

The thermal expansion coefficient of a material is the increase in length that it undergoes when its temperature is raised by a given increment. The thermal expansion of a polyethylene sample depends on two factors: the relative proportions of the ordered and disordered regions and the orientation of crystallite c axes with respect to the direction in which the expansion is being measured. Disordered regions exhibit substantially greater expansion than crystalline regions, due to an inherently greater degree of freedom of movement. The expansion of a crystallite is sensitive to the axis along which the measurement is being made. The c-axis dimension of the crystalline unit cell of polyethylene is essentially constant because the carbon–carbon bond length and its dihedral angle are relatively independent of temperature. In contrast, the a and b axes can expand by increased separation of adjacent linear sequences. Naturally, the expansivity of a sample increases with temperature as a greater proportion of the sample is converted from the ordered to the disordered state. Table 10 lists the thermal expansion coefficients of various types of polyethylene, selected polyolefins and engineering plastics, and some common nonpolymeric materials.

Table 10   Thermal Expansion Coefficients of Various Types of Polyethylene, Selected Polymers, and Some Common Nonpolymeric Materials

Material	Thermal expansivity [10^{-6} in./in./°C)]
High density polyethylene	60–110
Linear low density polyethylene	70–150
Polyethylene ionomer	100–170
Low density polyethylene	100–220
Very low density polyethylene	150–270
Ethylene-vinyl acetate copolymer	160–200
Polystyrene ("crystal")	50–85
Poly(methyl methacrylate)	50–90
Poly(vinyl chloride) (unplasticized)	50–100
Acetal (polyoxymethylene)	50–112
Acrylonitrile-butadiene-styrene (ABS)	60–130
Poly(ethylene terephthalate)	65
Polycarbonate	68
Polytetrafluoroethylene	70–120
Nylon 6,6	80
Nylon 6	80–83
Polypropylene (isotactic)	81–100
Copper	16.6
Glass	55
Water	71

Source: Refs. 97 and 100.

7. Transitions

Polyethylene samples exhibit a number of transitions in the solid state that are associated with small-scale motions in the various phases of the semicrystalline morphology. These transitions manifest themselves as changes in physical properties, which are principally revealed by oscillatory measurements. The nature, and even the number, of the various transitions in polyethylene have been widely debated, but it is generally agreed that they are three in number. It is standard practice to designate solid-state transitions in descending order from the melting temperature using letters of the Greek alphabet. Thus the three transitions found for polyethylene are the α, β, and γ transitions. The γ transition—which is widely accepted as corresponding to the glass transition—is always present and can be found in the range of $-130°$ to $-100°C$. In contrast, the α transition is found in a broad range of temperatures, normally between $10°C$ and $70°C$. The β transition, which occurs in the vicinity of $-20°C$, is not manifested by all samples.

   a. Glass Transition (γ Transition). The glass transition, also known as the glass–rubber transition, is a phenomenon observed in all synthetic polymers that contain a noncrystalline component. It is associated with a relatively abrupt change in the degree of freedom experienced by chains in the disordered region. Thus the chain segments comprising the disordered regions of a polymeric sample exhibit very little freedom of motion below its glass transition temperature (T_g), whereas above the T_g chain segments are free to move to a limited extent. Associated with the transition is an increase in the free volume of the system; i.e., the density of the sample begins to decrease at a faster rate as a function of temperature. Generally, the principal manifestation of the glass transition is a change in the deformation mechanism from glasslike (brittle) below T_g to rubbery (ductile) above it. As a rule of thumb,

$$T_g = (0.5-0.8) \times T_m$$

where

T_g = the glass transition temperature in kelvin
T_m = the peak melting temperature in kelvin

Historically the glass transition of polyethylene has been assigned to a wide variety of temperatures [33,34]. It was variously associated with the β or γ transition at temperatures ranging from $-20°C$ to $-140°C$. Careful work has strongly linked it with the γ transition, falling in the vicinity of $-110°C$ to $-130°C$ [35]. At this temperature the thermal expansivity of a variety of polyethylene samples exhibits a distinct increase and the storage modulus, measured by dynamic mechanical analysis, falls abruptly. These phenomena are illustrated in Figures 37 and 38, respectively.

   The location of the T_g of polyethylene—or any other semicrystalline or

Properties of Polyethylene

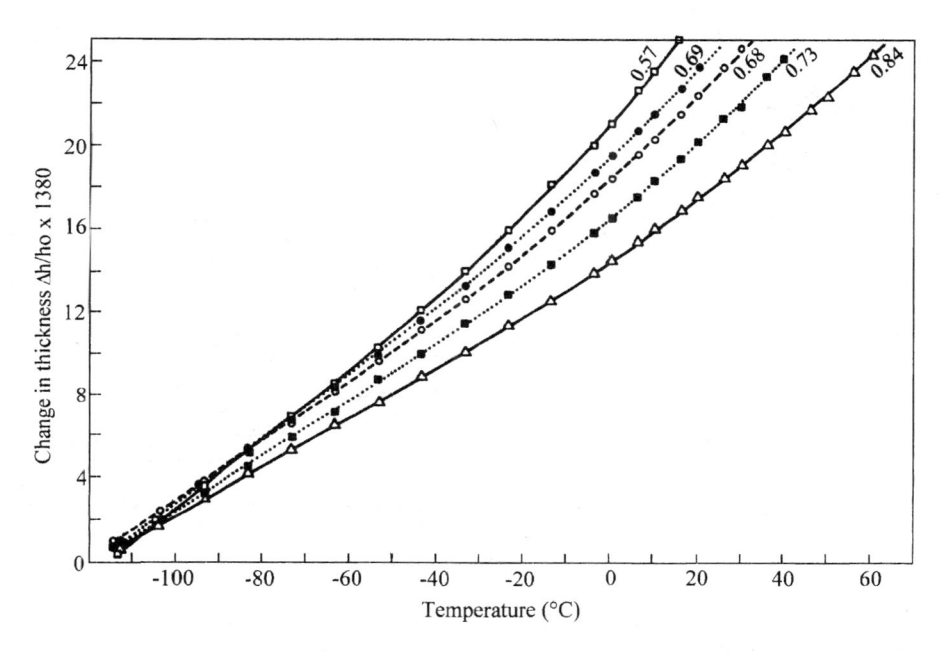

Figure 37 Thermal expansion as a function of temperature for high density polyethylene samples with different degrees of crystallinity. (From Ref. 35.)

amorphous polymer—depends on the testing procedure by which it is determined. In general, the more rapid the test, the higher the temperature at which the T_g will appear. Thus, increasing the frequency in dynamic mechanical analysis from 3.5 to 110 Hz can raise the observed T_g by approximately 20°C [36].

The glass transition of polyethylene occurs at such low temperatures that it is very rarely encountered in commercial applications. this effectively means that polyethylene samples remain in the ductile state at all service temperatures.

The intensity of the glass transition is strongly correlated with the fraction of disordered material in a sample. This is illustrated in Figures 37 and 38, where it can be seen that the rate of change of density as a function of temperature and the decrease in the storage modulus are inversely related to the degree of crystallinity. The dependence of the magnitude of the glass transition on the fraction of the disordered phase is taken as evidence that the physical changes that take place at the transition temperature do so in the disordered regions. The precise modes of chain motion that become allowed at the glass transition are unclear, but they probably involve the cooperative motion of short-chain sequences on adjacent molecules.

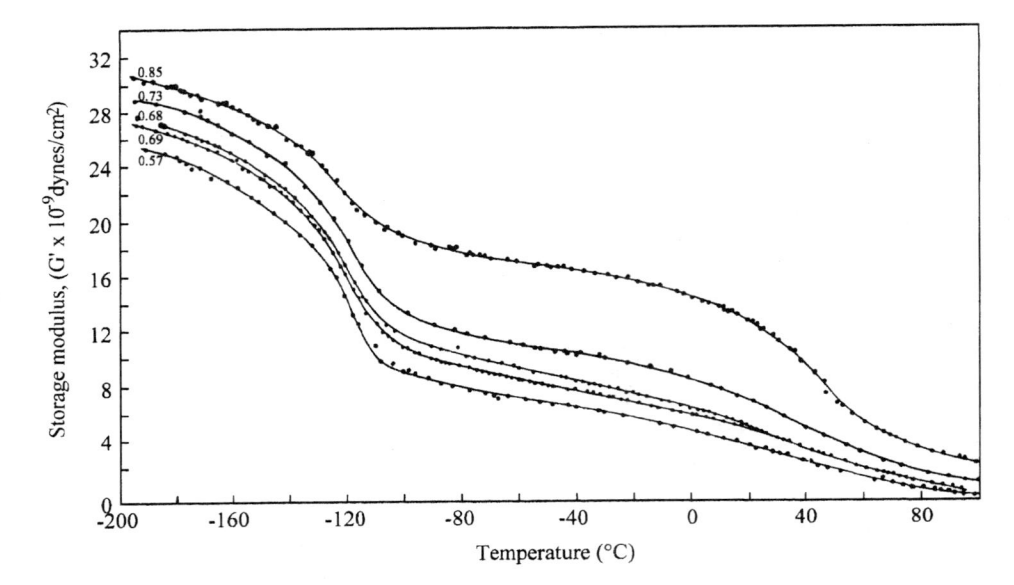

Figure 38 Storage modulus as a function of temperature for high density polyethylene samples with different degrees of crystallinity. (From Ref. 35.)

  b. α Transition. The intensity of the α transition in polyethylene is generally much less than that of the glass transition. Its appearance has been recorded at temperatures ranging from approximately −10°C to 120°C, depending upon sample preparation and measurement technique. The temperature at which the α transition occurs (T_α) has been correlated with the degree of crystallinity of a sample [37], but it has been shown to be primarily dependent upon the average crystallite thickness [38,39]. The relationship of T_α to crystallite thickness is particularly strong in the range from 50 to 200 Å; at thicknesses greater than this, T_α shows relatively little change [38]. This relationship is illustrated in Figure 39.

  The strong relationship between T_α and crystallite thickness is convincing evidence that the motions associated with the α transition take place in the crystalline regions of polyethylene. It has been suggested that the α transition is linked to the occurrence of the rotation or partial rotation of short molecular sequences in crystallites [34]. From a practical standpoint, no commercially relevant physical phenomena have been conclusively linked to the α transition.

  c. β Transition. The β transition in polyethylene is somewhat elusive; it is routinely observed in branched samples but is not present in all linear samples. This apparent anomaly has been satisfactorily explained by Popli et al. [38],

Properties of Polyethylene

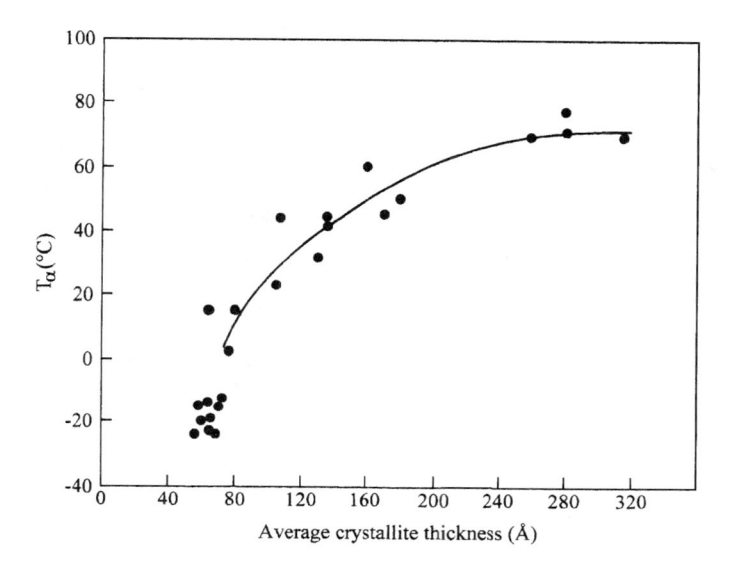

Figure 39 Plot of T_α as a function of average crystallite thickness. (From Ref. 38.)

who observed that the intensity of the transition was related to the fraction of the material contained in the partially ordered crystallite interface. This relationship is illustrated in Figure 40. It can readily be seen that the intensity of the β transition increases as the fraction of interfacial material increases. As a general rule it can be said that the β transition is observable when the interfacial regions make up more than 10% of the sample. Unlike the α transition, the β transition is found in a relatively narrow range of temperatures, as illustrated in Figure 40. What little variation exists in T_β can be explained by differences in measurement conditions and the errors associated with the evaluation of broad transitions. From a commercial point of view, the β transition is of little importance, because it does not manifest itself as a change in any of the more important physical properties of specimens.

D. Barrier Properities

The barrier properties of polyethylene might more precisely be termed its permeability properties, because polyethylene, in common with all other semicrystalline and glassy polymers, is permeable to some degree to most liquids, gases, and vapors. This said, it should be noted that the permeability of polyethylene to water is sufficiently low that it may effectively be considered to be impermeable on the relevant time scale. The barrier properties of polyethylene play an impor-

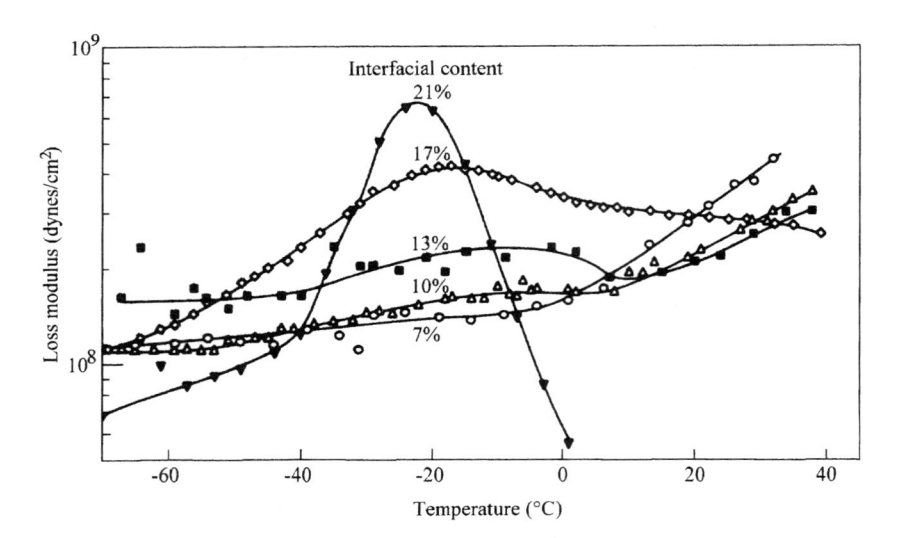

Figure 40 Plot of loss modulus at 3.5 Hz as a function of temperature for polyethylene samples with various levels of interfacial content. (From Ref. 38.)

tant role in defining its status as a desirable material in a number of fields. Apart from the obvious packaging applications, polyethylene is called upon to act as a barrier in such diverse uses as pipes, geomembranes, and housewares. In the latter three examples, polyethylene is considered to be essentially impermeable to the materials with which it comes into contact; thus it is basically in the area of packaging that permeability is important.

Permeability may be viewed as either a positive or negative factor depending upon the circumstances. If it is required that the contents of a package remain uncontaminated, then high permeability is naturally undesirable. However, if it is desired that the gaseous product of a reaction escape, high permeability may be desirable. An example of the latter is the outward diffusion of the respiration products from fruits and vegetables that have been packaged before they are fully ripe. The permeability of polyethylene to various molecules may also be advantageous by allowing the slow release of incorporated molecules such as fragrances or medications. Other beneficial applications of permeability include the diffusion of lubricants, incorporated during fabrication, to the surface of a moving part and the blooming of surface modifiers in films to aid in cling or to prevent blocking. For many applications, however, the diffusion of small molecules through polyethylene is an undesirable trait.

The components making up the barrier materials in a package serve two major purposes: They keep the contents in, and they keep contaminants out. Failure to meet the first requirement can result in preferential loss of certain compo-

Properties of Polyethylene 187

nents by the process known as "scalping." Failure to meet the second requirement can result in tainting of the contents. Scalping of specific molecules occurs because the chemical and physical characteristics of certain molecules allows them to permeate polyethylene more readily than others. Examples of scalping include the loss of fragrance molecules from toiletries, such as shampoos, soaps, and toothpaste, and the loss of essential oils from herbs, spices, and foodstuffs in general. The loss of even minute quantities of specific organoleptic molecules can change the whole character of a product. (Organoleptic molecules are those that are sensible to the human olfactory and gustatory senses.) In the case of toiletries and pharmaceutical products, the essential oils in the fragrances and flavorings can be some of the most expensive ingredients; manufacturers are therefore very interested in finding ways to contain such molecules and hence reduce their costs. Molecules that infuse a package may either taint the contents directly, causing an organoleptic change, or react with the contents to indirectly produce the same end result. Examples of tainting include the contamination of food with additives from the plastic packaging, the infusion of organic molecules from nearby products, and the oxidation of package contents by infused oxygen. Both tainting and scalping can render foods or other products unusable.

Two specific permeabiltiy constants should be highlighted, these being polyethylene's very low permeability to water and its relatively high permeation by oxygen. The first makes polyethylene highly desirable in food packaging to prevent dehydration, while the second detracts from its application to some packages by permitting the ingress of oxygen to products that may be readily oxidized.

The fact that polyethylene does not provide an adequate barrier to a number of key molecules (or whole families of substances) means that it is often used in conjunction with other polymers or materials that make up for this deficiency. In the packaging industry it is common to use multilayer films that combine the merits of two or more polymers. Multiple layers are not restricted to films; they are also found in bottles, tubs, and other containers. Polymer properties of interest to the packaging industry are not limited to barrier properties; polymer components also serve to physically protect their contents, act as a window to view the product, and provide a surface that can be decorated by the packager to make the contents more appealing to the consumer.

In addition to molecules diffusing through polyethylene, another important consideration is that of additives diffusing from the polyethylene itself, either contaminating other materials or changing the properties of the polyethylene. The latter can be particularly important when the migrant molecules are antioxidants and the polymer is in an oxidative environment. The loss of antioxidants from polyethylene products exposed to the elements is a major factor in some types of premature failure. Failure of certain products, such as electrical cables, can be linked to the loss of stabilizer; hence diffusion is an important process [40].

The diffusion of small molecules through polymers is a complex process about which much has been written. Descriptions given in this section are of a general nature, highlighting the molecular and morphological characteristics that control permeation. For more detailed information and a mathematical discussion of the various theories regarding polymer permeation, the reader's attention is directed to the references cited in the text. The terms "penetrant," "migrant," and "diffusant" are all used to describe molecules that permeate polymers. In the following discussion the term "migrant molecule" is used exclusively.

The permeation of a polymer substrate takes place in four stages. Initially the migrant molecules must make intimate contact with the surface of the polyethylene. Gases and vapors must adsorb on the surface of the polymer, and liquids must physically wet it. Second, the migrant molecules must dissolve into the polymer. The third stage is diffusion of the molecules down a concentration gradient, transporting them to the opposite side of the substrate. Finally, the molecules must leave the surface of the polymer, either by evaporation or by absorption into another substance.

Migrant molecules may be roughly divided into two categories: permanent gases such as oxygen, carbon dioxide, and methane and other, more complex molecules such as polar, aromatic, and aliphatic compounds. The permeation of polyethylene by permanent gases takes place according to classical Fickian mechanisms. Larger or more complex molecules diffuse through polyethylene in a somewhat more complicated manner, their rate of diffusion being dependent upon their size, shape, and specific interactions with polyethylene. Detailed discussions of the diffusion processes of small molecules through polymers can be found in the works of Comyn [41], Rogers [42], and Doong and Ho [43].

From a general point of view, the permeability of polyethylene to a migrant species is determined by kinetic and thermodynamic factors according to

$$P = D \times S$$

where

P = permeation coefficient (also known as permeability)
D = diffusion constant
S = solubility coefficient

The units of permeability are expressed in terms of the amount of migrant passing through a film of unit thickness per unit area per time per pressure difference, for example,

$$\frac{cm^3 \cdot mm}{cm^2 \cdot sec \cdot cm\ Hg} \quad \text{or} \quad \frac{cm^3 \cdot mil}{cm^2 \cdot 24\ hr \cdot atm}$$

Diffusion is the rate of passage of a migrant molecule through a unit area under the influence of a concentration gradient. In its simplest form the diffusion coefficient depends only on temperature, not on time or concentration, but in cases where migrant molecules interact with the polymer this relationship breaks down. Diffusion takes place when a dissolved molecule moves from one location (or "hole") to another. For this to occur, the new hole must be of sufficient size to accommodate the molecule and the intervening path must be sufficiently wide to permit its passage. Thus diffusion takes place by a series of jumps through the polymer matrix. Holes are transient, forming and collapsing as the chain segments in the disordered regions vibrate and rotate according to their degree of thermal excitation. Both the concentration and the size of holes are functions of the free volume of a system and hence are temperature-dependent. The more numerous and the larger the holes, the faster diffusion can take place.

For a molecule to pass from one hole to the next it must overcome an energetic barrier that may be considered in terms of an apparent activation energy. The apparent activation energy is dependent upon the diameter of the migrant molecule; the larger its diameter, the larger the opening it requires and hence the larger the new surface area that must be created and the greater the energy barrier. In the case of small molecules such as simple gases, the step size of diffusion will be of the order of the diameter of the molecule, but more complex molecules diffuse by steps that are a fraction of their molecular length. Thus long molecules may be considered as filling a number of adjacent holes, moving incrementally when a new hole opens at one end or the other. The diffusion of long molecules through the disordered regions of polyethylene may be likened to the "reeling in" of polymer chains during crystallization—but with a much smaller driving force.

The characteristics of the migrant molecules that control their diffusion rate through polyethylene are principally their size, shape, and chemical structure. In general, small molecules diffuse faster than large ones, short molecules diffuse more rapidly than long ones, and more streamlined molecules (such as linear alkanes) pass more readily than bulky ones (such as branched alkanes).

As may be readily deduced from the preceding equation, if the solubility of a molecule in polyethylene is low, then its permeability will be proportionally reduced compared to other molecules having the same diffusion coefficient. To a first approximation, the less polar the nature of a molecule, the more readily it will dissolve in polyethylene and hence the greater will be its permeability. Thus polyethylene is a good barrier to polar molecules because of their low affinity for the nonpolar polymer. Polyethylene barriers are penetrated most readily by those organic molecules whose solubility parameters are similar to that of the polymer. The degree of permeation of polyethylene by organic solvents, in increasing order, is alcohols, acids, nitro derivatives, aldehydes and ketones, esters, ethers, hydrocarbons, halogenated hydrocarbons.

An increase in the size of chemically similar migrant molecules leads to an increase in solubility, but this is accompanied by a proportionally greater drop in the diffusion coefficient. The net result is a decrease in the permeability coefficient. In the extreme case, an infinitely long alkane is perfectly compatible with polyethylene but will not be able to diffuese out of the system.

As for polyethylene properties, morphology is the major factor affecting permeability. Both diffusion rate and solubility decrease as the degree of crystallinity increases and hence the permeability falls. The solubility of migrant molecules in polyethylene is directly proportional to the fraction of disordered material, irrespective of differences in crystallite organization and molecular branching [44]. Two factors determine the diffusion rate, these being the degree of crystallinity (or rather the fraction of disordered material) and the arrangement of lamellae. The well packed atoms of the crystalline regions are impenetrable to migrant molecules and thus reduce the effective volume in which diffusion can take place. Crystallites also delineate the paths along which migrant molecules can move. Partially ordered interfacial regions, having a density approaching that of the crystalline regions, are also essentially impervious and further limit the paths of migrant molecules. Thus diffusion takes place primarily in the disordered regions of the polymer. Michaels and Parker [44] and Michaels and Bixxler [45,46] demonstrated that the diffusion of permanent gases through polyethylene was much slower than could be accounted for in terms of a simple distribution of isolated impermeable spherical domains in a permeable network. They introduced the concept of tortuosity, whereby the paths along which migrant molecules must pass between crystallites was far from being a direct course from one surface of the polymer to the other. They postulated that lamellae formed channels between which migrant molecules must pass on their way through the polymer. As may be readily imagined, if the lateral planes of lamellae are arranged parallel to the plane of a polyethylene film, the path that migrant molecules must follow will be very convoluted. In contrast, in the case of samples displaying a transcrystalline morphology, wherein columnar or rodlike aggregates of lamellae are formed with their lateral planes normal to the surface of polyethylene, ''channels'' of disordered material may be formed that conduct migrant molecules directly into the material. It is not the complexity of the route so much as its length that decreases the permeability.

In addition to tortuosity, Michaels and coworkers [44–46] also considered the effects of chain constraints on the movements of the disordered chains that would be required to form holes through which the migrant molecules could pass. Adjacency of disordered chain segments to the immobile crystalline regions was thought to reduce their freedom of movement and hence slow diffusion. Thus they concluded that diffusion would take place most slowly in high density polyethylene samples that had been crystallized in such a manner as to develop lamellae with very large lateral dimensions and thin intercrystalline zones. Accordingly,

Properties of Polyethylene

$$D = D^*/\tau\beta$$

where

D^* = diffusivity in (totally) amorphous polyethylene
τ = geometric impedance (tortuosity) factor
β = chain immobilization factor

Such principles can be used to guide the development of new resin grades and morphologies. However, despite our advanced knowledge of polyethylene morphology it is not possible to determine the tortuosity or chain immobilization factors solely on the basis of morphological parameters. These factors must still be evaluated experimentally.

The molecular weight and branching of polyethylene resins have little effect on their barrier properties, except as they pertain to the degree of crystallinity and other morphological factors. Chemical modifications to polyethylene alter its permeability by changing its chemical interactions with migrant molecules. Modifications invariably introduce increasing levels of polar species and thus raise the solubility of polar compounds, increasing permeability accordingly. In cases where specific chemical interactions bind a migrant molecule tightly to one location, holes may be effectively blocked, making them inaccessible to other diffusing molecules. Hence the rate of diffusion will be decreased.

Cross-linking polyethylene reduces chain segment motion in the disordered regions, and thus diffusion becomes more dependent upon the size and shape of the penetrant molecules. The chain immobilization factor will have the greatest effect on penetrants that have a large molecular volume.

Orientation affects the permeability of polyethylene by changing its degree of crystallinity, lamellar organization, and molecular constraints. Strain-induced crystallization may increase the degree of crystallinity, while noncrystalline regions become better packed as a function of orientation [47,48], both of which reduce the free volume of the system and hence reduce its permeability. Not withstanding, the effects of orientation are complex, and contradictory results have been reported [49].

To a first approximation, diffusion will increase as

Polyethylene characteristics	Migrant characteristics
Degree of crystallinity decreases	Polarity decreases
Lateral dimensions of lamellae decrease	Molecular volume decreases
	Branching decreases
Molecular constraints decrease	Molecular weight decreases
	Chain length decreases
	Flexibility increases

The rate of diffusion of migrant molecules will increase dramatically if the polymer film contains physical defects such as pinholes, cracks, crazes, or voids. Even the smallest of defects can provide a path for the migration of molecules. This characteristic can be exploited, as in "breathable" films in which a multitude of interconnected microscopic voids are intentionally created. Such films are readily permeable to individual molecules, i.e., those in the vaporous state, but are impermeable to liquids because surface tension is too great to permit the entry of liquid into the small holes. Such voids can be created when a polyethylene film containing a high loading (more than approximately 30%) of an incompatible inorganic filler (such as finely divided chalk or talc) is stretched. Interconnected microscopic voids form around the filler particles during orientation.

External conditions also play a role in determining the permeability of polyethylene. The rate of diffusion increases as the concentration gradient of the migrant molecule in the polymer increases, i.e., as the pressure differential across the barrier increases. If the penetrant interacts with the polymer, its diffusion coefficient will be pressure-dependent. Elevated temperatures increase chain segment rotational and vibrational motion and thus increase the size and concentration of holes required for diffusion. Thus diffusion can be thought of as a thermally activated process, with the diffusion coefficient obeying an Arrhenius relationship. Accordingly, permeability increases with temperature according to

$$P = P_0 \exp\left[\frac{-E_p}{RT}\right]$$

where

P_0 = intrinsic permeability
E_p = apparent activation energy
R = the gas constant
T = absolute temperature

Thus the permeation of polyethylene by various migrant molecules decreases logarithmically with the reciprocal of the absolute temperature [44]. Figure 41 illustrates the effect of temperature on the permeation of oxygen through various polyethylene films. The diffusion rate of migrant molecules through glassy and semicrystalline polymers in general declines markedly when the temperature of the polymer is reduced below its glass transition temperature. In the case of polyethylene, with its extremely low glass transition temperature, this is not normally a factor of much importance.

The permeability of polyethylene to a wide variety of migrant molecules has been determined by many authors using a variety of techniques. A thorough review covering the diffusion of organic molecules through polyolefins is that by Flynn [50], which is exhaustively referenced. In reading such reviews it is

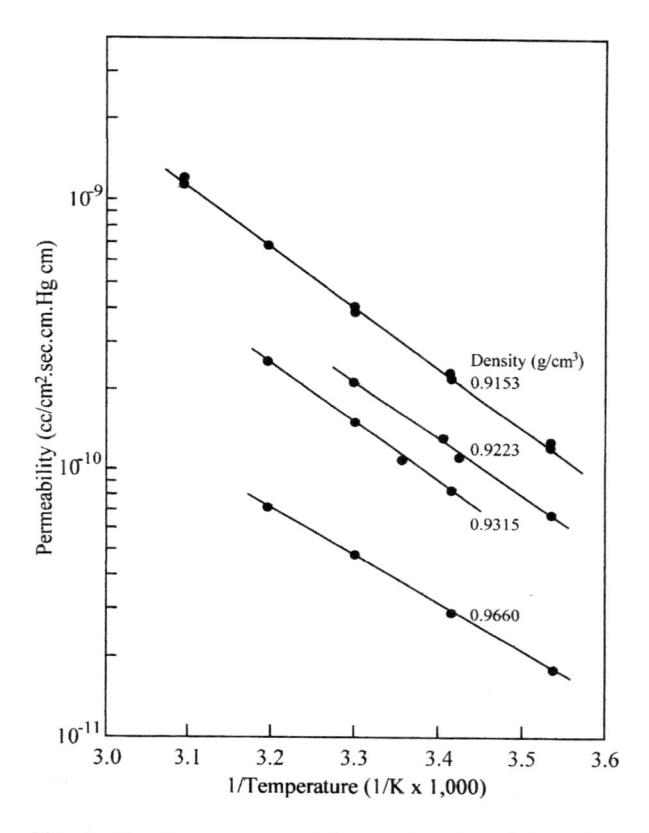

Figure 41 Oxygen permeability as a function of the reciprocal of the absolute temperature for polyethylene films of various densities. (From Ref. 44.)

striking how little agreement there is between different authors regarding diffusion constants. This fact should be kept in mind when such data are reviewed. The diffusion of migrant molecules in a specific system should be measured, rather than calculated, if it is critical to know its precise value. With this in mind, Table 11 is provided as a guide to the relative permeability of polyethylene and selected other polymers by various types of penetrant molecules.

E. Surface Contact Properties

The surface contact properties of polyethylene are those arising from its contact with materials with which it is in relative motion. The two principal consequences of contact are wear and friction. Wear is the phenomenon whereby the surface

Table 11 Permeability of Selected Molecules Through Various Types of Polyethylene and Selected Polymers

	Permeability [(cm^3·mm)/(sec · cm^2·cm Hg)]			
Polymer	N$_2$ 30°C	O$_2$ 30°C	CO$_2$ 30°C	H$_2$O[a] 25°C
High density polyethylene	2.7	10.6	35	130
Low density polyethylene	19	55	352	800
Poly(vinylidene chloride)	0.0094	0.053	0.29	14
Poly(ethylene terephthalate)	0.05	0.22	1.53	1,300
Nylon 6	0.1	0.38	1.6	7,000
Poly(vinyl chloride)	0.4	1.2	10	1,560
Polystyrene (crystal)	2.9	11	88	12,000
Polypropylene (isotactic)	—	23	92	680

[a] 90% relative humidity.
Source: Ref. 101.

of polyethylene is physically removed or permanently changed by contact with another substance. It can take one of three forms: erosion, abrasion, or cavitation. Erosion is caused by abrasive materials borne in a fluid medium, such as wind-blown sand. Abrasion is the consequence of one surface sliding against another, such as a piston in a cylinder. Cavitation is the effect of voids collapsing on a surface, such as those formed by the rotation of a propeller in water. (The latter is restricted to a very small number of cases [51]; as such it is of limited relevance and will not be discussed further.) Friction is the resistive force that occurs when two bodies in contact are moved relative to one another. The wear and frictional properties of polyethylene are important because polymeric products come into contact with surfaces in virtually every application. Contact properties may be of limited importance, as in the case of shopping bags and geomembranes, or crucial, as in the case of transfer lines, bushings, and prosthetic joints.

Polyethylene products exhibit excellent wear resistance, especially when made from ultrahigh molecular weight linear resins, in which form it surpasses all but the most specialized of polymers in their neat state (some filled polymers can exhibit superior wear resistance, but this is more a function of the filler than of the polymer). The coefficient of friction of high density polyethylene is very low, on a par with all but the most slippery of polymers. It rises as a function of molecular weight and increased branching levels.

Polyethylene's good balance of wear resistance and low friction suit it for a multitude of applications. It is particularly useful as an unlubricated bearing material. Such applications include those inaccessible for routine maintenance,

Properties of Polyethylene

where lubricants could be a source of contamination or where chemical inertness is required. Specific uses include cable guides, business machine bushings, food processing and pharmaceutical handling machinery, and prosthetic joints.

1. Abrasion Resistance

The wear resistance of polyethylene in its natural state is unsurpassed by any other unmodified polymer resin, the resistance being particularly high for articles fabricated from high density and ultrahigh molecular weight polyethylenes. The mechanisms by which the wear of polyethylene takes place are related to its deformational properties described earlier in this chapter. In practice, wear occurs by either erosion or abrasion. Erosion takes place when an abrasive material in a fluid medium impinges on a surface. Erosion normally involves particulate abrasives such as sand, but liquid droplets may also cause erosion (as in the case of rain striking airplanes in flight). Abrasive wear, in which solids rub against each other, can occur by two mechanisms. The first involves cutting and tearing of the polyethylene surface by sharp asperities on the counterface, while the second involves viscoelastic shearing due to adhesive forces. Both types of interactions are also active during erosion and contribute to friction. A secondary component of wear is the change of dimensions caused by creep. Much effort has been expended to determine the mechanisms of abrasive wear. Such results can be generalized to include erosion; samples that are resistant to abrasion also resist erosion. The wear of polyethylene surfaces can be disadvantageous from a number of points of view; physically it can result in a loss of fit between adjacent surfaces, such as a piston in a cylinder, or abrasion may mar the surface finish of a product, reducing its aesthetic appeal.

Abrasive wear in polyethylene occurs when the surface of a sample is removed by contact with a counterface with which it is in relative motion. The surfaces of the polymer and the counterface are always rough to some extent, either by design or due to the inescapable consequences of fabrication. Thus there are always asperities that protrude above the level of the surrounding surface. It is these asperities that make contact and are sites for ductile tearing failure. Asperities may be sharp and incisive, as in the case of those found on inorganic counterfaces, such as stainless steel and emery paper, or rounded and deformable, as in the case of those found on polymer surfaces. Sharp asperities cut and scour surfaces; smooth ones act by adhesion to viscoelastically shear the surface.

In measuring abrasion resistance, the regime of contact between the polymer surface and the counterface is very important. In some cases a new portion of the counterface is being continuously presented to the polymer surface, such as when the polymer follows a spiral path on the counterface; in others the polymer describes a circle on the counterface, so that it runs in a track over which it passes repeatedly; while in still other cases a linear path is traversed back and forth repeatedly. In the latter two cases the surface and counterface can "bed in" to each other, and the nature of the wear will change as a function of time and distance traversed.

The effects observed when a circular path is initiated are similar to those encountered when a fresh counterface is constantly being presented to the polymer surface. The asperities of the counterface either mechanically snag the polymer surface or adhere to it. In the first case the polymer can be cut or gouged. When circumstances permit, minute portions of the polymer surface are severed, and the debris may be left behind as loose particles. In the second case adhesive forces cause the two faces to bind, whereupon the softer one will undergo shearing of its uppermost layers. In the extreme case, layers can be sheared from the polymer and left adhering to the counterface. As the polymer continues to slide against the counterface the processes of gouging and adhesion continue, abrading the polymer surface. The volumetric wear rate (W_v), in terms of volume per energy (e.g., mm^3/Nm or in.^{3}/cal) is defined as

$$W_v = V/dL$$

where

V = volume of polymer removed
d = distance moved
L = load

The volumetric wear rate is also known as the "specific wear rate," "abrasion factor," "coefficient of wear," or the "K factor." The wear resistance of a sample is the reciprocal of its wear rate. The volumetric wear rate is related in a complex fashion to the morphology of the sample, the sliding speed, the roughness of the counterface, the specific heat capacity and heat conduction properties of the two faces, and the temperature at which the test was conducted. On a simpler basis, all other factors being constant, wear rate is inversely proportional to the product of the ultimate tensile strength and the elongation at break of a polymeric sample and directly related to its impact strength and fracture toughness.

When two surfaces start to slide against one another, steady state is not immediately achieved. Both the surface of the polymer and the counterface may be modified over a period of time. Abrasion removes the original polymer surface with its inherent properties, and the new face may be annealed by frictional heating or suffer thermal or oxidative degradation, resulting in chain scission or chemical modification. The counterface may also suffer abrasive wear (especially when the polymer is filled), or it may become coated with a film abraded from the polymer surface. Thus it takes some time before a steady rate of wear is achieved.

When the surface and counterface make repeated contact along a defined circular or linear track, the debris that is removed from the polymer surface may become trapped between the faces and alter the nature of the abrasion process. In this case a transfer film, consisting of thin layers of polyethylene, builds up on the counterface until it is entirely covered with polymer and the contact becomes essentially one of polyethylene against polyethylene. The material that composes the transfer film is removed from the polymer surface by the processes of severance or adhesion described above. The debris is then drawn into fibrils between the moving faces and smeared to form a film that is laid down on the counterface. The transfer film is built up of numerous small overlapping layers, which can adhere tenaciously to stainless steel counterfaces [52]. The formation of the transfer film may be enhanced by slightly elevated temperatures that facilitate the orientation and smearing of the polyethylene debris. Thus a transfer film forms faster at 30°C than it does at 15°C [52]. The polyethylene that makes up the transfer film is highly oriented and thus normally has a higher degree of crystallinity than the polymer surface from which it was removed [53]. Before the counterface is completely covered with polymer the wear rate changes in response to the proportion of the surface covered. Once a continuous transfer film is achieved, the wear rate stabilizes at a reduced level equivalent to that of polymer against polymer. Over a period of time the transfer film on the counterface may itself be worn away, but it is continuously replenished by material abraded from the polymer surface.

The wear resistance of polyethylene increases as the molecular weight increases and the level of branching decreases. Thus, under typical use conditions, ultrahigh molecular weight polyethylene—which is invariably unbranched—shows wear resistance superior to that of high density polyethylene, which in turn is more wear resistant than low density and linear low density polyethylenes, neither of which is commonly used as a bearing surface. The specific wear rate of high density polyethylene as a function of average molecular weight is shown in Figure 42.

The molecular characteristics of the polyethylene resin from which an article is fabricated play a role secondary to the roughness of the counterface with which it come in contact [53].

The wear rate of ultrahigh molecular weight polyethylene is essentially independent of irradiation at levels used to sterilize medical prostheses, i.e., 5 Mrad and below [54]. At higher levels of irradiation, such as those that result in significant levels of cross-linking, wear rate is increased substantially [55]. However, if the polyethylene exceeds its crystalline melting temperature because of an increase in sliding speed or applied load, the cross-linked polyethylene will exhibit a lower wear rate than the uncross-linked material [55].

For a given sliding speed, wear rate is approximately proportional to the applied load up to the point at which frictional heating melts the polymer surface. In the case of a circular or reciprocating motion, the molten polymer film may serve as a lubricant, decreasing the wear rate, or as a viscous brake, increasing

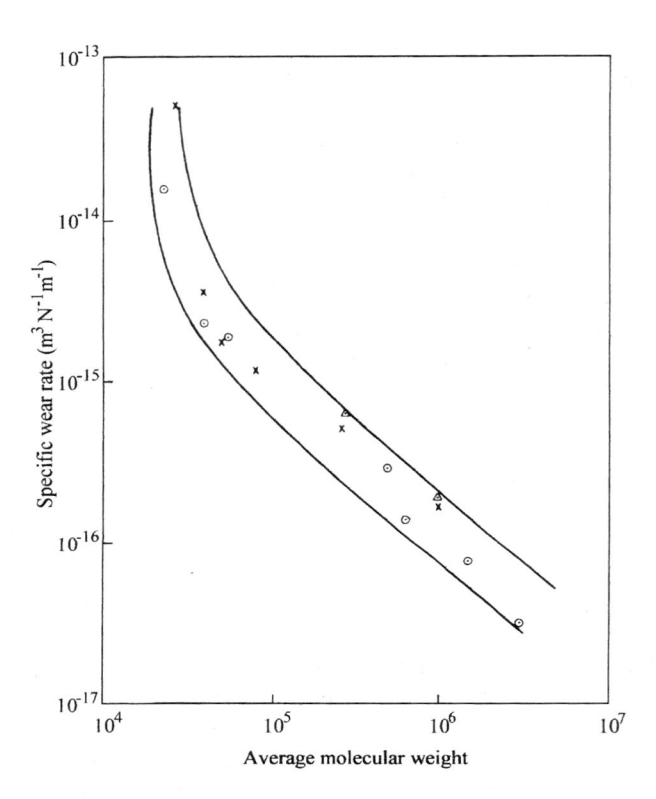

Figure 42 Specific wear rate as a function of molecular weight for linear polyethylene samples sliding against a steel counterface. (From Ref. 58.)

the wear rate. The precise effect is a function of the molecular weight of the polymer. Low molecular weight resins form lubricating films. Molten ultrahigh molecular weight polyethylene is so viscous that frictional forces are increased relative to the solid material. This results in increased heat input, thus melting more of the solid polymer and increasing the wear rate. When a polymer surface is constantly encountering a fresh counterface, melting inevitably increases wear rate because the molten polymer film is left behind on the counterface.

For a given load, the wear rate of polyethylene is approximately independent of the sliding rate up to a critical speed, beyond which frictional heating melts the surface of the polymer. In the case of ultrahigh molecular weight polyethylene, this results in a dramatic increase in the wear rate, as shown in Figure 43. This effect depends on the specific heat capacity and heat conduction of the counterface. The more heat the counterface can absorb and conduct away from

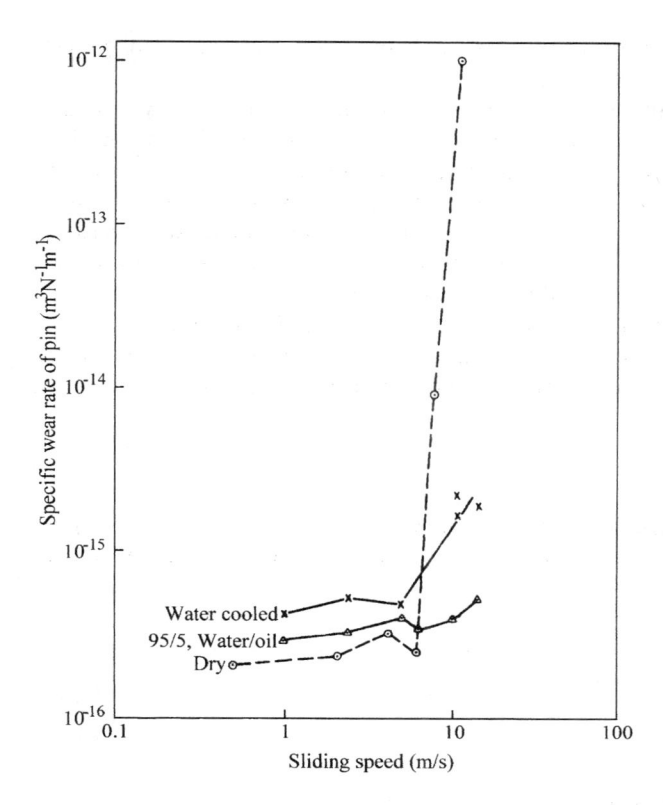

Figure 43 Specific wear rate as a function of sliding speed for ultrahigh molecular weight polyethylene sliding against a steel counterface. (From Ref. 58.)

the mating surfaces, the lower will be the frictional heating and the greater the sliding speed that can be achieved before the melting temperature is reached. Thus the wear rate of polymer against polymer may be higher than polymer against metal, because the metal is a better conductor. Cooling the counterface will retard the onset of increased wear, while an overall increase in temperature will have the reverse effect. In the case of a lower molecular weight resin, an increase of the sliding speed may reduce the wear rate due to the formation of a lubricating film.

As would be expected, the roughness of the counterface is an important factor controlling wear rate, especially under conditions in which the polymer continuously makes contact with fresh counterface surface. Such conditions prevail during the bedding-in period before a transfer film is fully formed and when the polymer describes a spiral trace on the counterface. The rougher the counterface, the greater will be the wear rate.

The wear rates of various polymers can be related to the energy of failure and hence to their cohesive energy density under certain conditions [56]. The cohesive energy density is a measure of the cohesion between adjacent chains in a solid and hence is a function of the molecular nature of the polymer. This relationship is found to hold true for a series of polymers subjected to abrasion by very rough surfaces as shown in Figure 44. However, when less rough counterfaces are encountered, this relationship breaks down.

The sliding wear rates of a number of polymer resins commonly used for dry bearing applications are listed in Table 12. These values are obviously dependent upon many factors and should be considered as comparative rather than absolute.

2. Friction

Friction is the resistive force encountered when two objects in contact slide against each other. The effect occurs under all circumstances, whether the relative

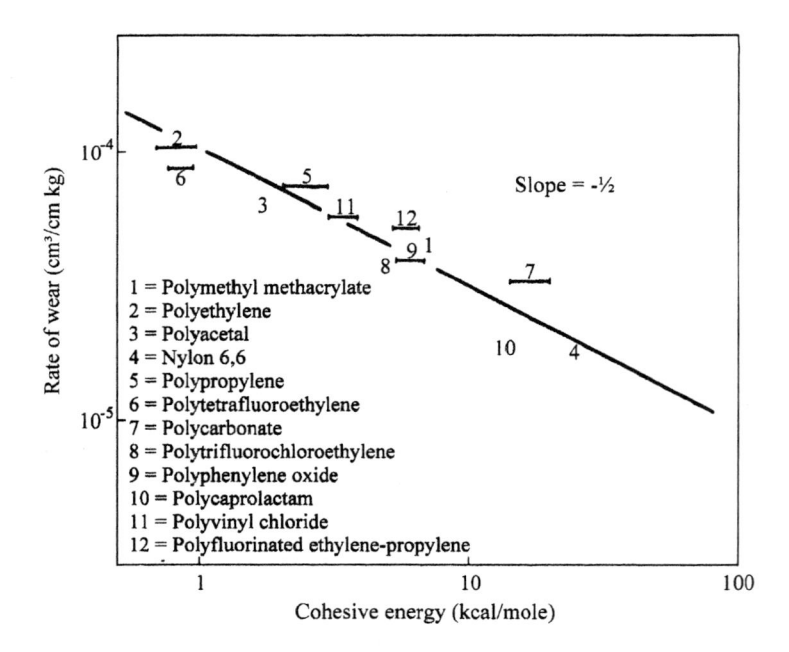

Figure 44 Rate of wear as a function of cohesive energy density for various polymers sliding against 100 grit sandpaper. (From Ref. 56.)

Table 12 Sliding Wear Rate of Various Types of Polyethylene and Selected Polymers Against Stainless Steel

Polymer	Abrasion against 180 grit abrasive belt (mm^3/cm^2)	Sliding wear rate against stainless steel [$mm^3/(N \cdot m) \times 10^{-6}$]
High density polyethylene	6.73	
Ultrahigh molecular weight polyethylene	7.35	0.51
Polypropylene (isotactic)	9.14	
Polycarbonate	26.64	
Acetal (polyoxymethylene)	14.81	11.81
Poly(ethylene terephthalate)	21.53	6.72

Source: Ref. 51.

motion is one of translation or rotation. Translation is exemplified by a polymeric piston ring sliding in a cylinder, and rotation by a metal rod rolling on the surface of a polymer film. The coefficient of friction (which is a dimensionless number) for a pair of surfaces is defined by Amonton's law:

$$\mu = F/W$$

where

μ = coefficient of friction
F = resistive force parallel to the direction of motion
W = normal force applied to the surfaces

In principle, the coefficient of friction is independent of the surface area of the two materials in contact. In the case of polymers the actual contact area is difficult to determine, because asperities of hard surfaces can indent the polymer while those of the polymer deform until the contact area is sufficient to support the applied forces.

The coefficient of friction developed between polyethylene and an adjacent hard surface is a function of various parameters related to material properties and testing conditions. The degree of branching and molecular weight are important, as are the surface roughness of the counterface and its thermal properties. Applied force, sliding rate, test temperature, duration of the test, and lubrication are also influential.

Friction is the sum of two components: external and internal energy absorption. The external component involves deformation and scarification of the surface and outer layers of the polymer, while the internal component involves hysteresis effects within the bulk of the polymer. Both external and internal components are active during sliding contact, whereas rolling contact principally involves internal effects.

The external component of friction results from processes very similar to those that occur during wear. There is an abrasive component active when rough surfaces make contact and an adhesive component active when nominally smooth surfaces make contact. Energy is absorbed by the formation of new surface area due to severance and gouging of the polymer surface by sharp asperities and the shearing deformation of surface layers brought about by adhesion at points of contact. On a molecular basis, the low coefficients of friction of high density polyethylene and polytetrafluoroethylene have been explained in terms of their smooth molecular profile [57–59]. The smooth molecular profile of these molecules facilitates shearing in the bulk near the surface and reduces mechanical interactions between the polymer surface and the transfer film.

Internal friction arises from damping effects that are caused by viscoelastic deformation occurring in the bulk. Such phenomena occur when adjacent segments of the polymer experience different compressive forces due to localized compression in the neighborhood of contacts made by asperities or on a larger scale in the general locale of the apparent area of contact. Internal friction may also be generated by compression at the leading edge of a sliding object and tension at its trailing edge. The overall deformations are a combination of compression, shear, and tension, depending on the forces active in adjacent segments of the polymer.

The primary factors that control the coefficient of friction of polyethylene are its molecular characteristics, while morphological factors play a secondary role. The molecular characteristics of greatest interest are those that control a sample's degree of crystallinity, i.e., branching levels and average molecular weight, the effects of which are addressed in Chapter 4. The spherulitic morphology of polyethylene plays an insignificant role in determining its frictional properties [59]. The internal component of the coefficient of friction of polyethylene decreases as the degree of crystallinity increases. Bulk modulus increases with crystallinity, and hence there is less shearing of adjacent segments with its accompanying energy damping. The external component of the coefficient of friction also decreases as the degree of crystallinity increases due to decreased abrasion resistance. In addition, the smooth molecular profile of high density polyethylene in comparison with that of low density and linear low density polyethylene reduces the frictional force. Thus the overall coefficient of friction of polyethylene increases as the molecular weight and branching levels increase [60], as illustrated in Figure 45.

Low levels of irradiation, such as those encountered in the sterilization of prosthetic joints prior to surgical emplacement, have little effect on the coefficient of friction of ultrahigh molecular weight polyethylene [61]. Higher radiation doses that result in significant levels of cross-linking increase the coefficient of

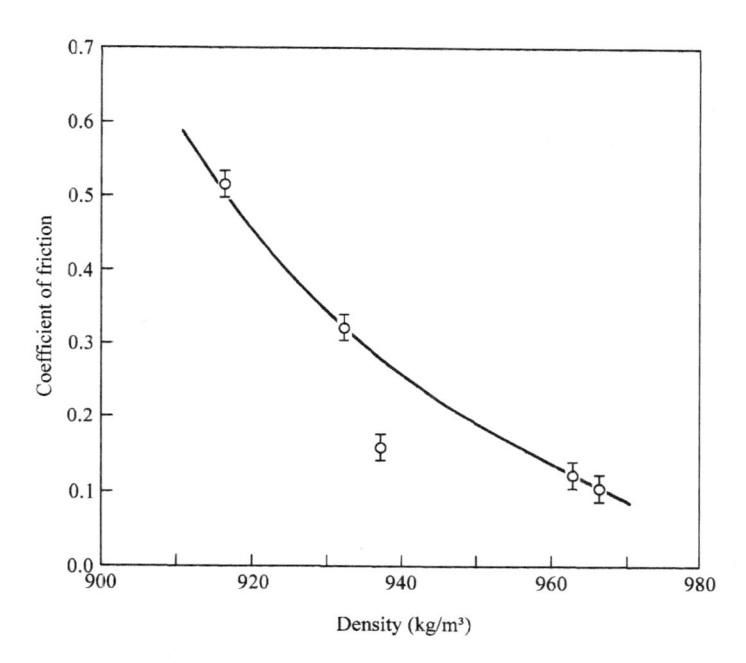

Figure 45 Coefficient of friction of polyethylene sliding against itself as a function of density. (From Ref. 60.)

friction somewhat, up to a maximum beyond which it falls back toward its original value [62].

Counterface properties that result in high abrasion of the surface of polyethylene increase the coefficient of friction. This is because abrasion requires energy absorption to deform the polyethylene morphology, create new surface area, and break polymer chains.

The coefficient of friction of polyethylene is a function of sliding speed. As the speed increases so does the coefficient of friction, up to a maximum, after which it falls [63]. The net result is a bell-shaped curve as a function of sliding rate, as shown in Figure 46. In a like manner, the coefficient of friction also passes through a maximum when plotted against temperature. The relationship between the coefficient of friction and temperature is so similar to that of the coefficient of friction and sliding speed that it is possible to develop a speed–temperature superposition curve in which a rise in temperature is equivalent to an increase in sliding rate and vice versa. This relationship is a consequence of the viscoelastic nature of polyethylene [64]. The initial rise in the coefficient of friction is due to increased energy absorbance by the viscous component, the

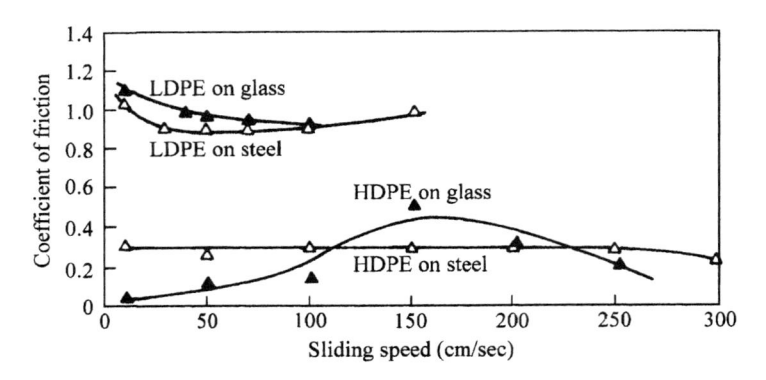

Figure 46 Coefficient of friction as a function of sliding speed for low and high density polyethylene against glass and steel. (From Ref. 63.)

decrease occurring when the elastic component becomes dominant. The elastic component comes to dominance because at high sliding speeds there is insufficient time for chains to slide past one another in a viscous manner, and at elevated temperatures because retraction of stretched chain segments is facilitated. The overall relationship between friction and sliding speed is somewhat complicated by frictional heating of the sample at higher speeds, which changes the temperature of the contact region. A decrease in the energy input required to permanently deform a sample at elevated temperatures also contributes to the decrease in the coefficient of friction as the temperature rises.

From the foregoing description it may readily be appreciated that coefficients of friction are dependent on so many interrelated and independent variables that it is not possible to assign exact values unless the system is precisely specified. Accordingly, the values given in Table 13 should be regarded as comparative only and not taken as having general application to all systems.

F. Optical Properties

There are three properties that define the principal optical characteristics of a polyethylene sample: its haze, transparency, and gloss. Haze is a function of light scattering; transparency is a function of unscattered light transmission, and gloss is dependent upon reflectivity.

> Haze is a measure of the incoming light scattered away from its original optical axis. A low percentage of haze indicates less light dispersion than a higher value. Two phenomena contribute to haze: internal and external (surface) light scattering. The first is a function of refractive index differences between adjacent regions within the sample, while the second is

Table 13 Coefficients of Friction of Various Types of Polyethylene, Other Polymers, and Selected Nonpolymeric Materials Sliding Against Themselves

Material	Coefficient of friction (against itself)
High density polyethylene	0.23
Low density polyethylene	0.5
Polytetrafluoroethylene	0.24
Polycarbonate	0.25
Poly(ethylene terephthalate)	0.25
Nylon 6,6	0.36
Polystyrene ("crystal")	0.38
Nylon 6	0.39
Poly(methyl methacrylate)	0.40
Poly(vinyl chloride) (unplasticized)	0.50
Polypropylene (isotactic)	0.67
Glass	0.9–1.0
Diamond	0.1
Steel	0.58

Source: Refs. 98 and 100.

a function of surface roughness. In general the internal haze, and hence the overall haze level, increases as a function of sample thickness.

Transparency is the ability of a sample to permit the direct transmission of light. In unpigmented samples, transparency is inversely related to haze. Pigments that absorb light can reduce transparency while not increasing the haze level proportionally. The relative transparency of a film or other sample can be gauged by trying to distinguish detailed features, such as text, through it. The further away from the sample that the text can be read, the higher is the transparency of the sample.

Gloss is a reflective phenomenon; in general, the smoother the surface, the greater will be its reflectivity and hence the higher will be its perceived gloss. The perceived level of gloss is not simply a function of specular (mirrorlike) reflection but also involves the distribution and intensity of the reflected light. To a first approximation, gloss is a function of viewing angle; the shallower the angle with respect to the surface plane, the more glossy the sample will appear. The apparent gloss of an otherwise matte surface when viewed at near grazing angles is known as "sheen."

Each of the three principal optical properties is related to the other two in a more or less complex manner.

In general, unpigmented thin films of low density polyethylene are quite transparent and relatively free from haze, while thin high density polyethylene films and thicker low density polyethylene films are translucent, that is, they transmit a certain amount of light but preclude a clear view of objects on the far side. Samples more than 1/8 in. thick, made from all but the very lowest density polyethylene resins, are opaque, blocking the transmission of virtually all light, even if they are unpigmented.

The optical properties of polyethylene are important in both thin and thick samples. Optical transmission properties are especially important for films, while reflective and coloration properties are of greater relevance to thick samples. Depending on the end use, polyethylene films may be used to either block or transmit light. In the former case, a pigment such as carbon black or titanium dioxide is incorporated into the film to scatter and absorb light. Films that are required to transmit light are generally unpigmented (natural). Examples of opaque films are those used for heavy duty garbage bags and agricultural ground cover. Translucent or clear films are exemplified by those used in food packaging and greenhouse covers.

When polyethylene is used in thicker applications such as bottles or household items, it is usually, but not always, pigmented. (Dyes, which are typically polar compounds, are rarely used in polyethylene because of their relatively low solubility.) In opaque applications the surface finish becomes more important, and efforts can be made to enhance or diminish the gloss depending upon the usage. Thus kitchenware is frequently molded against polished surfaces to improve gloss, while bottles may be blown into matte or polished molds depending upon the preference of the designer. When gloss is important in pigmented items, some degree of light penetration into the bulk is desirable because it results in a higher perceived color saturation. When the color saturation of a pigmented sample is augmented by surface clarity it is sometimes described as having a "see-through" appearance.

Haze is the visual evidence of two types of inhomogeneity: local anisotropy on a microscopic scale in the interior of the sample and surface roughness. Each class of inhomogeneity contributes independently to the total haze, but the different classes often stem from common causes. In general, both types of haze become more apparent with increased crystallinity. The two types of haze can be distinguished by wetting the surface of the polyethylene sample with a liquid that has a similar refractive index (approximately 1.51–1.54). Such treatment virtually eliminates surface haze by smoothing over surface irregularities; any haze still apparent can be attributed to internal scattering. The most common factor influencing internal haze is the morphological arrangement of the crystalline and disordered phases, which have different refractive indices. At the level of crystallites and in the intervening disordered zones, little diffraction occurs because lamellar and interlamellar thicknesses are typically an order of magnitude smaller than the wavelength of light. However, at the spherulite level, scattering occurs due to diffraction of light rays as they traverse the spherulitic boundaries. Such scattering is greatest when the range of spherulitic diameters approximates the visible wavelengths of light and the spherulites occupy about half the volume of the sample [65]. Thus the internal haze of a sample can pass through a maximum during the crystallization process when the peak scattering conditions are met. High density polyethylene generally exhibits greater internal haze than low density polyethylene because its range of spherulitic sizes more closely matches the wavelengths of visible light.

Rapid quenching of high density polyethylene from the melt reduces the haze level because the average spherulitic diameter is reduced, due to the higher nucleation density that occurs as a consequence of lower crystallization temperatures. Conversely, annealing low density polyethylene can increase its transmission of light [66]. The crystallization rate of polyethylene is so rapid and it is so readily nucleated by any extraneous matter (such as dust and catalyst residues) that spherulite size cannot be reduced significantly by the deliberate addition of a nucleating agent. To a first approximation, internal scatter increases linearly with the thickness of the sample. This relationship is complicated by variations in cooling rate throughout the thickness of the sample due to the relatively low heat transfer properties of polyethylene.

Another important factor influencing internal haze is the orientation of the ordered and disordered phases and the molecules that constitute them. Such orientation inevitably occurs during processing. The refractive index of polyethylene crystallites depends on the angle of view with respect to the crystal axes. Uniaxial or biaxial orientation preferentially aligns the crystalline axes, often in a complex manner, and thus the measured refractive index is dependent upon the viewing direction and the detailed morphology of the sample. In such cases the haze may be either increased or decreased, depending upon the exact molecular architecture and the processing conditions. The outcome of orientation during the crystallization process is further complicated by the effects on the spherulitic morphology and the degree of crystallinity. Thus it is not possible to generalize regarding the effects of orientation on haze.

Orientation of samples at room temperature can greatly affect their haze by a process known as "stress whitening," which occurs due to the formation of a multitude of microvoids. The voids provide a myriad of polymer/air interfaces that scatter light profusely. The resulting material is normally opaque with an attractive silvery sheen. Stress whitening can be healed by heating the sample up to a temperature approaching its DSC peak melting temperature.

It is very difficult to eliminate internal haze in high density polyethylene resins because the wavelengths of light span a range that overlaps broadly with that of the spherulitic diameters. The short-chain branching in linear low density polyethylene reduces the level of crystallinity and the size of the spherulites and consequently reduces the level of haze relative to that of unbranched resins. Further reduction of density levels, due to increased branching levels, either in resins made by high pressure processes (low density polyethylene and ethylene-vinyl acetate copolymers) or in those that incorporate very high levels of comonomer (very low density polyethylene) further reduces the incidence of internal haze. By similar reasoning, increased molecular weight reduces internal haze due to a decrease in the size of spherulites and the level of crystallinity. Irradiation of molten polyethylene also reduces internal haze in samples crystallized therefrom, due to restrictions placed upon molecular motion by cross-links.

Surface or external haze—also known as "grain"—is a function of the microscopic surface roughness. There are three principal factors affecting surface roughness: rheological properties of the molten polyethylene, formation of spherulites (and other supermolecular structures), and the surface against which the melt solidifies. In the case of film blowing, only the first two mechanisms are active and the surface haze is dependent only on the molecular characteristics of the resin and the processing conditions [67,68]. Surface roughness often initially appears on the molten film as it emerges from the lips of the annular die used to form the tube in film blowing. These irregularities tend to heal as the molten polymer is drawn away from the die. Such melt roughness is a function of resin molecular weight distribution and the processing conditions. The greater the proportion of very high molecular weight chains, the more severe will be the surface roughness and the resulting haze. Surface haze is strongly correlated with the ratio of the z-average molecular weight to the weight-average molecular weight ($\overline{M}_z/\overline{M}_w$). The relationship of haze to molecular weight distribution for a series of linear low density polyethylene resins, in which the surface haze is the major contributor to the overall value, is shown in Figure 47. The appearance of surface roughness can also be reduced by intensively mixing the melt prior to extrusion [68]. Thus the haze of low density polyethylene film can be reduced by repeatedly extruding it prior to film blowing, as shown in Figure 48. Further details of the formation of surface roughness under processing conditions are discussed in Section IV.

The second major cause of surface roughness is density fluctuation caused by incomplete crystallization. As polyethylene crystallizes, spherulites and other supermolecular structures form that have an average density greater than that of the surrounding material. This applies tension to the surrounding noncrystalline regions. When these higher density and stressed regions occur close to the surface of a film, microscopic sinkholes form that mar the surface, contributing to its roughness. On a much smaller scale, but by a similar process, surface roughness also occurs due to the impingement of bundles of lamellae with the surface [68]. The dimensions and concentrations of the crystalline regions and spherulites determine the severity of the sink marks on the surface.

The surface roughness of polyethylene is also strongly influenced by the

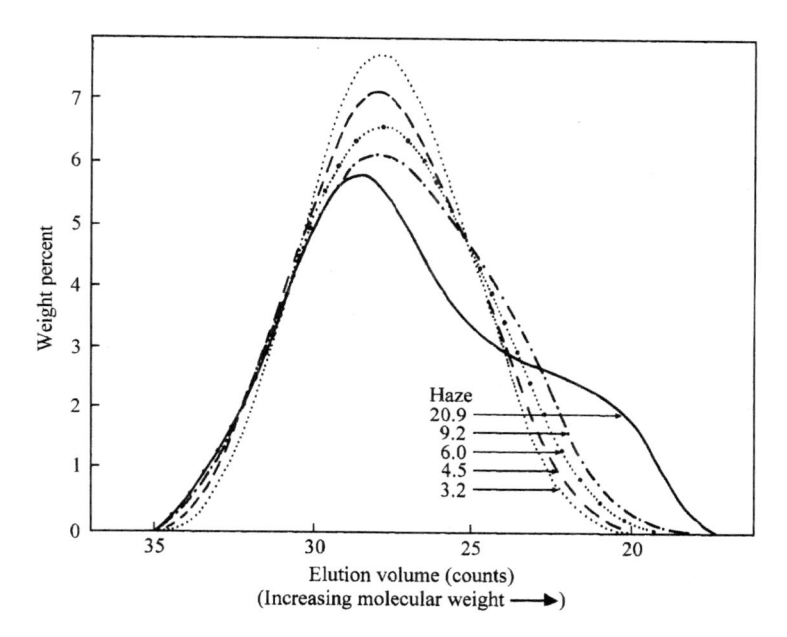

Figure 47 GPC curves (molecular weight distribution) of a series of low density polyethylene resins showing their correlation with haze levels. (From Ref. 68.)

nature of the surface against which it crystallizes. The smoother the molding surface, the lower its surface roughness. In the case of polyethylene films, an excellent surface finish can be achieved by casting molten polymer onto the surface of a highly polished roller in the chill roll casting process. Likewise, injection or blow molding dies may be polished to improve the surface appearance of parts molded in them.

In addition to the haze intrinsic to the molecular nature of polyethylene, it is possible to increase the haze level by accidentally or deliberately incorporating another material with a different refractive index that forms separate domains with dimensions on the order of the wavelength of light. The blooming of foreign materials to the surface of polyethylene to form droplets can also contribute to surface roughness.

Low haze is generally desirable from a packaging standpoint, because it permits a clear view of the contents. However, film strength is compromised when more transparent low density polyethylene resins are used in preference to high density polyethylene; in addition, the barrier properties are inferior. A compromise must therefore be found between optical and mechanical properties.

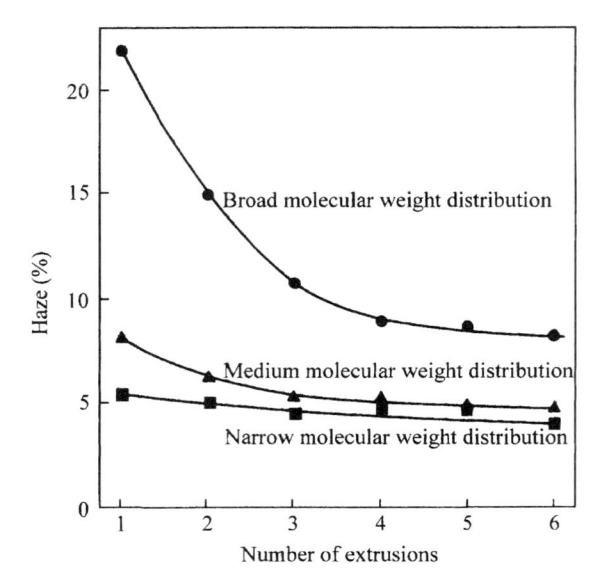

Figure 48 Effect of repeated extrusion prior to film blowing on the haze of low density polyethylene films. (From Ref. 68.)

Gloss is directly related to surface smoothness and as such is inversely correlated with surface haze. The factors that influence external haze are also active in controlling gloss.

In general, the refractive index of isotropic polyethylene samples tends to increase as the density increases. This is reasonable given that the refractive index of the relatively dense crystallites is significantly greater than that of the noncrystalline regions. Low density polyethylene typically has a refractive index of 1.51, while high density polyethylene has a refractive index of about 1.54.

III. ELECTRICAL PROPERTIES OF SOLID POLYETHYLENE

The electrical properties of pure polyethylene are governed by the negligible polar component of the carbon–carbon and carbon–hydrogen bonds that connect its constituent atoms. The absence of free electrons in the structure of polyethylene results in it being an excellent insulator, and the lack of polarizability of its bonds endow it with a general inertness to the effects of electrical fields. For these two reasons polyethylene finds extensive use as an insulator, primarily in the wire and cable industry. Despite its intrinsically desirable electrical properties,

Properties of Polyethylene 211

polyethylene is not totally immune to the effects of electrical fields and currents. Under the influence of high electrical stresses, trace amounts of polar molecules, such as catalyst residues and water, and polarizable bonds, such as those contained in carbonyl and vinyl groups, reduce polyethylene's electrical inertness. Under the influence of high electrical stress, polyethylene gradually deteriorates, both chemically and physically, reducing its effectiveness as an insulator.

A. Resistance and Capacitance

In common with the majority of synthetic polymers, polyethylene has no free electrons with which to conduct electricity. It is therefore a good electrical insulator. Another desirable characteristic of polyethylene is that its carbon–carbon and carbon–hydrogen bonds exhibit negligible polar character, thus making them essentially inert to electrical fields. This desirable combination of electrical properties makes polyethylene an excellent choice of material for a wide range of applications in which electrical resistance and inertness are required. Polyethylene finds extensive use in the areas of wire and cable coating and in a wide variety of other electrical applications, including terminal strips, electrical housings, and capacitors. The principal electrical characteristics of polyethylene can be defined in terms of its resistivity, permittivity, dissipation factor, dielectric strength and arc resistance. The first three characteristics are important at low electrical stress, while the latter two are more important at high electrical stresses.

Resistivity may be defined in terms of the bulk or surface conduction of current. In both cases it is a measure of the resistance to electrical flow exerted by the material. The bulk resistivity is largely a factor of the intrinsic nature of polyethylene and its various additives, while surface resistivity is strongly influenced by superficial contamination. Bulk resistance depends on thickness and is inversely proportional to cross-sectional area. It is defined in terms of volume resistivity, which is the resistance of a cube of a material, typically 1 cm or 1 m per side, quoted in terms of ohms per cubic centimeter or ohms per cubic meter. The bulk resistivity of polyethylene is not simply a function of its chemical structure; it may be reduced by contaminants such as antioxidants, catalyst residues, and water. The molecular characteristics of a polyethylene insulator (except those incorporating polar comonomers) are principally important with respect to morphology, which influences secondary physical characteristics such as melting temperature and abrasion resistance. For instance, high density polyethylene and cross-linked polyethylene are commonly used when higher service temperatures are likely to be encountered. Bulk resistivity is affected by temperature and to a much lesser extent by humidity, pressure, and sample morphology. Electrical resistance decreases as a function of increasing temperature:

$$\rho = \rho_0 \exp\left[\frac{\Delta E}{2kT}\right]$$

where

ρ = resistivity at temperature T
ρ_0 = limiting resistivity at low temperature
k = Boltzmann's constant
ΔE = energy gap between filled and unfilled electronic orbitals
T = absolute temperature

Surface resistivity is defined as the resistance between two electrodes that form opposite sides of a square and is quoted in units of ohms per square. The surface resistivity of polyethylene is much more susceptible to contamination than bulk resistivity, being especially sensitive to the presence of moisture, which reduces it considerably. The volume resistivities of polyethylene, selected polymers, and other materials are given in Table 14.

The dielectric constant (K), also known as the "(relative) electric permittivity" or the "electric inductive capacity," is a measure of the electrical inertness

Table 14 Volume Resistivities of Various Types of Polyethylene, Other Polymers, and Selected Nonpolymeric Materials

Material	Volume resistivity (ohm-cm at 50% relative humidity and 23°C)
Ethylene-vinyl acetate copolymer	2×10^8
Low density polyethylene	$>10^{16}$
High density polyethylene	$>10^{16}$
Nylon 6	10^{12}–10^{15}
Nylon 6,6	10^{14}–10^{15}
Poly(methyl methacrylate)	$>10^{14}$
Polystyrene ("crystal")	$>10^{14}$
Acetal (polyoxymethylene)	10^{15}
Poly(vinyl chloride) (unplasticized)	10^{15}
Polytetrafluoroethylene	10^{16}
Polypropylene (isotactic)	10^{16}
Acrylonitrile-butadiene-styrene (ABS)	1×10^{16}–5×10^{16}
Polycarbonate	2×10^{16}
Poly(ethylene terephthalate)	3×10^{16}
Aluminum	2.8×10^{-6}
Copper	1.7×10^{-6}
Graphite	65×10^{-6}
Silicon	10
Germanium	46

Source: Ref. 102.

Properties of Polyethylene

of a material to an applied electric field. It strongly depends on the polarizability of the dielectric material; the lower the polarizability of the constituent bonds, the lower the permittivity. Due to the negligible polar character of its carbon–carbon and carbon–hydrogen bonds, polyethylene has an extremely low interaction with electric fields and hence a very low electrical permittivity. The dielectric constant of a material is defined as the ratio of the capacitance of a capacitor in which it serves as the dielectric material to that of an identical capacitor in which the insulator is replaced by a vacuum. As with the electrical resistance of polyethylene, the dielectric constant is affected by temperature and humidity, an increase in either resulting in an elevated permittivity. The dielectric constants of polyethylene and selected polymers are given in Table 15.

The dielectric constant of polymers is approximately inversely proportional to the logarithm of the volume resistivity as illustrated in Figure 49.

The dissipation factor (D), also known as "tan d," is the ratio of the energy lost to that stored in an alternating electrical field. Low values are desirable, as they indicate efficient insulation, i.e., low power losses due to conversion of electric energy to heat, and are particularly important at high frequencies.

$$D = \frac{I_r}{I_c}$$

Table 15 Dielectric Constants of Various Types of Polyethylene and Other Polymers

Polymer	Dielectric constant (at 1 MHz)
Low density polyethylene	2.25–2.35
High density polyethylene	2.3–2.35
Ethylene-vinyl acetate copolymer	2.6–3.2
Polytetrafluoroethylene	2.1
Polypropylene (isotactic)	2.2–2.6
Poly(methyl methacrylate)	2.2–3.2
Polystyrene ("crystal")	2.4–2.65
Acrylonitrile-butadiene-styrene (ABS)	2.4–3.8
Poly(vinyl chloride) (unplasticized)	2.8–3.1
Polycarbonate	2.92–2.93
Poly(ethylene terephthalate)	3.37
Nylon 6,6	3.4–3.6
Nylon 6	3.5–4.7
Acetal (polyoxymethylene)	3.7

Source: Ref. 102.

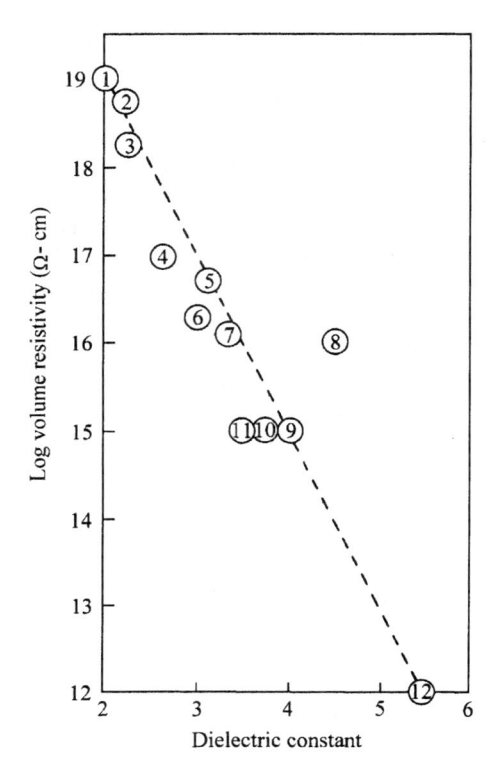

Figure 49 Correlation of volume resistivity with dielectric for various polymers. 1, polytetrafluoroethylene; 2, polyethylene; 3, polychlorotrifluoroethylene; 4, polyphenylene oxide; 5, polysulfone; 6, polycarbonate; 7, polyimide; 8, poly(vinylidene chloride); 9, nylon 6,6; 10, nylon 6; 11, epoxy resin; 12, polyester resin. (From Ref. 110.)

where I_r = dielectric absorption (loss factor) and I_c = dielectric constant (permittivity). This relationship is of the same form as dynamic-mechanical behavior, the induced field in the dielectric lagging behind the applied electric field.

The power factor (PF) is the ratio of the power dissipated to the wattage. In the case of good insulators, such as polyethylene and most other polymers, it is closely equivalent to the dissipation factor, and the two terms are often used interchangeably. The dissipation factor has been shown to decrease as a function of increasing crystallinity [69]. The dissipation factors of polyethylene and various selected polymers and other materials are given in Table 16.

As the voltage applied to an insulator is increased, there comes a point at which a catastrophic breakdown of electrical resistance occurs. The voltage at

Table 16 Dissipation Factors of Various Types of Polyethylene and Other Polymers

Polymer	Dissipation factor (at 1 MHz)
Low density polyethylene	>0.0005
High density polyethylene	>0.0005
Ethylene-vinyl acetate copolymer	0.03–0.05
Polystyrene ("crystal")	0.0001–0.0004
Acetal (polyoxymethylene)	0.0048
Polypropylene (isotactic)	0.0005–0.0018
Poly(vinyl chloride) (unplasticized)	0.006–0.019
Acrylonitrile-butadiene-styrene (ABS)	0.007–0.015
Polycarbonate	0.01
Poly(methyl methacrylate)	0.02–0.03
Poly(ethylene terephthalate)	0.0208

Source: Ref. 102.

which such a breakdown occurs is known as the "dielectric strength" or "breakdown voltage." It is normally quoted as a voltage gradient, e.g., volts per mil. The higher the value of dielectric strength, the more useful the material is as an insulator, especially at high electrical stresses. Dielectric strength varies with the thickness of the test piece, temperature, humidity, etc. Thin pieces (a few mils thick) often have higher dielectric strengths in terms of volts/mil than thick ones (>1/8 in.). Increased temperature and humidity decrease dielectric strength, while increased crystallinity and spherulite diameter increase it [69,70]. Dielectric strength is also time-dependent; paths along which current can flow take a finite time to develop. The aging of a conductor in an electric field tends to reduce its dielectric strength according to the strength of the field and various material properties that affect the growth of conductive paths [71]. Time to failure decreases as temperature and voltage increase [72]. The dielectric strength of polyethylene is influenced by the presence of contaminants such as catalyst residues, moisture, and dirt, by voids, and by polarizable species such as carbonyl or hydroxyl groups. In most cases, inhomogeneities within the chemical or physical structure of polyethylene reduce its dielectric strength; exceptions to this are the presence of antioxidants and electron traps, which reduce the rate of oxidation of the polymer. The dielectric strength of a polyethylene resin can be enhanced by increasing its purity and molding it under an inert atmosphere. In practice, dielectric strength is determined by ramping the applied electric field at a given rate until breakdown occurs.

The arc resistance of a material is the length of time its surface can be subjected to an electric arc (discharge) before it breaks down and conducts current. Failure generally occurs when a conductive line of carbonaceous material is formed by thermal decomposition. Arc resistance is normally measured in seconds. There is no correlation between the arc resistance of a polymer and its chemical composition. Values quoted in the literature for a given polymer vary widely. As would be expected, arc resistance is surface-dependent, decreasing when a sample is contaminated with moisture, grease, mica, etc. The arc resistance values for polyethylene and a variety of other polymers are listed in Table 17.

In the case of electrical insulation it is desirable that the dielectric constant and dissipation factor be as low as possible, while the material exhibits high resistivity, dielectric strength, and arc resistance. Polyethylene meets these requirements admirably at a very reasonable cost and thus is widely employed in electrical applications.

High molecular weight linear (high density) polyethylene and cross-linked polyethylene resins are used almost exclusively in high voltage applications typified by those in a power grid. The principal reason for selecting these two types of resins is their resistance to temperature effects. High voltage applications tend to generate substantial amounts of heat, which is dissipated slowly through the necessarily thick layers of insulation. An increase in temperature decreases the dielectric strength of polyethylene, especially when crystallites begin to melt. The crystallites found in linear polyethylene are less susceptible to melting than those found in branched resins because they are usually thicker. Thus, the effect of temperature on the dielectric strength of linear polyethylene is less pronounced

Table 17 Arc Resistance of Various Types of Polyethylene and Other Polymers

Polymer	Arc resistance (sec)
Low density polyethylene	135–160
Polycarbonate	10–120
Acrylonitrile-butadiene-styrene (ABS)	50–85
Poly(vinyl chloride) (unplasticized)	60–80
Polystyrene ("crystal")	60–80
Poly(ethylene terephthalate)	130
Polypropylene (isotactic)	136–185
Polytetrafluoroethylene	>200
Poly(methyl methacrylate)	No track
Acetal (polyoxymethylene)	129

Source: Ref. 102.

than on branched polyethylene. In situations where temporary current overloads can generate temperatures in excess of the crystalline melting range, cross-linked polyethylene will retain its dimensional integrity whereas uncross-linked resins would liquefy. In addition to their desirable thermal properties, high molecular weight and cross-linked polyethylene resins are also less susceptible to the phenomenon known as "treeing" than are lower molecular weight resins. This is presumably due to the high concentrations of tie chains found in high molecular weight and cross-linked resins.

Under less stringent conditions the choice of polymers available for insulation is much broader. In low voltage usage, considerations regarding a polymer's ultimate electrical properties may be outweighed by other physical characteristics and the cost of resin. Thus at low voltages polyethylene resins of all types come into competition with a variety of other polymers including polystyrene, poly (vinyl chloride), and polypropylene.

B. Treeing

When polyethylene insulation fails under the influence of high voltage electric fields it does so by the formation of conductive paths known as "trees." Trees are so named because they consist of a series of microscopic cavities fanning out in a dendritic pattern from a single point known as the "trunk" or "root." The point of origin, as might be expected, is invariably an inhomogeneity, either within the insulation or, more commonly, at its surface or interface with another material. Inhomogeneities, such as voids, contaminants, notches, or protrusions, act as electrical stress concentrators. Treeing occurs almost exclusively in the insulation of high voltage cables that make up the electric power grid. Trees come in a variety of configurations, from tall and slender, like poplar trees, to widely spreading, like oak trees. In all cases the overall growth habit is parallel with the applied electric field. The exact configuration depends upon a variety of factors, both internal and external, including the semicrystalline morphology of the insulator, the strength of the electric field, whether it is alternating or direct in nature, mechanical stresses on the insulator, and the presence of water or other fluids. Treeing is not unique to polyethylene; it is observed in a wide range of polymers. An optical micrograph of an electrically induced tree is shown in Figure 50.

Trees form most commonly in the presence of water, in which case they are termed "water trees." Other categories include "electrochemical," "electrical," and "sulfide" trees. Water trees consist of channels that are open and contain water in wet environments, but which close up when the insulator dries out. Upon rewetting, the channels reopen in their original locations. Electrochemical trees are water trees that contain chemical residues not originating from the polyethylene resin. It is generally thought that electrochemical trees form in the same

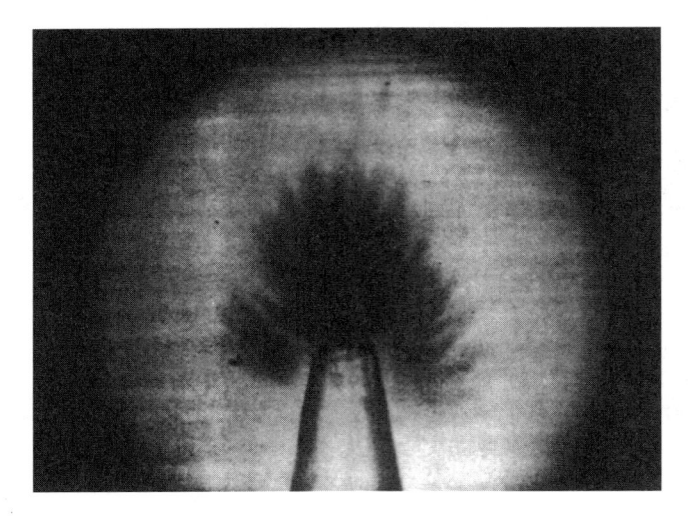

Figure 50 Optical micrograph of water tree. (From Ref. 74.)

manner as water trees, their cavity walls subsequently being chemically modified by the action of dissolved ionic chemicals. Electrical trees principally occur at field stresses much higher than those required to form water trees; they do not require the presence of moisture. The walls of the cavities composing electrical trees are largely carbonaceous, indicative of the high electric fields involved in their formation. When a water tree penetrates deeply into an insulator, the electric field experienced at its tip may be sufficient to initiate an electrical tree. Sulfide trees are relatively rare and contain high levels of contaminants, such as sulfur, from the environment. In most cases trees are rooted at the surface of electrical insulators, but they may initiate within the bulk of the material and grow both inward and outward, in which case they are termed "bow-tie trees." Trees that originate at a polymer surface are sometimes referred to as "vented trees" to distinguish them from bow-tie trees. Water trees account for the vast majority of trees, and the following discussion is restricted to their characteristics.

Water trees principally occur in buried high voltage electrical supply cables that experience wet conditions. The phenomenon was first recognized in the late 1960s after the practice of burying power lines became widespread [73]. The presence of water trees in polyethylene insulation surrounding an electrical conductor drastically reduces its breakdown voltage and can result in premature failure. The formation of water trees can take from weeks to years, depending upon the prevailing conditions. Their mechanism of formation has much in common with environmental stress cracking [71,74,75]. As such their formation is most facile in cable insulation that is subject to mechanical stress—both internal and external—in the presence of stress cracking agents. Polyethylene resins with substantial proportions of low molecular weight material and low concentrations of tie chains are most susceptible to treeing.

Water trees originate at locations of enhanced electrical stress, such as voids, notches, or protrusions into the insulation from the semiconductive sheath surrounding the conductor. When water is present at these sites it experiences a variety of electrically induced forces, such as Maxwell forces, electro-osmosis, and electrophoresis, that tend to force its molecules into the polymer. Preexisting voids, crazes, cracks, and spherulitic boundaries act as conduits by which water can penetrate. Once water infiltrates such features its molecules experience enhanced electrical stresses that increase its pressure, driving it further into the polyethylene or causing it to act as a wedge to enlarge the original conduit. The expansion of the original conduit to form a water-filled cavity causes the formation of crazes in advance of its tip, very similar to those created in environmental stress cracking. The electric field proceeds to drive water into the newly formed crazes, thus propagating the tree. Ionic solutions interact more strongly with electric fields than pure water and are thus more effective agents in the promotion of water treeing [76]. The newly created channels follow the path of least physical resistance, generally propagating through regions that have a low degree of crystallinity, few tie chains, or a poorly organized lamellar structure. Thus, penetration occurs most readily along spherulitic boundaries where branched and low molecular weight materials are concentrated. For this reason high molecular weight linear polyethylene resins and cross-linked polyethylene are more resistant to the propagation of trees than low density polyethylenes. Branching of trees occurs when propagating channels encounter regions in which growth is favorable in more than one direction.

Water trees initiate and grow more rapidly in the oscillating electric fields generated by alternating currents than they do in static fields [74]. Increasing the electrical stress generally, but not always, causes an acceleration in the growth rate of water trees [77]. Under the influence of higher frequencies, water trees initiate more rapidly [71], and the trees that form tend to have a more elongated habit than those grown at lower frequencies. Elevated temperatures increase the growth rate of water trees in most cases [77]. The rate at which water trees grow varies with time; initially they may grow at rates of the order of 1 µm/hr, but subsequently the growth rate falls by an order of magnitude [74]. It has been observed that bow-tie trees tend to stabilize over a period of time and rarely penetrate the insulation completely, whereas vented trees are far more likely to cause electrical breakdown [71].

Mechanical stresses present in cable insulation can promote the initiation and propagation of water trees. Such stresses typically arise as a result of uneven cooling during extrusion or when cables are subjected to tight bends. Thick layers of insulation can develop internal stresses as high as 37 MN/m^2 due to uneven cooling [71]. Regions of high stress are prime locations at which crazes form by the process of low stress brittle failure. The crazes so formed readily permit the infiltration of water and hence act as sites for the initiation of water trees. Regions of insulation that are under tension, i.e., those on the outside of a bend, show a higher propensity to form trees than do unstressed regions or those that are under compression. Thermal stress may also play a part in developing the network of cavities. Strong electric fields can generate high temperatures and associated temperature gradients within an insulator due to dielectric losses. This can result in uneven expansion of the insulator that generates mechanical stresses which facilitate the formation of crazes, hence promoting the initiation and growth of trees.

In addition to the physical effects of water penetration, enhanced electrical stress can increase the oxidation rate of polyethylene. Oxidation increases the polarity of the surface of cavities and locally enhances the effect of the electric field, increasing the growth rate of trees. Chemical oxidation can be retarded by the incorporation of antioxidants and electron-trapping compounds [78]. The presence of water can leach certain antioxidants from polyethylene, compounding the problem of water treeing [79].

Judicious choice of polyethylene resin and processing conditions for high voltage applications can greatly reduce the rate of tree initiation and growth. Beneficial processing conditions include predrying the polyethylene resin prior to extrusion and processing under an inert atmosphere, such as nitrogen, to reduce thermal oxidation. The use of ultraclean resins that extrude very smoothly minimizes the effect of contamination and reduces the number of protrusions from the semiconductive sheath into the insulator. The use of resins that extrude with an ultrasmooth surface can reduce the occurrence of water trees by as much as an order of magnitude [80].

IV. RHEOLOGY OF MOLTEN POLYETHYLENE

The properties of molten polyethylene are of great importance to the production of finished goods owing to the fact that all conversion processes require polyethylene to pass through the molten state. In its molten form, polyethylene is subject to a variety of deformational forces that influence its resulting solid-state properties, within the limits imposed by its molecular characteristics. Orientation frozen into the final product is especially important and is responsible for the anisotropic nature of most finished items. The rheological characteristics of a molten polyethylene resin determine which forming techniques are applicable and hence the range of final products that can be made from it. Conversely, processing techniques require polyethylene resins having rheological properties that fall within a given range.

Properties of Polyethylene

Molten polyethylene, or a melt of any thermoplastic, does not behave as a liquid in the classical sense, i.e., as a Newtonian liquid. This is to say, the viscosity of molten polymers depends on the shear imposed upon them. As a rule the viscosity of molten polymers decreases as the rate of shear increases, i.e., they are shear thinning. Some types of polymers do not conform to this rule, but polyethylene is not in this group.

Molten polyethylene is extremely viscous, up to many orders of magnitude more so than water or other low molecular weight liquids. In this state it displays certain elastic properties and is termed a viscoelastic liquid. Molten polyethylene is deformable, but when the deforming force is removed it tends to recoil toward its original dimensions. The viscosity and extent of the elastic recovery are functions of the entanglement of the molecules, which in turn depends on the molecular weight distribution and degree of branching of the resin. For example, the elastic component of molten ultrahigh molecular weight polyethylene is so much greater than its viscous component that it cannot be processed by ordinary techniques involving viscous flow.

On a conceptual basis, viscoelastic liquids, such as polymer melts, can be considered to behave according to the Maxwell model. In this model a deforming force acts upon a spring and dashpot arranged in series as illustrated in Figure 51. When a tensile force is applied to this system the spring instantaneously elongates followed by gradual displacement of the piston according to the resistance imposed by the dashpot. When the force is released, the spring retracts, but the dashpot retains a permanent set. Modeling the properties of a real polymer melt requires a large number of such pairs of springs and dashpots exhibiting a range of elastic constants and resistances, arranged in parallel as illustrated in Figure 52. On a phenomenological basis, the springs correspond to the elastic constants of segments of the polymer chains between entanglements and the dashpots correspond to the entanglements that control the rate at which molecules can slide past one another. Similar models can be envisaged for shear and compressive deformation. For comparison, a viscoelastic solid can be visualized as a spring and dashpot arranged in parallel according to the Voigt model, as illustrated in Figure 53. When a tensile force is applied for a period of time and then released, the system gradually returns to its initial state. The feature that distinguishes a viscoelastic liquid from a viscoelastic solid is that under the application of a constant force the liquid will deform indefinitely whereas the solid tends toward a finite deformational limit.

Much has been written on the mathematical and theoretical aspects of polymer rheology; see, for instance, Ferry's classic work on the subject [81] and other works listed in the bibliography. Although of great academic interest, precise mathematical models are applicable only to relatively simple polymeric systems and flow fields. Doubtless, as our understanding of molecular interactions improves, the time will come when precise mathematical models will be applica-

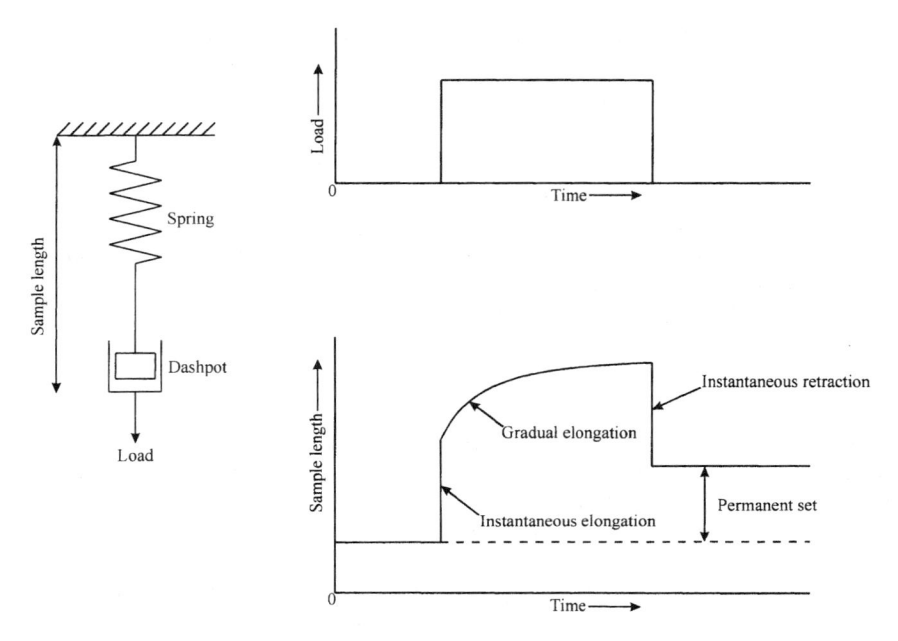

Figure 51 Maxwell model of a viscoelastic liquid.

ble to real polymer systems. When all is said and done, theoretical and laboratory-scale predictions are of little value if a particular resin cannot be processed economically on a commercial scale. It is sometimes said that the most important piece of rheological testing equipment is a processor's production line. Accordingly, the discussion of polyethylene rheology that follows is limited principally to the phenomenological aspects of the field.

A. Melt Viscosity

1. Zero Shear Viscosity

The viscosity of all thermoplastic melts is non-Newtonian, i.e., the viscosity is a function of the shear rate at which it is tested. For this reason great care must be taken to define deformational conditions when discussing viscosities. For purposes of comparison the viscosities of polymers are frequently quoted in terms of their apparent viscosity at zero shear rate. Zero shear viscosity is not a directly measurable value, but it can be obtained by extrapolation from observed viscosities over a range of finite shear rates.

The zero shear viscosity of a polymer is a function of various parameters,

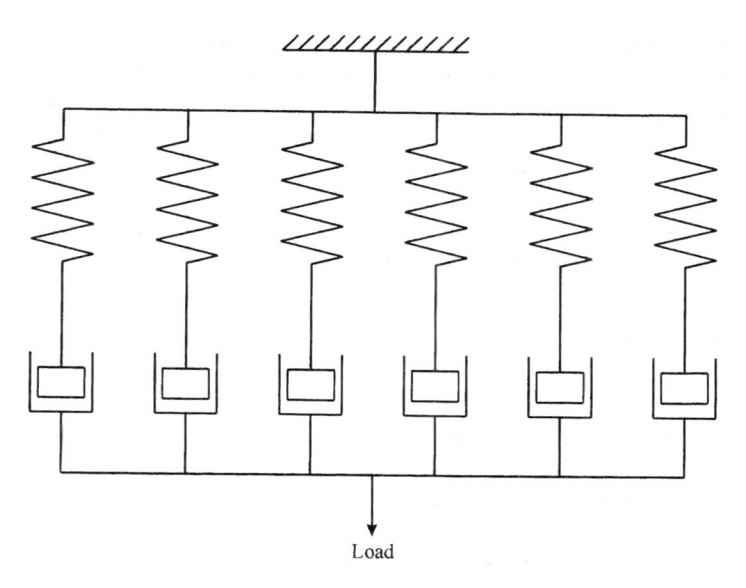

Figure 52 Generalized Maxwell model.

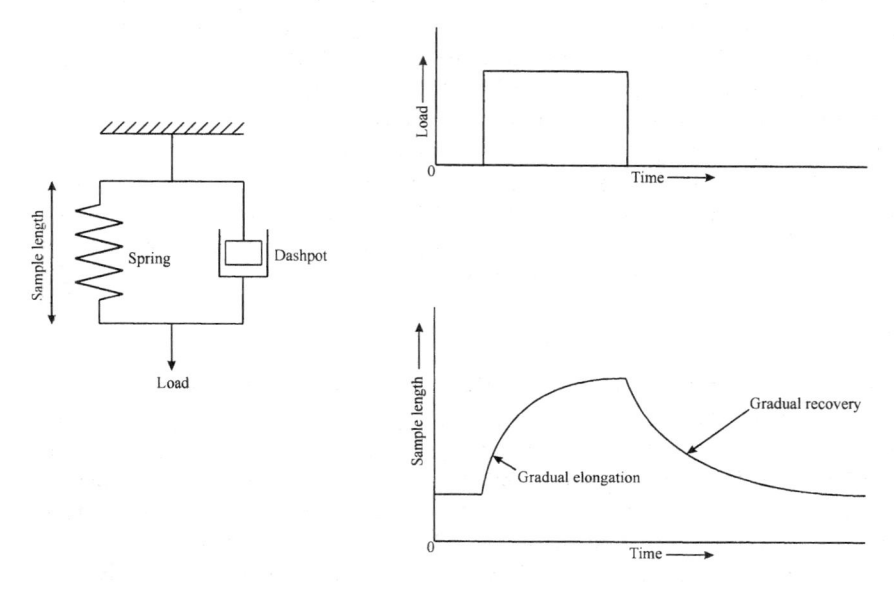

Figure 53 Voigt model of a viscoelastic solid.

both intrinsic and external. The primary intrinsic parameter for a high density polyethylene resin is its average molecular weight, molecular weight distribution playing a secondary role. For branched polyethylene resins, the degree and type of branching are also very important. The external parameter of principal interest is the temperature, with pressure being of lesser significance under conditions commonly encountered.

For unbranched polymers in general there is a common relationship between zero shear viscosity and average molecular weight:

For $\overline{M}_v < M_c$, $\mu_0 = K\overline{M}_v$
For $\overline{M}_v > M_c$, $\mu_0 = K\overline{M}_v^{3.4}$

where

$\overline{M}_v$ = viscosity-average molecular weight ($\overline{M}_v$ lies between $\overline{M}_n$ and $\overline{M}_w$)
M_c = critical molecular weight of entanglement coupling
μ_0 = zero shear viscosity
K = a constant for a given polymer and temperature

These relationships hold good for a wide range of linear polymers [82]. The value of the critical molecular weight of entanglement coupling (M_c) is approximately twice the molecular weight between entanglements (M_e) in the quiescent melt. The critical molecular weight for a given polymer is principally a function of the stiffness of its backbone and the molecular weight of the monomers of which it is composed. Thus, polyethylene with its low monomer molecular weight and highly flexible backbone exhibits a lower critical molecular weight than other polymer. The exponent 3.4 is derived experimentally, being the average of the observed values for a large number of linear polymers.

The effect of average molecular weight on the zero shear viscosity for a series of high density polyethylene resins of increasing molecular weight is illustrated in Figure 54. The change of slope as a function of molecular weight is readily seen at a molecular weight of approximately 3400.

2. Melt Index

The melt index (MI)—also known as the "melt flow index" (MFI)—of a polyethylene resin refers to the rate at which it extrudes from a capillary die under a standard set of conditions. The method by which it is determined is described in Chapter 6. The melt index of a polyethylene resin depends on its molecular characteristics, primarily average molecular weight, molecular weight distribution, and branching characteristics—short chain versus long chain, concentration, and distribution. The melt index reflects the average dimensions of the molecules in a resin and their entanglements with one another. From a commercial point

Properties of Polyethylene

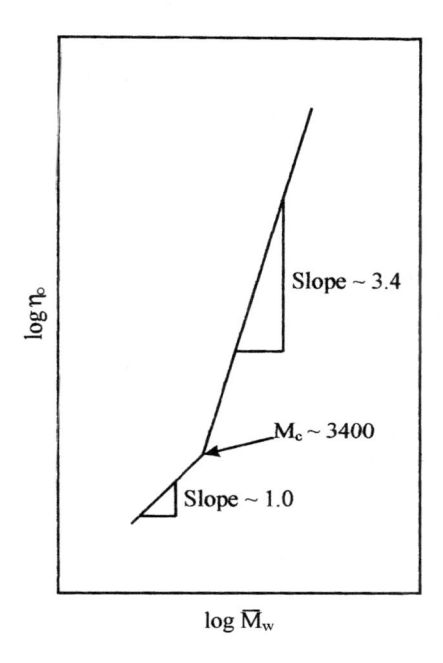

Figure 54 Schematic plot of zero shear viscosity as a function of average molecular weight for linear polyethylene.

[Plot shows $\log \eta_o$ vs $\log \bar{M}_w$ with Slope ~ 1.0 below $M_c \sim 3400$ and Slope ~ 3.4 above.]

of view, melt index is used as a rudimentary guide to flow characteristics in converting processes. Caution must be exercised when considering melt index, as it does not take into account changes of viscosity as a function of shear rate and hence may not accurately reflect the response of a resin subjected to varying levels of shear in processing equipment. This said, melt index is one of the most widely quoted descriptors used to define the characteristics of polyethylene resins.

The melt index of a polyethylene resin is sometimes equated with its weight-average molecular weight, there being an approximately inverse relationship between the two values. Care must be exercised when applying such relationships because they hold true only for series of resins that have very similar molecular characteristics, e.g., high density polyethylene resins made with the same catalyst system or high pressure products from a given reactor. There is no universal relationship between the melt index and molecular weight applicable to all resins.

The various conversion processes for manufacturing finished goods from base resin require different ranges of melt index for optimum performance; i.e., a balance of material properties and processability. Some typical ranges are listed in Table 18. These values should not be considered to cover all cases; specialized processes or the need for unique properties may require that resins outside these ranges be used.

3. Viscosity as a Function of Shear Rate

For a given polyethylene resin, the relationship between its measured viscosity and the applied shear rate depends on its molecular characteristics. All polyethylene resins are shear thinning. The general characteristics of the relationship of viscosity to shear rate are shown in Figure 55. Both viscosity and shear rate are plotted on logarithmic scales, reflecting the wide range of values encountered in commercial processes. Theoretically, the value of melt viscosity is predicted to level off at extremely high shear rates, but in practice melt instability sets in prior to this.

The change in the viscosity of polymer melts as a function of shear rate reflects changes in molecular entanglement. At zero shear rate polyethylene molecules in the melt adopt configurations that approximate random coils. The concentration of entanglements between molecules is thus determined by the average backbone length and degree of long-chain branching. For high density polyethylene resins, viscosity is directly related to the molecular weight. Linear low density polyethylene resins, with their short branches, follow a similar relationship to high density polyethylene. Low density polyethylene resins, with their long-chain branches, have a higher concentration of intermolecular entanglements and thus exhibit a higher viscosity than linear low density polyethylene or high density polyethylene for a given molecular weight. Low density polyethylene resins produced in autoclave reactors, having a higher concentration of long chain branches than those made in tubular reactors, have the highest zero shear viscosities at equivalent molecular weights.

Table 18 Approximate Ranges of Melt Indices Used in Commercial Conversion Processes

Conversion process	Melt index range (g/10 min)
Blow molding	0.05–2
Film blowing	0.05–2
Profile extrusion (pipes, etc.)	0.2–3
Film casting	2–5
Rotational molding	2–10
Injection molding	5–120
Extrusion coating	15–20

Properties of Polyethylene

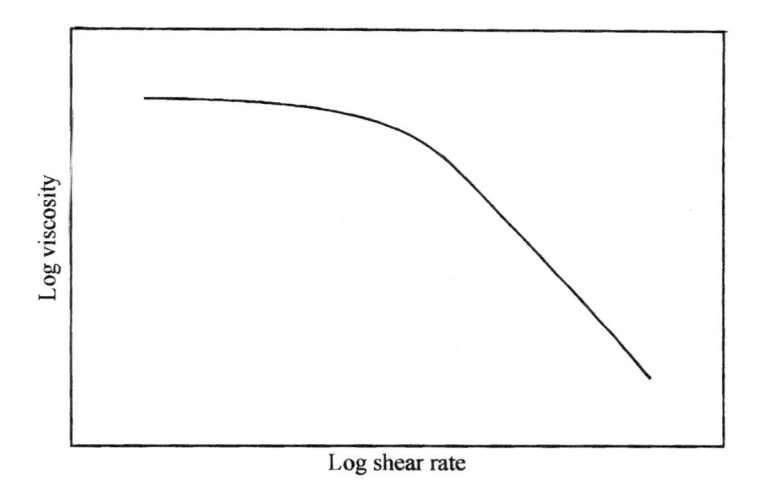

Figure 55 Schematic plot of melt viscosity as a function of shear rate for a typical polyethylene resin.

The response of the various types of polyethylene resins to increasing shear is a function of their degree of long-chain branching and molecular weight distribution. The lowest levels of shear thinning are exhibited by linear polyethylene resins that have a narrow molecular weight distribution. Resins with a higher proportion of long molecules, i.e., those with a broad molecular weight distribution, exhibit a greater reduction of viscosity as a function of shear rate than those having a similar average molecular weight but a narrower molecular weight distribution. Long-chain branched resins show a greater sensitivity to shear rate than linear ones because branched molecules have a more compact molecular profile at high shear rates than linear ones and hence have fewer entanglements to impede flow [83]. The effect of branching and breadth of molecular weight distribution on the shear thinning of polyethylene resins is shown schematically in Figure 56. The precise responses of viscosity to shear rate depend on specific molecular characteristics.

The viscosity of molten polymers (μ) decreases as the temperature increases. For Newtonian fluids and polymer melts over a limited range of temperature, the relationship of viscosity to temperature may be approximated by the Arrhenius equation

$$\mu = Ae^{E/RT}$$

where A = a constant, E = apparent energy of activation, R = the gas constant,

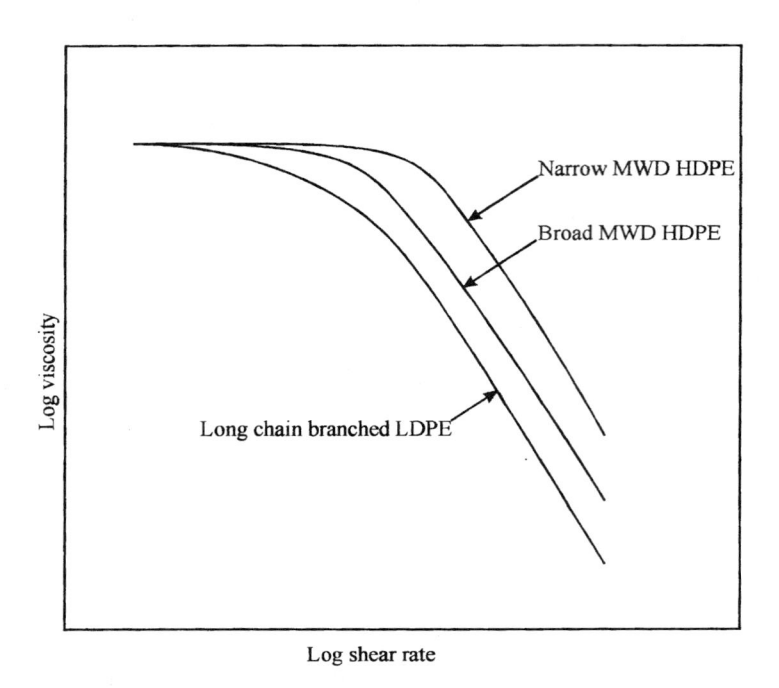

Figure 56 Schematic plot of melt viscosity as a function of shear rate for various types of polyethylene resin.

and T = absolute temperature. In practice a better fit is sometimes obtained from the empirical equation [84]

$$\mu = ae^{-bT}$$

where a and b are experimentally determined constants.

A large increase of pressure results in increased viscosity of polyethylene. This is thought to be due to decreased mobility of the molecular chains enforced by the decrease in the free volume of the system [85]. The observed viscosity is an exponential function of the applied pressure. Increasing the pressure on low density polyethylene from 2000 to 25,000 psi results in an increase in viscosity by a factor of 5 [86].

The viscosity of polyethylene melts can be lowered if the concentration of entanglements is reduced. Such a result can be effected by various strategies. The viscosity of molten ultrahigh molecular weight polyethylene may be greatly reduced if it has been previously precipitated from dilute solution. Dissolving polyethylene in a large excess of solvent greatly reduces the overlap between adjacent random coils, thereby decreasing the number of intermolecular entanglements along the length of each chain. When the polyethylene chains are precipitated by a rapid reduction of temperature they have insufficient time to reentangle. The net result is a material with relatively few chain entanglements in comparison to melt-crystallized samples. With few entanglements to impede the flow of molecules, the viscosity is greatly reduced in comparison to that of regular molten ultrahigh molecular weight polyethylene [87]. A similar phenomenon is observed when polyethylene melts are subjected to prolonged shear. Molecules become oriented to such an extent that they disentangle from their neighbors and the measured viscosity falls. If untangled resins remain in the quiescent melt state for a sufficient length of time, the molecules slowly reentangle and the original melt viscosity is restored. The greater the molecular weight of the chains, the longer it takes them to reentangle.

The viscosity of an ultrahigh molecular weight polyethylene resin that has been extruded and pelletized prior to testing is often substantially higher than that of nascent granules (i.e., those that are formed during polymerization). This is because monomers pass directly from the mobile phase to the solid state during polymerization, being added to the end of a chain that is essentially immobile. Without the opportunity to move freely, the newly formed chains cannot entangle. Subsequent extrusion in the melt affords the chains the opportunity to reentangle, and hence the viscosity rises.

4. Solution Viscosity

From a commercial fabrication point of view, the properties of polyethylene in solution are of limited interest. The only commercial process that uses dissolved polyethylene is the gel spinning of ultrahigh molecular weight polyethylene fibers. In its bulk state, molten ultrahigh molecular weight polyethylene is too viscous to be processed at any reasonable rate. However, by dissolving the polymer in an appropriate solvent at high temperature, the concentration of entanglements can be reduced sufficiently that the molecules can be highly extended. Rapid removal of the solvent from the drawn gel leaves highly oriented fibers that have extremely high elastic moduli.

The viscosity of concentrated solutions of polyethylene ($>\sim1\%$ w/v) is highly dependent upon the level of intermolecular entanglement. Thus solution viscosity is a function of molecular weight and chain branching. At very dilute concentrations ($<\sim0.1\%$), viscosity is a function of the average size of the random coils adopted by the chains. For purposes of comparison, the viscosities of polyethylene in solution are extrapolated to zero concentration to yield the intrinsic viscosity (applicable to a given solvent at a given temperature). For linear resins the intrinsic viscosity is approximately related to the average molecular weight according to the Mark–Houwink equation

$$\eta_0 = K\overline{M}_v^\alpha$$

where

η_0 = intrinsic viscosity
$\overline{M}_v$ = viscosity-average molecular weight
K, α = Mark–Houwink coefficients (empirically derived for a given solvent and temperature)

Numerous values of K and α are available in the literature that often yield very different values of $\overline{M}_v$ for a given value of η_0 [88].

B. Melt Elasticity Effects

When an isotropic melt is deformed, the chain segments between entanglements become aligned to a degree dependent upon the applied force. Entanglements act as transient cross-links, impeding the slippage of chains past one another when they attempt to return to a thermodynamically more favorable random coil configuration. When the deforming force is released, the partially aligned chain segments retract, causing macroscopic recovery of the molten sample toward its original dimensions. The extent of elastic recovery depends upon the duration and magnitude of the applied force and the rate at which chains can slip past entanglements. The rate of slippage is a function of the degree of orientation and the frictional resistance to chain movement. The resistance to chain movement is controlled by the degree of entanglement and the nature of the branching (if any). The degree of intermolecular entanglement is a function of the molecular weight and the long-chain branch concentration. Long-chain branches and large bulky ones are a greater hindrance to chain segment movement than short linear alkyl branches. Thus, small deformations of short duration are more recoverable than large ones of long duration, and higher molecular weight and long-chain branched samples are more elastic than lower molecular weight linear ones.

The overall elastic character of a polyethylene resin plays a major role in determining its processing characteristics. Resins with a high melt strength are required for such processes as film blowing and blow molding, which involve relatively low shear rates. A high critical shear rate (beyond which extrusion is unstable) is required for high speed extrusion and injection molding in which shear and elongational effects are pronounced. The production of highly oriented fibers requires a resin with a high breaking stretch ratio.

1. Die Swell

One of the classic examples of polymer melt elasticity is the phenomenon known as "die swell," in which an extrudate swells upon leaving a die, resulting in a product with a larger cross section than the die opening. As the extrudate exits the die, the constraints controlling its profile are removed and its dimensions are free to change according to its degree of orientation and entanglement. As molecules retract from the oriented state toward a random coil configuration, their shape changes from that approximating an ellipsoid toward that of a sphere. If there are no entanglements between molecules, the change of aspect ratio for each molecule will occur independently and its center of gravity will remain fixed in relation to its neighbors, in which case the profile of the extrudate will remain constant. In practice, molecules are entangled to some extent, and when the change of molecular aspect ratio occurs the molecules act in unison to cause an overall decrease in the length of the extrudate with a corresponding expansion of its melt at the instant it exits the die. The degree of die swell is expressed as the ratio of the cross-sectional area of the extrudate to that of the die opening. A schematic illustration of die swell is shown in Figure 57.

Orientation during extrusion can occur in two ways: by shear or by elongational flow. The former is typified by liquid flowing in a tube of constant cross section, the orientation being dependent upon the melt viscosity and shear rate, which in turn depends on the flow rate and the tube dimensions. Flow and shear rates are not constant over the cross section of the die; the flow rate is greatest at the center of the die, and the shear rate is highest at the walls. Elongational flow occurs when the cross-sectional area of the flow channel decreases; this elongates the profile of the molecules and increases their velocity. In most practical cases the effect of elongational orientation is greater than that of shear flow, orientation that occurs as the melt enters the parallel portion of the die relaxing somewhat under the lesser influence of shear flow. The longer the parallel region of the die, the greater will be the relaxation of the molecules and the lower their degree of orientation. Thus, a polyethylene resin extruded from a die with a large elongation flow at its entrance and a short length will exhibit greater die swell than the same resin extruded from a longer die or one with less elongation at its entrance. As a general rule, decreasing the length of time that a polymer spends in the parallel region of a die will increase its die swell.

For a given resin and die, the extent of die swell increases as the output rate increases, up to the critical shear rate at which "melt fracture" or "slip/stick" occurs (these phenomena are discussed in the following section). For a given resin at a fixed shear rate, die swell decreases as melt temperature increases. This is due to the higher potential for chain relaxation at elevated temperatures. Resins with long-chain branches exhibit less die swell than linear ones of similar molecular weight because their molecular profile is more compact and less prone to deformation during extrusion. The degree of die swell is very sensitive to the presence of long molecules; thus samples with a broad molecular weight distribution exhibit greater die swell than those with a narrow distribution. Samples that have been precipitated from solution prior to extrusion exhibit decreased die swell because their concentration of entanglements is lowered by the precipitation process.

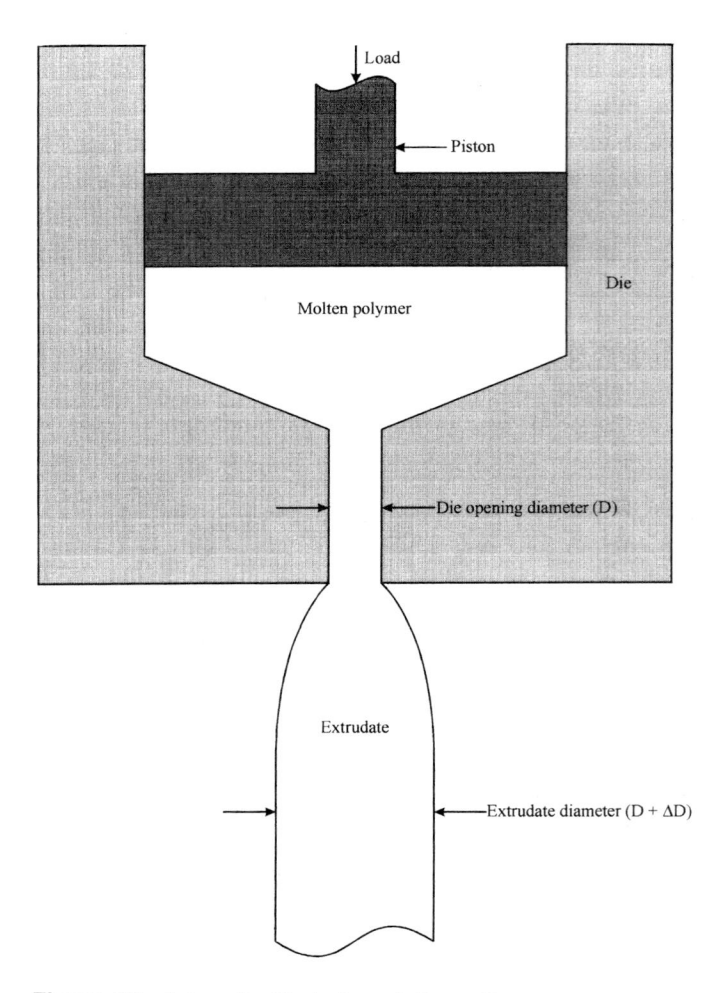

Figure 57 Schematic illustration of die swell.

Die swell has important consequences in the extrusion of products that require precise dimensions. The profile of the die, its output rate, the extrusion temperature, and the viscosity profile of the resin must all be balanced to give a product that has the correct dimensions and material properties. The problem is compounded in complex dies required to form extrudates with intricate profiles, such as window frames. Complex dies can generate different levels of shear in different regions, resulting in a range of die swell values within the same sample.

2. Ultimate Tensile Phenomena of Melts

a. Orientation of Melts Within Dies. Molten polyethylene samples are not infinitely deformable in the manner of low molecular weight liquids such as water. As the degree of orientation increases, the alignment of chain segments between entanglements improves to the point at which no further orientation is possible. At this juncture, if chains cannot slip past the entanglements quickly enough to relieve the applied stress, the entanglements will act as cross-links. When deformation within a die exceeds this limit, one of two things may occur: The adhesion of the molten polymer to the walls of the channel may fail and the flow mechanism will alternate between shear and plug flow (known as "slip/stick"), or chain segments may break and the melt will fracture. The term "melt fracture" is used here specifically to describe the process of melt failure that is caused by the severance of chain segments within the confines of a die. In practice, the critical shear rate beyond which melt fracture or slip/stick phenomena occur depends on the molecular characteristics of the sample and its shear history. Molten polyethylene samples that contain a high concentration of entanglements experience melt fracture at lower shear rates than those with lower concentrations.

Melt fracture—also known as "elastic turbulence"—occurs when the tensile forces experienced by a fully aligned chain segment between entanglements exceed the force required to break the polymer backbone. This requires that the frictional force exerted by entanglements also exceed the force required to break the polymer backbone. When an individual chain segment breaks, the load that it previously supported is transferred to its neighbors, which may in turn fail if they are also highly oriented. Melt fracture may be present on a localized basis, occurring in a highly oriented portion of a sample but not in other less oriented regions within the same cross section. In practice, melt fracture most commonly occurs at the walls of dies, where the shear rate is greatest, or at the entrance to dies, where the elongational effects are the most severe. The onset of melt fracture occurs at some critical shear rate characteristic of a resin's molecular characteristics and shear history. Polyethylene resins with high concentrations of entanglements exhibit melt fracture at lower shear rates than those with lower concentrations. The critical shear rate increases as molecular weight decreases, the degree of long-chain branching increases, and the molecular weight distribution broadens. Resins that have been subjected to procedures that reduce the entanglement concentration, such as prolonged shearing or solvent precipitation, exhibit elevated critical shear rates.

When melt fracture occurs, the opposing edges of the break retract to relieve the orientation. Less oriented material from adjacent regions is forced into the gap by the pressure on the melt. In the case of melt fracture occurring in the highly oriented skin of a polymer at the wall of a die, the infill comes from the core of the channel. When fracture occurs at die entrances due to elongational flow, the effect is more complex. At the entrance to a die, eddies may form as the polymer flow is compressed. It is these eddies that supply the infill material when fracture occurs at the die entrance [89]. The net result of either of these events is a product with a skin composed of materials that have experienced different shear histories. These processes are illustrated schematically in Figures 58a and 58b. During extrusion the occurrence of melt fracture may be accompanied by the sound of tearing from within the die [90]. The onset of melt fracture is accompanied by an abrupt change in the rate of polymer output. Up to the critical shear rate, the output rate is a smooth function of the extruder screw speed, but with the onset of melt fracture the output suddenly increases and may fluctuate wildly.

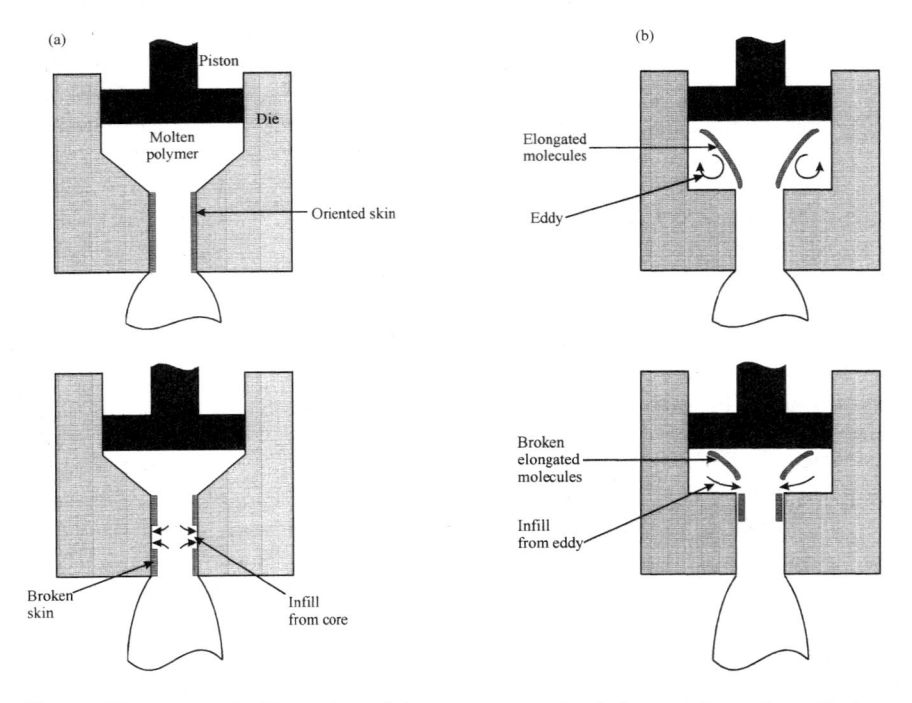

Figure 58 Schematic illustration of the occurrence of melt fracture (a) at die walls due to shear effect and (b) at die entrance due to elongational effects.

Melt fracture may also occur during the filling of injection molds, in which case the fracture typically occurs at the leading edge of the flowing melt, where the elongational flow is greatest. It manifests itself as periodic fluctuations of surface roughness, which are sometimes referred to as "flow marks."

Slip/stick occurs when the shear rate at the die wall exceeds the adhesive force of the melt to the surface. When this occurs, the melt jerks forward as a plug, relieving the pressure behind it and allowing the oriented chain segments to recoil somewhat. Once the pressure is relieved the rate of movement of the polymer slows and it re-adheres to the die wall. Shear flow resumes until once again the shear rate exceeds the critical value [91]. The effect is also known as "spurting" due to the erratic polymer output associated with it. During slip/stick flow the pressure within the die fluctuates and the polymer output is unsteady, both of which may vary periodically or erratically. The effects of stick/slip are closely related to those of melt fracture.

The results of melt fracture and slip/stick are most commonly observed during extrusion, the effects being manifest as a nonuniform extrudate. The nonuniformity may take the form of periodic fluctuations of the cross-sectional area (sometimes referred to as "bamboo"), helices, rough, highly erratic extrudate profiles, and, in extreme cases, fragmentation of the extrudate. Some of the manifestations of slip/stick and melt fracture are illustrated schematically in Figure 59.

Polyethylene extrudates sometimes exhibit "sharkskin" at shear rates below the onset of slip/stick. As the name implies, the surface of the extrudate feels rough, but there are no gross profile irregularities or fluctuations of polymer output. Its occurrence appears to be more closely related to extruder output than to shear rate. The onset of sharkskin can be delayed by raising the melt temperature. Unlike slip/stick and melt fracture, the average molecular weight of the resin is of less importance than the breadth of its molecular weight distribution, a narrow distribution enhancing the effect [91].

Under certain circumstances it is possible to achieve stable flow at shear rates in excess of the critical shear rate [92]. This phenomenon occurs for polyethylene resins with a weight-average molecular weight in excess of 400,000 when the output velocity is increased well above that required to exceed the critical shear rate. This behavior is explained in terms of the formation of a liquid crystal (hexagonal?) mesophase. It is postulated that this occurs in resins with a very high molecular weight because their molecules are amenable to being highly oriented.

Polyethylene can crystallize during extrusion if the conditions are favorable. This phenomenon, known as "shear-induced crystallization," occurs when highly oriented melts are extruded at temperatures at or below their melting points (93).

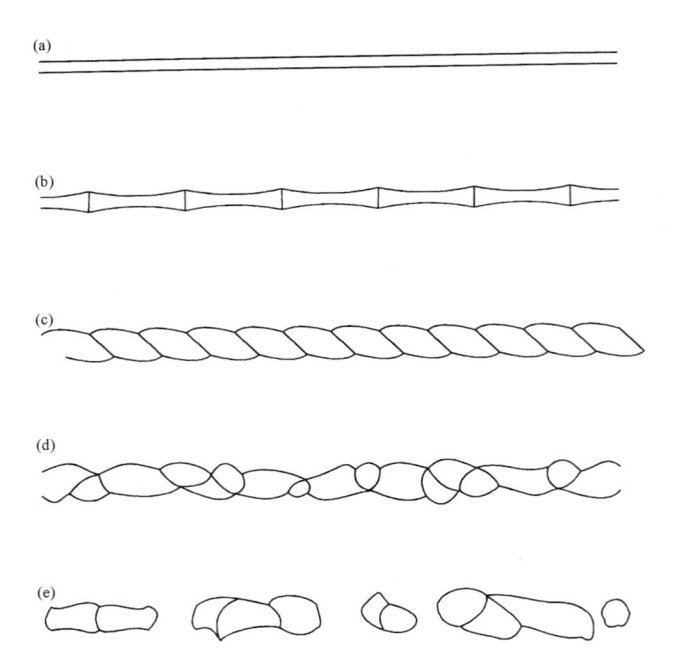

Figure 59 Extrudate profiles before and after the onset of slip/stick and melt fracture. (a) Smooth extrudate below the critical shear rate; (b) periodic fluctuations of extrudate profile (bamboo); (c) helical extrudate; (d) rough, irregular extrudate; (e) fragmented extrudate.

b. Orientation of Melts Outside Dies. When molten polyethylene is drawn outside the confines of a die it may rupture due to an insufficiency of entanglements to prevent catastrophic slippage of chains past one another or the breaking of chain segments between entanglements. Adequate melt strength and drawability are crucial to the economic success of several conversion processes, including film blowing, blow molding, and fiber spinning.

The extensibility of a polyethylene melt prior to rupture is known as its "breaking stretch ratio" (BSR), being the ratio of its length at rupture to its original length. The breaking stretch ratio decreases with increased resistance to chain slippage due to branching, higher average molecular weight, and broader molecular weight distribution. High orientation within a die reduces the external drawability of the melt.

The melt strength (MS) at rupture of molten polyethylene increases with increased levels of entanglement and orientation. Thus it increases with increased molecular weight and long-chain branching and higher shear rates in the die. For

Table 19 Summary of Effects of Molecular Characteristics on Melt Rheological Properties

Property	As average molecular weight increases	As molecular weight distribution broadens	As degree of long chain branching increases
Zero shear viscosity	Increases	—	Decreases
Shear dependence of melt viscosity	—	Increases	Increases
Melt index	Increases	Decreases	Decreases
Die swell	Increases	Increases	Decreases
Critical shear rate	Decreases	Decreases	Decreases
Breaking stretch ratio	Decreases	Decreases	Decreases
Melt strength	Increases	Increases	Increases

a given melt index, the melt strength of low density polyethylene is greater than that of high density polyethylene and linear low density polyethylene, which are fairly similar. Low density polyethylene manufactured in an autoclave has a higher melt strength than comparable tubular materials because of its more dendritic molecular architecture (94). Ionomers exhibit unusually high melt strengths and resistance to puncture (95). To a first approximation, the melt strength of an extrudate is related to its die swell ratio (96). As a general rule, the melt strength of a polyethylene resin is inversely related to its breaking stretch ratio.

C. Summary of Melt Properties

The general characteristics of molten polyethylene resins as a function of molecular characteristics are summarized in Table 19.

BIBLIOGRAPHY

Brydson JA. Flow Properties of Polymer Melts. New York: Van Nostrand Reinhold, 1970.
Dealy JM, Wissbrun KF. Melt Rheology and Its Role in Plastics Processing: Theory and Applications. New York: Van Nostrand Rheinhold, 1990.
Ferry JD. Viscoelastic Properties of Polymers. 3rd ed. New York: Wiley, 1980.
Kausch H H. Polymer Fracture. Berlin: Springer-Verlag, 1978.
Word IM. Mechanical Properties of Solid Polymers. New York: Wiley, 1983.
Williams JG. Fracture Mechanics of Polymers. Chichester UK: Ellis Horwood, 1984.

REFERENCES

1. P Smith, PJ Lemstra. J Mater Sci 15:505, 1980.
2. A J Peacock, L Mandelkern. J Polym Sci Polym Phys Ed 28:1917, 1990.
3. R Seguela, F Rietsch. J Mater Sci Lett 9:46, 1990.
4. J C Lucas, M D Failla, F L Smith, L Mandelkern, A J Peacock. Polym Eng Sci 35:1117, 1995.
5. V Lorenzo, JM Pereña, JM Fatou. Angew Chem 172:25, 1989.
6. D R Rueda, J Martinez-Salazar, F J Baltá-Calleja. J Mater Sci 20:834, 1985.
7. S Hashemi, JG Williams. Polym 27:384, 1986.
8. A D Channell, E Q Clutton. Polymer 19:4108, 1992.
9. T M Liu, W E Baker. Polym Eng Sci 32:944, 1992.
10. L C E Struik. Polymer 30:799, 1989.
11. J M Crissman. Polym Eng Sci 31:541, 1991.
12. J M Crissman. Polym Eng Sci 29:1598, 1989.
13. D T Raske and J E Young. Eur Patent Appl 0 436 520 A1, 1991.
14. X Lu, R Qian, N Brown, G Buczala. J Appl Polym Sci 46:1417, 1992.
15. X Lu, N Brown. J Mater Sci 25:411, 1990.
16. I D Peggs, D S Carlson. Brittle fracture in polyethylene geosynthetic membranes. In: ID Peggs, ed. Geosynthetic Microstructure and Performance, ASTM STP 1076, Philadelphia: Am Soc Testing and Materials, 1990, pp. 57–77.
17. Y-L Huang, N Brown. J Mater Sci 23:3648, 1988.
18. Y-L Huang, N Brown. J Polym Sci Polym Phys Ed. 28:2007, 1990.
19. D B Barry, O Delatycki. J Polym Sci Polym Phys Ed 25:883, 1987.
20. X Lu, A McGhie, N Brown, J Polym Sci Polym Phys Ed. 30:1207, 1992.
21. N Brown, X Lu, Y-L Huang, R Qian. Makromol Chem Makromol Symp 41:55, 1991.
22. S-W Tsui, RA Duckett, I M Ward. J Mater Sci 27:2799, 1992.
23. Y-Q Zhou, N Brown, J Polym Sci Polym Phys Ed 30:477, 1992.
24. J B Howard, In: E Baer, ed. Engineering Design for Plastics. Reinhold, New York: 1964.
25. H R Brown. Polymer 19:1186, 1978.
26. L Mandelkern, J G Fatou, R Denison, J Justin. J Polym Sci Polym Lett 3:803, 1965.
27. L Mandelkern, M Glotin, R A Benson. Macromolecule 14:22, 1981.
28. L Mandelkern, GM Stack, P J M Mathieu. Anal Calorim 5:223 1984.
29. L Mandelkern. Crystallization of Polymers. New York: McGraw-Hill, 1964.
30. U Leute, W Dollhopf, E Liska. Colloid Polym Sci 254:237, 1976.
31. F A Quinn, L Mandelkern. J Am Chem Soc 80:3178, 1958.
32. H S Bu, W Aycock, B Wunderlich. Polymer 28:1165, 1987.
33. R F Boyer. Rubber Chem Technol 36:1303, 1963.
34. N G McCrum, B E Read, G Williams. Anelastic and Dielectric Effects in Polymeric Solids. New York: Wiley, 1965, pp. 343ff.
35. F C Stehling, L Mandelkern. Macromolecules 3:242, 1970.
36. J G Fatou, L Mandelkern, A J Peacock. Unpublished observations.

37. C R Ashcraft, R H Boyd. J Polym Sci Polym Phys Ed 14:2153, 1976.
38. R Popli, M Glotin, L Mandelkern, R S Benson. J Polym Sci Polym Phys Ed 22: 407, 1984.
39. N Alberola, J Y Cavaille, J Perez. J Polym Sci Polym Phys Ed 28:569, 1990.
40. B B Pusey, M T Chen, W L Roberts. Proc 20th Int Wire and Cable Symposium, 1971, p. 209.
41. J Comyn, ed. Polymer Permeability. London: Elsevier Applied Science, 1985.
42. C E Rogers. In: E Baer, ed. Engineering Design for Plastics. New York: Reinhold, 1964, p 609.
43. SJ Doong, W S W Ho. Ind Eng Chem Res 31:1050, 1992.
44. A S Michaels, R B Parker. J Polym Sci 41:53, 1959.
45. A S Michaels, HJ Bixxler. J Polym Sci 50:393, 1961.
46. A S Michaels, H J Bixxler. J Polym Sci 50:413, 1961.
47. H Sha, I R Harrison. J Polym Sci Polym Phys Ed 30:915, 1992.
48. A Peterlin. Pure Appl Chem 39:239, 1974.
49. N E Schlotter, P Y Furlan. Polymer 33:3323, 1992.
50. J H Flynn. Polymer 23:1325, 1982.
51. H Böhm, S Betz, A Ball. Tribol Int 23:399, 1990.
52. T A Blanchet, F E Kennedy. Tribol Trans 32:371, 1989.
53. K Marcus, A Ball, C Allen. Wear 151:323, 1991.
54. W R Jones, W F Hady, A Crugnola. Wear 70:77, 1981.
55. K Matsubara, M Watanabe. Wear 10:214, 1967.
56. J P Giltrow. Wear 15:71, 1970.
57. J I Kroschwitz, ed. Polymers: An Encyclopedic Sourcebook of Engineering Properties. New York: Wiley-Interscience, 1987, p. 21.
58. J C Anderson. Tribol Int 15:43, 1982.
59. C M Pooley, D Tabor. Proc Roy Soc (Lond) A329:251, 1972.
60. S A Karpe. ASLE Trans 25:537, 1982.
61. W R Jones, W F Hady, A Crugnola. Wear 70:77, 1981.
62. K Matsubara, M Watanabe. Wear 10:214, 1967.
63. K Tanaka, Y Uchiyama, In: L H Lee, ed. Advances in Polymer Friction and Wear, Part B. New York: Plenum, 1974, p 499.
64. K A Grosch. Proc Roy Soc (Lond) A274:21, 1963.
65. M B Rhodes, R S Stein. J Polym Sci 45:521, 1960.
66. A M Hindeleh, M Al-Haj Abdallah, NS Braik. J Mater Sci 25:1808, 1990.
67. N D Huck, P L Clegg. SPE Trans 1:121, 1961.
68. F C Stehling, C S Speed, L Westerman. Macromolecules 14:698, 1981.
69. S Grzybowski, E Robles, O Dorlanne. IEEE Int Symp Elect Insul 1982, p 287.
70. S M Kolesov. IEEE Trans. EI-15:382, 1980.
71. E Ildstad, H Bårdsen, H Faremo, B Knutsen. IEEE Int Symp Elect Insul 1990, Toronto, p 165.
72. B S Bernstein, W A Thue, M D Walton, JT Smith. IEEE Trans 7:603, Toronto, 1992.
73. T Tabata, T Fukuda, Z Iwata. IEEE PAS-91:1354, 1972.
74. C T Meyer, J C Fillipini. Polymer 20:1186, 1979.
75. Y Poggi, V Raharimalala, JC Filippini. Polymer 32:2980, 1991.

76. Y Poggi, J C Filippini, V Raharimalala, Polymer 29:376, 1988.
77. M T Shaw, S H Shaw. IEEE Trans Elect Insul EI-19:419, 1984.
78. Y D Lee, P J Phillips. J Appl Polym Sci 40:263, 1990.
79. R J Densley, S S Bamji, A T Bulinski, J-P Crine, IEEE Int Symp Elect Insul 1990, Toronto, p 178.
80. N M Burns. IEEE Int Symp Elect Insul 1990, p 272.
81. J D Ferry. Viscoelastic Properties of Polymers. 3rd ed. New York: Wiley, 1980.
82. G C Berry, T G Fox. Adv Polym Sci 5:262, 1968.
83. J P Hogan, C T Levett, RT Werkman, SPE J 23(11):87, 1967.
84. J A Brydson. Flow Properties of Polymer Melts. New York: Van Nostrand Reinhold, 1970, p 47.
85. R C Penwell, R S Porter. J Polym Sci Polym Phys Ed 9:463, 1971.
86. R F Westover. Polym Eng Sci 6:83, 1966.
87. K A Narh, A Keller. J Mater Sci Lett 10:1301, 1991.
88. J Brandrup, EH Immergut, eds. 3rd ed. Polymer Handbook. New York: Wiley-Interscience, 1989.
89. E B Bagley, H M Birks. J Appl Phys 31:556, 1950.
90. J P Tordella. J Appl Phys 27:454, 1956.
91. J J Benbow, P Lamb. SPE Trans 3:7, 1963.
92. A J Waddon, A Keller. J Polym Sci Polym Phys Ed 28:1063, 1990.
93. J A Brydson. Flow Properties of Polymer Melts. New York: Van Nostrand Reinhold, 1970, pp 54, 55.
94. F P Mantia, D Acierno. Polym Eng Sci 25:279, 1985.
95. W F Busse. J Polym Sci A-2 5:1249, 1967.
96. K Wissbrun. Polym Eng Sci 13:342, 1973.
97. W A Kaplan, ed. Modern Plastics Encyclopedia '98. New York: McGraw-Hill, 1997, p B-52.
98. D W Van Krevelen. Properties of Polymers. 3rd ed. Amsterdam: Elsevier, 1990, p 707.
99. R A Isaksen, S Newman, RJ Clark. J Appl Polym Sci 7:515, 1963.
100. R C Weast, ed. Handbook of Chemistry and Physics. Boca Raton, FL: CRC, 1976.
101. R J Ashley. In: J Comyn, ed. Polymer Permeability. London: Elsevier, 1985, pp. 269–308.
102. C A Harper, ed. Handbook of Plastics and Elastomers. New York: McGraw-Hill, 1975.
103. R Popli, L Mandelkern. J Polym Sci Polym Phys Ed 25:441, 1987.
104. G Capaccio, T J Chapman, IM Ward. Polymer 16:469, 1975.
105. M A Kennedy, A J Peacock, L Mandelkern. Macromolecules 27:5297, 1994.
106. M A Kennedy, A J Peacock, MD Failla, JC Lucas, L Mandelkern. Macromolecules 27:1407, 1995.
107. A J Peacock, L Mandelkern, R G Alamo, J G Fatou. J Mater Sci 33:2255, 1998.
108. C F Popelar, C H Popelar, V H Kenner. Polym Eng Sci 30:577, 1990.
109. X Lu, N Brown. J Mater Sci 26:612, 1991.
110. D W Van Krevelen. Properties of Polymers. 3rd ed. Amsterdam: Elsevier, 1990, p 333.

6
Characterization and Testing

I. INTRODUCTION

The methods used to characterize polyethylene resins and products are numerous and varied in their principles of operation and range of properties investigated. Due to the close relationship between molecular characteristics, morphology, and final product attributes, a given technique may provide information about many aspects of a sample's characteristics. Thus, differential scanning calorimetry can yield information regarding the degree of crystallinity and distribution of crystallite thicknesses as well as data that can be used to predict elastic modulus and heat distortion temperature. No single technique can furnish a comprehensive picture of a polyethylene sample, the raw data most usefully being interpreted in light of the results from other experiments. Due to the large number of characterization techniques available—many of them highly complex—limitations of space permit only a brief description of each. In this chapter emphasis is placed on analytical methods that are singular to polymers.

The characterization and testing of polyethylene samples may be divided into five parts: molecular characterization, melt rheological analysis, solid-state morphological characterization, physical property determination, and electrical property testing. Molecular characterization principally involves determination of the molecular weight and compositional characteristics of a resin. Melt rheological measurements analyze the response of molten polyethylene to deformational forces, reflecting the molecular weight and branch distribution of a resin. Solid-state characterization seeks to determine the morphology of a specimen, which reflects its molecular characteristics and the method by which it was prepared. Physical property determination measures sample characteristics that are relevant to the attributes of the end product, such as elastic modulus, tear resistance, and electrical resistivity.

II. MOLECULAR CHARACTERIZATION

The molecular characterization of polyethylene resins is primarily aimed at determining their molecular weight and compositional distribution. The most widely used method for determining molecular weight distribution is size exclusion chromatography. Various spectroscopic methods, such as Fourier transform infrared or nuclear magnetic resonance spectroscopy, are used to determine the type and average concentration of short-chain branching. The distribution of short-chain branches as a function of molecular weight is determined by chromatographic separation techniques, the most common of which is temperature rising elution fractionation. The characterization of long-chain branching is typically addressed by a combination of size exclusion chromatography and either light scattering or viscometry. When taken together these techniques provide a clear understanding of the molecular nature of a polyethylene resin. From the molecular characteristics of a resin, together with some knowledge of processing techniques and structure–property relationships, it is possible to estimate the properties of final products made therefrom.

A. Molecular Weight Determination

The basis for most molecular weight determination techniques is the measurement of the size of random coils in very dilute solution. In theory this can be done by a variety of methods including light scattering, osmometry, viscometry, thermal field flow fractionation, sedimentation, and size exclusion chromatography. Fractionation methods can provide information with respect to the complete molecular weight distribution of a sample, whereas other methods provide only a single number corresponding to some moment of the distribution. By far the most common method for determining polyethylene molecular weights is size exclusion chromatography. The existence of long-chain branching in a sample complicates matters, and a combination of two or more techniques must be applied if an accurate molecular weight distribution is to be determined.

1. Size Exclusion Chromatography

Size exclusion chromatography (SEC) [also widely known as gel permeation chromatography (GPC)] is based on the premise that molecules in solution adopt random coil configurations with hydrodynamic volumes that increase as a predictable function of their molecular weight. By separating the molecules according to their random coil dimensions, a molecular weight distribution plot can be generated.

The principle of size elution chromatography is illustrated schematically in Figure 1. A dilute solution of polymer is pumped through a column packed with microscopic beads, the surfaces of which are riddled with pores whose range of sizes encompasses that of the polymer random coils. The largest molecules can only diffuse into a small fraction of the pores and are quickly eluted from the column. Progressively smaller molecules find a larger fraction of pores accessible and are thus impeded in their passage through the column in proportion to the numbers of pores available for them to enter. Thus the smaller the molecule, the longer it will take to pass through the column. The concentration of the solution eluting from the column is recorded as a function of time. With suitable calibration, a plot of the molecular weight distribution can be generated. From the distribution plot the various moments of the molecular weight, $\overline{M}_n$, $\overline{M}_w$ etc., are determined according to the equations listed in Chapter 1. In practice, of course, there are many complicating factors.

Size elution chromatography of polyethylene is normally performed on a commercial instrument equipped for computer-controlled data collection and analysis. These instruments are operated at high temperature (approximately 130–145°C), generally using 1,2,4-trichlorobenzene as the solvent (other solvents used include decalin, tetralin, and trichloroethylene). The concentration of solution is typically around 0.1% w/v, requiring only about 10 mg of polymer. The column packing consists of porous beads of a cross-linked polymer (typically based on polystyrene) approximately 5–20 µm in diameter. The pore sizes are conventionally rated in angstroms, ranging from 500 to 10,000,000 Å in steps equating to approximately one order of magnitude. It is practicable to use a series of columns that are each of a single pore size, a number of identical "mixed bed" columns (with a range of pore sizes), or some combination of single pore size and mixed bed columns. Invariably, columns packed with beads having the largest pore sizes are somewhat "fragile" and thus have a shorter lifetime than columns packed with beads having smaller pore sizes. Depending upon the range of pore sizes selected, linear polyethylene molecules with a range of molecular weights from approximately 500 to several million can be separated.

The most commonly used type of detector measures differential refractive index (DRI). The refractive index of the eluting solvent is assumed to be a linear function of the weight concentration of polyethylene in solution (this relationship actually changes somewhat as a function of molecular weight). Other detectors in use measure infrared or ultraviolet absorbance at a given wavelength, the scattering of a laser beam, or the viscosity of the solution.

The calibration of size elution chromatography instruments is a complex affair, and many methods—of varying levels of sophistication—are in use. The two main reasons that size elution chromatography calibration is so complex are the existence of long-chain branching on some resins and the lack of an adequate series of narrow molecular weight polyethlene fractions for instrument calibration.

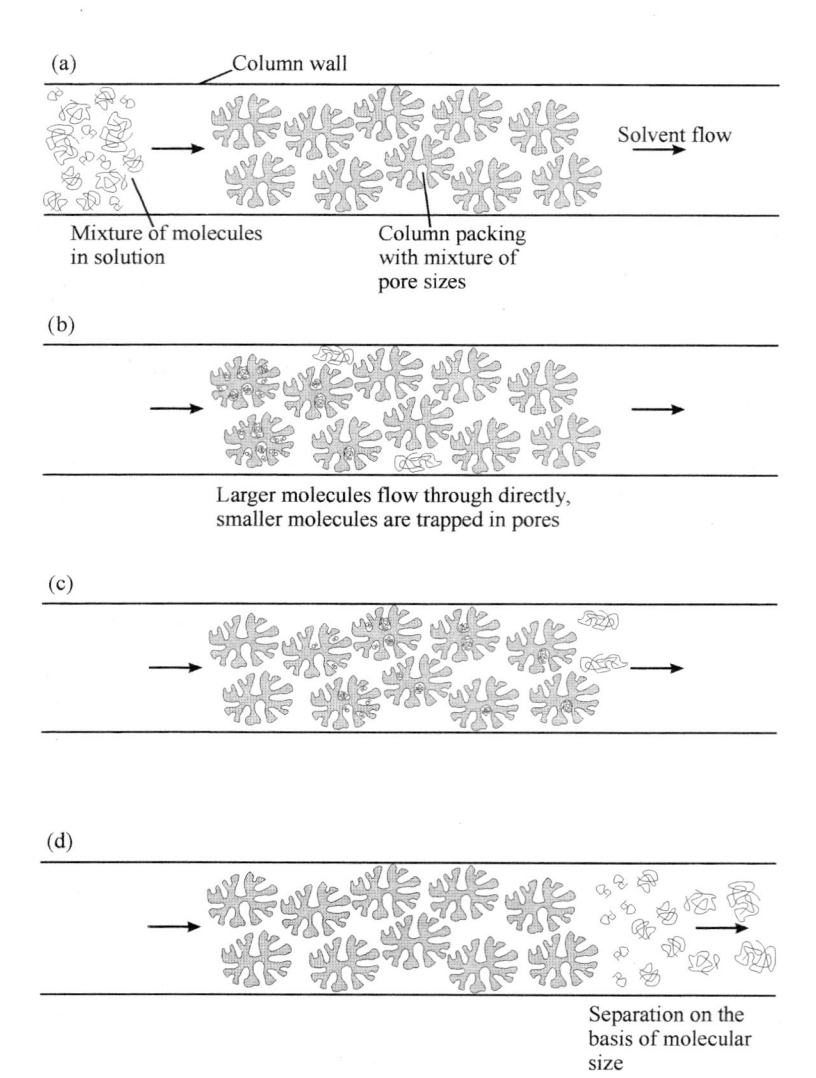

Figure 1 Schematic illustration of the principles of size exclusion chromatography. (a) Injection of dilute polymer solution; (b,c) progressive separation based upon hydrodynamic volume; (d) elution of the separated molecules from the column.

The presence of long-chain branching (LCB) on certain types of polyethylene seriously complicates the process of obtaining accurate molecular weights. Polyethylene molecules containing long-chain branches have smaller hydrodynamic volumes than linear molecules with identical molecular weights; thus they elute at longer times (the same can be said for short-chain branches, but the effect is much less severe). The discrepancy in elution times increases as a function of the branching complexity. This can give rise to serious errors in the determination of the molecular weight distribution when a single detector is used. This problem is especially important for low density polyethylene but also applies to high density and linear low density polyethylene samples that contain low levels of long-chain branching. In order to accurately determine a molecular weight distribution it is necessary to use a combination of a concentration-sensitive detector, such as a differential refractometer, and a detector sensitive to random coil dimensions, such as a viscometer or a light-scattering detector [1,2]. Branched samples yield anomalously low viscosities or scattering intensities compared to linear samples. The combination of data from the two detectors provides a branching function (g) applicable to the whole sample. A branching function of 1.0 indicates linearity of the sample's molecules, while lower values are indicative of long-chain branching—the lower the value, the more highly branched the sample. In practice, the degree of long-chain branching and the lengths of the branches themselves vary as a function of molecular weight. The assessment of short-chain branching distribution is addressed in section II.C.

In principle it is possible to obtain molecular weight distribution data for linear polyethylene samples using only a concentration-sensitive detector if the relationship of elution time to molecular weight is known. Ideally one would calibrate the instrument with monodisperse linear polyethylene standards whose molecular weights had been determined by a primary technique. In practice, anionically polymerized narrow molecular weight polystyrene standards are used to generate a calibration curve that is converted to the use of polyethylene by the application of the Mark–Houwink–Sakurada equation and corresponding coefficients according to

$$K_1 M_1^{\alpha 1} = K_2 M_2^{\alpha 2}$$

where

K_1, α_1 = Mark–Houwink constants for polystyrene
K_2, α_2 = Mark–Houwink constants for polyethylene
M_1 = molecular weight of polystyrene
M_2 = molecular weight of polyethylene

The drawback with this so-called "universal" calibration method is that the Mark–Houwink coefficients are somewhat dependent on molecular weight and are thus only approximately applicable to the full range of molecular weights of interest. Various values of coefficients for both polystyrene and polyethylene are available in the literature for a range of solvents and temperatures [3]. Unfortunately, the application of different pairs of coefficients can result in very different calculated molecular weight distributions for any given polyethylene sample.

An alternative method of calibrating a size elution chromatographic instrument that uses only a concentration-sensitive detector is to use a broad molecular weight sample for which the molecular weight distribution is accurately known. Such a linear polyethylene standard is available from the National Institute of Standards and Technology (in the United States). However, this standard does not cover a sufficient range to encompass the complete range of commercial polyethylene resins.

The report from size elution chromatography analysis usually contains a listing of the various average molecular weights and their ratios to one another, a branching index (where applicable), a plot of the distribution of the mass of polymer as a function of molecular weight, and, optionally, a plot of the raw signal as a function of elution time and a "slice report" (distribution data divided into a number of "slices" of equal time duration), listing the elution time of each slice, its calculated molecular weight, the mass of polymer in the slice, and the cumulative mass percentage eluted.

When utilizing data from size elution chromatography it should always be borne in mind that results depend upon the method of calibration. Variation is also introduced by different methods of sample preparation, the type of columns used, the type of detector (or detectors), the solvent type, and the temperature of elution. "Round robin" experiments run between a series of laboratories on identical samples can yield a wide variety of results unless a single experimental regime is rigidly employed. Typically, light-scattering detectors accentuate the contribution of higher molecular weight fractions, while differential refractive index detectors overemphasize the lower molecular weight ones. Due to the tailing of the molecular weight distribution at its high and low ends the $\overline{M}_n$, $\overline{M}_z$ and $\overline{M}_{z+1}$ averages show more variation than the $\overline{M}_w$ average.

As a rule of thumb, the repeatability of $\overline{M}_w$ determination for a given sample using a standard set of data collection and analysis conditions should be better than $\pm 2.5\%$. Variation of $\overline{M}_n$ and $\overline{M}_z$ under the same conditions should be less than $\pm 5\%$. The $\overline{M}_{z+1}$ may vary by as much as $\pm 10\%$. In absolute terms, the accuracy of the various molecular weight averages may be as much as double the error in repeatability.

The principle of size elution chromatography can be used to prepare polyethylene fractions on a large scale. This requires extremely large columns, which may be several orders of magnitude larger than those used for analytical scale size elution chromatography.

An unusual application of size elution chromatography is the determination of lamellar thicknesses in high pressure crystallized polyethylene [4]. Nitric acid etching was used to remove all but the crystalline molecular stems, which were

Characterization and Testing

subsequently dissolved and analyzed by size elution chromatography to yield their molecular weight and extended length. Comparison with direct lamellar thickness measurements made by electron microscopy revealed substantial agreement between the two techniques.

2. Viscometry

A viscometer can be used in molecular weight determination either on its own or, more commonly, in combination with one or more other detectors as a part of a size elution chromatograph. The determination of molecular weight from viscosity is based on the observation that the viscosity of a dilute polymer solution depends on its concentration and the size of its molecules. Viscometry used as a stand alone technique provides a single molecular weight known as the viscosity-average molecular weight ($\overline{M}_v$). When incorporated into a size elution chromatographic system, viscometry can provide information on the overall molecular weight distribution of a resin.

Viscometry encompasses a number of methods by which viscosity is determined from the flow properties of a solution in a capillary tube. Either the rate of flow or resistance to flow can be measured. The former is most commonly used as a stand-alone technique, while the latter is principally used for on-line detection in size elution chromatography. Both require that the polymer solution be at an accurately known temperature in the region of 135°C.

As a stand-alone method, the viscosity of a dilute polyethylene solution is normally measured using a capillary viscometer. A schematic representation of one such viscometer is shown in Figure 2. Detailed methodology for the use of such a capillary viscometer is found in ASTM D 1601.

A solution of precisely known concentration is introduced into the viscometer and maintained at the desired temperature in a thermostated oil bath. Pressure is used to force the solution from the reservoir up through a capillary tube into the index bulb. When the pressure is released, the solution flows back down through the capillary. Its rate of flow is determined from the time it takes the meniscus to traverse the distance between two index lines etched on the tube. Replicate measurements are made, and the average is taken. The viscosity is calculated from the rate of flow. The solution is then diluted with an accurately known volume of solvent, and the process is repeated. The viscosity is measured at a minimum of five dilutions, and the viscosity at infinite dilution, [η] (known as the intrinsic viscosity or the limiting viscosity number), is determined by extrapolation. From the limiting viscosity number, the viscosity-average molecular weight can be calculated according to the Mark–Houwink–Sakurada equation,

$$[\eta] = K\overline{M}_v^\alpha$$

where K and α are the Mark–Houwink constants for the specific solvent, temperature, and molecular weight range [3].

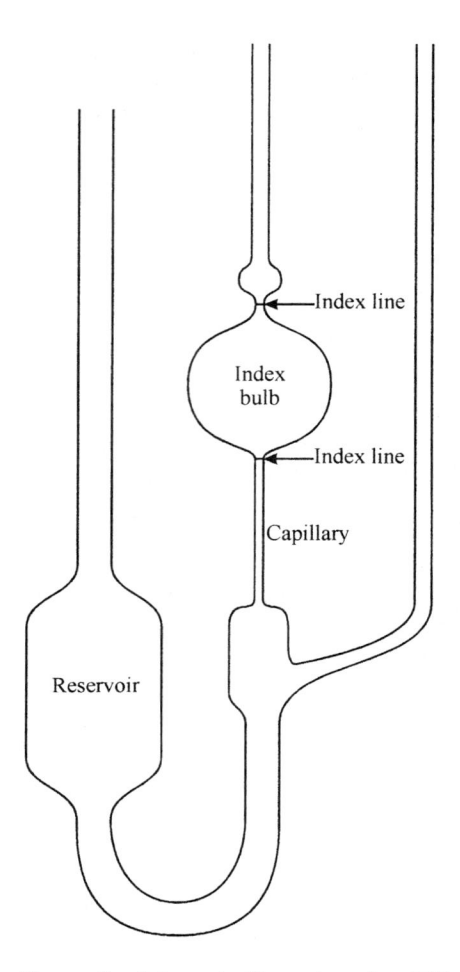

Figure 2 Schematic illustration of an Ubbelohde capillary viscometer.

As may well be imagined, this type of viscometry is tedious and fraught with the possibility of technical error. Its primary use lies in the analysis of polyethylene resins with ultrahigh molecular weights, for which size elution chromatographic columns with a suitable range of pore sizes are not currently available.

A number of authors have proposed alternative methods for calculating [η] from a single viscosity measurement at known concentration (see, for instance, the work of Raju and Yasseen [5] and references cited therein).

Characterization and Testing

The second major use of a viscometer is as a size-sensitive detector used in conjunction with a concentration-sensitive detector on size elution chromatographic equipment. In this application, which functions on-line, the resistance to eluent flow exerted by a capillary tube is measured. The viscosity is determined from the pressure drop along the length of a capillary or the pressure differential in comparison with flow through a reference capillary. The viscosity of the solution is calculated according to Poiseuille's law. In both cases, relatively small pressure changes are involved, requiring sophisticated transducers for their accurate determination. Systems that employ a single capillary require a pulse dampener to reduce the noise generated by the solvent pump. The molecular weight of the eluting polymer fraction is determined from the viscosity of the eluent, combined with the concentration of the polymer provided by another detector and the Mark-Houwink coefficients.

3. Light Scattering

Light scattering is the only primary method of determining the weight-average molecular weight of a polyethylene resin in the range of molecular weights applicable to most commercial resins. The distribution of intensities of scattered light as a function of angle is dependent upon the size, spatial separation, overlap, and concentration of polymer molecules in solution. As with viscometry, light scattering can be used either as a stand-alone technique or, more commonly, as a component of a size elution chromatographic system.

A very dilute solution of polyethylene with an accurately known concentration is placed in a cell at a temperature sufficient to maintain the polymer in solution. A beam of laser light is directed into the cell, and the scattering intensity at one or more angles is measured. The average molecular weight of the sample is calculated from the observed distribution of scattering intensities.

4. Membrane Osmometry

Membrane osmometry is a classic method for determining the absolute molecular weight of a solute in solution. It is based on the principle that when a compound in solution is separated from pure solvent by a semipermeable membrane (permeable to the solvent but not to the solute) the difference in chemical potential across the membrane generates a flow of solvent across the membrane. The pressure that must be applied to neutralize the flow is the osmotic pressure. Osmotic pressure is a function of the molar concentration and number-average molecular weight of the solute, increasing with the former and decreasing with the latter. Comparison of the calculated molar concentration of the solution with its weight concentration yields the molecular weight of the solute.

Membrane osmometry is rarely applied to commercial polyethylene resins because its accuracy is inversely related to molecular weight. The technique becomes unusable for molecular weights in excess of approximately 20,000. The lower extreme of the molecular weight determination range is set by the pore size of the semipermeable membrane. A further drawback is the relatively long time required to make a single molecular weight determination (up to 30 min). In the case of polyethylene, membrane osmometry is of real use only for determining the absolute number-average molecular weight of low molecular weight species extracted from commercial resins or in the analysis of ethylene-based waxes.

5. Vapor Pressure Osmometry

Vapor pressure osmometry (VPO) is another classic method of determining number-average molecular weights, based upon the colligative properties of a solute in solution. When a polymer solution is exposed to a saturated atmosphere of the pure solvent it will absorb solvent in an attempt to balance the chemical potentials of the solution and the solvent vapor. In the process, the temperature of the solution will rise as it absorbs the heat of condensation of the extra solvent. As the temperature of the solution rises its vapor pressure also increases until it matches that of the surrounding saturated solvent vapor (which is at a slightly lower temperature). By measuring the rise in temperature of a droplet of solution placed on a thermistor, the molar concentration of the solution can be calculated. Comparison of the molar concentration of the solution with its weight concentration yields the number-average molecular weight of the solute.

Although vapor pressure osmometry provides an absolute measure of the number-average molecular weight, it is rarely applied to commercial polyethylene resins because it is accurate only up to a molecular weight of approximately 25,000. Its uses are similar to those of membrane osmometry.

B. Spectroscopy

The role of spectroscopy in polyethylene resin characterization is to determine the types and concentrations of chemical species present in the material. The chemical species detected in a polyethylene resin may be directly attached to the polymer molecules or part of independent smaller molecules. Chemical moieties incorporated into the polymer molecules include olefinic branches, comonomers, unsaturation, and various oxidative products. Independent molecules that commonly occur include antioxidants, catalyst residues, and processing aids. Due to the enormous variety of chemical species that find their way into polyethylene, either deliberately or adventitiously, a detailed description of their qualitative and quantitative determination is beyond the scope of this book. Those wishing to explore this area more deeply are directed to the works of Haslam et al., Koenig, and Snyder listed in the bibliography.

Characterization and Testing

1. Infrared Spectroscopy

Infrared (IR) spectroscopy [invariably Fourier transform infrared spectroscopy (FTIR)] is used to determine a variety of molecular characteristics of polyethylene. It may be used to identify and quantify various additives and chemical groups attached to the polyethylene backbone. Additionally it can provide information with respect to solid-state morphology, which is addressed in Section IV.

To investigate the molecular nature of a polyethylene sample by infrared spectroscopy it is first necessary to reduce it to a state through which infrared radiation can pass. This generally involves compression molding a thin film of the sample at high temperature. Solvent casting is also feasible but is rarely practiced. The film is mounted in the spectrometer in the conventional manner, and a spectrum is recorded. Various specific absorbances may be measured with respect to the sample thickness or an internal calibration peak to identify the presence and determine the concentration of various species. Short-chain branches [6], methyl groups [7], unsaturation [8], and oxygenate species are among those commonly assayed. Many of the additives commonly used with polyethylene can be identified by their distinctive "fingerprints" found at characteristic wavelengths.

2. Nuclear Magnetic Resonance Spectroscopy

Nuclear magnetic resonance (NMR) spectroscopy is based on the fact that atoms with an odd number of protons in the nucleus can be induced to resonate in an applied alternating magnetic field. In the case of polyethylene, the hydrogen and carbon-13 (^{13}C) atoms respond to the applied field. The precise frequency (termed the chemical shift) at which a specific nucleus resonates is a function of its atomic type, its interaction with other resonant atoms in its immediate environment, and the extent of shielding by electron clouds. By analyzing the resonance frequencies, the types of chemical environment and the distribution of resonating atoms within them can be ascertained. The intensity of each NMR peak is directly proportional to the number of atoms that contribute to it. In practice this means that both qualitative and quantitative information regarding chemical species can be obtained. NMR spectroscopy of polyethylene can be carried out in the solid, molten, or solution state. It is in solution that the greatest detail regarding molecular structure is revealed.

a. Hydrogen Nuclear Magnetic Resonance. The spectral position of the peaks in hydrogen (proton) NMR spectra are characteristic of the environment of the atoms and thus can be used to qualitatively identify both chemical species directly attached to the polyethylene molecules and the presence of various additives. In addition, the intensity of each of the peaks is directly proportional to the number of atoms that contribute to the peak. This permits the quantitative determination of the concentration of the various species identified.

    b. Carbon-13 Nuclear Magnetic Resonance. In general, ^{13}C NMR provides more detailed information than proton NMR regarding molecular structure. This is primarily due to the greater range of chemical shifts experienced by ^{13}C atoms. ^{13}C NMR is frequently used for the determination of the type and concentration of side chains in branched polyethylene. In addition, it can provide a limited amount of information with respect to the distribution of branches.

    The ^{13}C atoms in the backbone of linear polyethylene and short olefinic branches of copolymers exhibit characteristic resonance frequencies. The relative abundance of the carbon atoms incorporated in the branches and backbone is quantitatively reflected in the intensities of their corresponding peaks and thus provides a method of determining the branch concentration. In addition, the presence of branches modifies the resonance frequencies of those carbon atoms in the backbone in the neighborhood of the branch. In practice this effect can be observed up to five backbone carbon atoms removed from the branch site. If two branches are separated by fewer than four ethylene units, the intervening carbon atoms will show characteristic frequency shifts. For a given copolymer incorporation ratio calculated from the relative intensities of the side-chain peaks, the statistical probability of randomly dispersed branches occurring in close proximity can be calculated. When the calculated probability is compared with the observed distribution, a measure of the randomness of copolymer insertion or branching can be obtained. This is of interest in calculating the distribution of branches due to "backbiting" in low density polyethylene and the effectiveness of various catalysts for the random incorporation of α-olefin comonomers in linear low density polyethylene. A limitation to this method is that olefinic branches containing more than six carbon atoms are indistinguishable. Side chains with six or less carbon atoms are referred to as short-chain branches (SCB), while those that are longer are termed long-chain branches (LCB).

    The collection of ^{13}C NMR spectra poses a number of experimental difficulties relative to proton NMR. The principal of these is the low abundance of ^{13}C atoms (1.1% of the total), which reduces the intensity of their signal to approximately 1/6000 of that of hydrogen. In practice this requires a more powerful (and more expensive) instrument for ^{13}C NMR spectroscopy and data collection times ranging from a few hours to several days per sample.

C. Composition Distribution Determination

Composition distribution (CD) refers to the distribution of concentrations of short-chain branches on the molecules of a polyethylene resin. It is analogous to molecular weight distribution, reflecting the range of branch concentrations found

Characterization and Testing

on the molecules of a sample. As would be expected, the overall level of branching and its distribution influence many sample properties, and thus their determination is a matter of some importance. Such measurements are most often made by temperature rising elution fractionation (TREF). Differential scanning calorimetry is also used to obtain a semiquantitative measure of composition distribution. Various alternative fractionation methods are also available and are used to a lesser extent.

1. Temperature Rising Elution Fractionation

Fractionation of linear low density and low density polyethylene resins on the basis of branch concentration is based on the fact that molecules with lower degrees of branching are preferentially incorporated into thicker crystallites than those with higher degrees of branching. The solubility of such crystallites is a function of their thickness, and thus fractionation can be effected if the crystallites can be systematically precipitated or dissolved as a function of their thickness. Preferential precipitation or dissolution of molecules can be achieved by changing the solvating power, pressure, or temperature of the solvent system. In practice the most effective separation is achieved by varying the temperature of the solvent rather than by changing its composition or pressure. The most widely used method for fractionation on the basis of the degree of branching is temperature rising elution fractionation, which involves both preferential precipitation and dissolution of the sample. The principles of this method are illustrated in Figure 3.

In temperature rising elution fractionation, a dilute solution of a branched polyethylene resin in an appropriate solvent, such as trichloroethylene at high temperature, is injected into a chromatographic column packed with an inert substrate such as sand or small glass beads. The temperature is then lowered at a controlled rate to precipitate the polymer onto the surface of the packing. The molecules with the largest separation between branch points crystallize first, in accordance with the principles outlined in Chapter 4. At successively lower temperatures molecules with higher degrees of branching crystallize and are laid down upon the previously precipitated layers. When crystallization is complete, the column packing is coated with a thin layer of precipitated polyethylene, such that there is a gradient of branching concentration decreasing from the outside inward. In many low density or conventional linear low density polyethylene samples, a small fraction comprising the most highly branched molecules may not crystallize at all, even at temperatures as low as 0°C. The polymer is then eluted from the column as a function of increasing solvent temperature. Uncrystallized molecules are eluted first, followed sequentially by molecules making up increasingly thicker crystallites. The polymer concentration in the eluent is measured, typically by an infrared detector, and recorded as a function of elution temperature.

(a)

Mixture of molecules in solution — Column wall — Inert column packing

(b)

Crystallizable molecules coat column packing

(c)

Separation on the basis of branching concentration

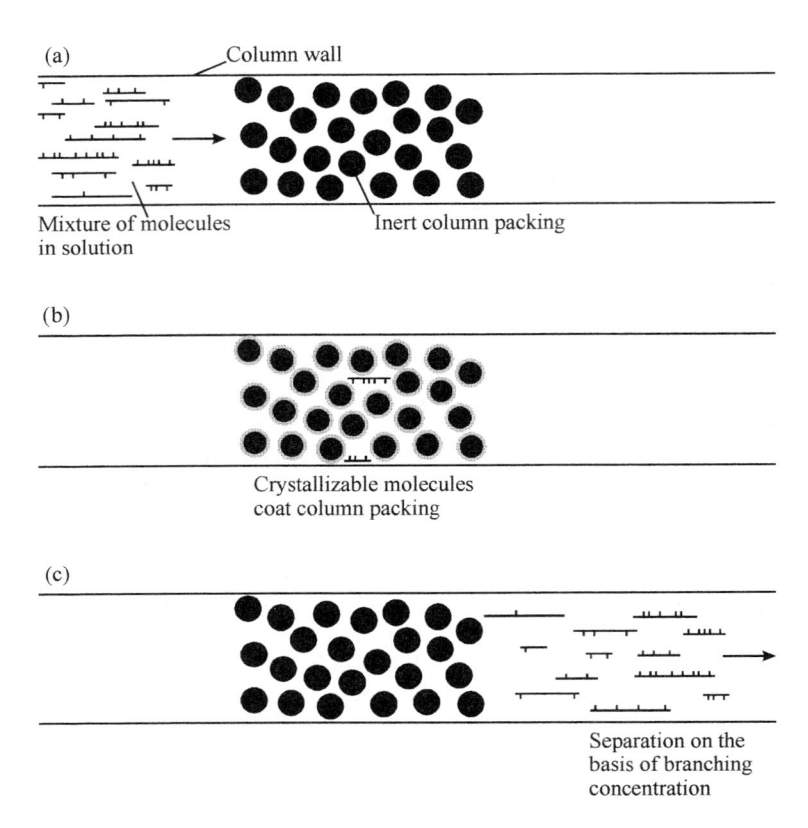

Figure 3 Schematic illustration of the principles of temperature rising elution fractionation. (a) Injection; (b) cooling/precipitation; (c) heating/elution.

The data from temperature rising elution fractionation are obtained as a distribution of polymer concentration as a function of elution temperature. These data can be converted into a composition distribution curve if the relationship between branch concentration and elution temperature is known. Such calibrations can be determined experimentally from well-defined narrow composition distribution samples, or they can be calculated theoretically [9]. A schematic representation of a composition distribution plot is shown in Figure 4.

2. Calorimetric Investigation of Composition Distribution

An approximate determination of the composition distribution of a polyethylene resin can be obtained by calorimetric investigation of its melting or dissolution

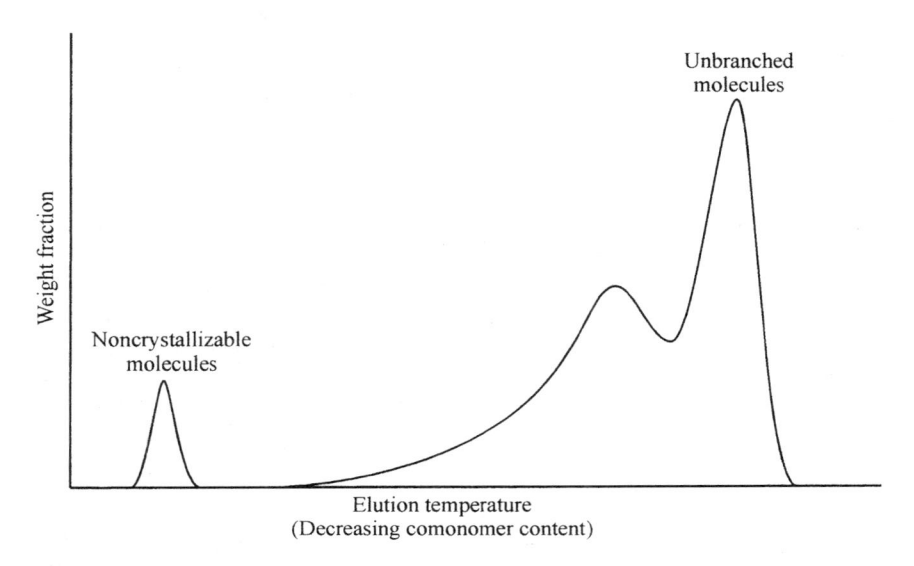

Figure 4 Schematic representation of comonomer distribution as a function of elution temperature, available from temperature rising elution fractionation.

characteristics. The melting or dissolution temperature of a crystallite is a function of its thickness. A plot of the heat required to melt the crystallites in a sample as a function of temperature reflects the distribution of lamellar thicknesses in the sample. Such data are available from differential scanning calorimetry (DSC), the principles of which are explained in a later section of this chapter.

When a branched polyethylene sample is slowly cooled from the melt, the distribution of lamellar thicknesses achieved is a function of the distribution of branch separations. By analyzing the melting characteristics of such a material, a qualitative understanding of the composition distribution can be obtained. For instance, conventional linear low density polyethylene resins typically show evidence of a bimodal distribution of melting temperatures as shown schematically in Figure 5. The relatively sharp peak at approximately 125–130°C corresponds to the melting of lightly branched molecules in thick lamellae, while the much broader, lower temperature peak corresponds to the melting of crystallites composed of more branched molecules. Several relationships have been proposed that purport to correlate crystallite melting temperature with lamellar thickness [10,11].

One of the drawbacks to investigating composition distribution by differential scanning calorimetry is its lack of resolution relative to temperature rising elution fractionation. This is caused by the hindrance to movement during crystal-

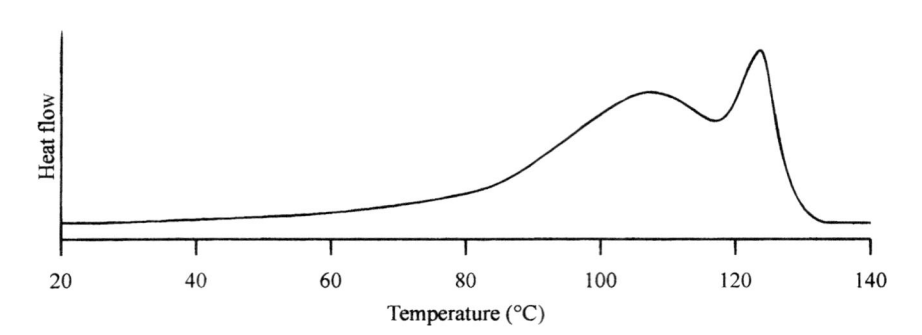

Figure 5 Schematic representation of differential scanning calorimetric trace of heat input as a function of temperature for a sample of conventional linear low density polyethylene.

lization imposed by the large number of entanglements present in the melt state. In solution, the concentration of entanglements is reduced and thus molecules are more likely to crystallize independently. A method of composition distribution analysis that takes advantage of this effect is solvated thermal analysis fractionation (STAF) [12]. In this method a sample of polyethylene is sealed in a differential scanning calorimeter sample pan with an excess of solvent. It is heated for sufficient time to dissolve the sample, then cooled slowly at a controlled rate. When reheated, the crystallites redissolve and a thermogram reflecting the composition is obtained.

3. Miscellaneous Fractionation Techniques

Fractionation of polyethylene has been accomplished by a wide variety of techniques other than size elution chromatography and temperature rising elution fractionation, with a multitude of variations thereon. The most notable of these less widely employed techniques include crystallization analysis fractionation [13], solvent gradient elution [14], successive solution fractionation (SSF) [15], continuous countercurrent extraction [16], high temperature thermal field-flow fractionation [17], supercritical fluid fractionation (SCF) [18,19], and high pressure Soxhlet extraction [20].

D. Cross-Fractionation to Determine Composition Drift

Most branched polyethylene resins are not monodisperse with respect to their degree of branching as a function of molecular weight. The distribution of branching concentration as a function of molecular weight is referred to as the

Characterization and Testing 257

composition drift. Typically, low density and conventional linear low density polyethylene resins have higher degrees of branching on their shorter molecules than on their longer ones. Analysis of composition drift requires cross-fractionation by a combination of molecular weight and compositional analysis methods. However one slices it, cross-fractionation is a time- and equipment-intensive technique.

In one variation of cross-fractionation, the output from a size elution chromatograph is sprayed onto a circular germanium wafer that is rotated slowly as the polymer elutes [21]. The solvent is flashed off at low pressure and high temperature to leave a thin layer of polymer in an arc on the wafer. The angular position of each element of the arc corresponds to a different molecular weight. The germanium wafer is then transferred to an infrared spectrometer in which it is rotated, and a series of spectra is recorded as a function of angular position. Thus the concentration of certain types of branching can be calculated at any given molecular weight. Overplotting the branch content information with that from the chromatograph reveals composition drift as a function of molecular weight. An example of this type of plot is shown in Figure 6.

In an alternative method the eluent from a temperature rising elution fractionator is collected as a series of fractions. These fractions are concentrated and subjected to size elution chromatography to determine their molecular weight

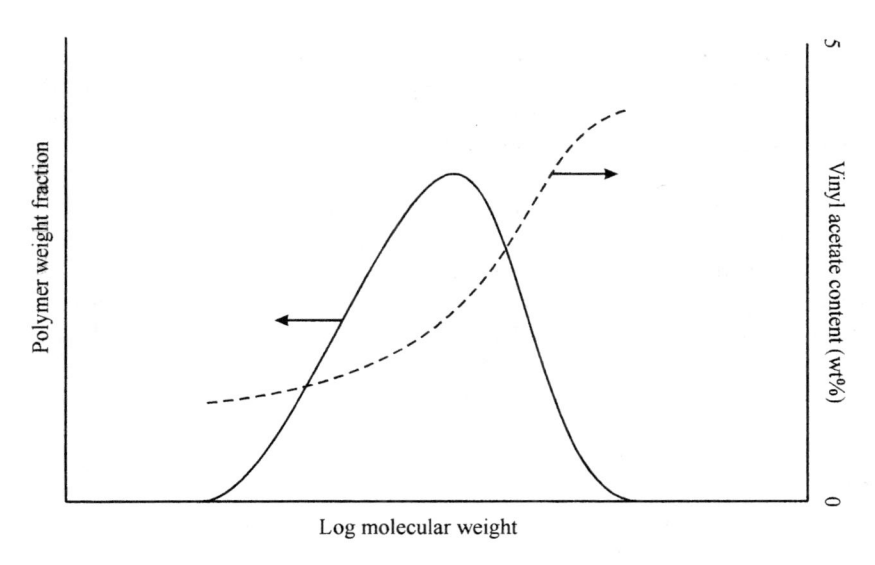

Figure 6 Schematic plot showing composition information overplotted with molecular weight distribution.

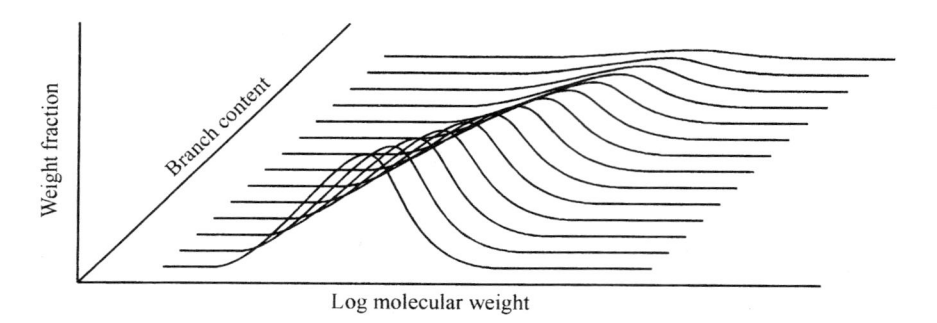

Figure 7 Schematic three-dimensional plot of molecular weight distribution as a function of branch content.

characteristics. The output from this analysis is available as a series of molecular weight distributions as a function of composition. These data can be presented in a contour plot or a three-dimensional plot as shown in Figure 7.

III. MELT RHEOLOGICAL CHARACTERIZATION

During its conversion from reactor product to its final form, a polyethylene resin inevitably passes through the molten phase at least once. While in the molten state it is subjected to various shear and elongation forces. It is the molten resin's reaction to these forces that determines the manner in which it can be processed and many of its final properties. (Conversely, the molecular characteristics of a resin must be tailored to a specific conversion process and the properties required of it.) Rarely is molten polyethylene allowed to reach thermodynamic equilibrium during commercial processing. Consequently the dynamic properties of molten polyethylene are far more important than the quiescent ones. This being the case, it is normal to characterize the properties of molten polyethylene in rheological terms—that is, under the influence of deforming forces.

The deformation of molten polyethylene may take place in either constrained or unconstrained systems. The former is exemplified by flow in a channel, such as that encountered in an extruder or an injection molding die. The latter involves elongation of the melt without the benefit of walls to constrain its shape, typified by film blowing or the drawing of an extrudate to form a fiber. Flow in a constrained system involves shear deformation whenever the polymer adheres to the walls. When adhesion to the wall fails, slip/stick or plug flow occurs, neither of which is desirable in the ordinary course of polyethylene processing. Elongational deformation is encountered when the cross section of the channel expands or converges. Flow in channels can involve both shear and elongational deformation, while that in unconstrained systems involves primarily elongational effects. Various analytical techniques are available to investigate both types of deformation. The various testing methodologies available are thoroughly reviewed by Whorlow (1980), Dealy (1982), and Collyer and Clegg (1988) (see bibliography), while a more comprehensive review of rheology as a science can be found in Dealy and Wissbrun (1990) and Ferry (1980).

A. Melt Flow Analysis

When the properties of a polyethylene resin are described, it is common to define the molten polymer's characteristics in terms of its melt flow rate under a standard set of conditions. Ideally the melt flow rate would be stated as a function of shear stress, but in practice a single value is normally all that is provided. The majority of analytical procedures involving the flow of polyethylene resins are carried out using instruments equipped with capillary dies. Many other types of rheometers exist, but their use in commercial situations is limited. The simplest type of capillary testing is melt indexing, which provides a single value characteristic of the resin. This method is a widely used technique suitable for routine analysis. Capillary rheometry is a more complicated technique that is highly versatile and can provide a detailed description of a resin's response to applied shear and some information with respect to its elongational properties. Rotational viscometry finds minor, but significant, use, generally at low shear stresses.

1. Melt Indexing

The melt index (MI)—also known as the melt flow index (MFI)—of a polyethylene resin is the weight in grams of polymer that extrudes from a standard capillary die under a fixed load, measured according to ASTM D 1238. The term "melt index" is limited to descriptions of polyethylene; "melt flow rate" (MFR) is the preferred term for all other polymers. The purpose of this measurement is to provide a value that reflects the ease of flow of a molten polymer. The melt index of a sample is primarily dependent upon its average molecular weight, but this relationship is strongly influenced by such factors as the molecular weight distribution and degree of long-chain branching. As the response of a molten polyethylene resin to applied shear depends on many molecular variables and the precise testing conditions used, the value of the melt index is of limited use in comparing resins. It is valid to make direct comparisons of melt index values of resins produced in identical reactors, using similar catalyst systems and polymerization conditions. In other cases relative melt indices should be used only as a preliminary guide when comparing the processing characteristics of different resins.

To determine the melt index of a polyethylene resin, a suitable mass of it (dependent upon the melt index) is charged into the barrel of a melt indexer (also known as an extrusion plastometer) preheated to 190°C. A weighted piston with a total mass of 2.16 kg is then placed atop the sample. A schematic illustration of the arrangement of a melt indexer is shown in Figure 8. The sample is allowed to preheat and melt for 6–8 min. During the time allowed for melting, a small quantity of polymer extrudes from the capillary die, of length 8.0 mm and diameter 2.0955 mm, that terminates the barrel. At the conclusion of the melting period, the extrudate is sliced off flush with the bottom of the die and a timer is started. The polymer is allowed to extrude for a preset period of time, after which it is severed and weighed. The melt index is the weight, in grams, of the extrudate multiplied by 10 divided by the extrusion time in minutes:

$$MI = \frac{\text{mass (g)} \times 10}{\text{time (min)}}$$

This corresponds to the weight of the polymer that would have extruded in a ten minute period. The precise methodology for determining the melt index is described in ASTM method.

The determination of the melt index is very sensitive to a number of factors that reduce its precision. These factors include temperature control; wear on the die, barrel, and piston; and operator inconsistencies. Data from ASTM D 1238 indicate a coefficient of variation within the same laboratory of 1.7–5.6%, with an interlaboratory coefficient of variation of 5–16%. In general, the coefficients of variation increase toward low and high values of melt index.

A crude measure of the shear sensitivity of a polyethylene resin can be obtained if it is extruded from a melt indexer under two different loads. It is conventional to make this determination using the standard load of 2.16 kg and one of 10.0 kg. The ratio of the mass of extrudate in 10 min at high load relative to that at low load is a dimensionless number known as the melt index ratio (MIR) or flow rate ratio (FRR). The higher the number, the greater is the sample's propensity to undergo shear thinning.

2. Capillary Rheometry

The purpose of capillary rheometry is to evaluate the rheological response of molten polyethylene resins to a wide range of shear rates. Primarily it is used to investigate the properties of melts under conditions of high shear akin to those found during processing. In particular, the relationship between melt viscosity and shear rate in ranges applicable to commercial molding processes can be obtained. Capillary rheometry is also useful in establishing the conditions under which melt instability (sharkskin, melt fracture, etc.) occurs. This information is valuable in determining the relative processability of resins.

Characterization and Testing

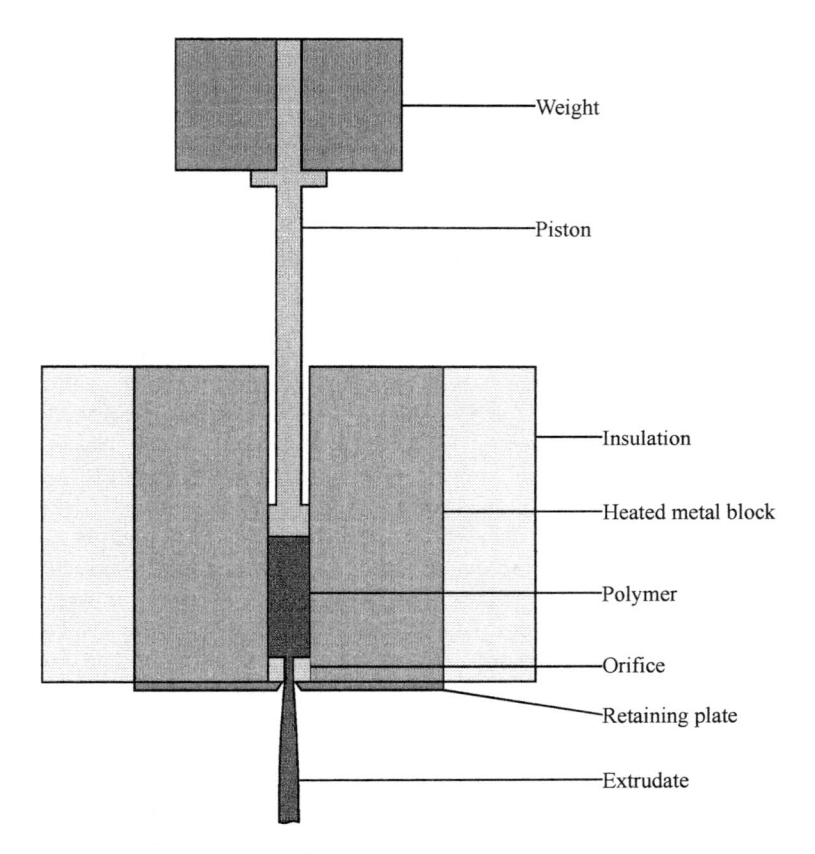

Figure 8 Schematic illustration of a melt indexer.

The basic geometry of a capillary rheometer is similar to that of a melt indexer. Molten polymer in a heated barrel is extruded through a capillary die under the influence of pressure exerted by a piston. Either the force applied to the piston or its rate of travel is controllable over a wide range, as are the temperature of the polymer and the dimensions of the capillary. The relationship between the force applied to the piston and its rate of travel reflects the response of the melt viscosity to the applied shear stress. In modern instruments, experimental control and final calculations are handled by computer. Additional refinements come in the form of interchangeable capillary dies of various lengths and diameters, the use of pressure transducers, and precisely controlled piston rates. Slit flow rheometers are also available but are far less common than capillary rheometers. The theory and practical aspects of capillary flow are extensively covered in the works cited in the bibliography.

Capillary rheometers are popular with those sections of the polymer industry that are interested in melt processing. The type of flow encountered in a capillary die is similar to that found in many commercial processes, and the shear rates accessible cover a wide range that includes those encountered during extrusion and injection molding. Shear rates of up to several thousand reciprocal seconds can be generated.

There are many variations on the basic theme of the capillary rheometer. Piston movement can be effected by gravity, pneumatic or hydraulic pressure, or various types of mechanical drives. Gravity-driven rheometers are generally limited to lower shear rates (higher shear rates would require the operator to handle unduly large weights). The greatest flexibility of experimental conditions is offered by hydraulic or servomechanical systems. Digital computer control permits either a constant driving force or constant speed to be applied to the piston. In practice the driving force on the piston is measured directly and the pressure in the barrel is calculated assuming negligible friction between the piston and the walls. The pressure drop along the capillary is calculated based on the assumption that the extrudate exits the die at ambient pressure. The melt is assumed to be incompressible, and the output rate is calculated from the velocity of the piston. To account for departures from ideal flow, corrections must be applied to allow for deviations from theory. The most prominent experimental factors that must be taken into account are convergent flow at the entrance to the die and pressure deviations approaching its exit. These effects can be determined if the experiment is duplicated using capillaries of identical diameters but different lengths. Deviations from theory also arise due to frictional (viscous flow) heating and pressure effects. High shear rates can give rise to nonuniform distributions of temperature within the die. Viscosity varies as a function of pressure; at high shear rates, this effect cannot be ignored, and corrections must be made to allow for it. In modern instruments all the calculations required by melt rheometry are handled by the computer that controls the experiment. Although capillary rheometry is primarily a high shear technique, the range of shear rates overlaps that available from rotational viscometers that operate at low shear. With appropriate selection of analysis software, the region of overlap can produce matching data, permitting a complete picture of viscosity as a function of a wide range of shear rates to be developed.

Capillary rheometry requires a fairly large sample for a complete analysis to be conducted. It would not be unusual to use several hundred grams of polymer during a thorough investigation. A potential problem in capillary rheometers is the degradation of the melt within the barrel during the course of the experiment. In the case of polyethylene, degradation can be largely avoided by the use of appropriate stabilizers and the exclusion of oxygen.

3. Drag Flow Rheometry

The measurement of the viscous response of molten polymer constrained between two surfaces moving relative to each other is used as a complementary method to capillary rheometry. Instruments of this general class are known as drag flow rheometers. For a number of practical reasons most of the instruments of this class involve the rotation of one surface in proximity to a stationary one.

A wide variety of rotational rheometers are available based upon a number of different testing geometries. Most of these are of little or no value in the study of polyethylene because the range of shear stresses attainable is too limited or they cannot handle the relatively high melt viscosity and elasticity of typical polyethylene resins. Two configurations are used in the study of the properties of molten polyethylene: the cone-and-plate and parallel plate geometries. The parallel plate geometry is also referred to as torsional flow. A schematic illustration of the cone and plate geometry is shown in Figure 9. The parallel plate geometry is similar with the exception that the cone is replaced by a plate. In these rheometers one surface is rotated at a known speed while the torque exerted on the other plate via the molten polymer is measured.

When unidirectional rotation is applied, such rheometers may be used to determine linear viscoelastic relationships at low shear rates, generally of less than 10 sec^{-1}, in which the viscous flow is Newtonian. The range of shear stress for which useful data are attainable is limited by the onset of flow instabilities at the polymer/air interface. Much higher shear rates can be accessed when an oscillatory motion is applied; this use is addressed in Section III.D. A distinct advantage of rotational rheometers over those that involve flow is the relatively small amount of polymer needed; often as little as one gram is sufficient. Such rheometers are useful when determining the long-term stability of resins, as the same sample can remain within the rheometer indefinitely. An operational advantage of the cone-and-plate and parallel plate configurations over most other rotational instruments is the ease with which they may be cleaned, as cleaning is a problem frequently encountered when testing molten polyethylene.

B. Melt Elongation Analysis

The analysis of polymer melts in an unconstrained system may be carried out in either a rigorous or a semiempirical manner. The former requires that the sample be deformed in a strictly uniform manner that requires specialized equipment and a high degree of devotion on the part of the investigator if it is to be carried out successfully. In practice, most laboratories that show an interest in the elongational properties of polymer melts rely on commercial instrumentation that is

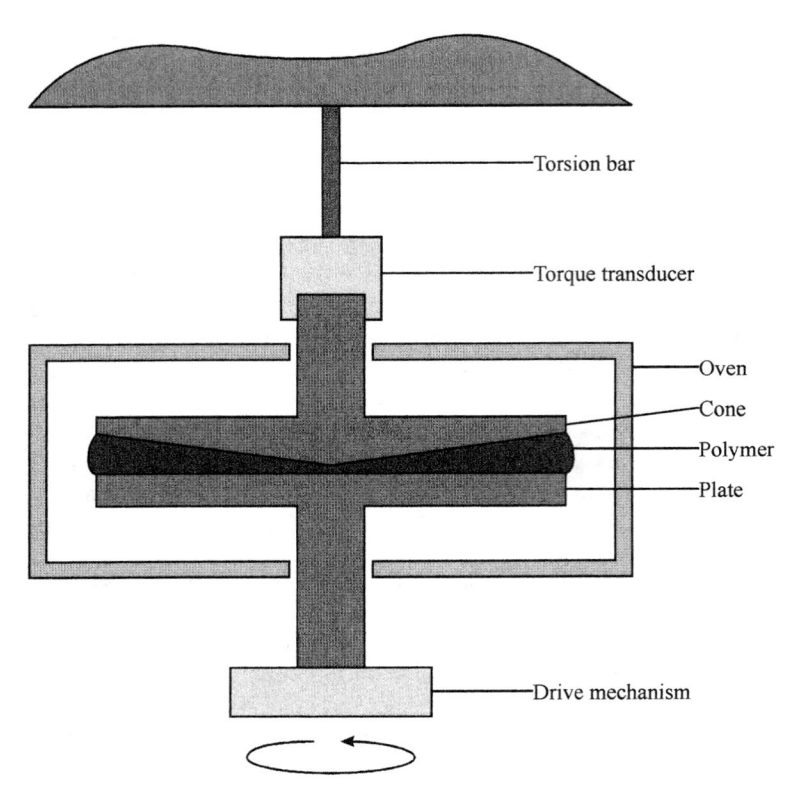

Figure 9 Schematic illustration of the geometry of a cone-and-plate rotational rheometer.

relatively simple and robust and permits the evaluation of less well defined criteria, such as melt strength or elongation at break.

1. Uniform Extensional Flow Analysis

The rigorous evaluation of the elongational properties of polymer melts requires that they be drawn in a highly controlled manner in such a way that the flow is uniform within the sample. This requires the use of extensiometers that are carefully designed and operated under conditions as nearly ideal as possible. Some of the problems that must be overcome are gripping the sample without introducing nonuniform deformation, supporting the sample in a nonintrusive manner, and maintaining a uniform sample cross section. A commercial instrument based on the design proposed by Meissner and Hostettler [22] is available. The general

Characterization and Testing

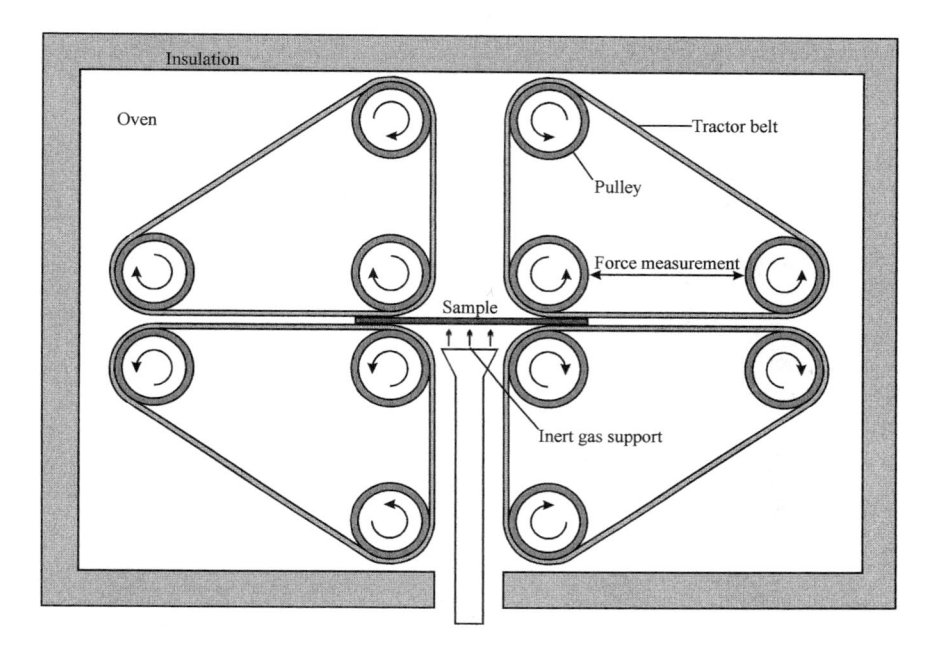

Figure 10 Schematic illustration of the geometry of Meissner and Hostettler's extensional rheometer.

configuration of this instrument is illustrated in Figure 10. The speed of the pulleys is programmed to maintain a constant strain rate in the sample. Stress is gauged from the force that must be exerted by the pulleys to maintain the required strain rate. As a general rule one would be unlikely to encounter such equipment in typical industrial laboratories. Some of the various techniques used for such analyses are described by Dealy (1982) (see bibliography).

2. Melt Strength Determination

From a practical standpoint it is often desirable to have some idea of the strength of a molten polymer and the degree to which it can be stretched as it exits a die. Such information is particularly useful with respect to the evaluation of resins used for film blowing and fiber spinning. The basic equipment used for such tests is relatively simple, consisting of a capillary rheometer to which is attached a filament windup device and some method for measuring the tension on the filament. The general configuration is shown in Figure 11. More sophisticated arrangements exist in which the extrudate is elongated within an environmental chamber that controls its temperature.

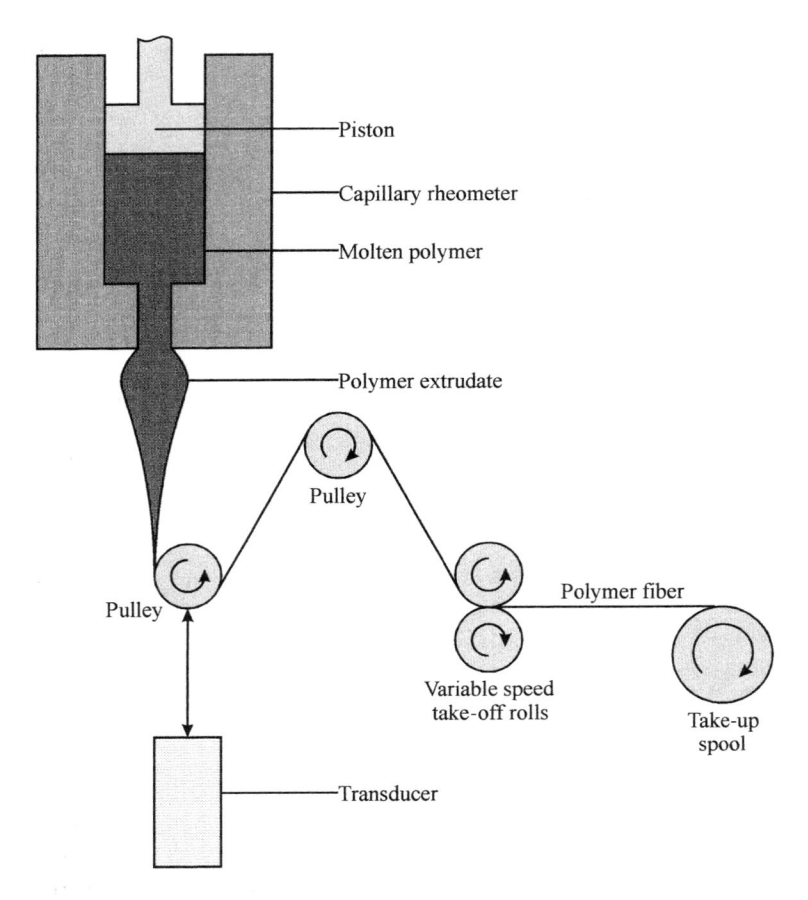

Figure 11 Schematic illustration showing the basic configuration for melt strength determination.

The maximum observed stress that can be applied to the fiber before it breaks is termed the "melt strength." The value of melt strength so measured is not an intrinsic property of the resin; it depends upon a large number of interrelated molecular and processing variables. As such, values of melt strength are used on a relative basis for the comparison of different resins under the same conditions or a single resin under a range of conditions.

In addition to obtaining the value of melt strength by this method, it is also possible to determine the extent to which the melt can be elongated. This involves measuring the diameter of the fiber after it has been wound onto the take-up spool.

3. Converging Flow in Melt Rheometers

Under certain circumstances, uniaxial elongational deformation can be approximated by convergent melt flow in the barrel of a capillary rheometer prior to the entrance of a die with zero length (i.e., a knife-edge hole). In this case the melt undergoes elongation parallel with the streamlines. Such converging flow does not result in uniform stretching at constant rate, so it is not strictly equivalent to elongational flow. However, with the use of appropriate analytical calculations, the elongational contribution to the overall melt deformation can be extracted [23].

C. Die Swell Measurement

Die swell occurs to a greater or lesser extent whenever molten polyethylene exits from a die into an unconstrained environment. Such conditions arise principally during extrusion, resulting in an extrudate with a cross-sectional area greater than that of the die. As discussed in Chapter 5, the degree of die swell depends on a large number of interrelated molecular and processing variables. It follows that there is no intrinsic value of die swell for any given resin.

Although there is no intrinsic value of die swell, comparative values may be obtained for different resins extruded under identical conditions or for a single resin extruded under various conditions. Such values are empirical, being characteristic of the sample and the conditions under which the measurement was made.

For the sake of convenience, most die swell measurements are carried out using dies of circular cross section, thus obviating the problem of changes in cross-sectional shape. In practice, melt rheometers are often used for this measurement, but it is also feasible to use laboratory-scale extruders equipped with appropriate dies. Die swell may be evaluated either on-line or off-line. The former involves evaluating the diameter of the molten polymer directly as it leaves the die and for a short distance thereafter. Methods of on-line measurement include the use of video imagery and scanning laser beams. Off-line measurement requires that the extrudate be rapidly quenched to maintain its cross-sectional area, which is evaluated subsequently on the cooled specimen. In either case the ratio of the diameter of the extrudate to that of the capillary provides a dimensionless value characteristic of the resin and its processing conditions.

The main problem that must be overcome in the evaluation of die swell is that of extrudate sag, which stretches the molten polymer, reducing its cross-sectional area. This problem may be eliminated or reduced by extruding the polymer into an oil bath with a temperature and density matching that of the melt or by making measurements on short lengths of extrudate for which the extensional mass is negligible.

Values of die swell are useful in screening polyethylene resins for suitability for various extrusion processes. The degree of die swell is indicative of the orientation of the melt achieved within the die. Such orientation may be desirable, as in the case of spun fibers, or undesirable if it adversely affects the shape of an extrudate for which a precise cross section is required.

D. Dynamic Mechanical Analysis of Melts

Dynamic mechanical analysis is used to determine the response of a polyethylene sample to an oscillating force. In its most general form a sample is attached to a pair of movable probes, one of which applies a sinusoidal oscillatory motion while the other measures the force transmitted by the sample. The temperature of the sample and the frequency of oscillation (ω) can be varied independently. The sample may be in either its solid or molten state. In the case of molten polyethylene, the sample typically takes the form of a disk sandwiched between a metal drive plate and a torque transducer. Rotation of the drive plate induces shear deformation within the sample, which is measured by the transducer. The basic configuration of the apparatus is similar to that of the cone-and-plate rotational viscometer shown in Figure 9. With appropriate modifications the same equipment can be used for both types of analysis.

Due to the viscoelastic nature of molten polyethylene, the sinusoidal motion experienced by the transducer is neither in phase with nor of the same amplitude as that of the driven plate. The observed torque is measured as a sinusoidal trace that lags behind that of the driven plate by a constant phase angle (δ). The observed sinusoidal trace can be considered to be the sum of two constituent sine waves, one in phase with the applied force and one out of phase. The amplitude of the in-phase sine wave is a function of the shear storage modulus of the sample (G'), while that of the out-of-phase sine wave is a function of the shear loss modulus (G''). The storage modulus is proportional to the average energy stored in a deformation cycle, while the loss modulus is proportional to the energy per cycle dissipated as heat. The measured moduli are functions of the shear rate (frequency) and temperature at which they are measured. Accordingly, such values are normally observed and plotted as a function of the logarithm of the frequency (ω, in rad/sec) at a constant temperature or as a function of temperature at a constant frequency. The former is known as a frequency sweep, and the latter, a temperature sweep. Naturally, equivalent compliance values can also be calculated.

The viscous component of the sample's response can be treated in a similar manner to its elastic component; h' and h'' are, respectively, the viscosity in and out of phase with the applied shear. The in-phase and out-of-phase viscosities are related to the loss and storage moduli according to

$$h' = \frac{G''}{\omega} \quad \text{and} \quad h'' = \frac{G'}{\omega}$$

When plotted as a function of decreasing frequency, the value of the in-phase viscosity approaches that of the steady-state flow viscosity (h_0) at very low frequencies.

IV. SOLID-STATE CHARACTERIZATION

The complete characterization of the semicrystalline morphology of polyethylene samples is presently beyond our capabilities. Although it is possible to define the supermolecular structure of a sample, the trajectories of the individual molecules comprising it remain unknown. Solid-state characterization of polyethylene, as currently practiced, is aimed at examining semicrystalline morphological features having dimensions in excess of approximately 20 Å. Three types of techniques are used: those that directly image morphological features, those that investigate the distribution of environments of constituent atoms, and those that measure some average property characteristic of a sample's morphology. An exhaustive evaluation of supermolecular structure is a complex and time-consuming affair that is rarely undertaken; it is more common to evaluate only those characteristics that directly influence the material properties of interest.

A. Microscopy

Microscopy of various types is widely used to visualize the structural characteristics of polyethylene samples. Techniques have been developed to obtain information regarding many disparate morphological features, including crystallites, spherulites, domain structure, surface roughness, and elemental composition. The importance of obtaining visual images of polymer morphology cannot be overvalued. The pictorial review of polymer morphology by Woodward (1989) listed in the bibliography illustrates the wealth of information available from optical and electron microscopy of various types.

At the level of magnification attainable by using optical methods, it is possible to identify features on the scale of spherulites and larger. Electron microscopy (EM) is used to obtain higher degrees of magnification, in which case features as small as crystallites can be resolved. The field of atomic force microscopy (AFM) is gaining popularity as a rapid method of obtaining detailed topographic and hardness maps of polymer surfaces at a similar level of resolution to electron microscopy. The subject of infrared microscopic spectroscopy is addressed in Section IV.C. Other specialized microscopic techniques that find limited use in the evaluation of polyethylene morphology include scanning transmission electron microscopy (STEM) and scanning tunneling microscopy (STM) [24,25].

1. Optical Microscopy

Optical microscopy (OM) covers a range of magnifications from approximately 5× to approximately 1000×. It is commonly used to study thin specimens, such as films and fibers, or sections cut from thicker samples. It is also used to examine the texture of surfaces and the overall shape of granules obtained directly from reactors. Thin specimens are generally studied by transmitted light, while incident light is used primarily for surface analysis. Optical microscopic techniques often suffer the drawback of a limited depth of field. This limitation is especially noticeable at higher magnifications.

At the lowest level of magnification, optical microscopy is often used as a screening or troubleshooting tool to investigate such features as surface irregularities on extrudates or "gels" and "fisheyes" in films and fibers. The term "gel" is used generically to describe any of an indefinite number of visually observed inclusions that differ in chemical or morphological structure from the bulk of the sample. Examples of gels include cross-linked polyethylene, inclusions consisting of other polymers, inorganic particles, plant fibers, and insect parts [26]. The term "fisheye" normally refers to localized thickened regions in films or fibers.

Due to its general chemical inertness, polyethylene cannot be readily stained for optical microscopy. The most useful tool for enhancing contrast in polyethylene samples is the polarizing filter. Polarizers may be used on either or both sides of the sample, with their planes parallel to one another, perpendicular to each other, or at any angle in between. The use of polarizers may be qualitative to enhance contrast or quantitative to measure the birefringence of samples. The various layers in coextruded films invariably exhibit different levels of birefringence. This makes it possible to distinguish them by light microscopy when viewed in cross section and to determine their thicknesses to an accuracy of approximately 0.5 μm.

In cases where polyethylene samples contain heterogeneous elements that provide optical contrast, it is possible to develop three-dimensional images using confocal optical microscopy. Such contrast is particularly well developed in the case of carbonaceous water trees [27]. Confocal microscopy generates a series of images corresponding to sections taken through the sample at incrementally greater depths. Computer-aided manipulation of the data can yield pairs of stereoscopic images along various optical axes or a rotatable pseudo-three-dimensional image on a computer monitor.

The size and shape of spherulites may be observed directly in thin films with the aid of polarizers. When viewed with crossed polarizers, each spherulite exhibits a characteristic "Maltese cross" pattern, its arms parallel with and perpendicular to the alignment of the polarizers. Photomicrographs of spherulites viewed through crossed polarizers are shown in Chapter 4, Figures 41 and 42.

With the aid of a heated microscope stage the melting and crystallization characteristics of thin films can be observed. Under suitable circumstances, fluxing of spherulites can be witnessed directly as their Maltese cross patterns dissipate. On a more general basis, loss of birefringence indicates melting. One specific use of this technique is in the analysis of coextruded films. The melting temperatures of each of the layers in a thin cross section cut from a coextruded film can be determined, facilitating their identification. Melting characteristics can also be investigated by recording light transmission through a sample as a function of temperature. As a sample melts, it loses birefringence and the image between crossed polars becomes increasingly dark. During crystallization, the rate of radial growth of spherulites can be measured as a function of temperature and time. The behavior of gels under the influence of rising temperature can provide clues to their nature.

Surface analysis includes observing extrudate roughness associated with flow instability and viewing fracture surfaces. Stereomicroscopes are especially suited to this type of analysis. Incident lighting at a shallow angle serves to enhance topographic contrast.

Interference contrast microscopy is a specialized type of transmittance microscopy used to examine the texture of relatively flat surfaces. It relies on the interference of light transmitted through the specimen at points separated by a distance less than the resolution limit of the microscope. The light passing through microscopic raised areas or indentations is out of phase with that from the rest of the surface, revealing the edges of such features as darker or lighter lines. Such features may consist of "pools" of low molecular weight material that have exuded to the surface, blisters, localized sink marks due to uneven crystallization, or voids beneath the surface. Careful adjustment of the height of the microscope stage to bring the nonplanar features in and out of focus permits their height or depth to be estimated.

2. Electron Microscopy

Electron microscopy permits the examination of morphological features ranging in size from lamellae up to spherulites. In most cases it requires that specimens be specially prepared to enhance contrast or stabilize their surface. With appropriate sampling and preparation techniques, electron microscopy can be applied to virtually any polyethylene sample regardless of its physical form or molecular characteristics. Scanning electron microscopy is used to view the surface morphology of specimens, while transmission electron microscopy is used to examine the fine structure of ultrathin sections. Electron microscopy of polyethylene is generally limited to the investigation of specimens at ambient temperature. The techniques involved in electron microscopy require no small degree of skill. Incorrect conclusions can be drawn if the preparation of samples, the recording of photomicrographs, or the analysis of images is carried out ineptly. Works describing the theory and application of electron microscopy to polymers are listed in the bibliography [Glauert (1973); Goldstein and Yakovitz (1975); Grubb (1982); and Michler (1996)].

a. Scanning Electron Microscopy. Scanning electron microscopy (SEM) is used to examine polymer surfaces. These surfaces may be those of a fabricated part, the fracture surface of a broken piece, or a surface revealed by microtomy. Scanning electron microscopy yields images with a very large depth of field compared to that of optical microscopy at equivalent magnifications. Most scanning electron microscopes use relatively high voltages to accelerate the electrons that strike the sample. The net result is that the electrons hitting the surface are highly energetic and impart much of their energy to the sample in the form of heat. Due to the nonconducting nature of polyethylene, unprotected surfaces rapidly heat up, causing thermal degradation of features by melting and cracking (due to localized expansion). Samples may also undergo reactions with the incoming electrons that directly change the sample morphology. Such effects are generically known as beam damage. Additionally, the high flux of electrons striking nonconducting surfaces builds up static charge on the sample, which deflects incoming electrons from their path, thus reducing the quality of the image. The higher the magnification, the worse these problems become. These troubles may be reduced by coating the surface of the sample with a conductive layer such as a layer of gold or carbon. The conducting layer has the disadvantage that it tends to obscure fine details on the sample. There is inevitably a trade-off between loss of detail and protection of the sample. The thicker the layer, the greater the obscuration of fine detail; typical conductive coatings are 50–200 Å thick.

In recent years instruments have become available that use low accelerating voltages, intense electron sources, and very sensitive electron detectors [28]. This technique is known as low voltage scanning electron microscopy (LVSEM or low voltage SEM). The use of such instruments minimizes beam damage and surface charging. This permits the examination of specimens without the need for the obfuscating conductive coating. Due to the lower penetration power of electrons in low voltage scanning electron microscopy and the fact that the original surface is being examined, its resolution is better than that of conventional scanning electron microscopy [29,30]. Even when examined by low voltage scanning electron microscopy, polymer samples are not immune to beam damage at higher magnifications during which the electron beam is concentrated on a small area for a long period of time. An additional benefit of low voltage scanning electron microscopy is its potential for differentiating domains within a specimen in terms of their chemical structure, distinguishing between various types of polymers in incompatible polymer blends. Despite their high cost, low voltage scanning electron microscopes are becoming widely used. In the description that follows it may be assumed that the techniques described apply to conventional scanning electron microscopy unless specifically stated otherwise.

In scanning electron microscopy it is often necessary to treat the surface of a sample to develop textural contrast. This is especially true for the surfaces of molded samples or those that have been obtained by microtomy. Such treatments may be physical or chemical. An example of physical treatment is exposing the surface to solvents such as xylene or cyclohexane to dissolve domains of molecules that are noncrystalline. This treatment is useful for examining such materials as rubber-toughened blends—providing that the rubber is not cross-linked. The resulting surface may exhibit a range of structures from isolated cratering, if the soluble domains were spherical and well dispersed, to fibrous if the soluble domains were highly elongated and interconnected. Chemical etching involves preferential digestion of the noncrystalline zones at the surface to reveal the underlying spherulitic structure or lamellar organization. Suitable chemical etchants include permanganic acid [31,32] and chlorosulfonic acid [33]. Chemical etching has the drawback that it has the potential for generating artifacts that may be mistaken for morphological features. Etching can also be effected by using an ionized gas (plasma) beam [34] or a combination of ion beam and chemical etching [35].

A method used to avoid beam damage and obscuration of details is that of surface replication [36,37]. A soft polymer film with a flexible backing is pressed into intimate contact with the polyethylene surface of interest. It is then peeled away from the polyethylene to provide an impression of the original surface. A thin coating of a conductive material, such as carbon, is applied to the surface of the impression. The polymer is then dissolved to reveal a conducting replica of the original surface that may be examined without fear of sample charging or beam damage. (Shadowing of the replica with heavy atoms, such as platinum, permits them to be viewed by transmission electron microscopy.)

The degree of magnification available to conventional instruments is limited by the resistance of the sample to beam damage and the obscuration of detail by the conducting coating. Magnifications of up to approximately $50,000\times$ are possible under favorable conditions. With low voltage scanning electron microscopy, magnification powers of up to $100,000\times$ are achievable, permitting the resolution of topographic features as small as 25 Å.

An ancillary technique sometimes associated with scanning electron microscopy is electron probe microanalysis (EPMA). This technique analyzes the characteristic X-rays that are emitted when the sample is struck by the electron beam. It is divided into two methods: energy-dispersive spectroscopy (EDS) and wavelength-dispersive spectroscopy (WDS). These techniques measure the energy and wavelength, respectively, of the emitted X-rays. Energy-dispersive spectroscopy collects all energy levels of interest simultaneously, while wavelength-dispersive spectroscopy scans the wavelengths of interest sequentially. Energy-dispersive spectroscopy is in essence a highly focusable elemental analysis technique used primarily as a qualitative tool. Wavelength-dispersive spectroscopy, on the other hand, is more quantitative. Energy-dispersive spectroscopy has better spatial resolution but poorer sensitivity to low levels of heteroatoms than wavelength-dispersive spectroscopy. In the field of polyethylene characterization, energy-dispersive spectroscopy and wavelength-dispersive spectroscopy find their main application in the analysis of nonpolymeric inclusions such as catalyst residues, catalyst supports, fillers, and antioxidants.

Microtomy is commonly used to expose the interior of samples. Diamond knives or freshly cleaved glass knives are required for the preparation of suitably smooth, undeformed faces. One of the problems associated with sample microtomy of polyethylene is the potential for destroying the features of interest by smearing the relatively soft surface. This problem is especially prevalent in samples with low crystallinity. To reduce surface deformation to a manageable degree it is necessary to cut samples at temperatures well below their glass transition temperature, i.e., below $-120°C$.

Scanning electron microscopy can be used to investigate the spherulitic morphology of samples. When microtomed surfaces are etched to reveal their morphology, the overall outlines of spherulites are visible, as is the general orientation of lamellar bundles within them. The individual lamellae can be observed by scanning electron microscopy only under specialized conditions. Low voltage scanning electron microscopy is sufficiently sensitive to detect the presence of lamellae at the surface of molded samples; however, the contrast and detail are not as great as those attainable by transmission electron microscopy.

Fracture surfaces are often viewed by scanning electron microscopy. Such surfaces may be those of parts that failed in service or testing; alternatively, specimens may be fractured at sub-glass transition temperatures to facilitate morphological analysis. Cleavage of polyethylene samples at low temperatures invariably occurs across the equators of spherulites or between them in the regions of low crystallinity.

The nascent morphology of polyethylene granules made in a reactor can be investigated by scanning electron microscopy. Competent analysis of such images can yield information regarding catalyst activity and polymerization conditions. Such analyses may be combined with energy-dispersive spectroscopy to aid in the identification of the class of catalyst used.

b. Transmission Electron Microscopy. Transmission electron microscopy (TEM) is used to investigate the morphology of ultrathin sections of polyethylene cut from larger samples. Using this technique it is possible to distinguish features as small as 20 Å. Typical magnifications used for polyethylene are in the range of 10,000×–250,000×. To develop sufficient contrast between crystalline and noncrystalline regions, the sample must be stained. This is carried out on the basis of the preferential introduction of heavy atoms into the noncrystalline regions. Two techniques are widely used for this purpose: chlorosulfonation [38] and ruthenium tetroxide staining [39–41]. Staining is carried out on bulk samples from which sections are cut with a diamond knife. The thickness of the sections depends upon the degree of resolution required; the thinner the section, the greater the resolution. In practice, sections are typically less than 1000 Å thick, often in the range of 600–800 Å. Needless to say, it requires more skill to cut thinner samples than thicker ones. Staining with ruthenium tetroxide has the added benefit of hardening the sample, making it less subject to deformation during microtoming.

When stained polyethylene sections are subjected to the beam of a transmission electron microscope, the regions containing the heavy atoms (the noncrystalline regions) scatter electrons more effectively than those containing only carbon and hydrogen (the crystalline regions). The saturation of each portion of the image is a function of the scattering power of the corresponding part of the sample. The darker the region, the greater the concentration of heavy atoms. Thus the micrograph consists of light and dark regions corresponding to the crystalline and noncrystalline regions, respectively. In practice, lamellae that have their a or b axes aligned perpendicular (or nearly so) to the plane of the section show up as white lines bordered by dark gray lines in a matrix of lighter gray. Examples of such images are shown in Chapter 4, Figures 12–15.

As an alternative to chemical staining, lamellar morphology can be investigated by examining thin replicas of ion or chemically etched fracture or microtomed surfaces [36,42].

As may well be imagined, a single image obtained by transmission electron microscopy (being that of a small cross section) is rarely representative of the sample as a whole. Thus it is necessary to examine numerous images from each sample, cut from mutually perpendicular planes, to build up an overall impression of the morphology. Preferably the images studied will cover a range of magnifications to ensure that any detailed structures are seen within the context of the large-scale morphology. Using transmission electron microscopy it is possible to study the arrangement of individual lamellae in relation to their neighbors and also the overall arrangement of lamellar bundles within the environment of the spherulite.

In addition to the qualitative interpretation of transmission electron micrographs, it is possible to quantify the thickness of lamellae [43]. As only those lamellae that intersect the plane of the section at angles close to the normal direction ($\pm$ 15–20°) were observed, it is possible to directly measure their thicknesses from a micrograph if the magnification is known. To ensure that a representative distribution of lamellar thicknesses is determined, it is necessary to make the appropriate measurements on a large number of lamellae observed in various regions of the sample. Such analyses, when carefully performed, yield distributions of lamellar thickness that agree well with determinations made by Raman longitudinal acoustic mode analysis [44,45].

3. Atomic Force Microscopy

Atomic force microscopy is used to obtain topological and local hardness maps of the surfaces of polymer samples. Images are built up as a series of pixels by tapping the sample with a finely tuned cantilever tipped with a pointed crystal as it rasters across the surface. Suitable crystals include silicon and silicon carbide. The local height at the surface or the distance that the tip penetrates the sample can be recorded individually or in combination as a function of the position of the stylus. Local height measurements are converted to topographic maps. Penetration measurements are converted to local hardness to generate a "phase" map that distinguishes between noncrystalline (soft) regions and (hard) crystallites. Local height or hardness is converted to a brightness scale; conventionally, lighter pixels correspond to higher or harder regions of the sample. Atomic force microscopy is capable of resolving features from as small as 25 Å up to several tens of micrometers. An example of a phase image of the surface of a linear low density polyethylene film is shown in Figure 12.

Atomic force microscopy can be used to investigate the surfaces of films and molded parts or fracture surfaces. A key advantage of this method is that no surface preparation is required. However, samples must be handled with care to avoid contamination by fingerprints and other substances or abrasive damage by contact with extraneous objects. Atomic force microscopy can provide information similar to that available from low voltage scanning electron microscopy of surfaces, but it does so at a higher resolution and can provide phase images in addition to topographic detail. To investigate the interior of a sample microtomy is necessary. This must be performed very carefully if useful information is to be obtained. In principle, microtomed surfaces can be etched to reveal further detail; great care must be taken to avoid artifacts. Methods and equipment for applying atomic force microscopy to polymer surfaces are still being refined, but much useful information has already been generated. Currently it is primarily a research tool, but its low cost and ease of use compared to electron microscopy ensure that it will gain increasing acceptance.

B. Scattering Measurements

Scattering measurements are used to determine various functions reflecting the distribution of environments in which the constituent atoms and molecules of the sample find themselves. The scale of the features probed varies from that of

Characterization and Testing

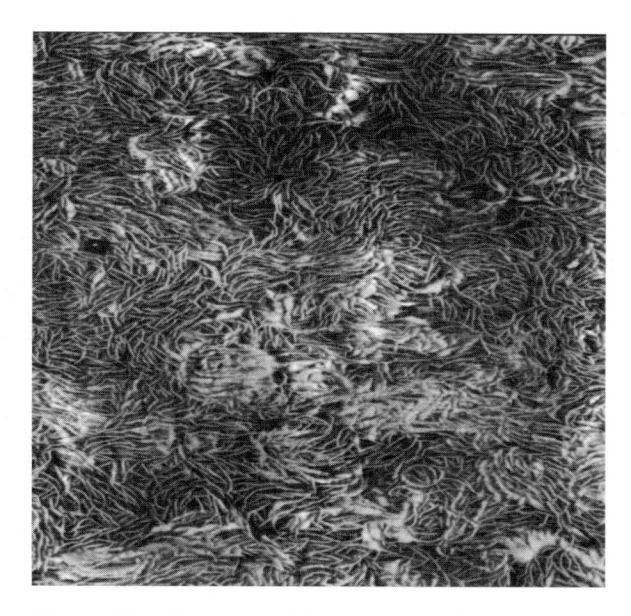

Figure 12 Phase map of the surface of a linear low density polyethylene film obtained by atomic force microscopy (courtesy of Exxon Chemical Company).

atomic spacings and unit cell dimensions through lamellar periodicity up to that of spherulitic radii. These are probed respectively by the techniques of wide-angle X-ray diffraction, small-angle X-ray and neutron diffraction, and small-angle light scattering. The distribution of scattering angles obtained by each technique is characteristic of the distribution of the periodicities of the features being probed, according to Bragg's law,

$$n\lambda = 2d \sin \theta$$

where

n = an integer
λ = wavelength of incident radiation
d = periodicity of scattering features
θ = scattering angle

All scattering experiments share the basic configuration illustrated in Figure 13. The output from a source of electromagnetic radiation or high energy particles is collimated to yield a beam of radiation that strikes a polyethylene sample, the intensity of the scattered radiation being recorded as a function of the scattering

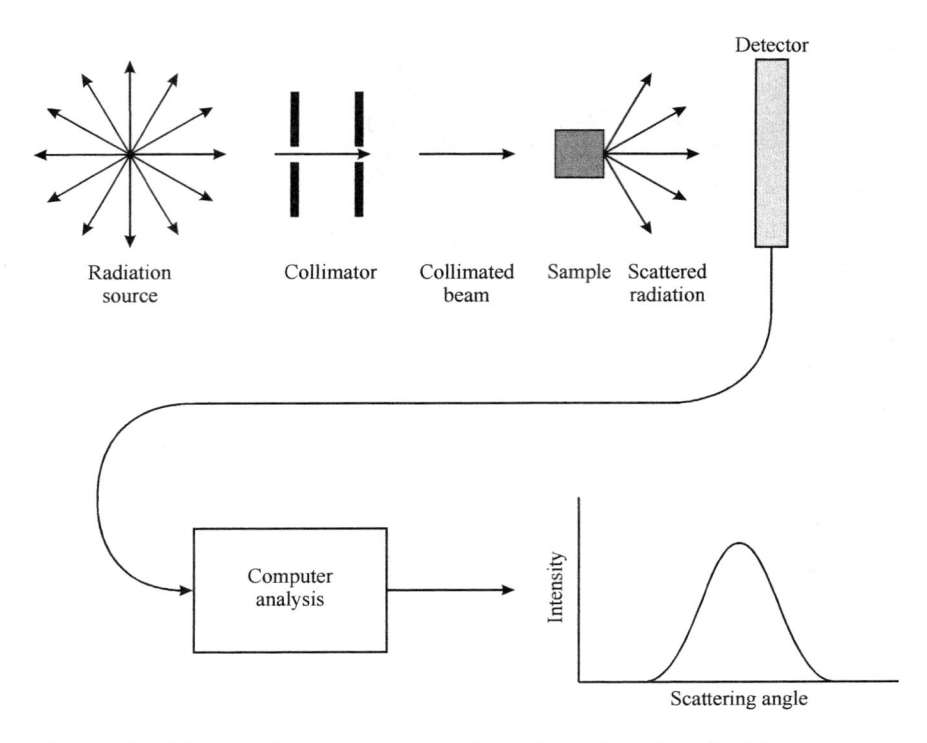

Figure 13 Schematic illustration showing the basic configuration of solid-state scattering experiments.

angle. Naturally, the nature of the radiation source, collimation system, and detector varies widely.

1. X-Ray Diffraction

X-ray diffraction (XRD) is one of the oldest techniques used to investigate the morphological character of polyethylene, dating back to the classic work of Bunn [46]. X-ray diffraction is a powerful technique that provides both qualitative and quantitative information with regard to many aspects of polyethylene morphology, including unit cell parameters, degrees of crystallinity, lamellar periodicity, and degrees of orientation.

a. Wide-Angle X-Ray Diffraction. Wide-angle X-ray diffraction (WAXRD), also known as wide-angle X-ray scattering (WAXS), refers to the scattering of X-rays over a range of angles (θ) from about 2° to 180°. When discus-

Characterization and Testing

sing X-ray diffraction it is customary to refer to the scattering angle in terms of 2θ. The precise range of scattering angles depends on the wavelength of the incident X-rays and the separation of the scattering planes. In the case of polyethylene irradiated with nickel-filtered copper K_α X-rays (a commonly used X-ray source), the most useful information is available in the range of 2θ angles from 5° to 50°. This corresponds to a range of atomic spacings on the order of 2 to 20 Å. This range spans the major planes in the crystalline unit cell and the intermolecular separations in the noncrystalline regions. The well-regimented arrays of atoms in the crystalline regions scatter the X-rays at discrete angles, whereas the noncrystalline regions give rise to diffuse scattering over a broad range of 2θ angles covering about 20°. A typical wide-angle X-ray diffraction pattern for an isotropic polyethylene sample of moderate crystallinity is shown in Figure 14. In addition to atomic spacings, the quantitative distribution of atoms between the crystalline and noncrystalline regions, i.e., the degree of crystallinity, can also be determined (details are given in Section IV.E.)

Variations on the basic configuration required for X-ray diffraction measurements abound. Scattered radiation may be collected in either transmission or reflection. The detector can take the form of a scintillation counter mounted on a goniometer, which swings in an arc, measuring the intensity of diffraction over a range of angles sequentially, or a detector that records the scattering intensity over a range of angles simultaneously. Such detectors may be photographic film or electronic position-sensitive devices (PSD) such as diode arrays and charge-coupled devices.

Analysis of the well-defined peaks due to scattering from the crystalline regions of polyethylene can yield precise information regarding the unit cell di-

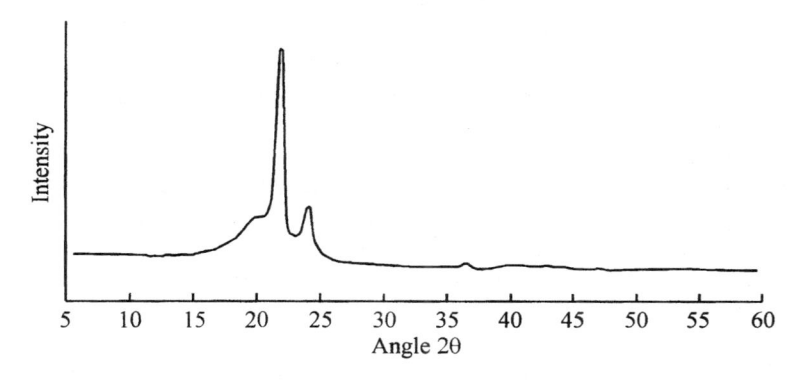

Figure 14 Plot of wide-angle X-ray scattering intensity as a function of 2θ angle for an isotropic linear low density polyethylene sample irradiated with copper K_α radiation.

mensions of the sample. It is possible to determine unit cell dimensions to an accuracy of four significant figures. Small changes in the dimensions of the unit cell can be linked to changes in crystallization conditions and the effect of comonomers, branches, and molecular weight. The diffuse band due to noncrystalline scatter can also be analyzed to yield information about the average density of the noncrystalline regions.

The width of the crystalline wide-angle X-ray diffraction peaks is a function of the dimensions and lattice perfection of the crystallites that give rise to the scattering. The larger and more regular the crystallites, the sharper will be the peaks. The intensity distribution of wide-angle X-ray diffraction peaks may be analyzed to yield quantitative measurements of crystallite thickness and various types of lattice disorder. Such methods can be applied to both isotropic and anisotropic samples [47].

Preferential crystallite alignment in oriented samples gives rise to anisotropic diffraction patterns. The diffraction pattern produced on a two-dimensional detector by X-rays scattered by an oriented film or fiber can indicate the degree and perfection of unit cell orientation within the sample. Isotropic specimens yield a series of concentric crystallite diffraction rings, whereas oriented specimens yield segments of rings, the length of each arc being inversely related to the degree of orientation. A more sophisticated technique for characterizing crystallite orientation in films is pole figure analysis (PFA). In this technique the angle between a scintillation counter and the incident radiation remains constant at a scattering angle corresponding to a major crystalline diffraction peak while the sample is rotated around two perpendicular axes. The measured intensity is plotted as a series of contours on a pole figure diagram that reflects the distribution of orientations of the unit cells in three dimensions.

The dimensions of the specimen required for wide angle X-ray diffraction depend on the information desired and the configuration of the equipment to be used. Reflectance measurements require a relatively large flat sample with lateral dimensions of more than approximately 1 cm and ideally a thickness in excess of 0.25 mm. Thinner samples may be used, but longer collection times are required. Thin films may be stacked closely upon one another to mimic a thicker specimen. Transmission experiments require a sample with lateral dimensions sufficient to block the entire cross section of the collimated beam. Sample thicknesses may vary from less than 0.05 mm to approximately 1 mm. Thick samples produce a greater scattering intensity (requiring shorter collection times), but resolution is lost due to secondary scattering of the diffracted radiation and imprecision in determining the sample-to-detector distance. Film samples for both transmission and reflectance should be of uniform thickness. Pole figure analysis requires a thin film or sheet with a diameter of approximately 20–40 mm. Fiber and yarn samples may be mounted individually in the path of a collimated beam or wound around a support. Powder samples may be encapsulated in a thin-walled

Characterization and Testing 281

glass capillary tube irradiated by a narrow beam of radiation to produce both reflectance and transmission scattering patterns.

The time required to collect wide-angle X-ray diffraction patterns varies greatly with the thickness of the sample, the intensity of the radiation, and the sensitivity of the detector. Position-sensitive detectors may provide sufficient information within a few seconds, while photographic film may require an exposure of many hours (or even days) to record the scattering pattern from a thin film or fiber. Scintillation counters typically scan at speeds on the order of 0.5–5°/min. When highly intense sources such as synchrotron radiation are combined with two-dimensional electronic detectors, data can be collected sufficiently fast that it is possible to obtain information regarding X-ray diffraction during real-time crystallization.

The presence of fillers may result in spurious sharp crystalline diffraction peaks (which may be used for identification purposes) or broad amorphous peaks that skew the diffuse polymer noncrystalline band.

b. Small-Angle X-Ray Diffraction. Small-angle X-ray diffraction (SAXRD), also known as small-angle X-ray scattering (SAXS), is used to investigate the periodicity of lamellar stacks within samples. The range of angles in which peaks occur depends upon the wavelength of the incident radiation and the periodicity of the lamellae. Typical 2θ angles for copper K_α radiation range from approximately 0.2° to 2°, corresponding to spacings of approximately 440 to 44 Å. Naturally, sophisticated equipment is required to accurately measure minute angular separations at such small diffraction angles. Either slit or pinhole collimation can be used to produce the extremely tight beam that must be employed. High power sources such as synchrotron lines typically use pinhole collimation, which obviates the problem of "smearing" that occurs with slit sources. The diffracted radiation pattern may be recorded on photographic film, for subsequent analysis with a scanning microdensitometer, or by an electronic position-sensitive detector. In both cases, mathematical manipulation of the data is required to remove background scatter and slit effects (if necessary) and to calculate the distribution of lamellar periodicities [48,49]. Such manipulation is routinely carried out by computer.

The lamellar periodicity calculated from small-angle X-ray diffraction can be used to estimate the average lamellar thickness if the degree of crystallinity of the sample is known. If one assumes that the sample is composed entirely of lamellar stacks, the approximate lamellar thickness is yielded by the product of the lamellar periodicity and the fractional crystallinity. Naturally, such estimates are somewhat crude, but they may be of use if it is not possible to obtain access to more accurate techniques such as Raman spectroscopy or electron microscopy.

Small-angle X-ray diffraction pole figures can be collected by a method analogous to conventional wide-angle X-ray diffraction pole figures [50].

2. Small-Angle Neutron Scattering

Small-angle neutron scattering (SANS) is a highly specialized technique used to investigate the average trajectory and distribution of molecular chains within a solid or molten sample. The material contrast required to scatter the neutron beam is provided by doping a small portion of the molecules of the sample with deuterium. Small-angle neutron scattering can be used to evaluate the mean-square radius of gyration of chains [51], to follow the changes of molecular profile during deformation [52], and to determine the degree of clustering of molecular stems in crystallites [53], i.e., the degree of adjacent reentry that occurs during the crystallization process. Another major application is the investigation of molecular segregation during the crystallization of blends.

The application of small-angle neutron scattering to the study of polymer morphology and crystallization is subject to numerous technical difficulties. One of the main drawbacks is the necessity that a portion of the sample be deuterated. To accomplish this it is necessary that deuterated species be prepared and blended into a protonated sample. When investigating a homogeneous sample, accurate experimentation requires that the deuterated molecules have the same molecular weight distribution as the protonated species. The preparation of well-matched samples can take place only on a laboratory scale (in part due to the high cost of deuterated compounds), which limits the quantities available for experimentation, effectively precluding the investigation of commercial conversion processes. Deuterium, having a higher molecular weight than hydrogen, introduces isotope effects into the sample even if the distributions of backbone lengths of the protonated and deuterated samples match perfectly. To reduce segregation based on crystallization temperature differences due to the isotope effect, it is common to quench samples rapidly from the molten state. This limits the types of crystallization experiments that can be performed. From an equipment standpoint, the production of neutrons suitable for small-angle neutron scattering requires a nuclear reactor, thereby eliminating its use as an analytical technique in commercial laboratories. The interpretation of small-angle neutron scattering data is not without ambiguity; the application of mathematical treatments based on different morphological assumptions and different ranges of scattering angles may result in very different interpretations being placed on the same data. Taken all in all, it is hardly surprising that this potentially powerful technique is not widely used.

3. Small-Angle Light Scattering

Small-angle (laser) light scattering (SALS or SALLS), also known as low-angle light scattering (LALS or LALLS), is used to investigate the supermolecular organization of lamellae within solid polyethylene samples. In its most common application it is used to identify the presence of spherulites and to determine their distribution of sizes and lamellar organization. Small-angle light scattering can also be used to follow the deformation of thin films and the growth of spherulites.

The basic equipment required to collect small-angle light scattering patterns is relatively simple, consisting of a source of polarized light (normally a low power solid-state laser), a stage on which to hold the sample, a rotatable polarizer, and a camera to record the scattering pattern. A schematic representation of the basic experimental configuration is shown in Figure 15. In most cases the plane of polarization of the incident light and that of the polarizer are perpendicular to each other, but other orientations are possible. The original description of this technique dates back to 1960 [54]; since then the equipment used has undergone relatively minor refinements. For more sophisticated experiments the sample can be supported on a hot stage (such as that used for optical microscopy), a video camera can be used to record patterns as a function of time or deformation, and a computer can be used to analyze the data.

The classic small-angle light scattering pattern obtained from spherulitic polyethylene films consists of a bright central point (due to unscattered light) surrounded by four distinct lobes in a "four-leaf clover" pattern. This type of scattering pattern is obtained from samples containing well-developed spherulites when the orientation of the polarizer and the incident light are perpendicular. An example of this pattern is shown in Chapter 4, Figure 43. The more highly developed the spherulites in a sample, the better defined will be the lobes of the pattern it produces. Randomly arranged lamellae give rise to circularly symmetric scattering patterns, while certain rodlike or sheaflike arrangements yield patterns with a fourfold symmetry with intense scattering in the vicinity of the center of the pattern [55].

The recording of small-angle light scattering patterns requires that the sample be in the form of a thin film with a thickness of the order of 25–250 µm. If

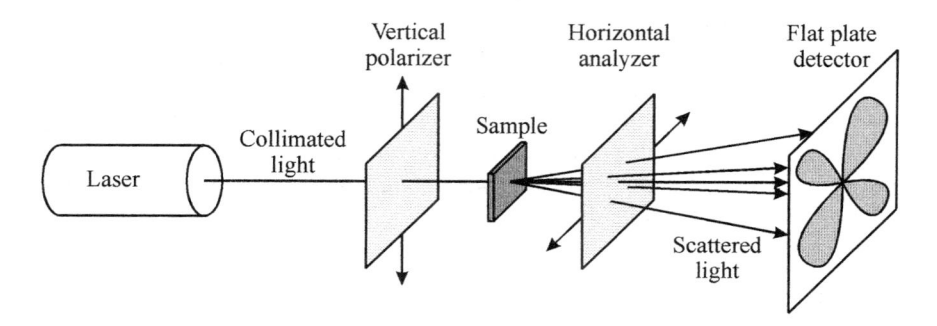

Figure 15 Schematic illustration showing the basic configuration required for small-angle light scattering experiments.

the supermolecular organization of a thick sample is to be examined, it is possible to shave off specimens of a suitable thickness using a razor blade or microtome. To reduce light scattering from uneven surfaces, specimens are sandwiched between a microscope slide and a cover slip with a drop of immersion oil to ensure good contact with the glass surfaces.

The azimuthal angle of light scattering at an angle of 45° to the planes of polarization of the laser beam and the polarizer depends upon the radius of the spherulites, according to the equation

$$R = \frac{\lambda}{\pi \sin (\theta/2)}$$

where

R = radius of spherulite
λ = wavelength of incident light
θ = azimuthal scattering angle

The most intense scattering in the lobe corresponds to the average size of the spherulites in the sample. The distribution of spherulitic radii may be obtained from the distribution of azimuthal scattering angles [56]. The distribution of spherulite sizes in isotropic samples is a function of molecular weight, branching, and crystallization conditions [57]. With the aid of a hot stage and a video camera the growth of spherulites can be monitored.

The change in the shape of spherulites during deformation can be followed by small-angle light scattering. During initial distortion of thin films, the familiar four-leaf clover pattern flattens out as the orientation of the lamellae within the spherulites changes. The extent of flattening reflects the degree of rotation of the individual lamellae. The pattern breaks down as the lamellae disintegrate during the yielding process.

Optical microscopy can be used as a complementary technique to small-angle light scattering for the determination of spherulite dimensions. However, in optical microscopy, when the diameter of the spherulites is substantially less than the thickness of the film it is difficult to distinguish between spherulites that overlap within the plane of the specimen. On the other hand, if a very thin section cut from a thicker sample is examined by optical microscopy, a relatively small proportion of the spherulites observed will contain the equatorial plane representative of the true spherulite dimensions.

C. Solid-State Spectroscopy

Solid-state spectroscopic techniques are used to investigate the vibrations of atoms and molecular segments within a sample and hence yield information re-

Characterization and Testing

garding its morphology. In polyethylene such information primarily relates to degrees of freedom of motion and molecular orientation. Quantitative estimates of degrees of crystallinity, orientation, and lamellar thickness can be obtained. The principal methods used to characterize solid-state polyethylene are nuclear magnetic resonance and vibrational spectroscopy, the latter being separated into infrared and Raman spectroscopy. Electron spin resonance (ESR) spectroscopy can also be applied to polyethylene, where its primary use is to elucidate the nature of additives rather than the overall morphology of the sample.

1. Vibrational Spectroscopy

Solid-state vibrational spectroscopy is divided into the techniques of infrared and Raman spectroscopy. Both techniques are sensitive to the vibrational activity of atoms with respect to their neighbors. These vibrations involve changes in bond length and angle, the characteristic frequencies of which can be measured by spectroscopy. The frequencies of such vibrations are controlled by the type of atoms involved, bond strengths, type of vibration (stretching, bending, twisting, etc.), and interactions with neighboring chemical species, both inter- and intramolecular. Infrared and Raman spectroscopy are complementary; infrared spectroscopy probes symmetrical vibrations that involve no change in the sign of the dipole moment during their motion, while Raman spectroscopy is sensitive to vibrations during the course of which the sign of the dipole moment changes. In solid-state vibrational spectroscopy the main aim is to analyze the effects of morphology on the motions of atoms, thereby investigating such parameters as degrees of ordering and orientation.

a. Infrared Spectroscopy. Solid-state infrared (IR) spectroscopy (normally Fourier transform infrared spectroscopy, FTIR) finds two principal applications in the morphological analysis of polyethylene, these being the determination of orientation and of local degrees of ordering within the sample. The latter involves the determination of the fractions of the sample that are crystalline (ordered), liquidlike (disordered), and interfacial (partially ordered). The use of vibrational spectroscopy for this purpose is addressed in Section IV.E. A variant of solid-state infrared spectroscopy involves the use of reflectance spectroscopy to investigate the nature of the layers of material closest to the surface of the sample.

All infrared-active vibrations have a directional component, as a consequence of which they will absorb only infrared radiation that is polarized parallel with their own vibrational axis. In the case of an isotropic or anisotropic sample placed in a randomly polarized infrared beam, this is of no consequence to the quantitative analysis of the various infrared bands. However, if the beam is polarized unidirectionally, only those molecular vibrations that have a matching directional component will contribute to the infrared spectrum. This phenomenon can be exploited to investigate the orientation of morphological features in film samples [58]. When the film is mounted in the spectrometer and its macroscopic orientation direction is known, spectra are recorded with the infrared beam polarized perpendicular to and parallel with the sample's orientation direction. Certain bands in the spectra, corresponding to crystalline and disordered regions, will be found to have different absorbance levels. The ratio of the absorbances, known as the dichroic ratio, of each of these pairs of bands in the two spectra yield an orientation function corresponding to either the ordered or disordered phase (or some combination thereof).

The use of conventional infrared spectroscopy requires that the sample be in the form of a thin film, of uniform thickness, preferably less than 125 µm thick, with lateral dimensions in excess of that of the cross section of the infrared beam, i.e., greater than approximately one in. in diameter. Thus blown and cast films, which constitute a large proportion of the production of polyethylene, are ideally suited to infrared spectroscopic examination. Other, thicker, products absorb too much of the infrared radiation to permit useful analyses to be performed. Films suitable for infrared spectroscopic analysis can be prepared from thick specimens by compression molding, but this inevitably destroys the original morphology. Samples that are heavily loaded with fillers such as carbon black or various minerals may not transmit sufficient infrared radiation for the collection of useful spectra. Infrared spectra with a high signal-to-noise ratio and a resolution of about 2 cm^{-1} can be recorded for typical film samples in 1 min or less. Greater resolution can be obtained by increasing the amplitude of oscillation of the interferometer mirror, which requires longer collection times. Unusually thick or thin samples may also require more time to obtain spectra with an acceptable signal-to-noise ratio. Infrared spectra can be obtained from samples at subambient or elevated temperatures with the use of appropriate heating or cooling cells.

In samples that are too thick for conventional infrared spectroscopy, or where it is desired to investigate the few micrometers of the sample closest to its surface, reflectance spectroscopy in the form of attenuated total reflectance (ATR) or grazing angle infrared spectroscopy can be used. Attenuated total reflectance relies on the phenomenon of total internal reflection observed when a beam of electromagnetic radiation is directed at the interface between two materials of disparate refractive indices at an incidence angle greater than the critical angle. In practice, the conditions of total internal reflectance are met when the infrared beam is projected through an infrared transparent crystal of high refractive index onto a sample of polymer in close contact with the crystal as shown schematically in Figure 16a. Suitable crystals can be made from germanium, zinc selenide, and many other materials. As the incident beam undergoes total internal reflection at the interface it interacts with surface layers of the polymer, which absorb some of the infrared radiation at their characteristic frequencies. If the crystal has a large length-to-thickness ratio and a reflecting upper surface, multi-

Characterization and Testing

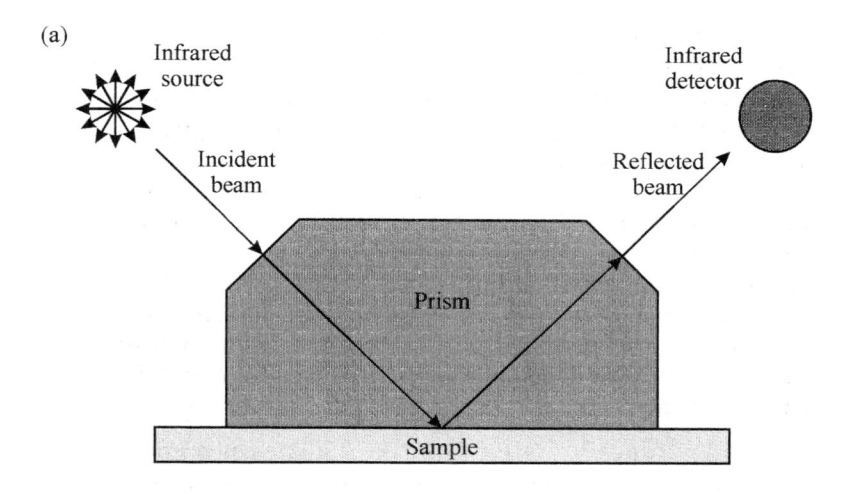

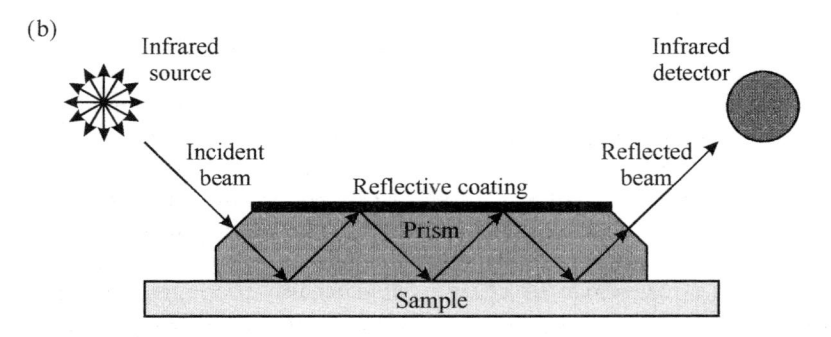

Figure 16 Schematic illustrations showing the optical configurations used for attenuated total reflectance infrared spectroscopy. (a) Single reflection; (b) multiple reflection.

ple reflections at the interface between the polymer and crystal can be achieved, as shown in Figure 16b. Multiple reflections have the advantage of enhancing the signal-to-noise ratio. The depth to which the infrared beam penetrates the sample is a function of the angle of incidence, the refractive index of the crystal, and the frequency of the radiation. The depth of penetration increases with decreasing angle of incidence, decreasing refractive index of the crystal, and decreasing frequency of radiation. By systematically decreasing the angle of incidence it is possible to record spectra from increasingly thick layers. Typical interaction depths for polyethylene are on the order of 0.5–3.0 μm.

The basic configuration of grazing angle infrared spectroscopy is similar to that of attenuated total reflectance, except that reflectance occurs at the interface between polymer and air at extremely high incidence angles. Under such conditions a high percentage of the beam is reflected after penetrating the sample to a depth of approximately 5–12 µm. The basic configuration of grazing angle infrared spectroscopy is shown in Figure 17.

When examining commercial samples it is often of great importance to identify small inclusions that give rise to gels in films or other unwanted effects. In such a circumstance, the infrared microscope is invaluable. Infrared microscopes focus an infrared beam onto a highly localized and precisely defined area of the sample, the spectrum of which is recorded. Under favorable conditions spectra can be recorded from inclusions with diameters of 20 µm and larger. Infrared microscopy is often used as a troubleshooting tool in conjunction with electron or optical microscopy.

  b. Raman Spectroscopy. When visible light strikes any material, a certain portion of it is absorbed—even by the most transparent of samples. Some of the photons absorbed by the sample excite its molecules into unstable virtual states that have energy levels higher than that of the original state by an increment equivalent to the energy of the exciting photon. When a virtual state decays, it normally decays to its original vibrational state, releasing a photon that has the same wavelength as the exciting radiation, i.e., elastic light scattering, known as Rayleigh scatter. In some cases, however, when the virtual state decays it decays to a vibrational state other than its original one, in the process of which it emits a photon with a wavelength different from that of the incident radiation, i.e., inelastic light scattering; this is the Raman effect. When the second vibrational state is higher than the original state, the scattering spectrum consists of Stokes lines at wavelengths longer than that of the exciting radiation. If the second vibrational state is lower than the original, the wavelengths of the inelastically scattered light are shorter than that of the exciting radiation and the spectrum consists of anti-Stokes lines. Typically the intensity of Rayleigh scatter is six orders of magnitude greater than that of the Stokes lines, which are in turn approximately three orders of magnitude more intense than the anti-Stokes lines. The frequency difference between the Rayleigh scatter and the Stokes lines may be determined

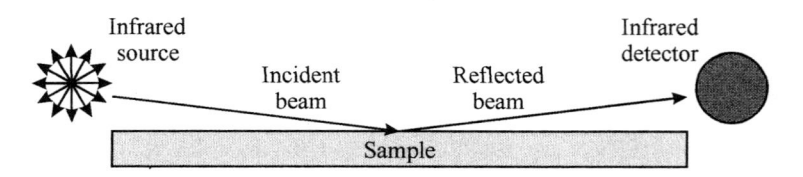

Figure 17 Schematic illustration showing the optical configurations used for grazing angle infrared spectroscopy.

by using a Raman spectrometer. The observed frequency shifts are a function of the vibrational characteristics of the sample.

Raman spectroscopy of polyethylene is divided into two clearly defined ranges: internal modes, which are characteristic of localized vibrations of small molecular subunits consisting of a few atoms, and longitudinal acoustic modes (LAM), corresponding to vibrations of extended molecular sequences arranged in the all-trans configuration. The internal mode frequencies of polyethylene are found in the range of 1000–4000 cm^{-1}, while longitudinal acoustic modes are found at frequencies ranging from approximately 5 to 250 cm^{-1}.

Raman spectroscopy requires a beam of highly intense monochromatic light, for which lasers are the ideal source. Common types of lasers in use include argon and krypton ion lasers. Three types of Raman spectrometers are available. Scanning instruments use one or more rotatable diffraction gratings to scan a range of frequencies sequentially, recording the intensity of scatter as a function of frequency, using a photomultiplier and a light-sensitive detector. Diode array instruments use a stationary diffraction grating, recording the scattering intensity over a range of frequencies simultaneously with a diode array detector. Fourier transform Raman spectrometers record spectra over a range of frequencies simultaneously using an optical system based on an interferometer. Any of these three types of instrument may be used to record internal mode spectra, but longitudinal acoustic mode spectra must be obtained with scanning type instruments—preferably ones with two or more gratings. (Dispersive spread from the highly intense Rayleigh line would overload the detection capabilities of diode array and Fourier transform instruments at low frequencies.) The time required to record an internal mode spectrum on a diode array or Fourier transform instrument may be two orders of magnitude shorter than the 1/2 hr or more required by a traditional scanning instrument.

One of the major difficulties encountered when recording Raman spectra of polyethylene on instruments that use diffraction gratings is that of fluorescence, which occurs over a broad range of frequencies, often swamping the weak Stokes lines. This problem may be alleviated by using a different laser frequency or by allowing the fluorescence to "burn out," by leaving the sample in the laser beam for an extended period of time (up to several hours). The use of Fourier transform spectrometers avoids this problem.

The standard optical configuration used to collect Raman spectra is one in which the scattered light is collected by a lens with its optical axis perpendicular to that of the incident laser beam. The sample is mounted at the intersection of the two optical axes, with its surface making an angle of approximately 45° to both. Thus, Raman spectra can be recorded from virtually any polyethylene sample, regardless of its nature. Specimens cut from films or molded items are held in the jaws of a three-way micropositioner, which is used to align the sample. Powdered samples can be contained in glass capillary tubes. Some care must be exercised that the highly intense laser beam does not heat the sample and hence change its morphology by melting or annealing it; this can be particularly troublesome for powdered samples or very thin films. The sample chamber of a Raman spectrometer is typically of sufficient size to accommodate large specimens or a variety of specialized cells. Hot and cold cells have been devised that can be used to record spectra from samples over a range of temperatures from $-196°C$ to $350°C$. It is even possible to record Raman spectra from molten polyethylene by using a specially designed extruder die mounted within the sample chamber [60].

a. Internal Mode Analysis. One of the principal applications of Raman spectroscopy to the study of polyethylene is the quantitative determination of degrees of ordering. Estimates of the fractions of the sample in the crystalline, partially ordered, and disordered phases can be obtained by analyzing the internal mode frequencies. This application is addressed in Section IV.E.

Raman-active vibrations exhibit a directional component that can be used to investigate macroscopic orientation in a manner analogous to that used for infrared dichroic measurements [61]. Similar levels of information can be obtained by either infrared or Raman measurements [62].

b. Longitudinal Acoustic Mode Analysis. Longitudinal acoustic modes—sometimes known as LA modes—are Raman peaks that occur at frequencies equivalent to mechanical acoustic vibrations. They occur when extended molecular sequences oscillate harmonically along their length. The frequency of such vibrations is determined by the elastic constant of the all-trans molecular configuration (parallel with the c axis of the unit cell) and the number of carbon atoms in the backbone of the extended segment. Such vibrations occur in the linear chain sequences that span the thickness of crystallites or in much shorter extended lengths found in partially ordered regions; the latter are known as disordered longitudinal acoustic modes (D-LAM). Analysis of longitudinal acoustic mode peaks can yield quantitative estimates of the distribution of extended chain lengths in the lamellae of a sample. The frequency of vibration of an extended molecular sequence is given by

$$\nu = \frac{n(E_c/\rho)^{1/2}}{2cL}$$

where

ν = frequency of vibration (cm^{-1})
n = order of vibration (1, 3, 5, etc.)
E_c = modulus of all-trans segment (2.9×10^{12} dyn/cm^2)
ρ = density of crystallite (1.00 g/cm^3)
c = speed of light (3.0×10^{10} cm/sec)
L = length of all-trans segment (cm)

The recording of longitudinal acoustic modes is not without difficulties, the most prominent of which is dispersive scatter of the Rayleigh line that invariably occurs within the instrument. The resulting "Rayleigh wing" overlaps the frequencies of the longitudinal acoustic mode, reducing the signal-to-noise ratio and introducing a sloping background from which the peak must be deconvoluted. The intensity of the Rayleigh wing can be reduced by decreasing the slit openings of the spectrometer or by using a triple-grating instrument rather than a double-grating instrument.

Deconvoluting the longitudinal acoustic mode from the Rayleigh wing yields a distribution of intensities as a function of scattering frequency from which the distribution of extended chain molecular sequences and hence lamellar thicknesses can be obtained. Three major difficulties are encountered during this process: deconvoluting the broad longitudinal acoustic mode peak from the Rayleigh wing, allowing for changes in the intensity of the Raman signal as a function of the frequency shift, and correcting for c axis tilt with respect to the thickness of lamellae. The first of these difficulties can be minimized by comparison of the shape of the Rayleigh wing with a reference spectrum obtained from a nonlamellar sample such as chalk. The effect of intensity changes as a function of scattering frequency can be removed mathematically [63,64]. The c axis tilt can be estimated from electron microscopy for a variety of crystallization conditions [65], typical angles falling in the range of 20–40°. The net result of such analyses is a distribution of lamellar thicknesses that favorably matches that measured directly by transmission electron microscopy of stained samples [66,67].

Within the phases of a polyethylene sample conventionally thought of as disordered, there exist localized regions of linear chain sequences. These submicroscopic regions of ordered material exhibit relatively high frequency Raman peaks known as disordered longitudinal acoustic modes. These peaks occur in the range of approximately 150–250 cm^{-1}, corresponding to extended sequences of approximately 10–20 methylene units [68,69].

2. Solid-State Nuclear Magnetic Resonance Spectroscopy

Solid-state nuclear magnetic resonance (NMR) spectroscopy can be used to investigate the distribution of degrees of freedom of the atoms in a sample. One of its principal applications to polyethylene samples is the determination of degrees of ordering. This application is described in Section IV.E. Solid-state NMR spectroscopy has also been used to investigate the longitudinal translation of chain segments through crystallites [70].

D. Thermal Analysis

The thermal analysis of polyethylene generally involves heating or cooling a sample at a controlled rate while monitoring some of its physical characteristics.

Changes in heat capacity are determined by using differential scanning calorimetry, weight changes by thermogravimetric analysis, and volume changes by dilatometry.

1. Differential Scanning Calorimetry

Differential scanning calorimetry (DSC) is a versatile technique used to determine thermal characteristics of polyethylene samples relevant to both real-life applications and fundamental morphological investigations. As its name implies, differential scanning calorimetry involves dynamic calorimetric analysis of a sample whose temperature is being ramped at a controlled rate. This is achieved by measuring the instantaneous heat capacity of a sample as a function of its temperature during heating or cooling. The results are presented in terms of heat flow as a function of temperature in a plot known as a thermogram. Endothermic and exothermic peaks respectively correspond to melting and crystallization processes, while step changes reflect material transitions, such as the glass–rubber transition. Quantitative information can be obtained with respect to both the temperature at which events occur and the heat flow associated with them. Differential scanning calorimeters can also be used to measure transitions involving heat transfer that occur at fixed temperatures, such as isothermal crystallization.

Two varieties of differential scanning calorimeters exist, both of which are capable of making accurate measurements on samples in the range of 1–20 mg. Figure 18 illustrates the basic features of the two types. In both cases specimens are encapsulated in small aluminum sample pans, which are placed in a sample chamber for comparison against an empty reference sample pan. In the first type (Fig. 18a), the flow of heat into the sample chamber via the sample support is kept constant while the temperature of the sample pan with respect to the reference pan is recorded. In the second variant (Fig. 18b)—known as the power-compensating type—the temperature of the sample and reference pans is ramped at a fixed rate while the relative heat flow required to effect the change in temperature is recorded. In both cases the temperatures of the sample and reference pans are determined to a precision of a few hundredths of a degree, while the flow of heat into the sample supports must be monitored and controlled to a similar precision. The net results of both methods are identical as far as the operator is concerned; each generates a precise plot of heat flow as a function of temperature.

As an alternative to a linear rate of temperature change it is possible to overlay a harmonic thermal oscillation over a linear temperature ramp. The net result is an average change of temperature equal to the linear temperature ramp with an observed temperature that fluctuates at a fixed frequency, each successive peak being a fixed increment in temperature higher or lower than the preceding one. The amplitude and frequency of the oscillation can be independently varied. This method was developed to distinguish reversible phenomena from nonrever-

Characterization and Testing

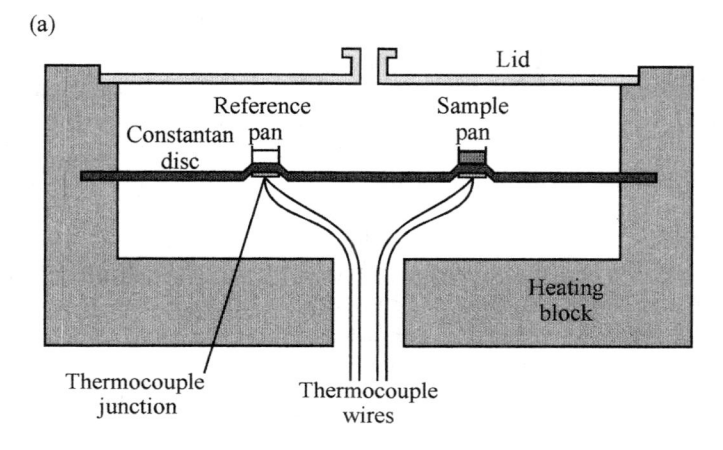

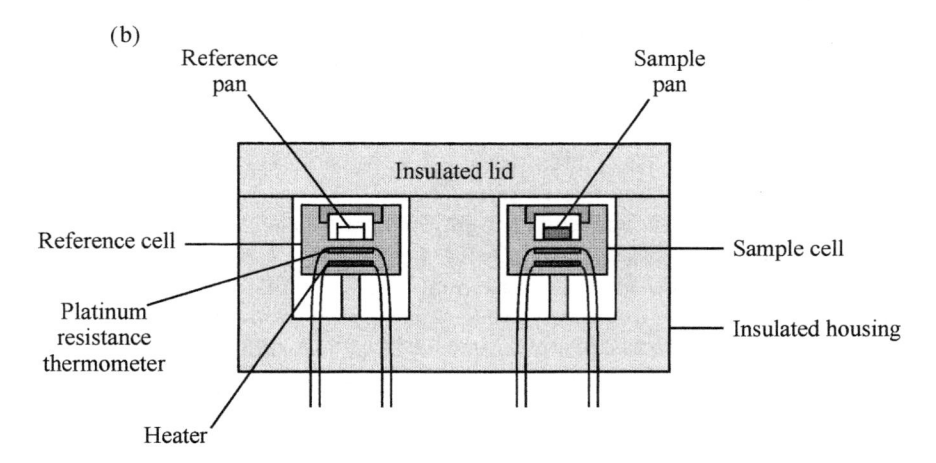

Figure 18 Schematic illustration showing the general configurations of the two types of differential scanning calorimeter sample chambers. (a) Constant heat flow into the sample chamber; (b) modulated heat flow to maintain specific temperature ramp.

sible phenomena. This is especially useful when a reversible step transition (such as the glass–rubber transition) is found in the vicinity of a nonreversible event, such as melting, crystallization, or chemical decomposition.

 a. Melting and Crystallization Temperature Determination. In principle, the melting temperature of a sample can be determined by recording its thermogram over a temperature range in which melting is known to occur and noting the position of the endothermic peak. In practice, things are not that simple, the main drawback being that the melting of a polyethylene specimen comprises the melting of innumerable crystallites, each with a melting temperature largely dependent upon its thickness. As there is a range of crystallite thicknesses, melting occurs over a range of temperatures. To complicate matters further, the temperature of any element within the sample depends on the thickness of polymer (which is an effective thermal insulator) intervening between it and the source of heat. Thus the temperature measured by the thermocouple may not represent that of the sample as a whole. As the processes of melting, crystallization, and heat conduction are all dynamic, the rate of temperature change influences the observed temperature. In practice, the faster the temperature is ramped and the thicker the sample, the larger will be the range of temperatures within the sample itself and the greater will be the lag between the average temperature of the sample and the registered temperature. One might suppose that the answer to this problem would be to use a thin sample and a very slow heating rate, but this is not so. If the temperature of a sample is raised too slowly, the sample is likely to undergo annealing or even melting and recrystallization, while decreasing the sample thickness leads to an increase of the signal-to-noise ratio. The net result of these circumstances is that there is no such thing as an absolute melting point or unique melting temperature of a polyethylene sample. Idealized thermograms representing the melting of high density, low density, and linear low density polyethylene are shown in Figure 33 of Chapter 5.

In practice, thermograms of polyethylene are frequently recorded at heating or cooling rates of 5–20°C/min using a sample size between 5 and 15 mg. For comparative work it is best to use samples with similar weights and thicknesses. Another factor influencing the accuracy of data is the calibration of the instrument, which is carried out by using samples of high purity metals or crystalline compounds. Calibration tends to drift with time, reducing accuracy. Ideally, comparative samples should be run consecutively on a freshly calibrated instrument.

Although melting occurs over a range of temperatures, the melting temperature is conventionally reported as being that of the endothermic peak maximum. The melting temperature of a sample may also occasionally be reported as the temperature at which the endothermic melting peak ends. The software on most commercial instruments permits the user to ascertain peak maxima to an unrealistically high precision of 0.01°C. If a given sample is run through a heating and cooling cycle several times, repeatability is likely to be ±0.25°C. For individually prepared specimens cut from the same sample, a precision of ±1°C could reasonably be expected. For samples run on different instruments—or even the same instrument over a period of weeks—it is unreasonable to expect a comparative precision of better than ±2°C.

The melting range and peak melting temperature of a specimen are functions of its thermal history. The shape of a melting endotherm reflects a specimen's distribution of lamellar thicknesses, which in turn depends upon its molecular characteristics and crystallization conditions, according to the principles discussed in Chapter 4. In practice it is common to record the melting thermogram of the same specimen twice, with an intervening crystallization step; this process is referred to as "first melt/recrystallization/second melt." The thermogram recorded during the first melt reflects the sample's original thermal history. Recrystallization under controlled conditions imposes a known thermal history. The thermogram recorded during the second melt may be used to compare samples that have all undergone the same recrystallization step; observed differences reflect variations in molecular characteristics.

The determination of the crystallization temperature of a sample is subject to many of the same considerations that apply to melting temperature determination. The crystallization temperature is normally reported as the temperature at which the exothermic peak maximum occurs but may also be reported as the temperature at which crystallization begins (the crystallization onset temperature). The observed crystallization peak temperature is always considerably lower (20°C or more) than the melting temperature observed subsequently for the same sample. The difference between the observed crystallization and melting peak temperatures increases as the rate of temperature ramp increases.

When reporting melting and crystallization temperature data, it is important to note the thermal history, sample size, sample configuration (powder, film etc), the heating or cooling rate, and whether the stated temperature was applicable to the peak maximum or the onset or conclusion of melting.

b. Crystallization Half-Time Determination. The temperature at which a sample crystallizes during cooling is useful for comparative purposes, but it is often more desirable to know the rate at which crystallization occurs at a fixed temperature. A measure of the latter is the crystallization half-time, which is the time it takes a sample to undergo half of the crystallization that it would ultimately undergo if left at a given temperature indefinitely. To make this determination with a differential scanning calorimeter, a sample is first heated to its molten state, after which it is rapidly cooled to the temperature of interest. Once at this fixed temperature, the flow of heat out of the sample caused by the crystallization exotherm is monitored. The integral of heat transfer as a function of time is recorded until the crystallization process is complete, i.e., heat transfer ceases. The crystallization half-time is the recorded time at which the heat transfer integral reaches half of its ultimate value. It is common practice to determine the crystallization half-time at a series of temperatures. At low crystallization temperatures there may be difficulty cooling the sample to the desired temperature before the onset of crystallization. This inevitably reduces the accuracy of the data from such temperatures.

Results may be quoted in terms of seconds or minutes, depending upon the crystallization rate. Crystallization rates involving half-times in excess of a few minutes are rarely of interest to commercial enterprises. The determination of half-times in excess of an hour with a differential scanning calorimeter is not very practical due to the difficulty of measuring the very low heat transfer rates involved; if such measurements are desired, dilatometry is a more practical alternative.

c. Heat of Fusion Measurement. The heat of fusion of a polyethylene sample is usually determined concurrently with its melting temperature characteristics. To do this, the area of the endothermic peak on the thermogram is measured in terms of energy per mass, normally calories or joules per gram (cal/g or J/g). It is assumed that the endothermic peak is due solely to the melting of the crystalline regions. The degree of crystallinity of the sample can be estimated by comparing the measured heat of fusion with that estimated for 100% crystalline polyethylene. This topic is further addressed in Section IV.E.

Although simple in theory, the measurement of heats of fusion suffers from some major practical drawbacks. The main problems stem from the selection of an appropriate baseline under the melting peak. It is rare that a sample yields a thermogram in which the baseline before the onset of melting extrapolates perfectly to that which is reestablished after melting is complete. Normally a straight line is drawn from the onset of melting to its conclusion, at points selected visually by an operator. As the baseline before and after the peak may be somewhat curved, the selection of the onset and conclusion of melting is subjective; no two operators consistently choose identical points. Selection of the onset of melting is further complicated by the fact that low density samples exhibit extensive low temperature tails. It is important that the temperature range scanned should be broad enough to allow a minimum of 20°C of baseline before and after the melting peak. Some of the problems associated with baseline selection are illustrated in Figure 19.

Errors associated with baseline selection, instrument calibration and sample weighing reduce the precision and accuracy of the measured heat of fusion. An accuracy ranging from about ± 2 cal/g for high density polyethylene to ± 4 cal/g for low crystallinity samples is typical.

2. Thermogravimetric Analysis

Thermogravimetric analysis is used to investigate the ultimate destruction of polymer samples at elevated temperatures by measuring mass loss as a function of temperature or time. The basic equipment consists of an extremely accurate balance suitable for determining differences in terms of micrograms and a temperature-programmable tube furnace flushed with an inert gas. The sample is mounted on a small metal weighing boat suspended from the arm of the balance

Characterization and Testing 297

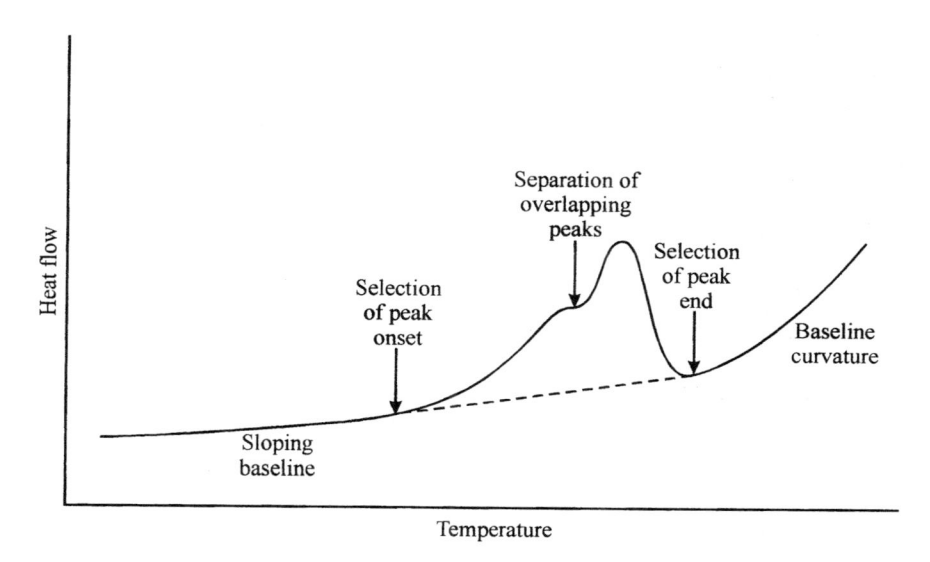

Figure 19 Thermogram of low density polyethylene illustrating some of the problems associated with the selection of the baseline under the peak.

that is inserted into the furnace. The temperature of the furnace may be programmed to increase at a constant rate or hold at a specified value. The weight loss of the specimen is recorded as a function of time.

Thermogravimetric analysis finds limited use in the field of polyethylene research. This is primarily due to the fact that most of the useful properties of polyethylene cease to exist at temperatures very much below that at which significant thermal decomposition occurs.

3. Dilatometric Analysis

Dilatometry is a classic technique that has largely fallen out of use in the field of polyethylene research. In the past it was principally used for the academic study of crystallization rates. A sample is prepared for dilatometry by being sealed into a small glass bulb attached to a long glass capillary tube. The system is evacuated, and mercury is introduced to surround the sample and partially fill the capillary. Changes in sample density are followed by measuring the change in the height of the mercury column in the capillary tube. To determine crystallization rates, the bulb of the dilatometer is immersed in an oil bath at a temperature adequate to melt the polymer, then transferred to an oil bath previously set for the temperature of interest. The progress of crystallization is tracked by observing the height of the mercury column as a function of time. Due to the extremely sensitive nature of dilatometry, oil baths with fluctuations of less than ±0.1°C must be employed.

Dilatometry can be used to follow the slow crystallization of polyethylene at high temperatures or secondary crystallization over extended periods of time. It can also be used to investigate density changes as a function of temperature. The sensitivity of dilatometry is such that changes in density of less than 0.0001 g/cm^3 can readily be detected. Measurements may be made over a period of days, weeks, months, or even years, providing that the oil bath in which a dilatometer is immersed is adequately thermostated.

E. Determination of Degree of Crystallinity

The degree of crystallinity of a polethylene sample reflects the quantitative distribution of chain segments between ordered and disordered regions. Many of the physical characteristics of polyethylene samples can be related to it in one way or another; such attributes include elastic modulus, yield stress, barrier properties, and mold shrinkage.

1. Methods

The degree of crystallinity of a polyethylene sample can be determined by a variety of methods, each of which measures a different property that is then related to the level of crystallinity via a particular set of assumptions. No two methods yield identical results for all samples, largely because each method probes different morphological structures. In addition, for a variety of reasons, most methods are not applicable to all samples. When comparing degrees of crystallinity the method by which the determination was made must always be taken into account. Ideally, more than one method of determination should be used. Some of the more commonly used methods are outlined below.

a. Density. Density is the one of the oldest, best established, and most widely used methods of determining the degree of crystallinity of polyethylene samples. Its importance is evident from the fact that commercial resins are divided into such categories as high density, low density, and linear low density polyethylene. The determination of the degree of crystallinity from density rests on two simple assumptions: that polyethylene samples consist of a two-phase morphology and that the density of each phase is uniform within the sample and consistent from one sample to another. The concept of two-phase morphology has been superseded by a more rigorous model in which a partially ordered transition zone separates the crystallites from the disordered regions (as discussed in Chapter 4). In addition, the unit cell dimensions, and thus the crystalline density, of polyethylene have been shown to vary with branch content [71]. Although both of the underpinning assumptions are true only to a first approximation, the degree of crystallinity is routinely derived from density. The density of polyethylene is sensitive to small changes in morphology and is easily measured to a high degree of accuracy. It can be directly measured by flotation methods, primarily density gradient column analysis and densimetry.

A density gradient column consists of a wide-bore glass tube filled with a mixture of liquids that has a smooth gradient from high density to low density as a function of height. This is achieved by filling the column with a pair of miscible liquids of different densities, the relative proportions of which gradually change during the filling process. Appropriate selection of the liquids results in a column having a range of densities that encompasses that of the polyethylene samples of interest. The density gradient is calibrated with glass floats of known density. If the column has been prepared correctly it will exhibit a linear density gradient in the range of interest. The density of a polyethylene sample can be determined by dropping it into the column, measuring the height at which it comes to equilibrium, and interpolating between the heights of the calibration beads above and below it. Alternatively a calibration plot of density versus the height of the calibration beads can be prepared, from which the density of a specimen floating at any height can be read directly. A common choice of liquids for the determination of polyethylene density is water and isopropanol. A drawback to this pair of liquids is a tendency toward incomplete wetting of samples; this can result in air bubbles on their surfaces, which invalidates measurements. This problem can be largely avoided if diethylene glycol or triethylene glycol is substituted for water. A variety of other pairs of liquids can also be used. A standard method for preparing density gradient columns and measuring the density of polymer samples can be found in ASTM D 1505. Under standardized conditions the density of a specimen can easily be determined to an accuracy of ± 0.0001 g/cm^3.

The weight percent degree of crystallinity of a sample can be readily calculated from its density if the densities of the crystalline and disordered regions are known [72].

$$X_D = \frac{1/\rho - 1/\rho_a}{1/\rho_c - 1/\rho_a} \times 100$$

where:

X_D = degree of crystallinity (%)
ρ = sample density
ρ_c = unit cell density
ρ_a = amorphous density

In practice, the value of density for the crystalline component is taken to be 1.000 g/cm^3 [73], and that of the disordered region, in the range 0.850–0.855 g/cm^3, with a commonly accepted value being 0.853 g/cm^3 [72]. The degree of crystallinity can be calculated to a precision of $\pm 0.1\%$, with an accuracy of approximately $\pm 2\%$.

The calculation of degrees of crystallinity from density is advantageous in that it is simple, precise, and applicable to samples of virtually any configuration, including pieces cut from fabricated parts and oriented materials. On the negative side, the procedure is somewhat slow, and it is not possible to determine the degree of crystallinity of filled samples, those containing voids, or those that contain significant quantitites of atoms other than carbon and hydrogen. Thus it is not applicable to ionomers or copolymers containing polar substituents.

The second common method of measuring density is densimetry. This is based on Archimedes' principle; i.e., a sample has a different weight when it is suspended in a liquid than when it is suspended in air. The difference in weight is a function of the density and mass of the sample and the densities of air and the liquid. The density of a sample can be determined according to

$$\rho = \frac{D_l W_a - D_a W_l}{W_a - W_l}$$

where

ρ = sample density
D_l = density of liquid
D_a = density of air
W_l = weight of sample in liquid
W_a = weight of sample in air

Under ideal conditions, the accuracy to which a sample's density can be measured is equivalent to that obtained from the density gradient technique. A standard method for measuring polymer density by densimetry can be found in ASTM D 792. Densimetry lends itself to automation [74]. As a method of calculating the degree of crystallinity, densimetry suffers from many of the same disadvantages as the density gradient method; however, it is much quicker, and, when automated, it is less subject to operator error.

b. Differential Scanning Calorimetry. The degree of crystallinity of a polyethylene sample can be calculated from its heat of fusion if one assumes a two-phase morphology, the noncrystalline portion of which does not contribute to the melting endotherm. The heat of fusion of a sample can be determined by differential scanning calorimetry, the principles of which were discussed in Section of IV.D.1. Comparison of the measured heat of fusion with that estimated for 100% crystalline polyethylene yields the fraction of the sample in the crystal lattice. In practice, the heat of fusion of pure crystalline polyethylene is taken to be in the range of 66–70 cal/g, with a commonly accepted value being 69 cal/g [75].

A major source of error in determining the degree of crystallinity by differential scanning calorimetry arises from the selection of the baseline under the endothermic peak. The problems associated with this procedure were discussed in reference to heat of fusion measurement in Section IV.D.1. Differential scanning calorimetry also suffers to some extent from poor sample to sample repeatability, which lowers its precision and accuracy. As with the determination of degrees of crystallinity from density, the presence of fillers invalidates this method. From a theoretical standpoint, the determination of crystallinity from a sample's heat of fusion relies on a simple two-phase model of morphology. These drawbacks and an uncertainty in the heat of fusion of 100% crystalline polyethylene limit the accuracy of this method to approximately ±5%.

On the positive side, differential scanning calorimetry has much in its favor. It can be used to determine the degree of crystallinity of a complete range of polyethylene samples, including those that contain polar groups, those that are oriented, and those that contain voids. In addition, much other useful information is also available from the same temperature scan.

c. Wide-Angle X-Ray Diffraction. Wide-angle X-ray diffraction is a time-honored but somewhat out-of-favor technique for the determination of degrees of crystallinity of polyethylene samples. According to this method the sample is scanned in a wide-angle X-ray goniometer, and the scattering intensity is plotted as a function of the 2θ angle, as shown in Figure 14. Alternatively, a two-dimensional "flat plate" detector can be used to collect data over a range of 2θ angles encompassing the complete azimuthal range. Quantitative decomposition of the diffraction pattern into its crystalline and noncrystalline components provides a measure of the degree of crystallinity. Theoretically the scattering pattern should be integrated over a complete range of 2θ angles from 0° to 180°. In practice a much narrower range (approximately 5–55°) is scanned and a pair of constants applied to the deconvoluted peak areas to compensate for the reduced angular range [76,77].

In its favor, wide-angle X-ray diffraction can be used to measure the degree of crystallinity of a complete range of polyethylenes, including some that contain fillers. In addition it can provide information about unit cell dimensions. Balanced against this, samples analyzed with a goniometer must be either a fine powder or unoriented and flat. Deconvolution of the scattering pattern may be subjective, and the model of morphology is two-phase. The accuracy of results is similar to that from differential scanning calorimetry, i.e., approximately ±5%. Experimental limitations largely preclude the application of wide-angle X-ray diffraction to manufactured items other than films, sheets, and fibers.

d. Vibrational Spectroscopy. The Raman spectrum of a polyethylene sample in the range of 1000–1600 cm^{-1} (the internal mode region) provides quantitative information with respect to all three phases of its morphology [78,79]. Peaks have been identified that can be assigned to the totally disordered and perfectly crystalline regions. When the area of these peaks is compared with that of an internal reference peak, the fractions of the two components can be calculated. The sum of the disordered and ordered components does not account for all the material; the difference is attributed to the interfacial region. Thus Raman spectroscopy can be used to assess the level of each of the three morphological regions. Other features in favor of this technique are its applicability to a complete range of polyethylene resins and the fact that sample size and shape are largely immaterial. On the negative side, it is time-consuming and is inapplicable to oriented samples and those containing certain additives. Deconvolution of the various peaks is prone to subjectivity unless a computer technique is employed [80,81]. The accuracy of this method depends largely on the signal to noise ratio of the spectrum and the validity of the calibration factors. Typical errors associated with the crystalline and disordered content are estimated to be ±3% of their values. The error associated with the measurement of the interfacial level is the sum of the errors of the crystalline and disordered regions.

Infrared spectroscopy can be applied to the quantification of the three morphological components of polyethylene in a manner similar to that used with Raman spectroscopy [82]. Many of the same advantages and disadvantages apply, with an added restriction that the sample must be in the form of film.

The restriction that samples must be isotropic largely precludes the application of vibrational spectroscopic methods to the measurement of the crystallinity levels of manufactured items.

e. Nuclear Magnetic Resonance Spectroscopy. Solid-state carbon-13 nuclear magnetic resonance spectroscopy can be used to determine the levels of all the components of the semicrystalline morphology of polyethylene [83–85]. When a pulse of radio-frequency electromagnetic radiation strikes a sample of polyethylene, a proportion of the carbon nuclei will be excited to an elevated energy level. The half-life of an excited state depends upon the freedom of movement of the atom and its neighbors; the more restricted the motion, the longer the half-life. The decay of the excited states to the ground state is observed as a function of time. The decay curve can be deconvoluted into three or more components corresponding to the crystalline and random regions and various regions of intermediate ordering. The strength of the signal associated with each of the regions yields a measure of the number of atoms that inhabit it.

Carbon-13 NMR is the most sophisticated method available for the determination of crystallinity levels. It is applicable to a complete range of polyethylene samples, requires no calibration, and is little affected by orientation, fillers, and

Characterization and Testing 303

voids. The major drawbacks to this method are the expense of the equipment and the long times required to collect data, sometimes in excess of 24 hr. Accuracy depends on the quality of the raw data, i.e., their signal-to-noise ratio, which in turn depends on the patience and workload of the operator. An accuracy of ±5% would be a reasonable expectation.

Similar methods are available that are based on solid-state proton nuclear magnetic resonance spectroscopy. Proton methods are less discriminating than the equivalent carbon-13 method, being based on a two-phase model of semicrystalline morphology. In their favor, they are several orders of magnitude faster than the carbon-13 method and the equipment required is much cheaper.

f. Ultrasonic Measurement. The ultrasonic determination of a sample's degree of crystallinity is based upon the observation that the speed of sound in a polyethylene sample is proportional to the sample density. The experimental arrangement requires a combined transducer/receiver, a flat sample, and a reflector to be immersed in a thermostated water bath. The transducer produces a train of pulses that are reflected from the various liquid/solid interfaces and recorded by the receiver, with and without the sample in place. From the time delay between the initial pulse and the various reflections it is possible to calculate the thickness of the sample and the speed of sound within it. The instrument must be calibrated with a series of known samples, the densities of which have been determined by an absolute method, i.e. flotation.

This method is fast, precise, and repeatable. Results are little affected by fillers. A major drawback is that it requires a relatively large, flat, defect-free isotropic sample, rendering it inapplicable to most fabricated items. Some degree of exactitude is inevitably lost in the calibration process, so the absolute precision and accuracy of the degree of crystallinity is somewhat lower than that of the flotation method against which it is calibrated.

2. Comparison of Methods

Whenever degrees of crystallinity are considered, it must always be kept in mind that values depend to some extent upon the manner in which they were obtained. Each method measures different material properties that are related to the sample morphology in different ways. Those methods that assume a two-phase model of morphology include the interface with either the crystalline or disordered region. A full comparison of the above techniques, applied to a common group of samples, is not available. Figures 20–23 compare the results of crystallinity determination from density, differential scanning calorimetry, and Raman spectroscopy for a variety of commercial and experimental resins [86–88]. In each plot a broken line is added to indicate the relationship that would be expected if the results were equivalent.

Degrees of crystallinity derived from differential scanning calorimetry and

Raman spectroscopy match quite well over the accessible range, as shown in Figure 20. However, the degree of crystallinity derived from density is invariably greater than that derived from either differential scanning calorimetry or Raman spectroscopy, as typified by the data in Figures 21 and 22, respectively. The increased scatter of the data derived from differential scanning calorimetry is indicative of the inherently greater error associated with this method relative to Raman spectroscopy. The difference between the density values and that of Raman or differential scanning calorimetry is inversely related to the level of crystallinity. The reason for this discrepancy becomes clear when the sum of the crystalline and interfacial regions determined from Raman spectroscopy is plotted against the degree of crystallinity determined from density, as shown in Figure 23. The excellent agreement indicates that the density method includes the interface in its measurement of crystallinity. From these data, it would appear that the density of the interfacial region is closer to that of the crystalline regions than to that of the disordered zone. This is in keeping with the view that the

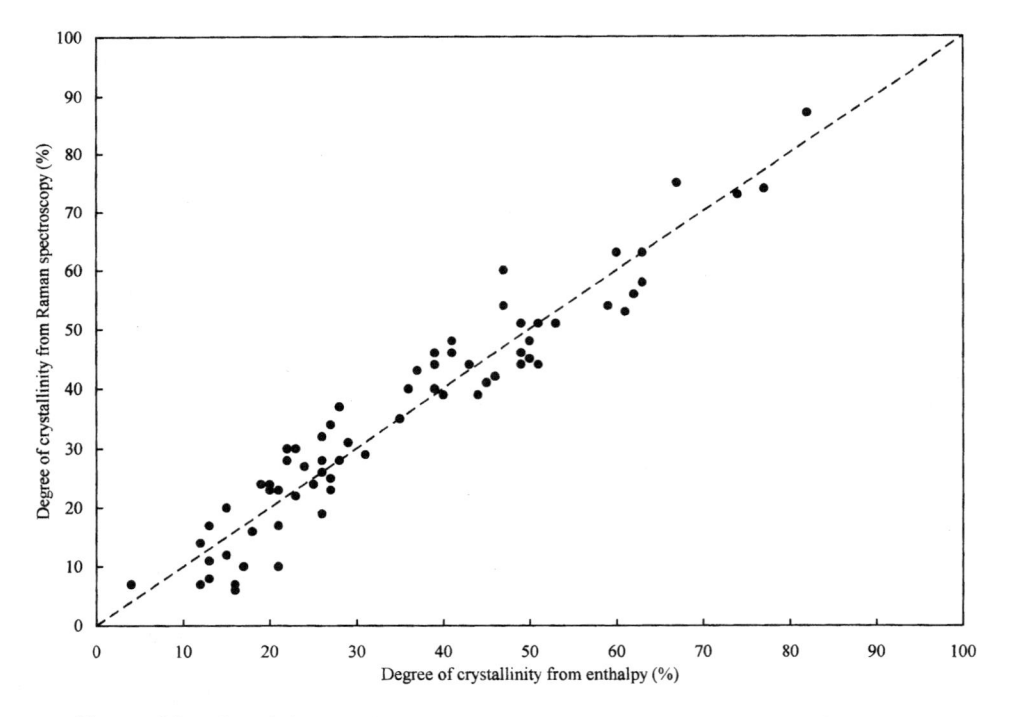

Figure 20 Plot of degree of crystallinity from Raman spectroscopy versus that from differential scanning calorimetry.

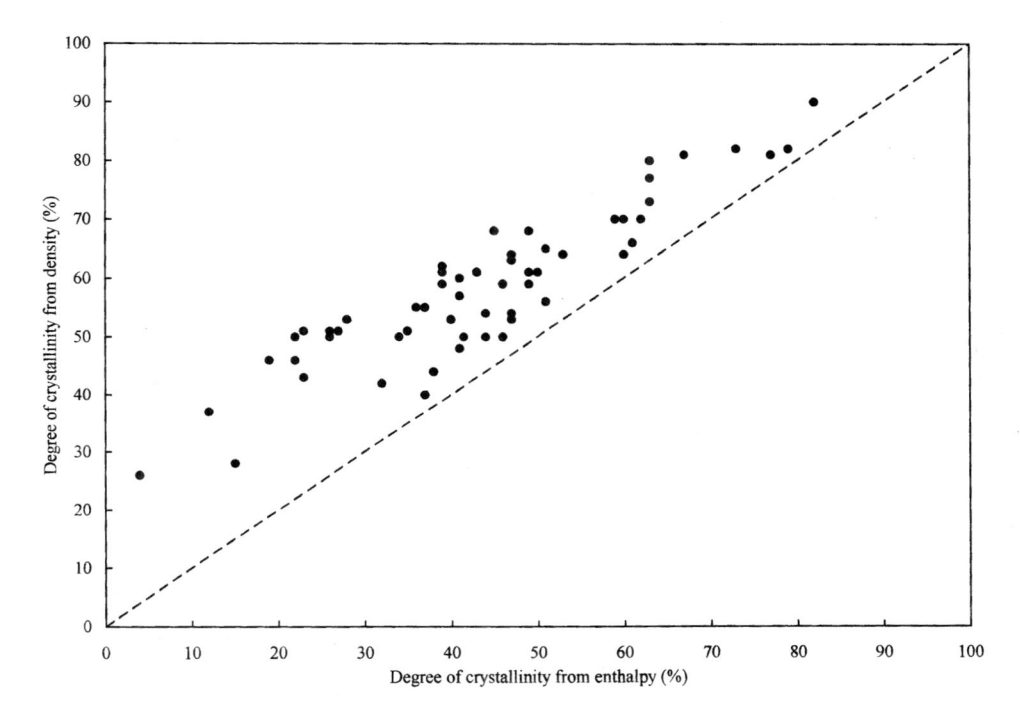

Figure 21 Plot of degree of crystallinity from density versus that from differential scanning calorimetry.

partially ordered region consists primarily of approximately parallel extended chains.

F. Characterization of Cross-Linked Polyethylene

Cross-linked polyethylene consists of a resin in which a substantial proportion of the chains are chemically bound together to form an insoluble network. The aim of cross-linking is to prevent the slippage of noncrystalline chain segments past one another. Cross-linking aids dimensional stability at temperatures in excess of the crystalline melting point and reduces creep and stress cracking. The properties relevant to the typical end use of such a material principally depend upon the proportion of the sample making up the network and the concentration of cross-linking sites that bind neighboring chains together. Gel content analysis addresses the first of these concerns, and swelling in hot solvent the second. For

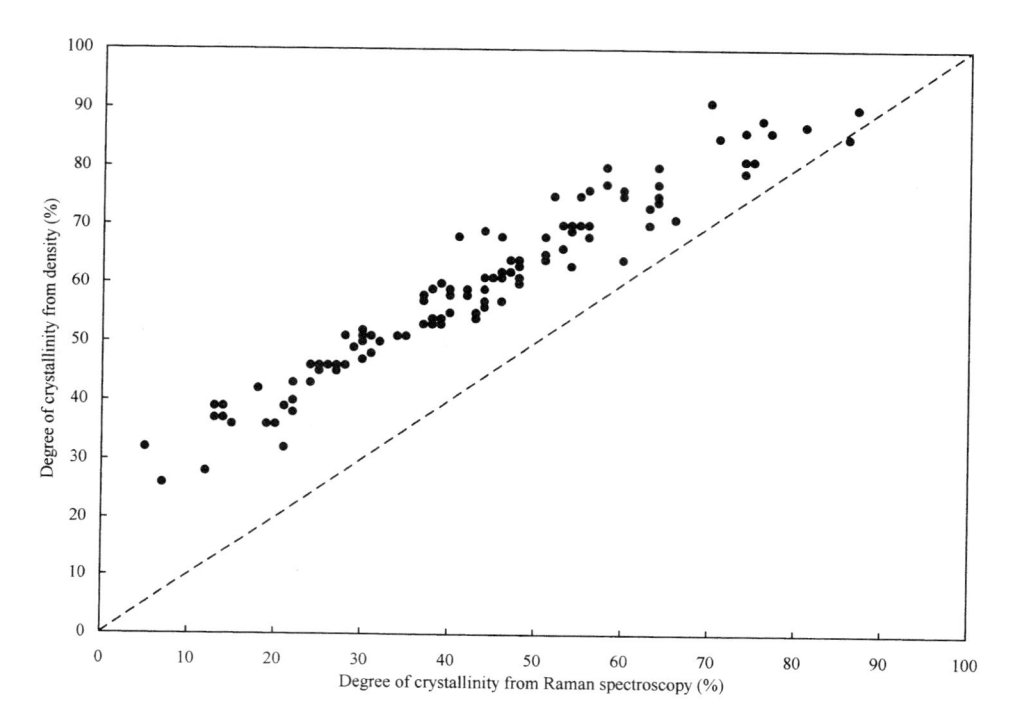

Figure 22 Plot of degree of crystallinity from density versus that from Raman spectroscopy.

most commercial purposes these tests are used to ensure that sufficient cross-linking has occurred for the end product to meet its service requirements

1. Gel Content Analysis

The goal of gel content analysis (GCA) is to ascertain what proportion of a cross-linked sample is bound into an insoluble network. This is determined by extracting a specimen in a hot solvent for a specified period of time and measuring its weight loss. Such tests are often used to determine whether the cross-linking reaction has occurred uniformly throughout the sample. They are also used routinely during the development of new product grades and processing conditions. A standard procedure whereby this measurement can be made is found in ASTM Method D 2765. Variations on this procedure abound in the scientific literature.

The essence of all gel content analysis procedures is to extract virtually all

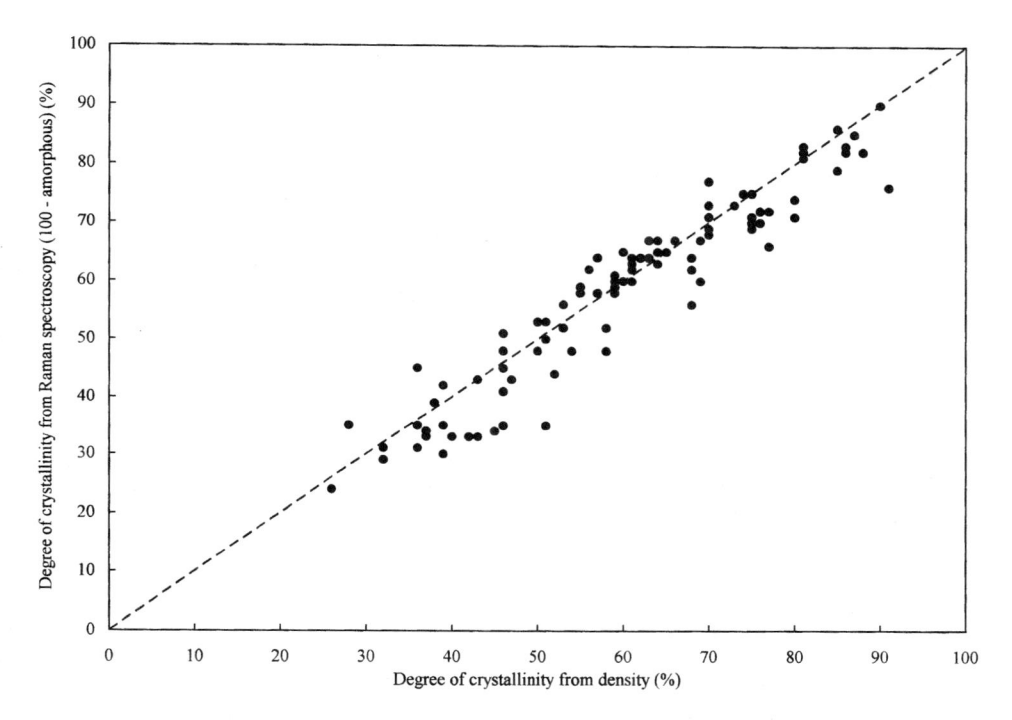

Figure 23 Plot of degree of crystallinity from Raman spectroscopy (core crystallinity plus interfacial contribution) versus that from density.

the uncross-linked material ("sol") from specimens of cross-linked polyethylene to leave behind the insoluble network ("gel"). To achieve this in a reasonable period of time requires specimens with a high surface-to-volume ratio, swollen to a high degree so that the sol can diffuse out. This translates to extraction of films, shavings, filings, or powder with a good solvent at high temperature. The preferred solvents are refluxing xylene or decahydronaphthalene (decalin). Samples that initially have a low surface-to-volume ratio must be prepared by milling, grinding, or microtoming to the appropriate degree. Milled or ground specimens should fall between 30 and 60 mesh. The time required to extract films or shavings is minimized if thicknesses less than 125 μm are used. A weighed amount of prepared specimen is encapsulated in a specimen holder contrived of folded and stapled 120 mesh stainless steel cloth. Films that are not prone to fragmentation in the boiling solvent do not need to be contained. Several duplicate specimens should be prepared, each with a starting mass of approximately 0.3–0.5 g. Specimens are fully immersed in refluxing solvent that has been well stabilized with an antioxidant, optionally with a nitrogen blanket. The volume of solvent should exceed that of the specimens by at least two orders of magnitude. The standard method calls for extraction in decalin for 6 hr or in xylene for 12 hr. Depending upon the nature of the samples under investigation, these time periods may be changed, providing that full extraction is achieved. It is also feasible to extract the specimens in a modified Soxhlet apparatus in which the extraction cup is surrounded by ascending solvent vapor. This arrangement maintains the specimens at a temperature very close to that of the boiling solvent. After the required period of time, the specimens are removed from the solvent and dried under vacuum to constant weight. The gel content of the sample is calculated according to

$$\text{Gel content (\%)} = 100 \times \frac{\text{final mass}}{\text{original mass}}$$

The degree of variation between duplicate specimens decreases as the gel content increases, ranging from $\pm 3\%$ at gel contents below 30% to $\pm 1\%$ at gel contents in excess of 90%.

2. Determination of Cross-Link Density

The degree to which a cross-linked polymer will swell when immersed in a solvent depends upon the polymer–solvent interaction parameter at the test temperature and the average molecular weight of the chain segments separating cross-links (effective chains). This relationship is defined by the Flory–Rehner equation [89,90].

$$V = -\frac{V_r + \mu V_r^2 + \ln(1 - V_r)}{V_o(V_r^{1/3} - V_r/2)}$$

where

V = concentration of effective chains (mol/cm^3)
V_r = volume fraction of polymer in swollen gel
μ = Huggins solvent–polymer interaction parameter
V_o = molar volume of solvent (cm^3)

$$V_r = \frac{1}{M_s\rho_p/M_p\rho_s + 1}$$

where

M_s = mass of solvent in gel (g)
ρ_p = density of polymer (g/cm^3)
M_p = mass of polymer in gel (g)
ρ_s = density of solvent (g/cm^3)

$$M_c = \rho_p/V$$

where M_c = molecular weight of effective chains (average separation of cross-links).

The Flory–Rehner equation in its original form does not take account of network imperfections due to the random distribution of lengths of effective chains or cilia consisting of chains that are bound to the network by only one end. To account for such imperfections in the network, Flory proposed the following modification [91], which was subsequently confirmed [92]:

$$M'_c = \frac{MM_c}{M + 2M_c}$$

where

M'_c = true molecular weight of effective chains
M = number-average molecular weight of resin prior to cross-linking

The applicable constants for cross-linked polyethylene in boiling xylene are

$\mu = 0.31$ [93,94]
$V_o = 139.3$ cm^3 at 140°C
$\rho_p = 0.806$ g/cm^3 at 140°C
$\rho_d = 0.761$ g/cm^3 at 140°C

The calculation of cross-link density is applicable only to samples from which the insoluble portion has been extracted. A standard procedure whereby this measurement can be made in combination with gel content analysis is found in ASTM method D 2765.

The ASTM procedure calls for the immersion of a minimum of two pre-weighed specimens of cross-linked polyethylene in xylene maintained at 110°C for 24 hr. Each specimen should weigh approximately 0.5 g. At the end of the extraction period, the specimens are removed from the solvent, exposed to a brief puff of air to remove the surface solvent, and quickly transferred to a pre-tared, stoppered weighing jar. The weight of the swollen specimens is determined, and then they are dried to constant weight in a vacuum oven. The final weight of the dried specimens is compared to their original weight to calculate their gel content. The final and swollen weights are used in the calculation of the swelling ratio. Duplicate determinations of the swelling ratio should fall within ±5% of the average value.

This procedure can be greatly accelerated if preextracted specimens are available. Preweighed, extracted specimens are fully immersed in gently refluxing xylene (at approximately 140°C) for approximately 2 hr, or until the specimens are fully swollen. Upon their removal from the xylene, the excess solvent evaporates from their surface almost instantaneously, and they are then transferred to a pre-tared, stoppered weighing jar. The weight of the swollen specimen is determined, and the molecular weight of the effective chains is calculated according to the preceding equations. Variation in the calculated molecular weight of effective chains (for a 5% variation in swelling ratio) varies from less than $\pm 10\%$ at swelling ratios below 3 to approximately $\pm 15\%$ at swelling ratios in excess of 10.

V. PHYSICAL PROPERTY TESTING

Physical property testing involves the determination of a sample's response to mechanical deformation under a variety of testing regimes. Tests may determine the relationship of deformation to applied force, stress as a function of applied strain, or the energy required to fracture a sample, or they may simply note a pass/fail result under specified conditions. Temperature, humidity, and other environmental factors may be varied to simulate end use conditions. Experiments may be performed on finished items for the purposes of quality control and product evaluation or on specimens specially prepared under standardized conditions for the purposes of fundamental research or comparative evaluation of competing resins. Experiments performed in the laboratory infrequently duplicate the conditions found in service. Laboratory results should therefore be treated as comparative, useful for ranking specimens prepared from different resins under standard conditions but not necessarily indicative of service performance.

A. Force Versus Deformation Measurements

Many physical property measurements involve deforming specimens under standardized conditions and recording the resulting stress generated in the sample. Such tests are routinely carried out on machines designed to measure the instantaneous force necessary to accomplish deformation at a fixed rate. The deformational force may take the form of tension, compression, shear, flexion, or torsion. Test specimens may be strips cut from film or sheet, molded or stamped "dogbones" (also known as "dumbbells"), fibers, rods, tubes, blocks, etc. With the exception of torsion experiments, the various testing regimes can usually be performed using a single instrument equipped with the appropriate grips and load cells. The configuration of a typical instrument set up for tensile testing is shown in Figure 24.

A sample of suitable dimensions (generally either a dumbbell or a strip) is firmly gripped at either end by a pair of jaws. One pair of jaws remains stationary while the other is moved at a predetermined rate, generally at constant speed

Characterization and Testing

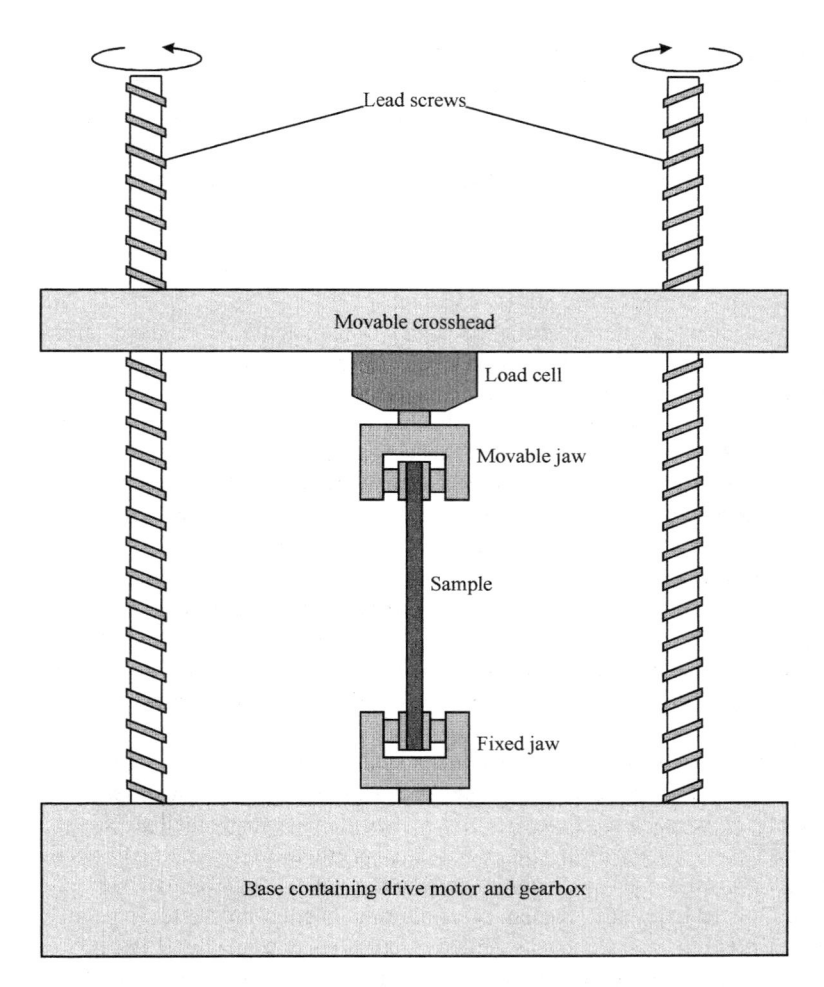

Figure 24 Schematic illustration showing the basic configuration of an instrument used to measure force as a function of tensile deformation.

but alternatively at some variable rate. The rate of deformation strongly influences the physical properties of the sample; suitable deformation rates applicable to different sample geometries and testing configurations can be found in the appropriate ASTM test method. The force required to deform the sample is monitored continuously by a load cell attached to one pair of jaws. The average strain in the sample is generally calculated based upon the measured displacement of the jaws relative to the original gauge length of the sample or the initial separation of the jaws. The gauge length of a dumbbell specimen is equal to the length of its narrow region in which the edges are parallel. When strips of uniform width and thickness are being tested, the gauge length is equal to the initial separation of the jaws. Ideally the strain in the sample would be measured directly using an extensometer (strain gauge). The load as a function of deformation or strain may be plotted directly, but it is more common to store the data on a computer for subsequent display and analysis. With appropriate modifications such instruments can be used to measure compressive, flexural, peel, and tear properties, the configurations of which are shown in Figure 25.

Variations on the basic testing configurations abound. Most instruments apply load vertically, but some are horizontal. Multiple sets of grips and load cells permit the simultaneous testing of duplicate specimens. Load cells come in various sizes, shapes, and sensitivities. The jaws that grip the sample may be mechanically, pneumatically, or hydraulically operated. Extensometers may be attached physically to the sample or track the movement of fiducial marks optically. Specimens may be enclosed in an environmental chamber, the interior of which may be heated or cooled, or filled with various media.

Mathematical analyses of the force versus elongation data yield a variety of results including initial modulus, yield stress and strain, and stress at break. Typically a minimum of five specimens are tested and the results averaged to provide the values to be quoted for the material. Some of the principal values of interest and their methods of calculation are discussed in subsequent sections.

When discussing mechanical testing, a clear distinction should be made between force versus elongation data and stress versus strain data. The terms ‘‘force’’ and ‘‘elongation’’ are independent of sample geometry; they are simply the force registered by the load cell of the instrument and the relative position of the jaws with respect to their original separation. Stress and strain are sample-dependent. The stress on any element of the sample is equal to the force experienced by the element divided by its effective cross-sectional area. If the cross-section of a specimen varies along its length, or the force experienced by a given element is different from that of its neighbors, the stress will vary accordingly, i.e., stress is not necessarily uniform along the length or across the width of a specimen. The strain, percent strain, and draw ratio for any portion of a specimen are defined as

$$\text{Strain} = \frac{\text{current sample dimension} - \text{original dimension}}{\text{original dimension}}$$

$$\text{Percent strain} = \frac{\text{current sample dimension} - \text{original dimension}}{\text{original dimension}} \times 100$$

$$\text{Draw ratio} = \frac{\text{current length}}{\text{original length}}$$

Characterization and Testing

(a) Compression

(b) Three-point bend

(c) T-peel

(d) Trouser tear

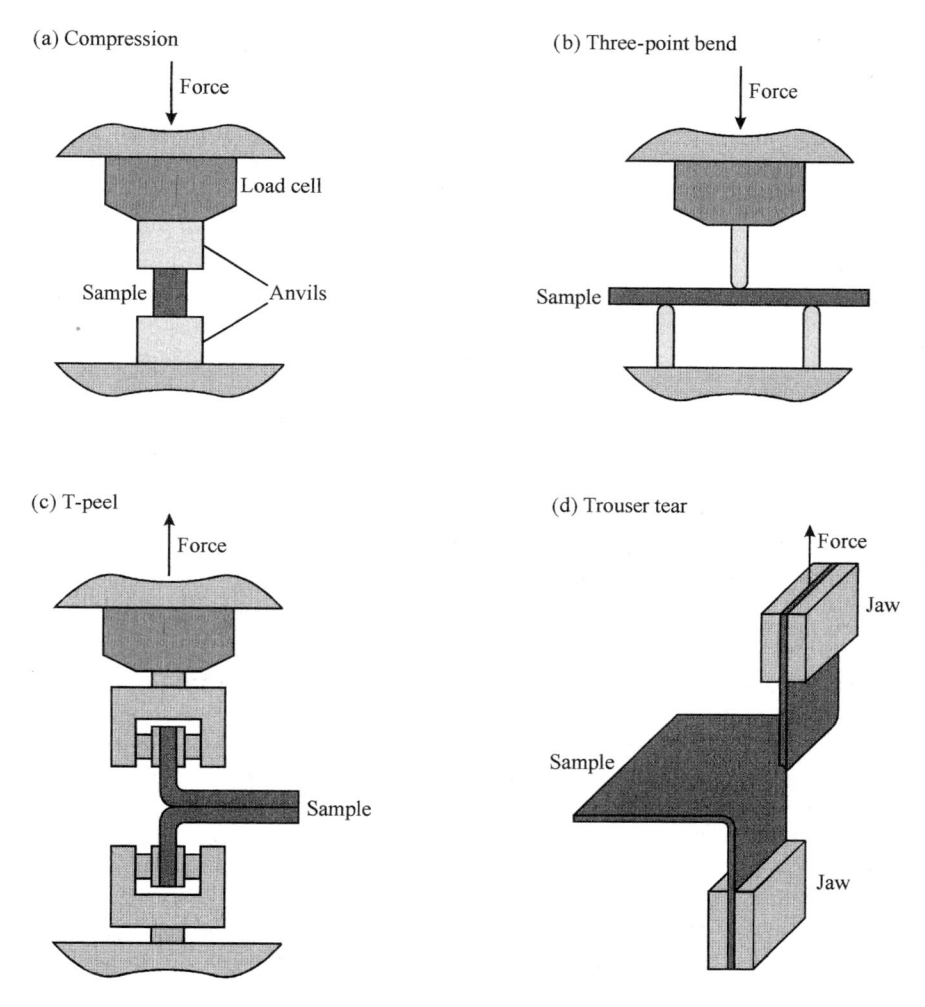

Figure 25 Schematic illustration showing the testing configurations used to measure. (a) Compressive, (b) flexural, (c) T-peel, and (d) "trouser tear" properties.

Polyethylene specimens rarely deform homogeneously; therefore, strain is rarely constant along their length during the deformation process. Bearing in mind that stress and strain vary within a polyethylene specimen, it is misleading to refer to the stress versus strain curve of a whole specimen, the use of the term "force versus elongation" being more accurate. (A good approximation to a stress versus strain curve can be generated by calcualting the instantaneous stress and strain for a single element during the deformation process [95].)

Samples of polyethylene available from typical conversion processes are rarely isotropic. When determining the properties of such samples it is crucial that the testing direction with respect to the orientation axes of the sample be controlled. It is common to make property determinations on two sets of specimens cut from anisotropic samples, one parallel with the principal orientation direction and the other perpendicular to it.

1. Tensile Testing

Tensile testing is one of the most common forms of physical testing performed on polyethylene samples. The data available from such testing are extremely informative with regard to both applications and fundamental knowledge. A generalized force versus elongation curve illustrating the major points of interest is shown in Figure 3 of Chapter 5. It should be recognized that, due to the variety of morphologies associated with polyethylene, there is no typical force versus elongation curve applicable to all samples. Figures 4–7 in Chapter 5 depict a series of tensile force versus elongation curves generated from samples with a wide range of molecular and morphological characteristics. Standardized methods applicable to the tensile testing of polyethylene are described in ASTM Methods D 638, D 882 and D 1708.

a. Elastic Modulus. The elastic modulus (also known as the initial or Young's modulus, E) of a sample is effectively a measure of its stiffness. Ideally it is defined as the stress required to effect unit strain in the elastic portion of the force versus elongation curve. In practical terms it is the slope of stress as a function of strain in the region prior to yielding. The difficulty with this definition in the case of polyethylene is that for most samples any applied strain is nonrecoverable to some extent. The net result of this is that there is rarely an initial linear portion of the force versus elongation curve that can be used to define the elastic modulus. Thus for most polyethylene samples there cannot be said to be a true elastic modulus. Obviously, for comparative purposes it is desirable to be able to quote a value for the initial modulus of a sample. Accordingly, various stratagems have been devised to obtain a value, four of which are illustrated in Figure 26.

To determine the tangent modulus, a straight line is drawn tangent to the curve; the slope of which is reported. Typically, the steepest part of the curve is selected; this is invariably the initial slope, in which case it is reported as the initial modulus. When the tangent is taken at a specific strain level, this value must also be reported. To determine the chord modulus, two points on the curve are selected and the slope of the line connecting them is calculated. When the initial point chosen is 0% strain, the value is reported as the secant modulus of the upper level, e.g., 1% or 2% secant modulus. In all modulus calculations it is assumed that deformation is homogeneous at low strains and that deformation

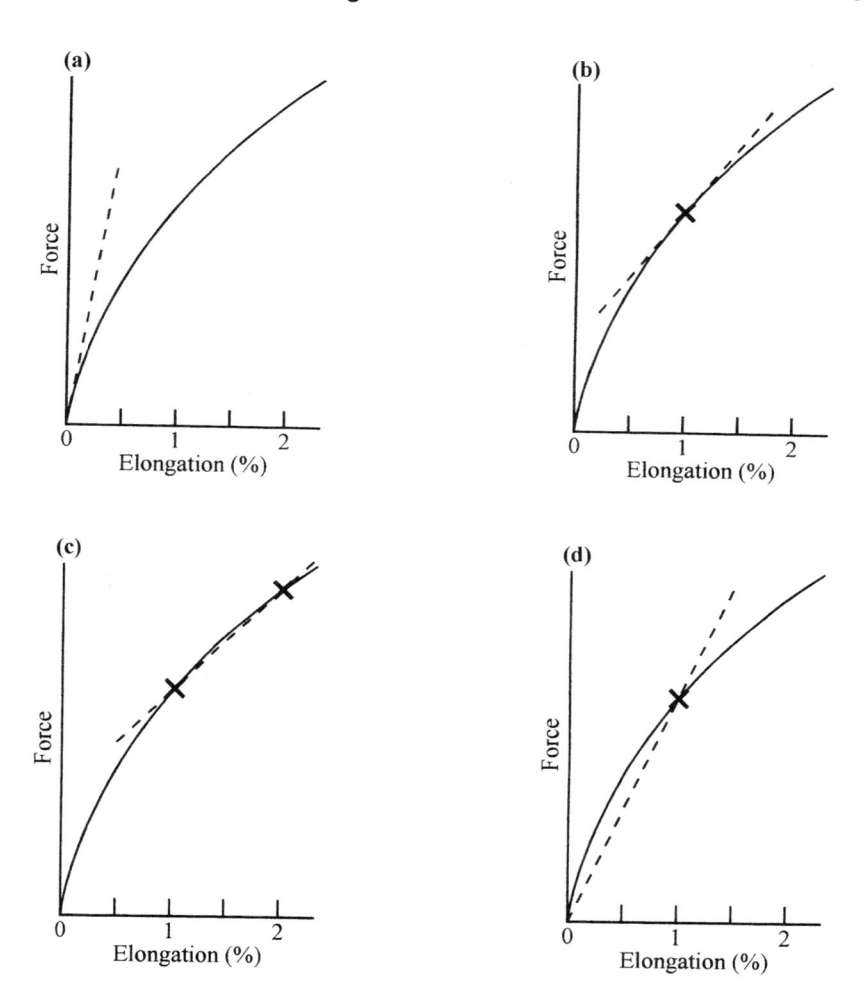

Figure 26 Schematic illustration of four of the principal methods used to calculate elastic modulus. (a) Initial modulus; (b) tangent (1%); (c) chord, (1–2%); (d) secant (1%).

occurs only in the gauge region of the sample. Ideally the strain used in the calculation of modulus is measured directly with an extensometer, which accurately measures the separation of two points on the sample during the deformation process. In the absence of an extensometer, strain is defined as the elongation of the specimen relative to its original gauge length. For a general discussion of various methods used for calculating modulus, the reader's attention is directed to ASTM Method E 111.

Due to the low strains involved in the calculation of initial modulus and the likelihood of curvature of the force–elongation curve, the reported values will be subject to some degree of error. With judicious selection of the calculation parameters these errors can be reduced, but differences will always exist between the values calculated by the different methods. The standard deviation for a series of specimens cut from a single sample should not exceed 5% for a given calculation method. Different calculation methods may provide values that are discrepant by 50% or more.

b. Yield Phenomena. A specimen is said to yield at the point beyond which applied strain is no longer fully recoverable. If the luxury of a true stress versus strain plot is available, the yield point can be obtained unambiguously using Considère's construction. In practice, the yield point is identified by one of various methods depending upon the shape of the force versus elongation curve and operator preference. Three such methods are illustrated in Figure 27. When a well-defined yield peak is present, the yield point is identified as the first maximum in the curve. When there is no stress maximum, the yield point may be defined as the point at which deviation from linearity exceeds a specified offset value. Alternatively, the yield point may be identified with a point on the force versus elongation curve corresponding to a defined change in slope. As would be expected, each of these methods provides a slightly different value of yield stress (σ_y, YS) and yield strain (ε_y, YE).

The yield stress of a specimen is conventionally reported as the force required to induce yielding divided by the initial cross-sectional area of the specimen (i.e., the engineering stress). If a specimen deforms extensively before the onset of yielding, the original cross-sectional area will be reduced significantly and the value reported as the yield stress will not accurately reflect the true stress at yield.

The strain at yield is the strain induced in the sample at the point of yield. Ideally it would be measured directly with an extensometer, but it is often calculated from jaw displacement, assuming that all deformation occurs within the gauge region (which is not necessarily true, especially for ductile samples).

The errors associated with the yield stress calculated by a given method should not exceed 5%. The relative values returned by the different calculation methods applied to a single sample will depend upon the calculation parameters chosen. In general, the value calculated from the yield maximum will be the largest.

Errors in the yield strain are typically greater than those for the yield stress, largely because relatively small extensions are involved. Errors of $\pm 5\%$ or more for a given calculation method can be expected. The relative values returned by the different calculation methods will vary according to the calculation parameters selected.

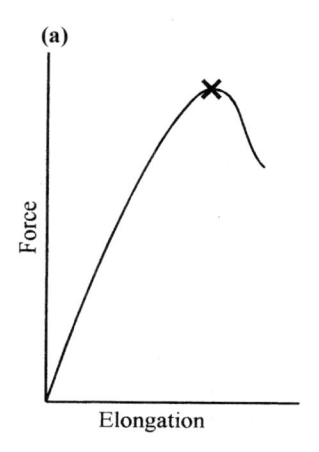

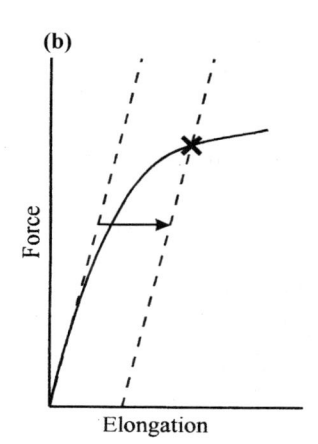

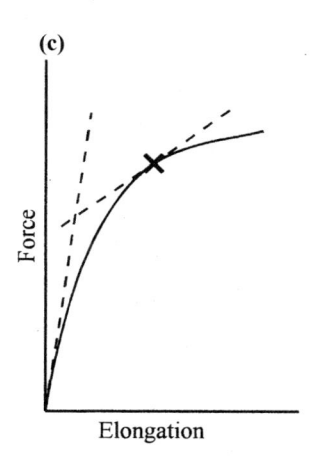

Figure 27 Schematic illustration of various methods used to calculate the yield point. (a) First maximum; (b) offset yield; (c) change of slope.

    c. Tensile Strength. The tensile strength (TS) of a specimen is conventionally reported as the maximum force measured during a tensile test divided by the original cross-sectional area of the specimen. In the case of polyethylene the maximum recorded force typically occurs just prior to break; tensile stress thus coincides with stress at break (σ_b, also known as the ultimate tensile stress, UTS). The conventional value does not take account of reduced cross-sectional area due to drawing. Consequently it is sometimes referred to as the engineering stress at break to distinguish it from the "true" stress at break (true ultimate tensile stress, TUTS), which makes allowance for the reduction of a specimen's cross-sectional area.

The tensile strength of specimens that strain harden (which includes most polyethylene samples) is strongly dependent upon their elongation at break. Bearing this in mind, standard deviations of 10–20% are realistic. Errors associated with the "true" ultimate tensile stress will be greater than those associated with the tensile strength because their value is compounded by variation of elongation prior to break.

   d. Elongation at Break. The elongation at break of a polyethylene sample is reported as the observed strain (ε_b), percent strain, or draw ratio (λ_b) that occurs immediately prior to sample failure. Ideally the strain at break would be measured directly with an extensometer. However, it is more common to assume that all deformation occurs within the gauge region of the sample and to calculate the elongation based upon crosshead displacement as a function of the initial gauge length. This can result in unrealistically high values of elongation when deformation occurs in portions of the specimen outside the gauge region. Elongation at break is also highly susceptible to the manner in which the specimen is gripped. Strips held between planar jaw faces frequently break at much lower strains than those held in "line grips." Premature break also reduces the observed tensile strength. The difference between planar jaw faces and line grips is illustrated in Figure 28.

The ultimate achievable strain for a specimen may be reduced by sample defects, overlaying its dependence upon material properties. Thus the variability associated with compression-molded specimens is likely to be higher than that found for molded specimens or those cut from commercially prepared films. Er-

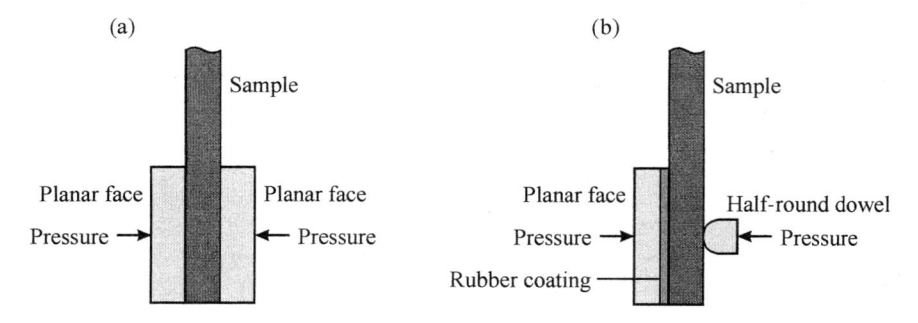

Figure 28 Schematic illustration of gripping methods. (a) Planar grip faces; (b) line grips.

rors in the range of ±10–20% could be expected once any abnormally low values are rejected.

e. Tensile Heat Distortion Temperature. The determination of heat distortion temperature by means of tensile testing is performed according to ASTM Method D 1637 on samples with a thickness of less than 0.06 in. (Thicker specimens should be tested in the flexural mode.) A load sufficient to develop a stress of 50 psi is applied to a strip specimen supported in an oil bath or oven. The relative strain in the sample is measured as its temperature is increased at a rate of 2°C/min. The heat distortion temperature is reported as the temperature at which the specimen exhibits a change in length, either elongation or shrinkage, equivalent to a strain of 2%. A minimum of two specimens should be tested and the results averaged. In the case of anisotropic samples, strips should be cut both parallel with and perpendicular to the principal axis of orientation.

A reproducibility of ±2°C is acceptable within a single laboratory, with errors of ±10°C between different laboratories. The results obtained from this test should be used for comparative purposes only. The temperatures so determined are not directly indicative of end use performance.

2. Compressive Testing

Compressive testing of polyethylene can be carried out on blocks, rods, or tubes. The compressive equivalents of tensile modulus, yield stress, yield elongation, and tensile stress can be calculated, but there is no equivalent to tensile elongation at break. The instruments used for such testing are similar to those used for tensile testing, with the exception that the jaws of those tensile testers are replaced by "anvils" that compress the sample. ASTM Method D 695 describes a regime of testing suitable for the determination of the compressive properties of plastic samples.

Compressive data of a limited kind can be obtained by using a constant-load device in which the deformation of the sample is measured as a function of time subsequent to loading. ASTM Method D 621 describes procedures suitable for this type of determination.

a. Hardness Testing. The relative hardness of polyethylene specimens may be determined with Rockwell or durometer hardness testers according to ASTM Methods D 785 and D 2240, respectively. The hardness values so determined are a function of both the elastic and viscous components of a sample's deformation. Such values are useful for comparative purposes, but they cannot be used to predict service performance. Relative rankings obtained for a series of samples by one type of test do not necessarily correspond to that of the other type of test.

The Rockwell hardness of a sample is calculated from the difference in the depth to which a steel sphere of standard size penetrates a specimen when impressed by two standard loads for a set period of time. The hardness of the sample determines the diameter of the steel ball and the two loads to be used. Different ranges are identified by one of a number of hardness scales. The harder the sample, the smaller the ball and/or the heavier the second load.

Durometer hardness testing determines the resistance of the sample to penetration by an indentor, which is a spring-loaded point initially protruding 1 mm from a flat surface. The profile of the indentor and the elastic constant of the spring to which it is attached define the type of durometer. In the case of polyethylene, durometer Type D is commonly used. When the flat surface is brought into firm contact with a block of polymer (minimum thickness 6.35 mm), the depth to which the indentor penetrates the specimen is inversely proportional to its hardness. After a fixed interval of time (commonly 10 secs), the hardness, in arbitrary units, is read directly from the scale.

3. Shear Testing

Shear testing of polyethylene can follow one of two different regimes that measure different aspects of the material's properties. The shear strength of a sample is determined by punching a hole in it according to the procedure described in ASTM Method D 732. Shear modulus is determined from a test that applies torsion to a specimen, following a procedure described in ASTM Method D 1043.

a. Shear Strength. The shear strength of a specimen is defined as the force necessary to shear one part of the specimen away from the rest divided by the area of the sheared face. A circular punch is used to pierce a sheet having a thickness of 0.05–0.5 in. A mechanical tester set up for use in the compressive mode is used. A circular or square specimen is clamped into a specimen holder, and the punch is forced through it at a constant rate of 0.05 in./min. The maximum force registered is divided by the circumference of the punch and the thickness of the specimen to yield the shear strength. The shear strength is reported as the average of a minimum of five specimens.

Repeatabiltiy within one laboratory should be less than $\pm 2\%$ with a variation between laboratories of approximately $\pm 5\%$. The shear strength determined by this method depends upon a variety of factors that may or may not be thickness-dependent; accordingly, it is possible to compare results only from samples of similar thickness.

b. Shear Modulus. The apparent shear modulus (apparent modulus of rigidity) can be obtained from a torsional test in which a twisting force is applied to a rectangular strip cut from a sample. The term "apparent" is used because this testing method involves both recoverable and nonrecoverable deformational components. The specimen is mounted in a pair of grips aligned along a common rotational axis, the lower set fixed and the upper rotatable via a pulley system. When a fixed torsional load is applied to the upper grips via the pulley system, the sample experiences a torsional force. The resulting angular strain is measured 5 sec after the application of load. The apparent shear modulus is calculated on the basis of the specimen dimensions, the load applied, and the observed angular strain. Details of this testing method and the calculations associated with it are available in the ASTM method.

4. Flexural Testing

The flexural testing of polyethylene can be performed in a two-, three-, or four-point bending mode, the configurations of which are illustrated schematically in Figure 29. Three-point bending produces a line of maximum stress directly beneath the central beam, whereas the four-point mode results in maximum stress in the region between the two central beams. The three- and four-point modes, described in ASTM Method D 790, are typically used for stiffer samples, while the two-point (cantilever beam) method, described in ASTM Method D 747, is used for more flexible ones.

The flexural deformation of polyethylene involves both recoverable and nonrecoverable components, precluding the determination of a true flexural modulus; accordingly, an apparent modulus is reported. A minimum of five specimens should be tested and their results averaged. For anisotropic samples, testing should be carried out both parallel with and perpendicular to the principal orientation direction. Reproducibility of results for all the modes of testing is of the order of ±5%.

a. Three- and Four-Point Bending. The instrumentation used in the three- and four-point bending modes is similar to that used for compressive testing. Flexural testing of polyethylene is normally carried out to limited deformations, generally to a degree no more than is necessary to define flexural yielding. Polyethylene samples rarely fracture in the flexural mode unless they exhibit very high degrees of crystallinity. The apparent flexural (bending) modulus is calculated from the force required to deform the specimen to some fixed extent (the apparent secant modulus) or from a tangent drawn from the steepest part of the force versus deformation curve (modulus of elasticity). Standard testing beam separations, deformation rates, and equations for calculating the apparent modulus and yield stress are listed in the ASTM method.

b. Cantilever Beam Testing. The cantilever beam flexural testing method determines the force required to bend a specimen through a series of angles. From the initial slope of the force versus angle plot, the apparent flexural modulus can be determined according to the formula in the ASTM method.

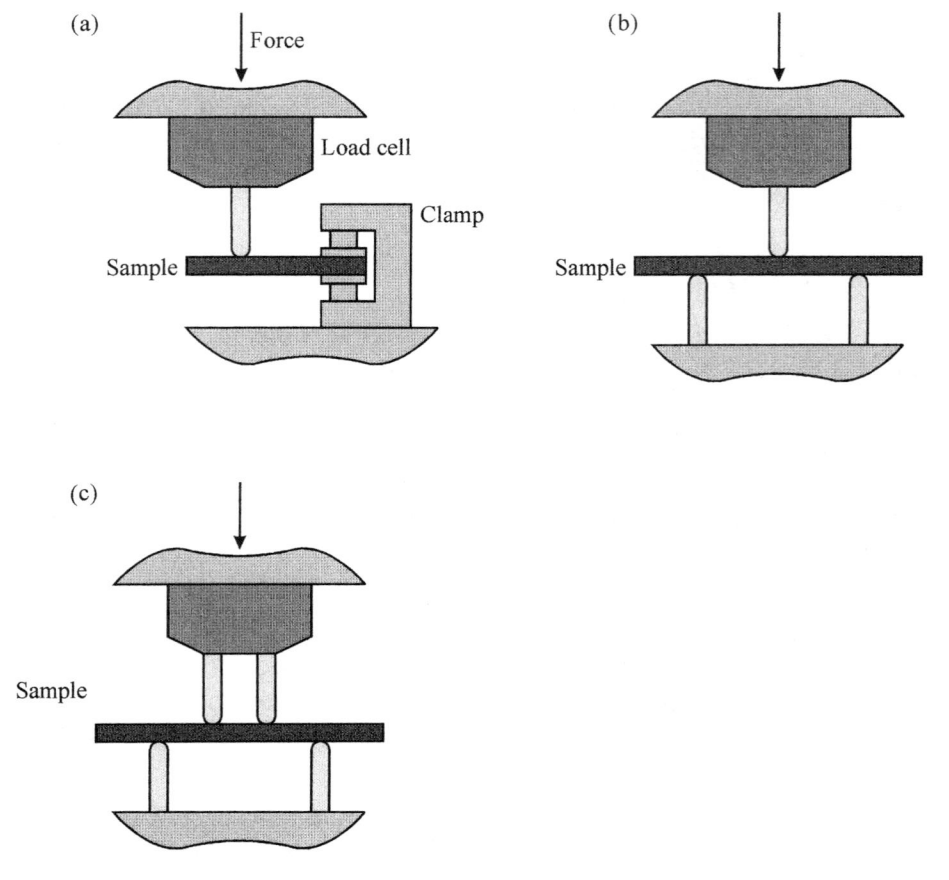

Figure 29 Schematic illustration of flexural testing configurations. (a) Two-point; (b) three-point; and (c) four-point bending.

    c. Flexural Heat Distortion Temperature. The flexural heat distortion temperature of a sample is determined according to ASTM Method D 648. Samples with thicknesses in the range 0.04–0.5 in. may be tested. (Thinner samples are tested according to the tensile heat distortion method). A specimen 5 in. long and 0.5 in. wide is supported horizontally on parallel bars 4 in. apart in an oil bath or oven. A load that results in a fiber stress of 66 psi is applied to its center while the temperature is increased at a rate of 2°C/min. The heat distortion temperature is reported as the temperature at which the sample deflects by 0.01 in.

Anisotropic samples should be tested both parallel with and perpendicular to the principal orientation direction.

The results of this test may be considered to be repeatable to a precision of ±5–10°C. The results obtained from this test should be used for comparative purposes only, the temperatures so determined not being directly predictive of end use performance.

5. Tear Strength

The determination of tear resistance is relevant to films and thin sheets of polyethylene. Two types of film tear strength are measured: initial resistance to tearing and resistance to tear propagation. Determination of resistance to tear initiation is specified in ASTM Method D 1004, while tear propagation resistance is measured according to ASTM Method D 1922 or D 1938.

a. Initial Tear Resistance. The determination of initial film tear resistance is performed using a tensile testing instrument. The geometry of the specimen specified by the ASTM method is such that stress is concentrated at a 90° notch. Specimens of the shape illustrated in Figure 30 are die cut from the film of interest. When the specimen is deformed at a constant rate, a tear initiates at the tip of the notch, and the applied force at which this occurs is noted. The initial tear resistance is reported as the force required to initiate rupture divided by the film thickness. The vast majority of commercially manufactured films are anisotropic; accordingly the initial tear resistance should be determined both

Figure 30 Specimen geometry used in the ASTM Method D 1004 determination of initial tear resistance in films.

parallel with and perpendicular to the principal orientation direction. A minimum of 10 specimens should be run in each direction and the results averaged. Because of the many factors involved during the tearing of films, it is not possible to directly compare the results of tests made on films that differ in thickness by more than 10%.

b. Tear Propagation Resistance. The determination of tear strength propagation resistance involves the extension of a preexisting razor cut through a film specimen. Two different methods of determination are routinely used. One effectively measures the energy absorbed in propagating a tear through a standard distance, while the other measures the load required to propagate the tear. The former uses a pendulum-type instrument and the latter a tensile tester. Both methods provide information that can be used to rank the tear propagation resistance of films.

The standard pendulum-type measurement of tear strength is known as the Elmendorf test. A schematic illustration of a typical instrument used in the pendulum-type measurement is shown in Figure 31. The geometry of the specimen, specified by the ASTM method, is illustrated in Figure 32. The specimen contains a precut slit and is so shaped that the tear must traverse a fixed distance

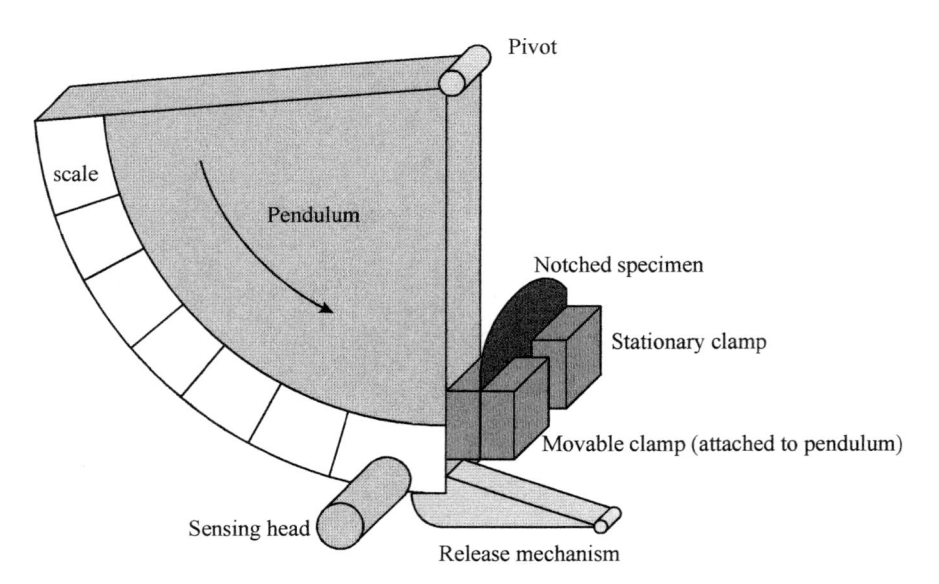

Figure 31 Schematic illustration showing the basic configuration of a pendulum-type tear propagation resistance instrument.

Characterization and Testing 325

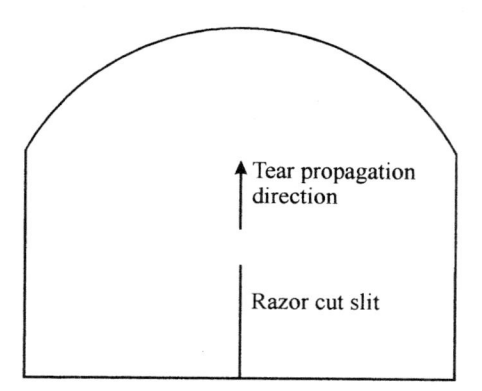

Figure 32 Specimen geometry used in the ASTM Method D 1922 determination of propagation tear resistance in films.

between the notch tip and the opposing edge of the film. The specimen is gripped in two clamps; one is stationary, and the other is movable, attached to a pendulum of known mass. When the raised pendulum is released, it swings down through an arc, taking the movable clamp with it and tearing the specimen. The energy absorbed by the propagation of the tear is determined by comparing the heights to which the pendulum rises with and without the specimen in place. A minimum of ten specimens should be run for isotropic samples and the results averaged. For anisotropic films a minimum of 10 specimens should be run parallel with and perpendicular to the principal orientation axis of the film. The tearing force is reported as the relative loss of energy multiplied by the rated mass of the pendulum. The thickness of the film is also reported, but comparative ranking can be made only between films having closely matched thicknesses.

The second type of tear propagation resistance test uses a tensile testing instrument. It is colloquially known as the "trouser tear test," so named for the specimen geometry specified by the ASTM method, which is illustrated in Figure 33. The tabs marked A and B are gripped in the opposing jaws of a tensile tester, which are then separated at a rate of 10 in./min. The force required to propagate the tear is measured by a load cell. If the resistance to tear propagation is relatively constant, the average force required to propagate the tear is reported. However, if specimens deform extensively prior to tear propagation, both the force required to initiate propagation and the maximum force observed are reported. As with the pendulum type of test, the thickness of the film is reported, with comparative rankings permissible for films having nearly identical thicknesses.

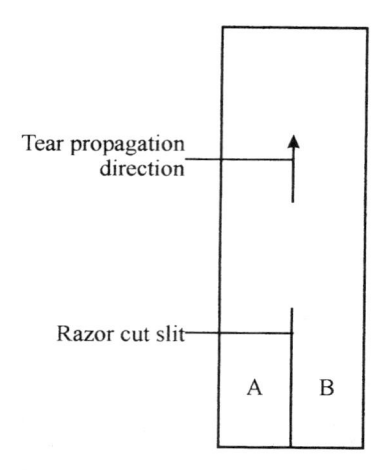

Figure 33 Specimen geometry used in the ASTM Method D 1938 determination of propagation tear resistance in films.

6. Creep Measurement

The measurement of creep involves the determination of a sample's response to applied stress over a prolonged period of time. Measurements can be made in tension, compression, torsion, and flexion, for periods extending from hours to years. Naturally, suitable instruments must be capable of maintaining their calibration for the duration of the experiment; this requires careful design and construction. The environment in which experiments are conducted must be precisely controlled to eliminate long-term fluctuations that may affect results. For a thorough analysis, a series of different loads should be applied at two or more temperatures; several specimens being tested under each set of conditions. To fulfill these requirements numerous instruments must be used to run experiments concurrently or a few instruments to run experiments sequentially. The former case is extremely capital-intensive, while the latter may involve inordinately long periods of testing. Clearly, creep measurement is not an endeavor to be undertaken lightly.

In each testing mode the same basic procedure is followed. A load is applied to the specimen under the appropriate conditions of temperature and chemical environment, and its deformational response is followed over a prolonged period of time. The applied load is much less than that required to induce yielding of the sample during standard mechanical property testing. The selection of appropriate experimental parameters requires a knowledge of how the creep data will be used. As creep data are principally used to predict reliability in end use

Characterization and Testing 327

applications, it is important to select testing conditions that permit the results of tests to be extrapolated over the desired time period. Chemical environments should be duplicated as closely as possible, and stresses and temperatures should be within the range over which time/temperature superposition can be applied.

Creep testing is generally performed under conditions of fixed load. Equipment is available that compensates for sample deformation to provide a constant level of stress. The appropriate load is applied to a sample rapidly and evenly, at which point a timer is started. One or more dimensions of the sample are monitored as a function of time, either on a continuous basis or intermittently, the interval between measurements being increased approximately logarithmically. Data can be reported in various fashions, the two principal ones being the length of time to reach a pre-determined failure criterion as a function of testing conditions, and deformation as a function of the logarithm of time under the various testing conditions. A value of "creep modulus" can be obtained by dividing the initial stress by the observed strain at some point in time. The creep modulus naturally varies as a function of time, depending upon the response of the sample to the testing conditions. Isochronous stress versus strain curves can be plotted in which the observed strain at a series of fixed observation times is plotted against a series of initial stress values. Such curves are useful for predicting strain as a function of stress and time. Recommended experimental conditions and testing procedures can be found in ASTM Method D 2990.

B. Impact and Puncture Resistance Determination

Many applications of polyethylene exploit the excellent toughness of many of the grades available. This toughness may be exhibited in a number of beneficial ways such as puncture resistance of films, drop strength of blown bottles, and impact resistance of molded items. The types of loads that polyethylene items are exposed to vary extensively; it is therefore desirable that toughness be measurable under a similarly broad range of configurations. To this end, numerous testing regimes have been developed to determine the toughness of specifically prepared samples or the ability of fabricated items to withstand specific hazards. The wide range of available tests is based on a few general procedures, modified appropriately to investigate the property of concern. The basic procedures are discussed in the following subsections.

1. Impact Beam Testing

The response of small beams of polyethylene to impact is highly dependent upon the nature of the specimen. Narrow beams typically deform by buckling or twisting, while thicker ones—especially those that have defects—rupture by the processes of crack initiation and crack propagation. To obtain reproducible results from beam testing it is important that specimens be thick enough to avoid buckling and twisting under impact. (Materials that are too flexible for impact beam testing should be tested according to the tensile impact methodology.) The impact strength of thick beams is highly sensitive to the presence of notches, either adventitious or deliberately introduced, which can drastically reduce impact resistance.

The impact testing of polyethylene beams is normally performed with a pendulum-type tester in which a weighted pendulum strikes and breaks a specimen. The energy required to rupture the sample is calculated from the reduction in swing height of the pendulum after it strikes the specimen. In the Izod configuration (also known as the cantilever beam test), a beam is mounted vertically, the lower part of it being clamped in a vise while the upper part is struck by the pendulum. In the Charpy configuration (also referred to as the simple beam test), the beam is mounted horizontally, its ends resting unclamped against supports while the center of the beam is struck. A schematic illustration of a pendulum type impact tester is shown in Figure 34, with the Izod and Charpy configurations being illustrated in Figure 35. Such testing equipment can be used to determine

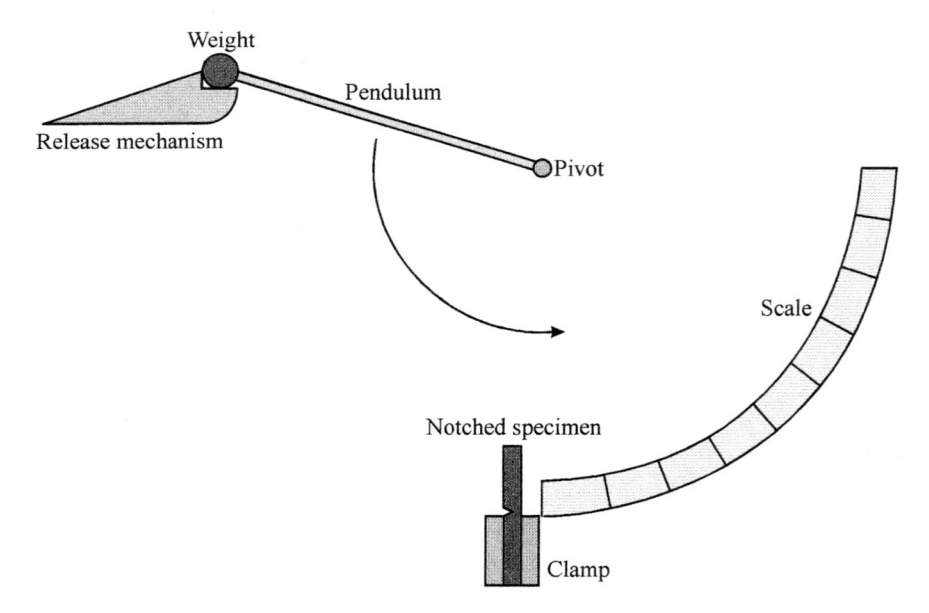

Figure 34 Schematic illustration showing the basic configuration of a pendulum-type beam impact resistance instrument.

Characterization and Testing

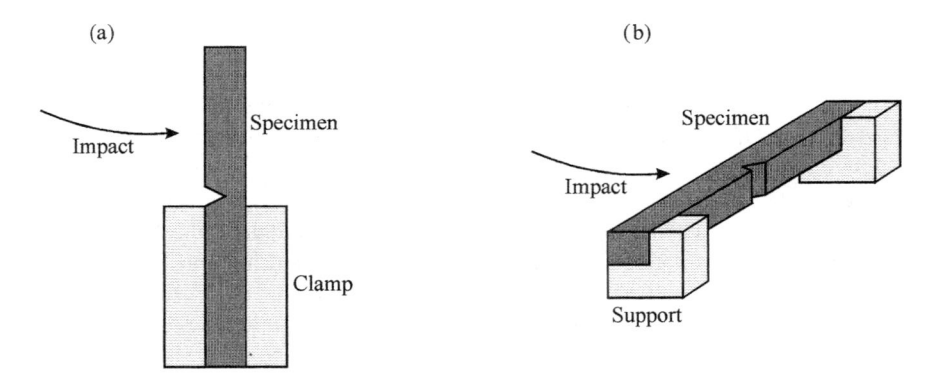

Figure 35 Illustration of (a) Izod and (b) Charpy beam impact testing configurations.

the impact resistance of both unnotched and notched beams, in accordance with ASTM Methods D 4812 and D 256, respectively.

The basic operation of the pendulum tester is identical in the Izod and Charpy configurations. A weighted pendulum is supported in the raised position while a specimen is mounted on the appropriate support or gripped in a vise. When the pendulum is released it swings down, striking and breaking the sample at its lowest point of travel and swinging up on the far side. The loss in height of the swing compared to that when no specimen is in place is a measure of the energy required to rupture the sample and propel the broken part to its final resting place. A minimum of five specimens are tested in any given configuration, and the impact resistance is reported as the average value in terms of energy absorbed divided by the width of the specimen. The nature of the break is also noted as "complete," "hinged," "partial," or "no-break" for each specimen. A hinged break is one in which the two halves of the specimen remain connected by a ligament that is too weak to support either part in the vertical position. A partial break is one in which the break extends more than 90% of the way through the specimen but the remaining ligament does not form a hinge. The no-break condition is an incomplete break that spans less than 90% of the thickness of the specimen. Tests can be performed on specimens at room temperature or those that have been conditioned prior to testing at any desired temperature, humidity, etc. It is common to run such tests at more than one temperature, frequently at ambient and a much lower temperature.

There is no direct relationship between the impact resistance of notched and unnotced specimens. In a series of samples the same general trend may be followed, but the actual rankings are liable to vary appreciabley. Errors associated with impact beam testing are of the order of ±5–15% for different operators using different equipment.

 a. Unnotched Testing. The testing of unnotched samples is normally carried out in the Izod testing configuration. A standard specimen 2.5 in. long, 0.5 in. wide, and 0.125 in. thick is clamped vertically so that one half of its length protrudes from the jaws of the vise. The specimen is struck upon its narrow face by a pendulum bob of known weight swung through a standard arc. Specimens of other dimensions may be used if necessary, but the results obtained with them cannot be directly compared to those obtained with specimens of the standard dimensions. Various pendulum weights are used, depending upon the toughness of the sample, but all strike the specimen in exactly the same spot with a striking head of standard dimensions. The energy absorbed to rupture the specimen is determined from the height to which the pendulum rises after striking the specimen.

 b. Notched Testing. The testing of notched samples can be carried out in either the Izod or Charpy testing configuration. Specimens having the standard length, width, and thickness are notched at the center of their span to a depth of 0.1 in. using a tool that leaves a notch with a tip radius of 0.01 in. The specimen is mounted in such a way that when struck the notch is subjected to a tensile opening force. In the Izod regime the sample is struck on the notched face, while in the Charpy regime it is struck on the opposite face. The weight of the pendulum is chosen with regard to the impact resistance of the sample.

 The sensitivity of a sample's impact resistance to notch radius can be determined by testing two sets of specimens with notch tip radii of 0.01 and 0.04 in., respectively. The notch sensitivity of the sample is reported as the difference between the impact resistance measured at the two notch radii divided by the difference between the radii.

 A measure of a sample's resistance to unnotched impact can be obtained by mounting notched specimens in the Izod configuration such that they are struck on the face opposite the notch, i.e., reversed from the normal mounting direction. The results obtained for unnotched impact resistance determined by reversing a notched beam do not always coincide with those determined directly for an unnotched beam.

2. Impact Plaque Testing

The impact resistance of injection-molded plaques or thick extruded sheets of polyethylene can be tested by means of a falling weight method. Two variants of a similar testing configuration are used; in one a tup (impactor) falling through a guide tube strikes a specimen directly, while in the other a falling weight strikes a separate impactor already positioned in contact with the specimen. In both cases

Characterization and Testing

the specimen is mounted or clamped on an annulus. A schematic illustration of a falling weight impact tester is shown in Figure 36. The energy required to crack or split the specimen is determined from the mass of the weight and the distance through which it falls. Standard testing conditions are to be found in ASTM Method D 3029. One of the geometries specified in the ASTM method (type GB) coincides with the conditions of the commonly used Gardner impact test.

In both the free-falling dart and the weight-striking impactor variants, the falling mass and the distance it drops may be altered within broad limits. The dimensions of the tup or impactor and the support ring are specified in the ASTM method. A projectile of known weight with a striking face of standard dimensions is dropped from a preselected height onto the center of the specimen. If the specimen fails, the weight of the dart is reduced by a fixed increment; if it remains intact, the weight is increased by the same increment. Alternatively, the distance through which the weight drops may be varied by a fixed increment. Failure is defined as a crack or split discernible with the naked eye. A fresh specimen is mounted in the instrument and the test repeated. This process is repeated using appropriately increased or decreased dart weights or heights until 20 specimens

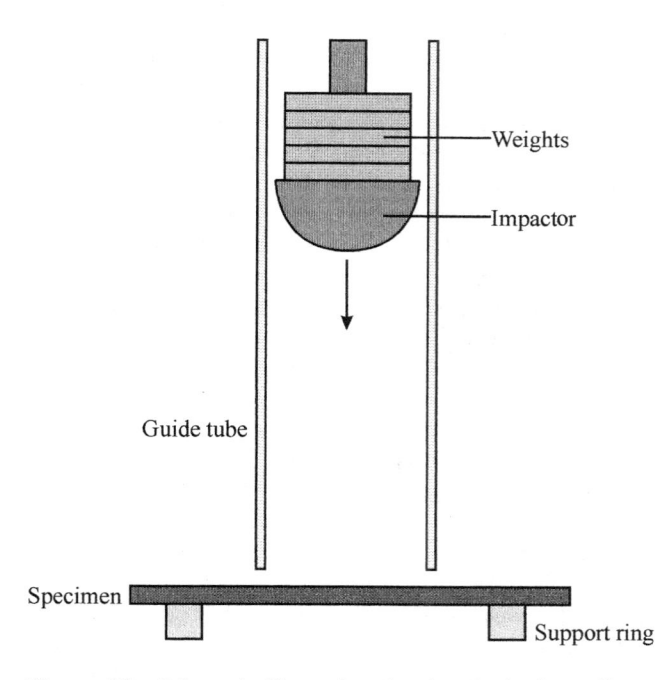

Figure 36 Schematic illustration showing the basic configuration of a falling weight impact resistance tester.

have been tested. This type of testing procedure is known as the Bruceton staircase or up-and-down method. The numbers of failed and intact specimens at each projectile weight (or height) are noted, and the impact strength is calculated using the procedure detailed in the ASTM method. Tests may be run on specimens over a wide range of temperatures. A stable thermal environment is important, because results are often highly sensitive to small temperature changes.

The results of impact plaque testing depend on the configuration of the support and tup employed. The results obtained with a given configuration are not directly comparable with those of other configurations. The data generated by this test are useful for ranking samples that have been tested under identical conditions and fail in a similar manner. Such data may not be directly relevant to end use applications.

3. Tensile Impact Testing

Tensile impact testing is used to determine the impact resistance of samples that are too thin or too flexible for impact beam or impact plaque testing but cannot be considered films. It involves strain rates that are intermediate between those used for impact beam testing and those used for conventional tensile testing. This procedure is performed according to ASTM Method D 1822.

A small dogbone specimen is securely clamped by its ends in a pair of grips, one of which is part of the weighted pendulum on a pendulum-type tester, while the other is part of a crosshead that initially moves in unison with the pendulum. When the pendulum is released it swings down until it reaches the bottom of its arc, at which point the movement of the crosshead is halted by impact against a rigid anvil. The pendulum continues to swing, applying tension to the sample, one end of which moves with the pendulum while the other remains stationary in the crosshead. If the pendulum has sufficient momentum, the sample will break. The energy absorbed by the specimen's failure is determined from the height to which the pendulum swings compared to that when no specimen is present. A schematic illustration of a tensile impact tester is shown in Figure 37.

Two types of dogbones are used; one is a waisted specimen in which the gauge length is effectively zero (type S), and the other has a short parallel gauge region (type L). When the latter breaks it does so with greater elongation than the former, which effectively gives it a higher energy to break. The strength of a specimen is calculated in terms of energy per unit cross-sectional area.

The reported tensile impact resistance is the average for a minimum of five specimens. Samples that are believed to be anisotropic should be tested both parallel with and perpendicular to the principal axis of orientation. As with other impact resistance tests, the results obtained from this method are not predictive of end use performance. No direct relationship exists between the results obtained

Characterization and Testing

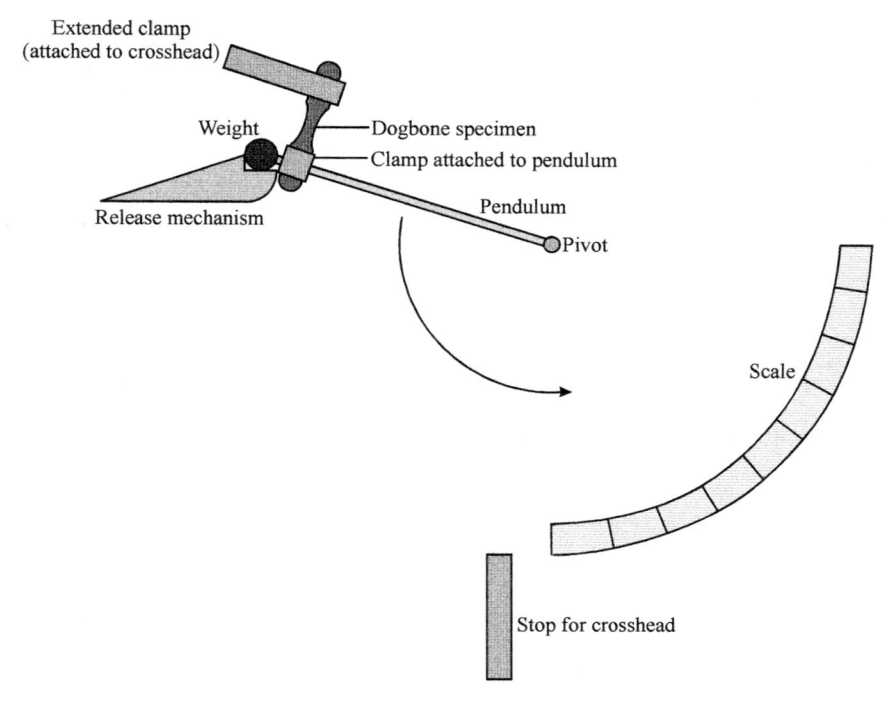

Figure 37 Schematic illustration showing the basic configuration of a tensile impact tester.

by the various testing procedures for a given material. All impact resistance results should be used for comparative ranking purposes only.

4. Film Puncture Resistance Testing

The puncture resistance of polyethylene films is measured by various procedures that employ a falling or swinging projectile to rupture a film. Two principal methods of measurement are used. The first determines the weight of a projectile, falling from a fixed height, that has a 50% probability of initiating rupture in a specimen. The second (with two variations) measures the energy absorbed during the puncturing of a specimen by a falling dart or swinging pendulum. These methods are described respectively in ASTM Methods D 1709, D 4272, and D 3420. Each method provides a relative ranking of the puncture resistance of films. The rankings provided by the different methods are not necessarily coincident; the first method determines the force required to initiate rupture, while the second requires the projectile to penetrate the specimen completely. Both methods are susceptible to variations due to surface contamination, film blemishes, wrinkles, and other sources of sample inhomogeneity. For any given method it is permissible to compare the thickness-normalized results of puncture resistance provided that there is less than 25% thickness variation between samples.

In the first method (which has much in common with impact plaque testing) a film specimen is gripped by an annular clamp in a horizontal plane. A projectile of known weight with a hemispherical striking face of standard diameter is dropped from a height of 26 in. onto the center of the specimen. Standard dimensions and weights are listed in ASTM Method D 1709. The weight of the projectile is adjusted according to whether or not the sample is punctured, according to the Bruceton staircase procedure outlined in Section V.B.2. Puncture is defined as the situation in which light can readily be seen through a crack or split in the film. The numbers of punctured and intact specimens at each dart weight are noted, and the impact strength is calculated using the formula detailed in the ASTM method.

If the energy to completely puncture a film is to be measured, either a falling weight or a pendulum device may be used. In the falling weight procedure, a dart of known weight with a hemispherical tip falls from a height of 26 inches onto a horizontal film gripped in an annular clamp. Standard dimensions and weights are listed in ASTM Method D 4272. The energy absorbed by the puncture process can be determined from the relative speed of the dart after it passes through the sample compared to its speed when no sample is in place. The speed of the dart is determined by an optical "speed trap." The puncture energy of a sample is reported as the average of a minimum of five separate penetration tests.

When a pendulum type of device is used, the film is gripped in an annular clamp mounted vertically. A pendulum of known mass with a tip of standard size is swung down through a standard arc to strike the center of the specimen. Standard dimensions and weights are listed in ASTM Method D 3420. The energy absorbed by the puncture process is determined by comparing the height to which the pendulum rises after penetrating the sample to a reading taken with no sample present. The puncture force is reported as the relative loss of energy multiplied by the rated mass of the pendulum. The puncture force is reported as the average of a minimum of five separated tests.

D. Dynamic Mechanical Analysis

The term "dynamic mechanical analysis" (DMA) is used loosely to describe a broad range of techniques that measure a sample's physical response to an applied oscillatory strain. Such experiments are used to determine the elastic and damping components of a sample's response to mechanical perturbation as a function of frequency, time, or temperature. The resulting information may be used directly

Characterization and Testing

to evaluate a material's suitability for a specific application or in the investigation of the various mechanical transitions of polyethylene. Experimental configurations included in the general category of dynamic mechanical analysis include shear, torsion, compression, tension, and flexion. A general method for determining the dynamic mechanical properties of polyethylene is given in ASTM Method D 4065.

The theory behind the experimental determination of the dynamic mechanical properties of solids has much in common with that of the dynamic mechanical analysis of melts. Samples in the form of strips, beams, fibers, or rods may be used. Such specimens may be subjected to oscillatory deformation in the form of tension, torsion, and—if they are sufficiently thick—flexion and compression.

Tension, torsion, and compression experiments can be conducted with equipment having the basic configuration shown in Figure 38. One end of the specimen is gripped in a jaw that can be driven with the appropriate oscillatory motion, while the other is gripped in a jaw attached to a transducer. Due to the viscoelastic nature of polyethylene, the sinusoidal motion of the driven jaw is not transmitted directly by the sample to the transducer. The stress measured by the transducer has a sinusoidal trace that lags behind that of the driven jaw by

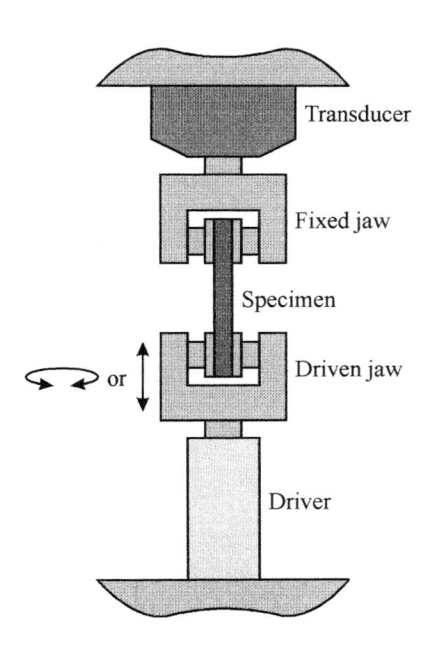

Figure 38 Schematic illustration showing the basic configuration for tensile, torsional, and compressional dynamic mechanical analysis of solid samples.

a constant phase angle (δ). The observed sinusoidal trace can be considered to be the sum of two constituent sine waves, one in phase and one out of phase with the applied force. The amplitude of the in-phase sine wave is a function of the storage modulus of the sample, while that of the out-of-phase sine wave is a function of the loss modulus. The storage modulus is proportional to the average energy stored in a deformation cycle, while the loss modulus is proportional to the energy per cycle dissipated as heat. The tensile storage and loss moduli, respectively, are given the symbols E' and E''. Those of shear (torsion) are known as G' and G'', while their compressive equivalents are K' and K''. Formulae for calculating the various values of interest are available from ASTM Method D 4065.

The measured moduli depend to a large extent upon the frequency and temperature of testing. It is normal to run a series of constant-frequency experiments as a function of temperature or a series of constant-temperature experiments as a function of frequency. The former is known as a temperature sweep, and the latter as a frequency sweep. Results are typically plotted as loss modulus, storage modulus, and tan δ as functions of temperature or frequency. The moduli and tan δ may be plotted as a function of time for samples that undergo a slow transition at a specific temperature and frequency.

Shear experiments can be conducted using a torsion pendulum device. In such experiments a horizontal beam is suspended from a rectangular strip of polymer, a rod, or a fiber as shown in Figure 39. A transient force is applied to

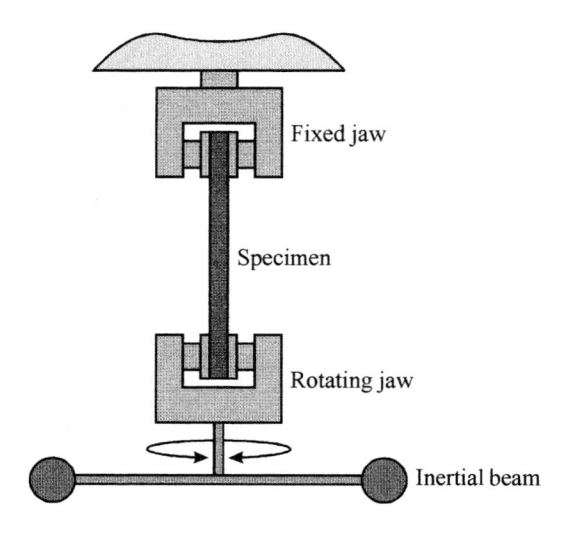

Figure 39 Schematic illustration showing the basic configuration of a torsion pendulum.

Characterization and Testing

the beam, twisting the sample. The amplitude and frequency of the resulting oscillations are then observed as functions of time. Mathematical analysis of the frequency, amplitude, and decay of the oscillations yields the storage and loss moduli.

Flexural experiments can be conducted using a three-point bending configuration. The flexural moduli are given the same symbols as the tensile moduli.

D. Stress Cracking Resistance Testing

Stress cracking of polyethylene is an insidious problem, normally occurring after a substantial delay subsequent to an item being placed in service. The conditions of physical stress under which it occurs are those that would not ordinarily result in immediate failure. It is therefore very useful to know a particular resin's propensity to undergo stress cracking. In actual use an item's tendency to stress crack is a function of a wide range of factors related to the molecular characteristics of the resin, molding conditions, and environmental pressures. Tests exist for the accelerated determination of a resin's intrinsic tendency to experience environmental or thermal stress cracking and also for the testing of fabricated items that emulate performance under critical conditions. Such tests, although not strictly predictive of a material's end use performance, provide valuable information that is useful in the selection of appropriate resins for specific applications.

1. Environmental Stress Cracking Resistance

The determination of a resin's resistance to environmental stress cracking is carried out under conditions that accelerate failure, either on compression-molded test pieces or on fabricated articles. Both types of tests involve exposure of test pieces to a powerful stress cracking agent under conditions of high stress at elevated temperatures. Such tests are useful for determining a sample's general resistance to environmental stress cracking but do not necessarily predict actual performance under service conditions. Procedures for testing a resin's intrinsic propensity to undergo environmental stress cracking can be found in ASTM Methods D 1693 and D 2552, the former being performed on bent specimens, the latter under conditions of tensile stress. Procedures for testing injection-molded pails, blown bottles, and extruded pipes are found in ASTM Methods D 1975, D 2561, and F 1248, respectively.

The bent strip method of determining environmental stress cracking resistance involves immersing severely bent specimens in a potent stress cracking agent at elevated temperatures and measuring the time that it takes for 50% of the samples to fail. Strips 1.5 in. long and 0.5 in. wide are cut from sheets of polyethylene compression-molded under standard conditions. Each strip is carefully nicked along its centerline, the nick being 0.75 in. long and penetrating approximately one fifth of the thickness of the specimen. Specimens are bent into a U shape with the nick on the outside and inserted into a metal channel that preserves this configuration. Each channel, containing 10 or more specimens, is placed in a bath of a stress cracking agent, used neat at 100°C or as an aqueous solution at 50°C. Suitable agents include various detergents, soaps, and organic solvents that do not appreciably swell the polymer. A commonly used agent is nonylphenoxy poly(ethyleneoxy)ethanol (otherwise known as Igepal CO-630). Samples are deemed to have failed when one or more visible cracks appear. The time it takes for 50% of the samples to fail is reported as the F_{50} value. Precise conditions can be found in ASTM Method D 1693. The interlaboratory variation associated with this test is quite high: A two sigma confidence limit of approximately 2.9 is reported in ASTM Method D 1693 [i.e., there is a 95% probability of obtaining a result that is within the range of (reported value)/2.9 to (reported value) $\times$ 2.9].

In the tensile environmental stress rupture resistance test, dogbone specimens are subjected to a constant stress while immersed in a bath of hot stress cracking reagent; the time it takes for 50% of the specimens to fail is reported. Small dogbone specimens of standard dimensions are cut from compression-molded sheets approximately 0.1 in. thick. Each of 20 specimens is mounted on a bracket that permits an individual loading that results in a stress of 8×10^5, 9×10^5, or 10×10^5 dyn/cm^2. The prestressed specimens are then immersed in a bath of neat nonylphenoxy poly(ethyleneoxy)ethanol at a constant temperature of 50°C. Samples fail by rupturing at their narrowest point. The time it takes for 50% of the samples to rupture is reported as the F_{50} value. Precise conditions can be found in ASTM Method D 2552. The interlaboratory precision of this test is approximately $\pm 15\%$.

Injection-molded pails with tightly fitting lids may be tested for environmental stress cracking resistance by filling them with a stress cracking agent and applying a substantial top load at an elevated temperature. A minimum of three pails are tested either under standard conditions, listed in ASTM D 1975, or under some user-specified conditions appropriate to the prospective end use. Under the standard conditions the pails are filled with a 10% aqueous solution of nonylphenoxy poly(ethyleneoxy)ethanol and placed in an oven at 130°F. An identical pail filled with sand or water is placed on top of the first along with an additional load selected according to the rated capacity of the pail. The number of days that elapse before a container starts to leak, from either the lid or the pail itself, is noted. Pails that have not failed after 60 days are considered to have surpassed the test criteria.

Blow-molded bottles are tested for environmental stress cracking resistance by placing bottles filled with a stress cracking agent in an oven at elevated temperature and noting the time it takes for leakage to occur. Bottles may be tested with only the inside in contact with the stress cracking agent, either at atmospheric pressure or subject to some internal overpressure, and with both the inside and exterior base exposed to the agent. ASTM Method D 2561 describes the testing of various types of bottles exposed to a variety of stress cracking agents appropriate to end use conditions or a standard bottle exposed to an aqueous solution of nonylphenoxy poly(ethyleneoxy)ethanol. The former procedure is useful for assessing specific end use applications, while the latter is useful for comparing the environmental stress cracking resistance of different resins. In each type of test a minimum of 15 bottles are tested, with the time to initial failure (F_i), time to failure of half the specimens (F_{50}), and time to failure of the last specimen (F_{100}) being reported. An interlaboratory study of bottles manufactured and tested under standard conditions in accordance with ASTM Method D 2561 suggested a two sigma confidence limit of approximately 1.6 [i.e., a 95% probability of obtaining a result that is within the range of (reported value)/1.6 to (reported value) $\times$ 1.6].

The environmental stress cracking resistance of extruded pipes can be determined by severely compressing short sections in the presence of a stress cracking agent at an elevated temperature and noting duration to failure. A pipe is cut into rings that have a width of 0.5 in. or 30% of its nominal outside diameter, whichever is larger. The outside of each section is nicked to a predetermined depth, the incision being made parallel with and midway between the cut edges. Sections are compressed in a jig until the minimum distance between the inner walls of the pipe is equal to the nominal wall thickness. The sections are then immersed in a 25% aqueous solution of nonylphenoxy poly(ethyleneoxy)ethanol at a constant temperature of 50°C. Specimens are inspected periodically for evidence of deterioration; failure is denoted as the presence of one or more visible cracks. Precise testing conditions are to be found in ASTM Method F 1248. Repeatability of results for sections cut from a given pipe are reported to be ±17% of the natural logarithm of the time in hours for the specimen to fail.

2. Thermal Stress Cracking Resistance

The resistance of polyethylene resins to thermal stress cracking is investigated by winding strips of polymer tightly around a mandrel and noting whether cracks appear at an elevated temperature. According to ASTM Method D 2951, strips of polymer 5.0 in. long and 0.25 in. wide are cut from 0.05 in. thick sheet compression-molded under standardized conditions. These strips are helically wound around a 0 25 in diameter mandrel, and their ends are firmly clamped. Nine such specimens are exposed to a temperature of 100°C in an oven for 1 week. The specimens are periodically inspected for evidence of cracking. Satisfactory samples do not exhibit cracking within this time period. The results of this "pass or fail" test are neither intended to predict end use performance nor to rank the thermal resistance of different resins.

E. Weathering Resistance

The study of weathering resistance is a two-part activity: exposing samples to a weathering environment followed by analyzing exposed versus unexposed samples for evidence of deterioration. The weathering step can consist of exposure of samples to the natural elements in an outdoor setting, accelerated outdoor testing in which mirrors are used to concentrate sunlight on samples, or exposure of samples to an artificial light source in a controlled environment, optionally in the presence of water in the liquid or vapor state. The type of tests involved in the second step depends on the particular properties that are of concern; any test that can be run on virgin materials can also be run on exposed samples. Weathering experiments are generally run for one of two diametrically opposite reasons, either to determine how well a sample resists exposure or to investigate how readily it degrades. The results of artificial accelerated testing do not always correlate with natural (unaccelerated) outdoor weathering experiments. Before specific materials are put to use in critical applications in which they are exposed to the elements, they should always be tested for outdoor weathering resistance using unaccelerated techniques.

Methodologies for investigating outdoor weathering are to be found in ASTM Methods D 1435, D 4364, and D 5272. The first relates to natural exposure, the second details accelerated testing procedures and the third relates to the testing of photodegradable plastics. The operation of artificial weathering instruments is described in ASTM Methods D 1499, D 4329, D 4364, D 5071, and D 5208, the latter two being relevant to the testing of photodegradable plastics. ASTM F 1164 describes a specific procedure for evaluating the deterioration of optical properties of film under conditions of biaxial stress and accelerated weathering.

1. Natural Outdoor Weathering

Outdoor weathering consists of exposing either specially molded test plaques or fabricated items to the elements. Naturally, with such a test there can be a large degree of variation depending upon typical climatic conditions at the test site, seasonal fluctuations, unique weather patterns, etc. Other effects, such as wind-blown sand, salt-laden air, and pollution fallout must also be considered. Test racks, manufactured from some noncorrodible material, hold a series of samples in such a way that none may overshadow its neighbors. Racks are typically arranged at some angle that best provides high exposure to sunlight; this may vary from horizontal to vertical depending upon the latitude and the nature of the sample. Specimens are normally arranged so that the side of interest faces the equator. When testing samples (often sheets or films) for their photodegradability, it is standard practice to mount samples at an angle of 5° from the horizontal as this simulates the conditions to which litter is typically exposed. Solar radiation at the test site is monitored with a pyranometer or ultraviolet radiometer mounted at the appropriate angle. Rainfall and temperature are also monitored. In order to obtain results that are predictive of end use performance, test conditions should simulate as closely as possible those encountered in actual use. Samples may be exposed for various lengths of time, individual specimens being removed at intervals while others are left for further exposure. Ideally some specimens should be exposed for at least a year to reduce the influence of seasonal fluctuations. After the allotted period of time the exposed specimens are tested in comparison to unexposed specimens to determine the extent of property deterioration. Where photodegradability is of concern it is normal to analyze molecular characteristics and tensile properties.

2. Accelerated Outdoor Weathering

Accelerated outdoor weathering is achieved by concentrating sunlight onto specimens by the use of a series of mirrors. The equipment used to achieve accelerated testing consists of a series of up to 10 mirrors affixed to motorized mounts arranged in such a way that each one forms a tangent to a parabolic curve, the locus of which is the sample rack. Throughout the day the mirror mounts are driven in such a way that they constantly reflect the sun's rays onto the specimens. Optionally, samples may be sprayed with water to simulate exposure in humid climates. To limit the effects of heating, the samples are cooled from the reverse side with the aid of powerful fans. Samples should be no more than 0.5 in. thick if cooling is to be effective. Most of the considerations that apply to natural outdoor weathering also apply to accelerated outdoor testing. Accelerated outdoor testing is useful in determining resistance to outdoor exposure where ultraviolet radiation is the principal cause of deterioration. Care must be taken that the higher temperatures involved do not unduly bias the results.

3. Artificial Weathering

The indoor weathering of plastics is conducted in enclosed cabinets that expose specimens to radiation from an artificial source, optionally in the presence of moisture. The machines used for this process are sometimes referred to as "weatherometers." The light source may be fluorescent tubes that emit ultraviolet radiation or a carbon or xenon arc. In no case does the spectrum of the artificial radiation exactly match that of natural sunlight. In the case of arc sources, the specimens are mounted in racks that encircle the source and revolve around it. The ASTM methodology requires that the interior of the cabinet be kept at a constant temperature of 63°C. Optionally the specimens may be sprayed with water at intervals to simulate rain and dew. Depending upon whether photodegradation or weathering resistance is of interest, the ASTM methodology defines different schedules of water exposure. For the determination of resistance to weathering, 18 min of water spray is used in every 2 hr period. When fluorescent tubes are used as the radiation source they are arranged in banks, the tubes being mounted vertically. Specimens are mounted on racks parallel to the banks of tubes. A trough of water is maintained in the bottom of the cabinet that can be warmed to a predetermined temperature. When the water in the trough is heated, vapor rises and condenses on the specimens. An alternating cycle of 4 hr of exposure to radiation followed by 4 hr of exposure to condensation is used. Different lengths of exposure time may be scheduled, but 720 hr is typically used when determining resistance to weathering.

F. Permeation Characteristics

The permeation of small molecules through polyethylene may be viewed as a positive or negative attribute depending on the application and the nature of the migrant molecules. However permeation is viewed, either good or bad, it is beneficial to know the rate at which it takes place. Permeation through polyethylene is most important as it applies to packaging products, principally films and bottles. Permeability resistance may be determined by one of several general methods. On a macroscopic scale, changes in weight, volume, or pressure can be measured. On a human level, changes in taste and smell can be evaluated. The first two types of methods are quantitative, while the last is qualitative. Each method has its own place but may be used in conjunction with others as conditions warrant.

1. Macroscopic Methods

Macroscopic methods of determining the permeability characteristics of packages or films involve the measurement of relatively large effects by such coarse techniques as infrared absorption differences and weight, volume, or pressure changes. Methods exist for determining both liquid and vapor transmission rates, the former being addressed by ASTM Method D 2684, the latter by a variety of techniques, including those described in ASTM Methods D 895, D 3985, D 1434, E 96, E 398, F 372, and F 1249.

a. Liquid Permeation. Liquid permeation rates can be determined for screw-top polyethylene bottles of standard dimensions or any other container that has a positive closure. The liquids involved may be any reagent, solution, or proprietary product of interest. The method is basically the same irrespective of whether the standard bottle or other type of package is tested. At least three containers are filled to their rated capacity with the liquid of interest. The containers are then sealed and accurately weighed before being placed in an environmental chamber for the duration of the test. The temperature and humidity conditions under which the test is conducted remain constant at predetermined levels. The containers are weighed at regular intervals (typically every 7 days) until their

Characterization and Testing

weight loss as a function of time stabilizes. For bottles of standard dimensions, the permeability factor (P_t) is calculated from

$$P_t = RT/A$$

where

R = mass loss per day
T = average bottle wall thickness
A = surface area of the bottle

For nonstandard containers, the average weight change as a function of time is reported.

The average interlaboratory coefficient of variation using the standard bottle is approximately 14%.

  b. Vapor Permeation. The determination of vapor permeation rates is performed using a two-part cell in which a barrier separates two chambers that contain different concentrations of the molecule in question. The migrant molecules tend to diffuse from the higher concentration to the lower one, and the rate at which they do so is measured. Many variations on this basic format exist to test various barrier configurations against a host of different molecules. The most widely measured vapor transmission rates are those of water and oxygen.

In one of the simplest tests used to determine water vapor permeability, a desiccant is placed inside a package that is subsequently sealed and accurately weighed. The package is placed in an environmental chamber, where it is exposed to predetermined constant temperature and humidity. The entire package is weighed on a periodic basis. Any increase in weight is attributed to permeated water that has been absorbed by the desiccant. The rate of water permeation is calculated once the absorption of water as a function of time has stabilized. The evaluation of the water vapor permeability of film or sheet may be performed by using it to seal the mouth of a dish. Either water or a desiccant may be sealed into the vessel prior to its exposure to a controlled environment. The mass of the dish is monitored until the rate of weight change as a function of time stabilizes. Dishes that contain water lose weight, while those that contain desiccant gain weight. The water vapor transmission rate (WVTR) is calculated as

$$\text{WVTR} = R/A$$

where

R = rate of weight change
A = area of barrier

The two methods yield results of a similar magnitude, but the calculated WVTRs for the desiccant method tend to be slightly lower.

An alternative method for evaluating the water vapor transmission rate of a flexible material is to use it as a barrier separating two chambers, one at high humidity and the other at low humidity. The rate of water vapor permeation is determined by measuring the increase in the concentration of water vapor in the low humidity chamber. Both absolute and comparative methods exist for calculating the rise in water vapor concentration, an infrared absorbance technique being commonly used. In comparative methods the diffusion cell is calibrated with a film of known barrier characteristics. The water vapor transmission rate is calculated as the mass of water permeating the barrier as a function of time and area.

The gas permeability characteristics of a film may be determined by using it as a barrier to separate two chambers that contain the gas at different concentrations. In the extreme case one of the chambers is evacuated. The rate of gas permeation across the barrier can be determined from changes in pressure or volume in the low concentration chamber or by analytical techniques that measure the concentration of the gas of interest in an inert carrier gas on the low concentration side of the barrier. When pressure is used to measure the permeation rate, the low concentration chamber is evacuated and the high pressure side is flooded with pure gas. The rise in pressure is monitored with a manometer, from which the rate of permeation can be calculated and thus the gas transmission rate. When the volumetric method is used, a pressure differential is maintained across the barrier and the movement of a slug of liquid in a capillary tube is followed to determine the volume of gas crossing the barrier.

The rate of permeation of oxygen through a flexible barrier may be determined by monitoring the rate at which it crosses the barrier. Two chambers are separated by the barrier film; one contains pure oxygen, and the other is continuously flushed with pure nitrogen. The concentration of oxygen permeating the barrier into the nitrogen is monitored with a coulometric detector. The magnitude of the electric current produced by the detector is proportional to the rate of oxygen permeation across the barrier. This procedure requires that the equipment be calibrated using a film with a known oxygen gas transmission rate.

2. Microanalytical Methods

Microanalytical techniques involve the isolation, identification, and quantification of minute quantities of molecules that may permeate into or out of a package or some other product. The isolation step may involve passing a highly purified carrier gas over a product that sweeps away molecules that have diffused from the product, to be condensed in the cold finger of a cryogenic trap. After a sufficient length of time the flow of carrier gas is halted and the trapped material injected into a gas chromatograph coupled with a mass spectrometer. The gas chromatograph separates the molecules according to their adsorption characteristics on the column packing. As each peak elutes from the chromatograph it is passed to the mass spectrometer, where its molecules are partially disintegrated, the molecular weight of the fragments being analyzed according to their molecular weights. Each organic molecule yields a characteristic fragment pattern ("fingerprint") from which it may be possible to identify the primary molecule. In addition to qualitatively identifying migrant molecules, their relative concentrations can also be measured.

3. Organoleptic Analysis

Organoleptic analysis involves the classification of the taste and smell of a product by a panel of trained human testers. This type of analysis is principally qualitative, but various degrees of odor and taste may be ranked. Virtually any problem involving the migration of noninjurious molecules that can be tasted or smelled is amenable to this type of analysis. Whenever possible, tests should be comparative, but it is permissible to subjectively rate individual products. The two principal properties investigated are contamination and scalping. In the former, the contents of a package are deleteriously affected by the inward migration of molecules that either taint the contents directly or chemically react to produce the same effect. Scalping is the process by which certain flavoring or fragrant molecules preferentially migrate out of the package, changing the organoleptic balance of the remaining contents.

In order for the testing panel to make valid judgments it is necessary to isolate the effect in question. The procedure followed depends upon the nature of the product under investigation. In the case of foodstuffs, this may simply require the panel to compare fresh material with that stored in the package in question for a given period of time under specific conditions. Members of the panel try to ascertain the nature of the differences in descriptive terms and rank them according to their relative severity. Contaminants with a concentration as low as a few tens of parts per billion may be distinguished under favorable circumstances. Skilled panelists can recognize the chemical families involved and may even be able to identify specific molecules. When molecules migrate out of the polymer itself, the sample may be placed in airtight glass jars or in contact with purified water for a given period of time. At the end of the conditioning period, the odor in the jar or the taste of the water is classified and given a comparative severity ranking.

The establishment and maintenance of a competent organoleptic testing panel is not a trivial undertaking. Such panels consist of at least half a dozen members who have been specially selected and trained. To maintain their skills, panel members must practice their art on a regular basis and participate in periodic refresher training. In addition, a room free of potential odoriferous contamination should be set aside solely for the purposes of organoleptic testing.

G. Optical Characterization

The optical characterization of polyethylene materials consists of three major analyses and two lesser ones. From a commercial point of view, the most important optical characteristics of polyethylene products are their haze, gloss, and transparency. The refractive indices of polyethylene specimens, which generally fall into a relatively narrow range, are of little consequence from a practical point of view except as they pertain to the measurement of optical birefringence.

A standard methodology for the evaluation of haze and luminous transmittance can be found in ASTM Method D 1003. The determination of reflection haze is addressed in ASTM Methods D 4039 and E 430, and the various types of gloss in ASTM Methods D 2457, D 4449, E 167, and E 430. The measurement of transparency is described in ASTM Method D 1746. Refractive index may be measured according to the methodology described in ASTM Method D 542. The large number of methods devoted to the evaluation of reflective properties is indicative of the difficulties associated with the evaluation of this complex, and often subjective, set of characteristics.

1. Haze Determination

The haze observed in a sample is that part of the transmitted light that is scattered away from the optical axis of the incident beam. The haze associated with a polyethylene sample is due to the scattering of light by discontinuities within the material and from surface irregularities. The former is known as internal (or occasionally intrinsic) haze, and the latter as surface or exterior haze. Internal and total haze can be measured directly; the value of surface haze is obtained by subtracting one from the other.

The haze of a specimen is measured on a device known as a hazemeter, the basic layout of which is illustrated in Figure 40. Hazemeters basically consist of a collimated light source that illuminates a sample of film or sheet mounted on a small opening in a sphere, the interior of which is coated with a highly reflective matte white paint. On the far side of the sphere, on the same optical axis as the collimator, is a light trap that intercepts undeviated light. The intensity of the light scattered inside the sphere is measured by a photocell mounted in its wall at right angles to the optical axis.

The determination of haze requires the measurement of scattered light by the photocell under four different conditions. The total incident light (T_1) is measured with no sample in place and with a reflectance standard blocking the light trap. The total light transmitted by the specimen (T_2) is determined with both the specimen and the reflectance standard in place. Instrumental scattering (T_3) is determined with the light trap unobstructed and no specimen in place. The light scattered by the specimen and instrument (T_4) is measured with the specimen in place and the light trap unobstructed. The total transmittance (T_t) is the ratio of total light transmitted by the specimen to the total incident light:

Characterization and Testing

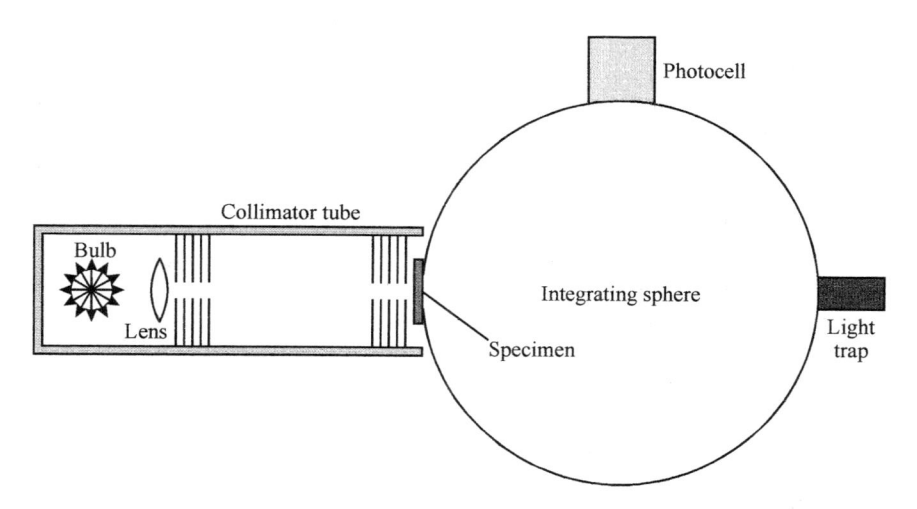

Figure 40 Schematic illustration showing the basic configuration of a hazemeter.

$$T_t = T_2/T_1$$

The percent haze is then calculated from

$$\text{Haze (\%)} = \frac{(T_4 - T_3 T_t)/T_1}{T_t} \times 100$$

Ideally, a minimum of three determinations should be made on separate specimens cut from the same sample and the average value reported. As the total haze of a specimen is highly dependent upon its surface characteristics, it is important that the surfaces of each specimen be free of blemishes, abrasion damage, and contamination. The precision of haze measurements is typically 0.1%, which is beyond the resolution of the human eye.

The contribution of surface haze to the total haze of a specimen can be removed if its surfaces are coated with a liquid (such as immersion oil) with a refractive index approximately matching that of polyethylene ($n \approx 1.5$). Haze determined on coated specimens is indicative of the internal scattering. The surface haze value can be obtained by subtracting the internal haze from the total haze measurement.

2. Reflective Property Determination

The reflective properties of a polyethylene sample are governed by a variety of surface characteristics and the direction from which observations are made. No single number or adjective can adequately describe the reflective properties of a given sample. The most readily appreciated component of reflection is specular gloss, which is the portion of the incident light reflected at an angle equal to the angle of incidence. Specular gloss depends strongly on the viewing angle, generally increasing as the angle of incidence increases. The key material properties that affect gloss are surface roughness, sample planarity, and any preferential orientation or periodicity of surface texture. Reflective haze is that portion of the incident light that is scattered from the surface at an angle oblique to specular reflection. Reflective haze is responsible for the diminishment in contrast of objects observed by reflection. The reflective characteristics of some polymeric surfaces are so complex that the most accurate way of distinguishing similar surfaces may be human observation. In such cases the samples of interest are observed side by side under various lighting conditions and at such angles that subtle differences are highlighted.

The specular gloss of a specimen is the ratio of the intensity of reflected to incident light measured at a specified angle of incidence. The basic configuration of the device used for this determination—known as a glossmeter—is illustrated in Figure 41. A beam of light is aimed at the surface at an incident angle i, and the beam that is reflected at an angle r is collected and its intensity measured by a photocell. In practice, the incident and reflected angles share the same value, which is set to 20°, 45°, or 60°. The specular gloss of the specimen is determined from the amount of light measured by the photocell relative to the incident intensity. Glossmeters are routinely calibrated with two or more standards for which the specular gloss is accurately known at the appropriate angle of incidence. Ideally, a minimum of three determinations should be made on different regions of the specimen and the average value reported. Repeatability should be within 1% of the average gloss value when using the same instrument or within 3% when using different instruments.

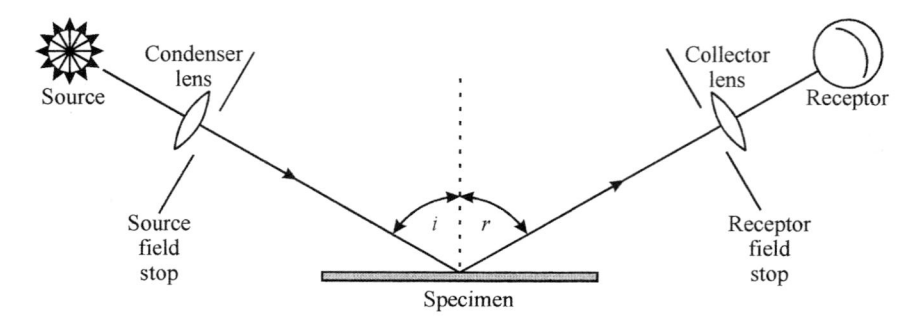

Figure 41 Schematic illustration showing the basic configuration for measuring specular gloss.

Characterization and Testing 349

When mounting samples it is imperative that they be held perfectly flat. In the case of thin films this may be achieved by using a vacuum mount that holds the sample against a planar backing plate. Care must be taken when handling samples to avoid marring or contaminating their surfaces. In the case of transparent films it is necessary to mount the sample against a matte black background to avoid complications introduced by reflection from the backing plate. When characterizing samples that show preferential alignment of surface texture, it is important to align specimens consistently.

A measure of the reflection haze of a specimen may be determined from the specular gloss values obtained at 20° (G_{20}) and 60° (G_{60}). The haze index H is equal to $G_{60} - G_{20}$. Repeatability by a single operator should be less than two haze index units.

Other values associated with gloss are measured with a goniophotometer. A goniophotometer shares the same basic configuration as the glossmeter described above, but it can also measure the intensity of reflected light when the incident and reflected angles are not equal. Goniophotometers can be used to measure specular gloss, distinctness of image gloss, and sheen. The reflection haze can also be measured with such instruments. In addition, the directionality of the various types of gloss can be determined when the specimen is mounted on a turntable.

Goniophotometers can be used to measure light scattered at various angles, either through slits mounted at specified viewing angles or with the aid of a photoreceptor that can be scanned through a range of angles. Angles of incidence of 20° and 30° are typically used. Specular gloss is measured using a method similar to that employed with the glossmeter; however, the angle over which the reflection is collected is limited to a narrow range, ±0.9° at an incidence angle of 20° and ±0.2° at an incidence angle of 30°. The distinctness of image gloss defines the clarity with which a reflected image is perceived. It is calculated from the intensity of the light (haze) scattered at an angle of 29.7° or 30.3° ($H_{0.3}$), with a slit width of ±0.07°, when the incidence angle is 30°, according to

Distinctness of image gloss = $100 \times (1-H_{0.3})$

Reflection haze is the portion of the incident light that is scattered at angles other than the specular angle. When the incidence angle is 30°, the reflection haze is measured at 28° ±0.2° or 32° ±0.2° (H_2) and 25° ±0.25° or 35° ±0.25° (H_5). When the incidence angle is 20°, the reflective haze (H_{20}) is measured at 18.1° ±0.9° or 21.9°± 0.9°. The directionality is defined as the ratio of H_2 across machine direction to H_2 with machine direction. The machine direction is defined as the direction at which the specular gloss is greatest.

The precision of the values determined from goniophotometers depends to some extent on the model used. Repeatabiltiy should be within 1% of the magnitude of the measured values when using the same instrument.

The reflective characteristics of samples exhibiting high gloss can be ranked visually. The samples to be evaluated are arranged side by side and illuminated by a fluorescent desk light over which a wire grid has been placed. The reflective characteristics are evaluated by observing the reflection of the grid at various angles of incidence and at various orientations of the samples with respect to the direction of incident light. Samples can thus be ranked for specular gloss, distinctness of image, reflective haze, sheen, directionality, surface topography, and texture.

3. Transparency Determination

The transparency of a sample is its ability to transmit light directly; it is complementary to haze. Transparency is measured in terms of the ratio of the intensity of undeviated light to that of an incident beam. A narrow beam of collimated light is shone perpendicularly onto a specimen, and the specularly transmitted light is measured by a photocell equipped with a narrow aperture ($0.1° \pm 0.025°$) to obstruct the forward-scattered component. Measurements of the intensity of the beam are taken with and without the specimen in place. The transparency is calculated as

$$\text{Transparency } (\%) = 100 \times \frac{I_t}{I_i}$$

where

I_t = intensity of transmitted light
I_i = intensity of incident light

Ideally, a minimum of three determinations should be made on separate speciments cut from the same sample and the average value reported. As with other optical characterization methods, it is important that the sample be mounted flat and wrinkle-free and not be subjected to physical or chemical abuse.

4. Refractive Index and Birefringence Measurement

The refractive index (n) of a sample reflects its ability to refract light at its interface with a material that has a different refractive index. The greater the difference in refractive indices, the greater will be the angular deviation of a ray of light at the interface. The refractive index of an isotropic sample is uniform, but anisotropic samples exhibit nonidentical refractive indices that depend on the angle at which they are measured. The absolute refractive index of a polyethylene sample is of limited interest from either a scientific or commercial standpoint, but differences between refractive indices determined along three mutually perpendicular axes are informative regarding the orientation of an anisotropic sample.

Characterization and Testing

In practice, the three axes in a film sample are defined by the principal fabrication direction (the machine direction), an axis perpendicular to this in the plane of the film (the transverse direction), and an axis perpendicular to the plane of the film (the normal direction), as illustrated in Figure 42. Birefringence (Δ_{ij}) in a given plane is the difference between refractive indices of the axes that define the plane:

$$\Delta_{zy} = n_z - n_y; \quad \Delta_{zx} = n_z - n_x; \quad \Delta_{yx} = n_y - n_x$$

Many physical properties of commercially prepared films depend strongly on the measurement direction relative to the principal fabrication direction. Thus, a quantitative measure of a film's birefringence can be very useful.

The refractive index of a film can be determined by applying Snell's law to the critical angle required for total internal reflection against a prism of known refractive index. Abbé refractometers can be used for this purpose following the technique described in ASTM Method D 542. To apply this method, the film must be sufficiently thin, flat, and soft to obtain good optical contact with the refractometer prism face. Alternatively, a prism–waveguide coupling instrument can be used [96]. The prism–waveguide method is amenable to automation and can be used to determine the refractive indices in the three principal directions and hence the optical birefringence. Birefringence can also be determined by measuring the optical retardation of a light beam passing through a film of known thickness. This is performed with an optical microscope equipped with polarizers [97].

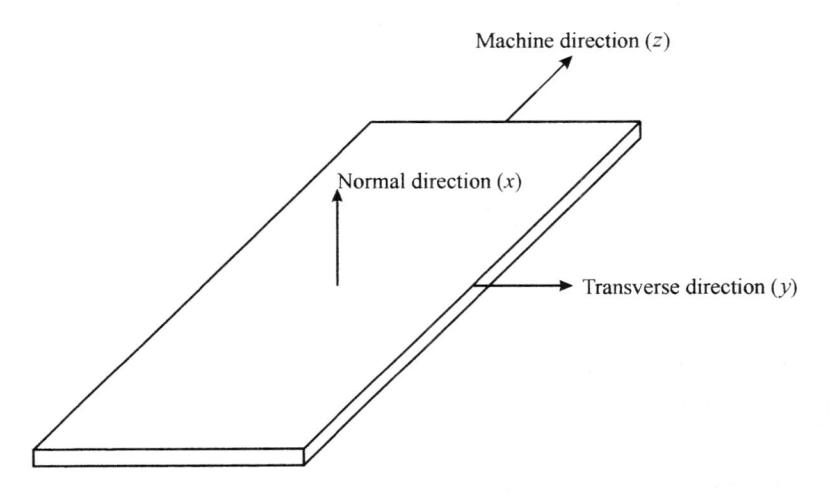

Figure 42 Schematic illustration showing the three principal axes in an oriented film.

H. Cling Measurement

The "blocking" of thin polyethylene films occurs when they are extremely smooth, flexible, or soft or contain a low molecular weight viscous component that can migrate to the surface. When one or more of these conditions is met, films that come into contact adhere to each other to a greater or lesser extent. Blocking leads to problems such as excessive force being required to unwind film from a roll and plastic bags that are hard to open. The degree of blocking (cling) between a pair of films can be evaluated by sliding a thin rod between them according to the procedure outlined in ASTM Method D 1893 or by applying sufficient tensile force perpendicular to their planes to separate a given area as described in ASTM Method D 3354.

The force required to progressively separate a pair of blocked films by drawing a thin rod between them is measured by using a tensile testing instrument. The basic configuration of the testing equipment is shown in Figure 43. The pair of films are hung from the upper grip with the principal fabrication

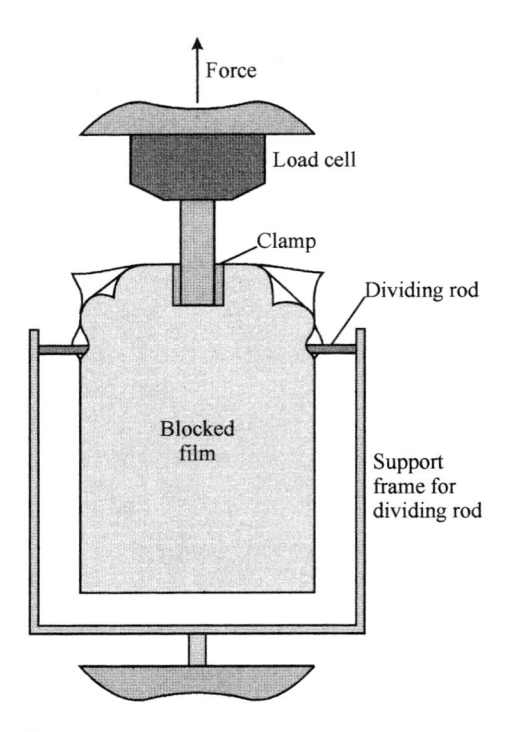

Figure 43 Schematic illustration showing the basic configuration for blocking force measurement by the drawn rod method.

direction vertical. A smooth aluminum rod with a diameter of 0.25 in. is inserted between the films and drawn downward at a speed of 5 in./min. The force required to separate the films is recorded by the load cell. The blocking force is reported as the average force exerted on the moving rod divided by the width of the film sample. For greatest accuracy, the film should be as wide as practicable. Four replicates should be run for each sample and the average value reported.

The blocking load required to separate a pair of films may be determined by the parallel plate method. This test consists of applying an increasing tensile force perpendicular to the plane of a pair of films with a standard area of contact. The basic configuration of the testing equipment is illustrated in Figure 44. A pair of blocked films 10 cm × 18 cm are placed between a pair of square aluminum blocks with planar faces measuring 10 cm × 10 cm. The upper block is suspended from one arm of a balance beam while the lower one is held immobile. The layers of film are separated where they overhang the blocks, and the protruding pieces are attached to the upper and lower blocks, respectively, by means of adhesive tape. A separating force is applied to the upper block by adding weight to the opposite end of the balance beam. A smooth increase in load is achieved

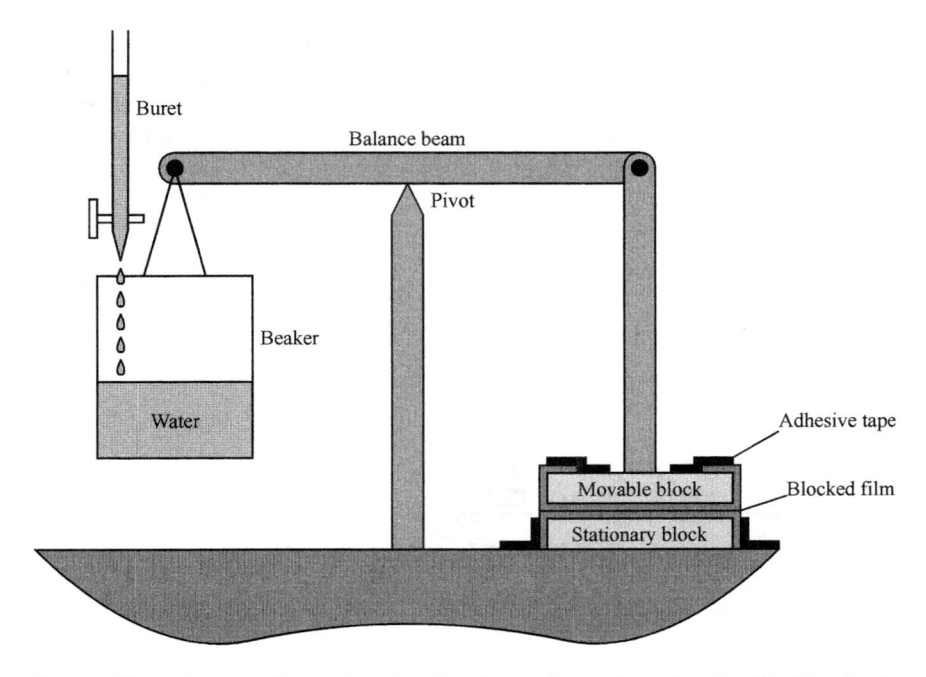

Figure 44 Schematic illustration showing the basic configuration for blocking load measurement by the parallel plate method.

by running water into a beaker from a buret at a rate of 90 mL/min. The addition of water is halted when the blocked film separates. The blocking load is taken as the weight of water that is required to separate the film less a tare value required to separate the aluminum blocks with no film present. Five replicates should be run for each sample and the average value reported. Duplicate results obtained by a single operator should vary by less than 20%. The variation between reported average values from different laboratories should not exceed 15%.

I. Coefficient of Friction Measurement

The coefficient of friction between two surfaces in contact is the ratio of the force required to slide them against each other relative to the perpendicular force that holds them in contact. In practice, the coefficient of friction depends on the nature of the surfaces, whether or not they are in motion, and, if so, their relative speed. The measurement of static and kinetic coefficients of friction is descried in ASTM Method D 1894.

The apparatus for measuring coefficients of friction basically consists of a flat plane and a sled, to both of which sheet or film can be attached. The basic configuration of the testing equipment is shown in Figure 45. Typically this equipment is mounted on a tensile tester. The sled and plane are translated at a fixed rate relative to each other while the force required to initiate and maintain motion is measured. If the coefficient of friction of a film or sheet sliding against itself is to be determined, a layer of it is taped to both the plane and the sled in

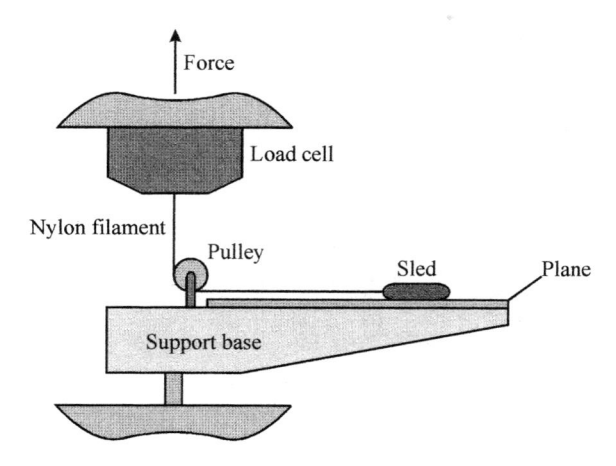

Figure 45 Schematic illustration showing the basic configuration of equipment used to measure coefficients of friction.

Characterization and Testing

such a way that only the film surfaces make contact. Other materials may be substituted, which may be attached to either the sled or the plane. Care must be taken that neither surface is contaminated or damaged during attachment. The orientation of materials with respect to the principal fabrication direction may affect the results, so a consistent orientation must be selected.

After carefully placing the sled upon the plane, crosshead movement is initiated. Once a small amount of slack is taken up in the filament connecting the sled to the load cell, force is registered. The force required to initiate motion and the average force required to maintain a constant relative speed of 150 mm/min are recorded. The static coefficient of friction (μ_s) and the corresponding kinetic value (μ_k) are calculated according to:

$$\mu_s = \frac{A_s}{B} \quad \text{and} \quad \mu_k = \frac{A_k}{B}$$

where

A_s = initiating force
A_k = average steady-state force
B = mass of sled

The coefficients of friction are reported as the average values determined from a minimum of five duplicates. For each run, fresh specimens must be used.

J. Abrasion Resistance Determination

Abrasion to the surface of polyethylene products may result in material loss or degradation of the surface finish. In the first case, structural degradation may occur, while the marring of the surface finish, although leaving the physical properties unchanged, may lead to a lack of aesthetic appeal. Resistance to abrasive material loss by bonded or loose abrasives may be determined according to the procedures described in ASTM Method D 1242, while the resistance of polymer surfaces to abrasive marring may be tested according to ASTM Method D 673.

The relative resistance of samples to abrasive loss against a bonded abrasive may be determined by subjecting them to excoriation with an abrasive tape. The basic configuration of the equipment used for this test is shown in Figure 46. Flat plaques of polymer, which have been accurately weighed, are attached to a continuous belt that is driven at a constant rate around a pair of pulleys. Successive plaques are subjected to scarification by an abrasive tape that is drawn between the specimens and a contact roll at a fixed speed. Contact is maintained between the abrasive tape and the plaques by means of a dead weight and a spring. The grade of abrasive tape and the number of cycles to which samples are subjected may be varied according to the predicted abrasion resistance of the

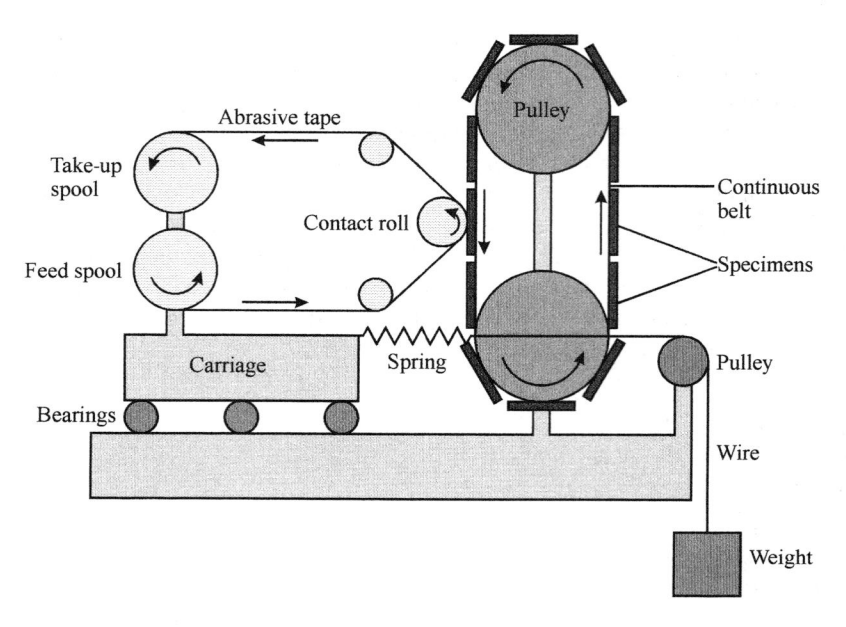

Figure 46 Schematic illustration showing the basic configuration of equipment used to measure abrasive loss by exposure to bonded abrasives.

various plaques attached to the belt. After a predetermined number of cycles, the plaques are removed from the belt and blown free of any extraneous matter. Their weight loss is measured, and the abraded volume is calculated based upon their density.

Comparison of samples abraded under identical conditions provides a ranking of their relative abrasion resistance. As there is no exact method of predicting abrasion resistance under one set of conditions based upon a different set, comparative rankings must be determined under identical conditions. One of the drawbacks to this test is that it is applicable only to relatively thick planar samples, effectively precluding its application to many fabricated samples.

The resistance of polymer samples to the effects of loose abrasive grit can be determined by trapping an abrasive between a flat sample and a hard surface that are in motion relative to each other. The basic configuration of the equipment used for this test is shown in Figure 47. Abrasive grit is steadily fed from the hopper onto the surface of a metal disk rotating at a fixed rate. An accurately weighed polymer plaque mounted in a sample holder is rotated against the surface of the disk. Thus the metal disk spins at one speed while the sample rotates around another axis at a different rate. A cam periodically lifts the sample to ensure a fresh supply of grit between the plaque and the disk. After a predeter-

Characterization and Testing

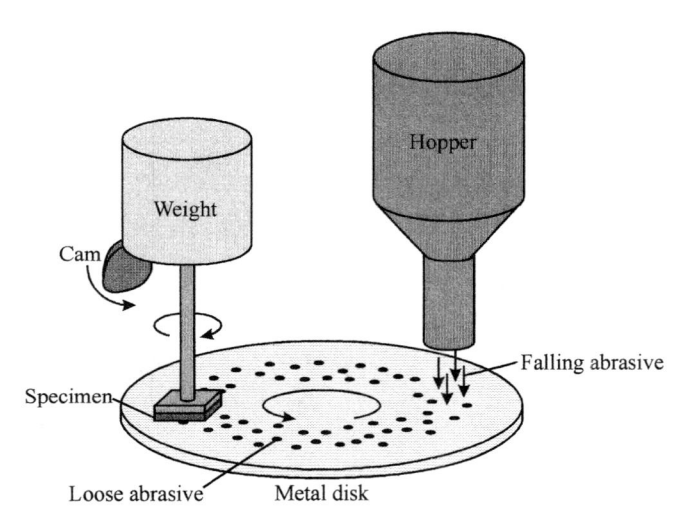

Figure 47 Schematic illustration showing the basic configuration of equipment used to measure abrasive loss by exposure to loose abrasives.

mined number of rotations, the sample is removed from the holder and blown free of any extraneous matter. Its weight loss is measured, and the abraded volume is calculated based upon its density.

Comparison of samples abraded under identical conditions provides a ranking of their relative abrasion resistance. This technique, like the one previously described, suffers from the drawback that it is inapplicable to many fabricated samples.

The resistance of polymer surfaces to marring by abrasive action may be determined by dropping a steady rain of abrasive grit onto an inclined surface. The basic configuration of the equipment used to perform this test is shown in Figure 48. Abrasive grit is released at a constant rate from the supply hopper to fall 25 in. down a guide tube, striking a sample that is inclined at an angle of 45° to the horizontal. The sample is removed after a predetermined weight of grit has fallen on it. It is then cleaned with a blast of compressed air, and its optical characteristics are compared with those of an unabraded specimen. In the case of polyethylene, loss of gloss or increase in surface haze are typically determined. The effects of such abrasion are similar to those encountered when airborne particles strike a moving or stationary surface. For comparative purposes it is possible to plot the change of a key optical characteristic as a function of the weight of abrasive to which it is subjected. The results of this test are used for comparative purposes only; they cannot be used to predict the absolute performance of products in end use applications.

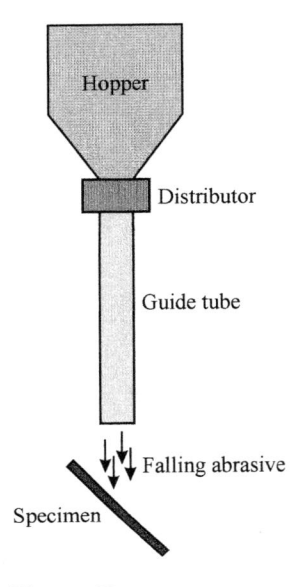

Figure 48 Schematic illustration showing the basic configuration of equipment used to measure mar resistance of polymer surfaces to free-falling abrasive grit.

VI. ELECTRICAL PROPERTY TESTING

The electrical testing of polyethylene materials seeks to determine the response of a sample to various types of electric fields. Variables include voltage, current (alternating and direct), contact or noncontact conditions, and various types of surface contamination. Methods of characterizing the electrical properties of polyethylene fall into two general categories: those that determine electrical characteristics predictive of end use performance and those that seek to rank materials with respect to one another. The first type is exemplified by the measurement of relative permittivity, the second by the determination of arc resistance.

A comprehensive list of the terms and definitions relating to the electrical properties of materials can be found in ASTM Standard D 1711. General methodologies describing the characterization of the electrical properties of polymeric sheet and film are described in ASTM Methods D 229 and D 2305, respectively.

A. Electrical Resistance

The electrical resistance of a material is defined in terms of the voltage that must be applied to a sample in order for it to conduct current at a given amperage.

Characterization and Testing 359

The resistance of a specimen is the ratio of the applied voltage to the current flowing through it. The overall resistance of a sample comprises bulk and surface components, which are respectively determined by its volume and bulk resistivity and its surface area and surface resistivity. The determination of electrical resistance properties of insulating materials is outlined in ASTM Method D 257.

The basic requirements for determining the overall, bulk, and surface resistance and resistivities of a sample are a pair of electrodes, a stable power supply, and meters for measuring the voltage drop and current flowing through the sample. A variety of experimental configurations based upon direct measurement or comparison against a known standard are available. Depending upon the sample configuration, which may be sheet, rod, tube, insulated wire, etc., various types of electrodes may be employed. Electrodes can take the form of tapered pins inserted through specimens, binding posts, conducting paint, liquids, metal bars, foil, or the wire comprising the conductor in a cable. When surface resistance is to be determined, two electrodes are applied to the same surface of a specimen. As surface resistance is highly dependent upon contamination, specimens should either be pristine or cleaned thoroughly prior to measurement—unless, of course, the effects of contamination are under investigation. For bulk resistance measurements, electrodes are applied to opposite sides of the specimen or inserted through its thickness. In each case the dimensions of the specimen and the electrodes and the separation between electrodes must be accurately known if the resistivity is to be calculated. Once the bulk or surface resistance of the sample has been determined, the resistivity can be calculated according to

$$r_v = R_v A / t$$

where

r_v = volume resistivity
R_v = measured volume reisistance
A = effective area of measuring electrode
t thickness of specimen

and

$$r_s = R_s P / g$$

where

r_s = surface resistivity
R_s = measured surface resistance
P = effective perimeter of measuring electrode
g = separation of measuring electrodes

With the many different measuring systems available it is not possible to make any general statements regarding the precision of such measurements.

B. Capacitive Properties

The effectiveness of any material used as a dielectric to separate the plates in a capacitor depends upon its polarizability in an applied electric field. The degree to which the insulator polarizes is termed its (relative) permittivity, dielectric constant, or electrical inductive capacity. Permittivity is defined as the ratio of the capacitance of a capacitor constructed using the insulator to an identical one in which the insulator is replaced with vacuum. When a capacitor is subjected to alternating current its dissipation factor becomes important. The dissipation factor, also known as tan δ, is the ratio of the energy lost to that stored when the capacitor is subjected to an alternating field. Low values, indicating low power losses due to conversion of electric energy to heat, are desirable and are particularly important at high frequencies. The characteristics of dielectric materials may be determined according to the procedures defined in ASTM Methods D 150 and D 1531. The techniques described in ASTM Method D 150 apply to the determination of the capacitive properties of a dielectric by direct comparison with an identical system in which the insulator is replaced by a vacuum. The procedures presented in ASTM Method D 1531 apply to the determination of capacitive properties using a cell in which the dielectric displaces a fluid with known dielectric properties.

For the direct determination of the permittivity of an insulator, a capacitor is constructed in such a way that its vacuum capacitance can be measured or calculated. Ideally, specimens take the form of film or sheet, but tubes can also be accommodated. Electrodes may consist of metal foil or plates, vapor-deposited metal, or conductive liquid. The dielectric of interest is sandwiched between the plates of the capacitor, and the capacitance and dissipation factor of the system are measured. The observed capacitive properties are compared against the vacuum characteristics calculated for the cell configuration, and the permittivity and dissipation factor of the insulator are calculated. Equations applicable to the various capacitor and electrode configurations can be found in the ASTM test method.

The capacitive properties of thin films and sheets can also be determined in a cell into which a specimen is placed between fixed parallel plate electrodes, thereby displacing a fluid of accurately known dielectric properties. The basic configuration of the test cell used for such measurements is shown in Figure 49. If the thickness of the specimen can be accurately measured, it is only necessary to determine the capacitance and dissipation factor of the cell with a single fluid separating the plates and with the specimen inserted between the plates displacing some of the fluid. If the average thickness of the specimen cannot be accurately measured, as in the case of extremely thin films, two different fluids must be

Characterization and Testing

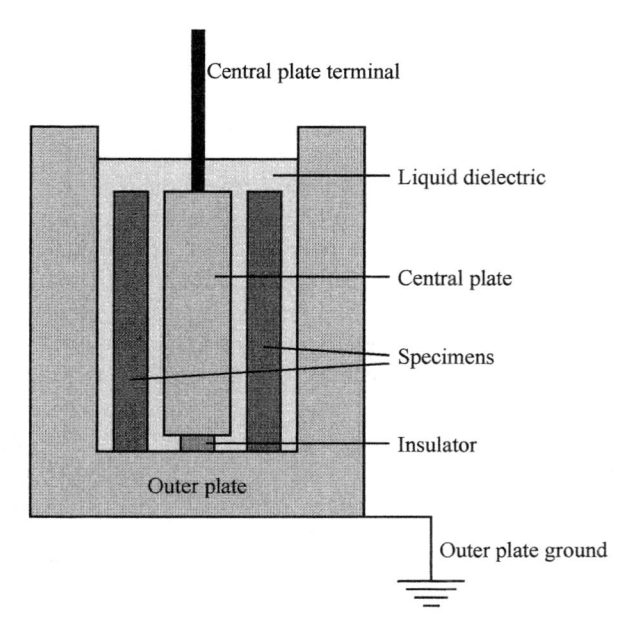

Figure 49 Schematic illustration showing a cross section of a test cell used to measure electrical capacitive properties by fluid displacement procedures.

used. Four measurements are made, one with each of the fluids alone in the cell and a matching pair in which the specimen displaces some of the fluid. The applicable equations for calculating the permittivity and dissipation factor of the sample can be found in the ASTM test method.

C. Dielectric Strength and Breakdown Voltage

When electrical insulators are subjected to a potential difference of increasing magnitude, there comes a point at which they physically break down and begin to conduct electricity. The dielectric strength of a sample is the voltage gradient at which this failure occurs. The (dielectric) breakdown voltage is the potential difference at which dielectric failure occurs under specified conditions. The determination of these two characteristics is described in ASTM Method D 149.

In practice, dielectric strength or breakdown voltage is determined by applying an electric field across an insulator and ramping it at a fixed rate until the sample fails. Failure is defined as puncturing of the sample with subsequent conduction of electricity. Test specimens can take many forms, ranging from end use products to sheet and film molded specifically for purposes of testing. Electrodes are attached to opposing faces of the specimen; their precise placement may be critical. When nonsymmetrical or end use products are to be tested, standard electrode positions should be adopted. Electrodes are preferably metal plates of a standard shape, size, and surface finish, but other types may be employed as conditions warrant. One of the electrodes (normally the larger if there is a difference in size) is grounded, and the other is attached to a controllable source of alternating voltage. The root mean square (rms) voltage is increased according to one of three general methods. In method A, the "short-time test," the applied voltage is ramped steadily from zero at a constant rate ranging from 100 to 5000 V/sec. The ramp rate is chosen such that failure occurs within 10–20 sec of the start of the test. Method B, the "step-by-step test," calls for the incremental increase of voltage as a function of time, the increment and interval between steps being constant. Increments may be in the range of 0.25–10 kV, with a typical time interval of 60 sec. The starting voltage is selected such that failure occurs within 4–10 steps. In method C, the "slow rate of rise test," the applied voltage is ramped steadily at a constant rate, starting from a predetermined voltage similar to that used in method B. The voltage is increased at a rate that approximates the average rate of increase used in method B. Breakdown should occur after a minimum of 120 sec, at a voltage not to exceed 1.5 times the initial value. Failure is deemed to occur when there is an abrupt increase in measured current, normally accompanied by a physical rupturing of the specimen that is frequently audible. In the case of the step-by-step test, the breakdown voltage is taken to be that of the last complete time interval before failure. For each sample a minimum of five specimens are run and the average breakdown value is reported. The dielectric strength is the breakdown voltage divided by the thickness of the specimen.

The coefficient of variation reported for a single operator is usually less than 9%. When different operators using different sets of equipment analyze a given material, the single-operator coefficient of variation is approachable if all the experimental variables are rigorously controlled.

D. Arc and Tracking Resistance

An arc occurs when current jumps the air gap between two electrodes. When the localized heating associated with arcing occurs in the proximity of a polymeric insulator, chemical reactions take place on the insulator surface that may result in the formation of a conductive path known as a "track." Tracks also form when surface contamination of an insulator results in the transmission of current across its surface. The formation of tracks on insulators is extremely detrimental to their performance. Several methods have been developed to measure the relative propensity of insulators to tracking under a variety of conditions that more or less attempt to duplicate the types of environments encountered in service. In

Characterization and Testing

the interest of obtaining results in a timely manner, the conditions under which testing is performed are generally much more severe than those associated with actual use. Dry arc resistance testing may be carried out according to ASTM Method D 495. The determination of the tracking resistance of contaminated materials is described in ASTM Methods D 2132, D 2303, and D 3638, which respectively describe testing regimes designed to ascertain the effects of dirt and mist, flowing electrolyte on an inclined plane, and electrolyte pooling on a horizontal surface.

When the dry tracking resistance of an insulator is to be determined, an arc is struck between a pair of electrodes with sharp edges that rest on the surface of a specimen. The length of time it takes for a conductive path to form and the arc to disappear into the specimen is recorded. Specimens may take the form of molded sheets or end use products. The electrodes may be either strips of stainless steel that have a sharp corner or tungsten rods whose ends have been polished at an acute angle to leave an elliptical face. Tungsten electrodes are used when samples are more resistant to the arc and stainless steel electrodes would erode during the course of the test. The electrodes are placed in contact with the surface of the insulator at a spacing of 0.25 in. A high voltage, low current arc is struck between the electrodes, and the time it takes for a track to form is recorded. A minimum of five specimens are tested, and the average is reported. The results available from this test should be used for comparative purposes only. The conditions of this test are rarely duplicated during end use; other factors, especially surface contamination, play an important role in determining service performance.

The comparative tracking index (CTI) of a specimen is determined by dripping an electrolyte onto its surface, with the drops falling between a pair of electrodes making contact with it. The comparative tracking index is the voltage required to produce failure after 50 drops of electrolyte have been applied. Chisel point platinum electrodes rest on the surface of a thick specimen (>0.1 in.), with a spacing between them of 4 mm. A low voltage (<600 V), low current alternating stress is applied between the electrodes while drops of electrolyte are allowed to fall between them. The electrolyte may be any liquid or solution of interest; a common choice is a 0.1% aqueous solution of ammonium chloride. The standard drop size is 20 mm^3, with a drop rate of two per minute. Failure is deemed to occur when an abrupt drop in resistance indicates the formation of a conducting path. The test is repeated at the same voltage using a fresh specimen (or isolated regions of the same specimen) five times, with the average number of drops required before failure being noted. The procedure is repeated using a number of different voltages. The average number of drops required before a track forms is plotted as a function of voltage, and the comparative tracking index is reported as the voltage required to produce failure at 50 drops of electrolyte. Intralaboratory variation for this method is in the region of ±25 V, with a value of up to

±50 V for interlaboratory comparisons. It should be noted that the comparative tracking index is an arbitrarily defined value and has no predictive value.

One of the most challenging environments that insulating materials experience is a combination of dirt and water. Insulators subjected to this combination in the presence of an electric field may fail by tracking or erosion in the vicinity of the electrodes. To determine the relative resistance of insulators to dust and fog, three parallel strip electrodes of brass or copper are placed in contact with the surface of an insulator. The outer electrodes are grounded, and the central one is attached to an alternating current supply. The surface of the insulator is coated with a layer of synthetic dust about 0.020 in. thick, and the surface is evenly wetted with a fine water spray emitted by a fog nozzle. When the central electrode is energized, arcs form between it and the adjacent electrodes. Power is maintained until the insulator fails, either by the formation of a track that conducts current or by erosion beneath the electrodes that punctures the specimen. A minimum of three specimens should be tested and the average time to failure and type of failure reported. The results of this test should be used for comparative purposes only.

The effects of contamination on the surface of insulators may also be determined by the inclined-plane liquid contaminant test. This test seeks to determine the resistance of samples to the effects of moisture contamination in the presence of dirt. In practice, exposure to moisture would be sporadic, but by using a continuous supply of electrolyte, failure due to tracking may be greatly accelerated. In this test a specimen 5 in. long by 2 in. wide is mounted at an angle of 45°. Electrodes are attached to its underside at each end, and a supply of electrolyte is pumped over the upper electrode and allowed to run down the underside of the specimen to the lower electrode, where it drips off. Various electrolytes may be used; a common choice is a 0.1% aqueous solution of ammonium chloride. Once a steady flow of electrolyte has been established, an alternating field, the potential of which may be varied from 1000 to 7500 V, is applied between the electrodes. The sample can be tested by the initial tracking voltage or time-to-track methods. According to the first method, the applied voltage is increased in steps of 250 V at intervals of 1 hr. The voltage at which a track is established between the electrodes is reported as the initial tracking voltage. In the second method, the time taken to establish a track 1 in. in length that grows from the lower electrode at a given voltage is reported as the time to track. In either case a minimum of five specimens are tested, and the average value is reported. As with the other methods of determining tracking resistance, the results of this test should be used for comparative purposes only.

E. Partial Discharge Resistance

Partial discharge (known as corona discharge when ionized gas glows) occurs at electrodes that are separated from one another by an electrical insulator when a

Characterization and Testing

high potential is applied. Insulators exposed to coronas for extended periods of time may deteriorate under their influence, eventually undergoing a breakdown of resistance. Failure may occur by one of several methods. Erosion of the insulator may reduce its thickness until it can no longer withstand the applied voltage, at which point rupture occurs. Corona discharge may promote the formation of electrical trees that gradually penetrate the thickness of the insulator. Alternatively, the surface of the insulator may become conducting due to chemical reactions induced by the electric field. The resistance of insulators to the effects of corona discharge may be determined according to the procedures described in ASTM Method D 2275.

The basic configuration used to determine resistance to corona discharge is similar to that used for testing dielectric strength. Specimens consisting of film or sheet are placed between pairs of electrodes across which an alternating potential is applied. Typically a single large flat electrode is placed under the sample while a number of smaller spherical or cylindrical stainless steel electrodes are placed on top. An electric field of a given strength is applied, and the time to failure at each of the small electrodes is measured. Voltage levels are chosen that are higher than the corona inception level but below that expected to cause failure within 24 hr. Electric fields should be strong enough to cause failure at some points within 30 days. It is common to use nine pairs of electrodes and report the fifth failure as the median value. Tests are run at a number of different voltage levels, and the results are plotted as time to (median) failure as a function of applied voltage or voltage stress (V/mil or V/mm). For samples that show high resistance to corona discharge, tests may be accelerated by increasing the frequency of the alternating field. The results of this test are not predictive of end use performance; they may only be used for comparative purposes to rank materials relative to one another.

F. Electrical Treeing

Electrical trees occur within thick insulators that are subjected to very high electric fields. In practice, such intense electric fields are generated at localized inhomogeneities within a sample that is exposed to strong electric fields. The evaluation of electrical treeing resistance may be performed according to ASTM Method D 3756. Currently there is no standard method for evaluating the resistance of polyethylene materials to the formation of water trees.

The resistance of polyethylene samples to the generation of electrical trees is determined by exposing them to the intense electric fields associated with sharp electrodes. Pairs of electrodes can be incorporated into samples during compression molding or inserted into samples that have been softened by warming them to an appropriate temperature below their crystalline melting point. The diameter of the electrodes can vary from 0.7 to 10 mm, diameters in the lower part of the range being most common. Each pair of electrodes consists of one with a hemispherical tip while the tip of other is sharpened to a point with a radius of 2.5 mm and an included angle of 30°. Electrode gaps of 2–12 mm are commonly used. For comparative tests it is important that the electrode gap be constant in all cases. Two types of tests are performed. In the first, known as electrical stress testing, various voltages are applied to different sets of specimens for 1 hr. In the second type, known as divergent-field voltage life testing, a single voltage is applied until a minimum of half of the specimens have failed. In electrical stress testing, the existence of trees originating at the needle point is determined by microscopic observation. The electrode gap used in this test is typically 6–12 mm. The fraction of specimens that develop trees is plotted against the applied voltage. The double-needle characteristic voltage (DNCV) is defined as the voltage at which it is predicted that 50% of the samples would fail within 1 hr. In divergent-field testing, specimen failure is defined as complete breakdown, i.e., current flows. Electrode gaps for this test are typically in the range of 2–6 mm. The time it takes for 50% of the samples to fail under the stated conditions is reported as the t_{50} of the material. The results of these tests are not predictive of end use performance; their value lies solely in ranking the electrical treeing resistance of samples under controlled conditions.

VII. ASTM METHODS

The following is a list of the various ASTM methods applicable to polyethylene referenced in the foregoing sections. It is generally organized according to the order in which the various topics are addressed in the chapter.

Definitions	
Standard terms	ASTM D 883
Packaging and distribution terminology	ASTM D 996
Electrical insulation terminology	ASTM D 1711
Ultrahigh molecular weight polyethylene molding and extrusion materials	ASTM D 4020
Dynamic mechanical testing definitions	ASTM D 4092
Stretch, shrink, and net wrap	ASTM D 4649
Mechanical testing terminology	ASTM E 6
Precision and bias	ASTM E 177
Thermal analysis standard definitions	ASTM E 473
Specimen preparation	
Conditioning	ASTM D 618
Polyethylene molding and extrusion specification	ASTM D 1248
Compression molding	ASTM D 1928

Characterization and Testing

Molecular weight determination

Solution viscometry	ASTM D 1601
Molecular weight determination by size elution chromatography	ASTM D 3593
Molecular weight determination by light scattering	ASTM D 4001

Melt rheological characterization

Flow properties	ASTM D 569
Flow rates with an extrusion plastometer	ASTM D 1238
Capillary rheometry	ASTM D 3835
Rheology by dynamic mechanical analysis	ASTM D 4440

Thermal analysis

Thermal expansion	ASTM D 696
Accelerated weight and shape change	ASTM D 756
Thermal shrinkage of thermoplastics	ASTM D 1204
Accelerated linear dimension changes	ASTM D 1042
Melting point with hot stage microscope	ASTM D 2117
Heats of fusion and crystallization by thermal analysis	ASTM D 3417
Transition temperatures by thermal analysis	ASTM D 3418
Heats of fusion by differential scanning calorimetry	ASTM E 793
Melting and crystallization temperatures by differential scanning calorimetry	ASTM E 794
Linear thermal expansion	ASTM E 831
Heat flow calibration of differential scanning calorimeters	ASTM E 968
Specific heat capacity	ASTM E 1269
Glass transition temperature by differential scanning calorimetry	ASTM E 1356

Density determination

Density by displacement	ASTM D 792
Density gradient column method	ASTM D 1505

Cross-linked polyethylene analysis

Gel content analysis	ASTM D 2765

Force versus deformation measurements

Modulus	ASTM E 111
Tensile properties	ASTM D 638
Tensile properties of films	ASTM D 882
Microtensile properties	ASTM D 1708
Tensile heat distortion temperature	ASTM D 1637
Degradation brittle point by tension	ASTM D 3826
Compression deformation	ASTM D 621

Force versus deformation measurements

Compressive properties	ASTM D 695
Rockwell hardness	ASTM D 785
Vicat softening	ASTM D 1525
Durometer hardness	ASTM D 2240
Flexural heat distortion temperature	ASTM D 648
Apparent bending modulus by cantilever	ASTM D 747
Flexural deformation	ASTM D 790
Shear modulus	ASTM E 143
Shear strength with punch tool	ASTM D 732
Torsion stiffness	ASTM D 1043
Tear initiation resistance of film	ASTM D 1004
Tear propagation resistance (Elmendorf)	ASTM D 1922
Tear propagation resistance (trouser tear)	ASTM D 1938
Thermal shrinkage of film	ASTM D 2732
Shrink tension of film	ASTM D 2838

Impact and puncture resistance determination

Notched cantilever beam impact resistance (Izod, Charpy)	ASTM D 256
Tensile impact	ASTM D 1822
Gardner impact	ASTM D 3029
Unnotched cantilever beam impact resistance (Izod, Charpy)	ASTM D 4812
Brittleness temperature by impact	ASTM D 746
Film brittleness temperature by impact	ASTM D 1790
Dart drop impact resistance of film	ASTM D 1709
Pendulum impact resistance of film	ASTM D 3420
Dart drop energy absorbance of film	ASTM D 4272
Creep rupture, various methods	ASTM D 2990

Dynamic mechanical testing

Dynamic mechanical properties	ASTM D 4065

Stress crack resistance determination

Resistance of plastics to chemicals	ASTM D 543
Environmental stress crack resistance of ethylene plastics	ASTM D 1693
Environmental stress crack resistance of injection-molded pails	ASTM D 1975
Tensile environmental stress rupture	ASTM D 2552
Environmental stress crack resistance of blow-molded containers	ASTM D 2561
Resistance to thermal stress cracking	ASTM D 2951
Environmental stress crack resistance of pipes	ASTM F 1248

Characterization and Testing

Weathering resistance

Outdoor weathering	ASTM D 1435
Operation of weatherometer (carbon arc)	ASTM D 1499
Operation of weatherometer (xenon arc)	ASTM D 2565
Operation of weatherometer (ultraviolet)	ASTM D 4329
Accelerated outdoor weathering	ASTM D 4364
Color stability indoors	ASTM D 4674
Operation of weatherometer (xenon arc) for photodegradable plastics	ASTM D 5071
Operation of weatherometer (ultraviolet) for photodegradable plastics	ASTM D 5208
Outdoor weathering for photodegradable plastics	ASTM D 5272
Weathering of transparent plastics	ASTM F 1164

Permeation characteristics

Water absorption	ASTM D 570
Water vapor permeability	ASTM D 895
Gas permeability	ASTM D 1434
Permeability of packages	ASTM D 2684
Oxygen transmission using a coulometric sensor	ASTM D 3985
Water vapor transmission	ASTM E 96
Water vapor transmission by dynamic method	ASTM E 398
Water vapor transmission by infrared spectroscopic methods	ASTM F 372
Water vapor transmission by modulated infrared spectroscopic methods	ASTM F 1249

Optical characterization

Refractive index	ASTM D 542
Haze and transmittance	ASTM D 1003
Film transparency	ASTM D 1746
Specular gloss	ASTM D 2457
Reflection haze of high gloss surfaces	ASTM D 4039
Birefringence and residual strain	ASTM D 4093
Visual evaluation of gloss differences	ASTM D 4449
Goniophotometry	ASTM E 167
Measurement of high gloss	ASTM E 430

Surface property testing

Film blocking	ASTM D 1893
Friction coefficients	ASTM D 1894
Blocking load of film	ASTM D 3354
Mar resistance	ASTM D 673

Surface property testing

Abrasion resistance of transparent plastics	ASTM D 1044
Abrasion resistance	ASTM D 1242

Electrical property testing

Alternating current dielectric strength and breakdown voltage	ASTM D 149
Alternating current loss and permittivity	ASTM D 150
Electrical insulation	ASTM D 229
Direct current electrical resistance	ASTM D 257
High voltage, low current arc resistance	ASTM D 495
Dielectric constant and dissipation factor	ASTM D 1531
Corona discharge	ASTM D 2275
Liquid-contaminated inclined plane tracking	ASTM D 2303
Tests for electrical insulation	ASTM D 2305
Resistance to dust and fog tracking	ASTM D 2132
Measurement of tracking	ASTM D 3638
Resistance to water treeing	ASTM D 3756
Tracking index using various electrodes	ASTM D 5288

BIBLIOGRAPHY

Alexander LE. X-Ray Diffraction Methods in Polymer Science. New York: Wiley-Interscience, 1969.
Azároff LV. Elements of X-Ray Crystallography. New York: McGraw-Hill, 1968.
Baltá-Calleja FJ, Vonk CG. X-Ray Scattering of Synthetic Polymers. Amsterdam: Elsevier Science, 1989.
Collyer AA, Clegg DW, eds. Rheological Measurement. London: Elsevier, 1988.
Cutler DJ, Hendra PJ, Fraser G. In: Dawkins JV, ed. Developments in Polymer Characterisation—2. London: Applied Science, 1980.
Dealy JM. Rheometers for Molten Plastics. New York: Van Nostrand Reinhold, 1982.
Dealy JM, Wissbrun KF. Melt Rheology and Its Role in Plastics Processing. New York: Van Nostrand Reinhold, 1990.
Ferry JD. Viscoelastic Properties of Polymers, 3rd ed. New York: Wiley, 1980.
Glatter O, Kratky O, eds. Small Angle X-Ray Scattering. London: Academic, 1982.
Glauert AM, ed. Practical Methods in Electron Microscopy. Amsterdam: North-Holland, 1973.
Glöckner G. J Appl Polym Sci Appl Polym Symp 45:1, 1990.
Goldstein JI, Yakowitz H, eds. Practical Scanning Electron Microscopy. New York: Plenum, 1975.
Grubb DT. in Bassett DC, ed. Electron Microscopy of Crystalline Polymers, from *Developments in Crystalline Polymers*-1. Barking, UK: Applied Science, 1982.

Haudin JM. In: Meeten GH, ed. Optical Properties of Polymers. London: Elsevier Applied Science, 1986, pp 167–264.
Haslam J, Willis HA, Squirrel DCM. Identification and Analysis of Plastics. 2nd ed. London: Heyden, 1972.
Hemsley DA. Microscopy Handbooks: The Light Microscopy of Synthetic Polymers. Oxford UK: Oxford Univ Press/Royal Microscopical Society, 1984.
Hendra PJ, Gilson T. Laser Raman Spectroscoy. London: Wiley-Interscience, 1970.
Kajiyama T, Tanaka K, Ge S-R, Takahara A. Prog Surf Sci 52:1, 1996.
Koenig JL. Chemical Microstructure of Polymer Chains. New York: Wiley, 1980.
Magonov SG, Whangbo M-H. Surface Analysis with STM and AFM. Weinheim: VCH, 1996.
McCrone WC, McCrone LB, Delly JG. Polarized Light Microscopy. Chicago: McCrone Research Inst, 1984.
Meeten GH, ed. Optical Properties of Polymers. London: Elsevier Applied Science, 1986.
Michler GH. J Macromol Sci Phys. B35:329, 1996.
Painter PC, Coleman MM, Koenig JL. The Theory of Vibrational Spectroscopy and Its Application to Polymeric Materials. New York: Wiley, 1982.
Randall JC. Polymer Sequence Determination—Carbon-13 NMR Method. New York: Academic, 1977.
Sawyer LC, Grubb DT. Polymer Microscopy. London: Chapman and Hall, 1989.
Snyder RG. J Chem Phys 27:1316, 1967.
Tung LH, ed. Fractionation of Synthetic Polymers. New York: Marcel Dekker, 1977.
US Patent 5,008,204, Apr. 16, 1991. FC Stehling.
US Patent 5,030,713, Jul. 9, 1991. L Wild, DC Knobeloch.
White JL. Principles of Polymer Engineering Rheology. New York: Wiley, 1990.
Whorlow RW. Rheological Techniques. Chichester, UK: Ellis Horwood, 1980.
Wild L. Adv Polym Sci 98:1, 1991.
Willis JN, Wheeler LM. In: Provder T, Barth HG, Urban MW, eds. Chromatographic characterization of polymers: hyphenated and multidimensional techniques, Adv Chem Ser Vol 247. Washington DC: American Chemical Society, 1995 pp. 253–263.
Willmouth FM. In: Meeten GH, ed. Optical Properties of Polymers. London: Elsevier Applied Science, 1986, pp 265–333.
Woodward AE. Atlas of Polymer Morphology. Munich: Hanser, 1989.
Yay WW, Kirkland JJ, Bly DD. Modern Size-Exclusion Liquid Chromatography. New York: Wiley, 1979.

REFERENCES

1. JL Ekmanis, RA Skinner. J Appl Polym Sci Appl Polym Symp 48:57, 1991.
2. S Pang, A Rudin. J Appl Polym Sci 46:763, 1992.
3. J Brandrup, EH Immergut, eds. Polymer Handbook. 3rd ed. New York: Wiley-Interscience, 1989, pp VII/4ff.
4. DC Bassett, BA Khalifa, RH Olley. J Polym Sci Polym Phys Ed 15:995, 1977.

5. KVSN Raju, M Yaseen. J Appl Polym Sci 45:677, 1992.
6. T Usami, S Takayama. Polym J 16:731, 1984.
7. ASTM Method D 2238.
8. J Haslam, HA Willis, DCM Squirrel. Identification and Analysis of Plastics, 2nd ed. London: Heyden, 1972, p 369.
9. E Karbashewski, L Kale, A Rudin, WJ Tchir, DG Cook, JO Pronovost. J Appl Polym Sci 44:425, 1992.
10. B Wunderlich, G Czornyj. Macromolecules 10:906, 1977.
11. A Wlochowicz, M Eder. Polymer 25:1268, 1984.
12. DR Parikh, BS Childress, GW Knight. ANTEC '91, p 1543.
13. B Monrobal. J Appl Polym Sci 52:491, 1994.
14. K Shirayama, T Okada, S-I Kita. J Polym Sci A 3:907, 1965.
15. P Schouterden, G Groeninckx, B Van der Heihden, F Jansen. Polymer 28:2099, 1987.
16. H Geerissen, P Schützeichel, BA Wolf. Makromol Chem 191:659, 1990.
17. L Pasti, S Roccasalvo, F Dondi, P Reschiglian. J Polym Sci Polym Phys Ed 33:1225, 1995.
18. JJ Watkins, VJ Krukonis, PD Condo Jr, D Pradhan, P Ehrlich. J Supercrit Fluids 4:24, 1991.
19. E Kiran, W Zhuang. Polymer 33:5259, 1992.
20. KM Scholsky. J Appl Polym Sci 47:1633, 1993.
21. AH Dekmezian, T Morioka, CE Camp. J Polym Sci Polym Phys Ed 28:1903, 1990.
22. J Meissner, J Hostettler. Rheol Acta 33:1, 1994.
23. FN Cogswell. Rheol Acta 8:187, 1969.
24. DH Reneker, J Schneir, B Howell, H Harary. Polym Commun 31:167, 1990.
25. KD Jandt, M Buhk, J Petermann, LM Eng, J Fuchs. Polym Bull 27:101, 1991.
26. GM Brown, DF Brown, JH Butler. Proc 47th Annual Meeting of the Electron Microscopy Society of America, 1989, p 362.
27. DM Shinozaki, PC Cheng, S Haridoss, R Mitchell, A Fenster. J Mater Sci 26:6151, 1991.
28. J Pawley. J Microsc 136:45, 1984.
29. JH Butler, DC Joy, GF Bradley, SJ Krause, GM Brown. Microscopy: The Key Research Tool. Woods Hole MA: Electron Microscopy Society of America, p 103, 1992.
30. JH Butler, DC Joy, GF Bradley, SJ Krause. Polymer 36:1781, 1995.
31. RH Olley, AM Hodge, DC Bassett. J Polym Sci Polym Phys Ed 17:627, 1979.
32. KL Naylor, PJ Phillips. J Polym Sci Polym Phys Ed 21:2011, 1983.
33. MM Kalnins, MT Conde Braña, UW Gedde. Polym Testing 11:139, 1992.
34. M Kojima, H Satake. J Polym Sci Polym Phys Ed 20:2153, 1982.
35. IG Voigt-Martin, EW Fischer, L Mandelkern. J Polym Sci Polym Phys Ed 18:2347, 1980.
36. DC Bassett, AM Hodge. Proc Roy Soc Lond A359:121, 1978.
37. PJ Goodhew. Specimen Preparation in Materials Science. 137ff, vol. 1, part I, Glauert AM, ed. Practical Methods in Electron Microscopy. Amsterdam: North-Holland, 1973.
38. G Kanig. Prog Colloid Polym Sci 57:176, 1975.

39. JS Trent, JO Scheinbeim, PR Couchman. Macromolecules 16:589, 1983.
40. H Sano, T Usami, H Nakagawa. Polymer 27:1497, 1986.
41. GM Brown, JH Butler. Polymer 38:3937, 1997.
42. DC Bassett, AM Hodge. Proc Roy Soc Lond A377:39, 1981.
43. DC Bassett. Makromol Chem Macromol Symp 69:155, 1993.
44. IG Voigt-Martin, R Alamo, L Mandelkern. J Polym Sci Polym Phys Ed 24:1283, 1986.
45. IG Voigt-Martin, GM Stack, L Mandelkern, AJ Peacock. J Polym Sci Polym Phys Ed 27:957, 1989.
46. CW Bunn. Trans Faraday Soc 35:482, 1939.
47. A Kawaguchi, S Murakami, M Tsuji, T Ohta. Bull Inst Chem Res Kyoto Univ 70:188, 1992.
48. W Ruland. Colloid Polym Sci 255:417, 1977.
49. YD Lee, PJ Phillips, JS Lin. J Polym Sci Polym Phys Ed 29:1235, 1991.
50. S Röber, P Bösecke, HG Zachmann. Makromol Chem Macromol Symp 15:295, 1988.
51. AT Boothroyd, AR Rennie, CB Boothroyd. Europhys Lett 15:715, 1991.
52. W Wu, GD Wignall. Polymer 26:661, 1985.
53. DM Sadler, A Keller. Macromolecules 10:1128, 1977.
54. RS Stein, MB Rhodes. J Appl Phys 31:1873, 1960.
55. J Maxfield, L Mandelkern. Macromolecules 10:1141, 1977.
56. T Pakula, M Kryszewski, Z Soukup. Eur Polym J 12:41, 1976.
57. L Mandelkern, M Glotin, RA Benson. Macromolecules 15:22, 1981.
58. B Jasse, JL Koenig. J Macromol Sci Rev Macromol Chem C17:61, 1979.
59. CSP Sung. Macromolecules 14:591, 1981.
60. AJ Peacock, PhD Thesis. University of Southampton, 1984.
61. DI Bower. J Polym Sci Polym Phys Ed 10:2135, 1972.
62. B Jasse, JL Koenig. J Macromol Sci Rev Macromol Chem C17:61, 1979.
63. RG Snyder, SJ Krause, JR Scherer. J Polym Sci Polym Phys Ed 16:1593, 1978.
64. RG Snyder, JR Scherer. J Polym Sci Polym Phys Ed 18:421, 1980.
65. IG Voigt-Martin, EW Fischer, L Mandelkern. J Polym Sci Polym Phys Ed 18:2347, 1980.
66. IG Voigt-Martin, R Alamo, L Mandelkern. J Polym Sci Polym Phys Ed 24:1283, 1986.
67. IG Voigt-Martin, GM Stack, L Mandelkern, AJ Peacock. J Polym Sci Polym Phys Ed 27:957, 1989.
68. RG Snyder, NE Schlotter, R Alamo, L Mandelkern. Macromolecules 19:621, 1986.
69. WL Mattice, RG Snyder, R Alamo, L Mandelkern. Macromolecules 19:2404, 1986.
70. K Schmidt-Rohr, HW Spiess. Macromolecules 24:5288, 1991.
71. PR Swan. J Polym Sci 56:409, 1962.
72. R Chiang, PJ Flory. J Am Chem Soc 83:2857, 1961.
73. R Kitamaru, L Mandelkern. J Polym Sci A-2 8:2079, 1970.
74. DG Moldovan. Adv Lab Autom Robot 6:284, 1991.
75. PJ Flory, A Vrij. J Am Chem Soc 85:3548, 1963.
76. PH Hermans, A Weidinger. Makromol Chem 44–46:24, 1961.
77. MR Gopalan, L Mandelkern. J Polym Sci Polym Lett 5:925, 1967.

78. GR Strobl, W Hagedorn. J Polym Sci Polym Phys Ed 16:1181, 1978.
79. M Glotin, L Mandelkern. Colloid Polym Sci 260:182, 1982.
80. L Mandelkern, AJ Peacock. Polym Bull 16:529, 1986.
81. C Shen, A Peacock, R Alamo, T Vickers, L Mandelkern, C Mann. Appl Spectry 46:1226, 1992.
82. H Hagemann, RG Snyder, AJ Peacock, L Mandelkern. Macromolecules 22:3600, 1989.
83. R Kitamaru, F Horrii, K Marayama. Macromolecules 19:636, 1986.
84. KJ Packer, JM Pope, RR Yeung, MEA Cudby. J Polym Sci Polym Phys Ed 22:589, 1984.
85. RR Eckman, PM Henrichs, AJ Peacock. Macromolecules 30:2474, 1997.
86. R Popli, L Mandelkern. J Polym Sci Polym Phys Ed 25:441, 1987.
87. L Mandelkern, AJ Peacock. Stud Phys Theor Chem 54:201, 1988.
88. AJ Peacock, L Mandelkern. J Polym Sci Polym Phys Ed 28:1917, 1990.
89. PJ Flory, J Rehner. J Chem Phys 11:512, 1943.
90. PJ Flory. J Chem Phys 18:108, 1950.
91. PJ Flory. Ind Eng Chem 38:417, 1946.
92. L Mullins. J Polym Sci 19:225, 1956.
93. QA Trementozzi. J Polym Sci 23:887, 1957.
94. LH Tung. J Polym Sci 24:337, 1957.
95. PD Coates, IM Ward. J Mater Sci 15:2897, 1980.
96. SS Hardaker, S Moghazy, CY Cha, RJ Samuels. J Polym Sci Polym Phys Ed 31:1951, 1993.
97. WC McCrone, LB McCrone, JG Delly. Polarized Light Microscopy. Chicago: McCrone Research Institute, 1984.

7
The Chemistry of Polyethylene

I. INTRODUCTION

Polyethylene by its very nature is relatively chemically inert. The small dipole moments associated with carbon–hydrogen and saturated carbon–carbon covalent bonds severely limit the types of reactions that polyethylene is likely to undergo. The introduction of unsaturation or the incorporation of various other atoms increases the probability of chemical reaction. In particular, the copolymerization of ethylene with polar comonomers results in polymers that undergo a wide range of reactions characteristic of the polar group. Commercially the most relevant categories of reactions undergone by polyethylene are chain degradation, cross-linking, oxidation, surface modification, and grafting.

Molecular degradation of polyethylene can follow a variety of paths, including chain scission, cross-linking, and the insertion of extraneous chemical moieties. Degradation of polyethylene can be caused by photo-oxidative, thermal, mechanical, or radiological processes. The changes in molecular weight associated with chain scission or cross-linking can radically affect rheological characteristics and the mechanical properties of the solid state. The balance between competing chain scission and cross-linking reactions depends upon many factors, including temperature, chemical environment, and the presence and effectiveness of stabilizers. Chain scission reduces the average molecular weight of the material, deleterious effects being observed in the ultimate mechanical properties of products, such as reduced tear strength and the onset of embrittlement. Oxidative reactions that introduce new chemical species can affect optical properties and the propensity to undergo electrical breakdown.

Cross-linking may be induced deliberately during the fabrication process to enhance certain properties that would otherwise be deficient. The principal aim of cross-linking is to improve high temperature structural integrity, i.e., to prevent viscous flow when the crystalline melting temperature is exceeded. Cross-linking can be effected by chemical means or by treatment with radiation. Chemical cross-linking may be subdivided into reactions that take place between the carbon backbone atoms of polyethylene chains and those that involve reactions of side groups grafted onto the backbone.

Surface modification of polyethylene is carried out principally to increase the surface energy of products that come into contact with liquids. Specifically, surface treatment aids the adhesion of paints, inks, and glues to polyethylene products.

Reactive side groups may be grafted onto the backbone of polyethylene to endow it with specific chemical properties. Grafting can provide sites for subsequent reactions, to improve miscibility with other polymers or enhance adhesion to various inorganic fillers.

Copolymerization of ethylene with polar comonomers results in such resins as ethylene-*co*-vinyl acetate, ethylene-*co*-vinyl alcohol, and ethylene-*co*-methacrylic acid copolymers. The polar side groups so incorporated may interact with each other to endow the product with specific physical properties, or they may be used as sites for subsequent chemical reactions. A major family of polymers falling into this category are "ionomers," which consist of ethylene-*co*-vinyl acid copolymers, the acid functions of which have been neutralized to form metal salts.

II. DEGRADATION AND STABILIZATION

A. Mechanisms of Degradation

The degradation of polyethylene takes place when a chemical reaction results in a detrimental change to the characteristics of a specimen. In common parlance, the term "degradation" is often taken to mean chain scission, but this is only one example of the many changes that can occur. (In this chapter the term is used in its more general sense.) Degradation includes embrittlement, the development of color, loss of clarity, an increase in the electrical dissipation factor, and changes in viscosity. It may be brought about by thermal, photic, mechanical, chemical, irradiative, or biological action. Degradation is generally viewed as an unwanted occurrence; however, once the useful life of an article has expired and no specific properties are required of it, environmental degradation (especially that due to biological processes) may be viewed as attractive. The subject of biodegradation is treated as a separate topic in Section II.C.

For the most part, degradation takes place in the presence of oxygen, resulting in its gradual incorporation into the polymer molecules by a series of autocatalytic reactions. The most common degradative environment encountered by polyethylene products is weathering in outdoor situations. Under these circumstances the primary factor is exposure to ultraviolet (UV) radiation, with thermal effects playing a secondary role. Prolonged exposure to high temperature in the absence of light is sufficient to cause degradation, but polyethylene's relatively poor mechanical performance at elevated temperatures ensures that such conditions are rarely encountered in normal service. Temperatures in excess of the crystalline melting point are inevitably encountered during processing, but exposure to such harsh conditions is generally short-lived, and generally little harm comes to the physical properties of adequately stabilized resins. Electrical properties, which are sensitive to small concentrations of contaminants, may be adversely affected by even short exposures to molding temperatures in the presence of oxygen. During the weathering process it is quite conceivable that temperatures sufficient to promote the photo-oxidation process will be encountered. Significant levels of high energy radiation and mechanical and chemical stresses are rarely encountered in common use, but may promote degradation of various types when they are present.

The oxidative degradation of polyethylene consists of four stages: initiation, propagation, branching, and termination. The principal reactions involved in each step are illustrated in Figure 1. The process as a whole is often referred to as "autoxidation" or "auto-oxidation." Many reactions other than those illustrated in this simplified scheme can also occur.

Pyrolysis occurs when thermal degradation takes place in the absence of oxygen at high temperatures. Conditions suitable for pyrolysis are rarely encountered unless a deliberate effort is being made to depolymerize polyethylene. The use of pyrolysis as a tertiary recycling technique is discussed in Chapter 10.

1. Autoxidation

a. Initiation. The primary event in the autoxidation of polyethylene is the generation of radical species. The radicals so produced may react with one another, with species from the same or different polyethylene molecules, or with various molecules absorbed in the resin. Radicals can be generated on polyethylene either by the abstraction of a hydrogen atom attached to the backbone or by cleavage of the backbone to yield terminal radicals. The latter is generally caused by severe physical deformation, either in the molten state under conditions of extreme shear or during rupture of solid samples. The breaking of carbon–hydrogen bonds is more frequently encountered, being brought about by chemical or radiation attack, which occurs more readily at elevated temperatures. A radical species, such as an oxyradical generated by the homolysis of a peroxide molecule, may abstract a hydrogen atom to yield an alkyl radical and an alcohol. Alternatively, high energy radiation may directly separate a hydrogen atom from the backbone. Cleavage of carbon–hydrogen bonds produces a free hydrogen atom that may be temporarily trapped (caged) in the vicinity of its complementary alkyl radical. Such caged radicals have the opportunity to recombine, regenerating the original bond. The probability that the radicals will escape the cage is the "chain

Initiation

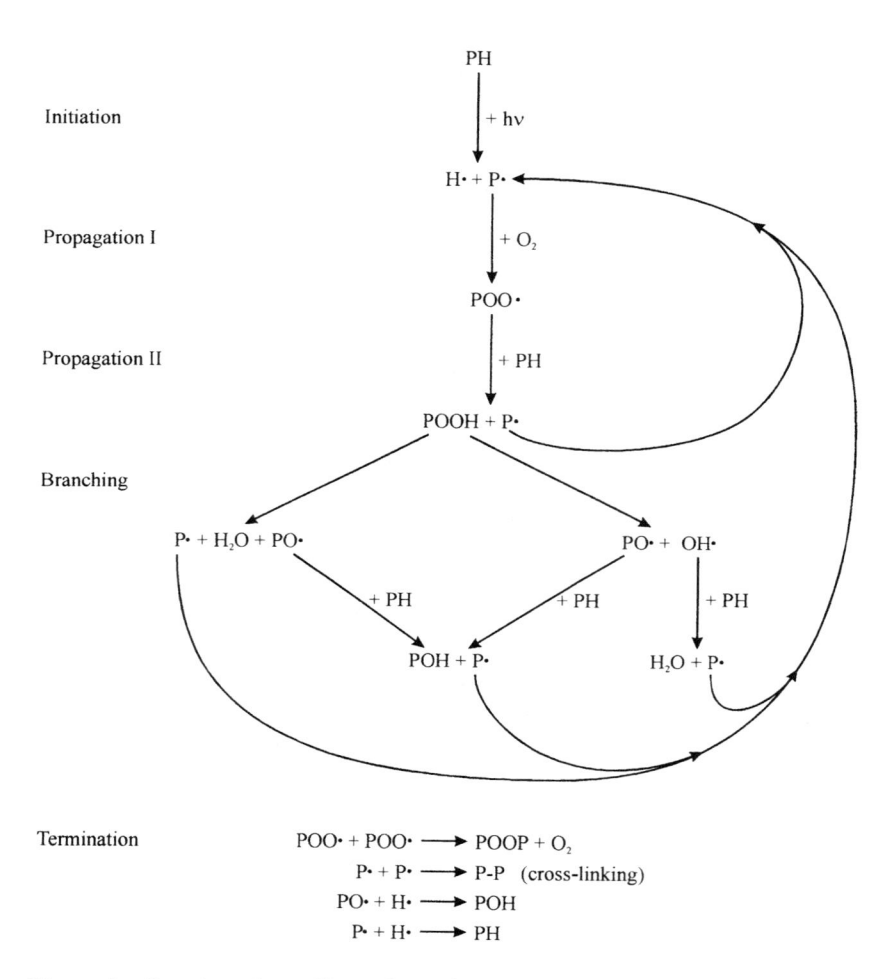

$$PH \xrightarrow{+h\nu} H\cdot + P\cdot$$

Propagation I

$$P\cdot \xrightarrow{+O_2} POO\cdot$$

Propagation II

$$POO\cdot \xrightarrow{+PH} POOH + P\cdot$$

Branching

$$POOH \longrightarrow P\cdot + H_2O + PO\cdot \quad \text{or} \quad PO\cdot + OH\cdot$$

$$P\cdot + H_2O + PO\cdot \xrightarrow{+PH} POH + P\cdot$$

$$PO\cdot + OH\cdot \xrightarrow{+PH} POH + P\cdot \quad \text{or} \quad H_2O + P\cdot$$

Termination

$$POO\cdot + POO\cdot \longrightarrow POOP + O_2$$
$$P\cdot + P\cdot \longrightarrow P\text{-}P \ (\text{cross-linking})$$
$$PO\cdot + H\cdot \longrightarrow POH$$
$$P\cdot + H\cdot \longrightarrow PH$$

Figure 1 Reaction scheme illustrating major reactions of the autoxidation process. P = polyethylene backbone.

generation probability,'' i.e., the effectiveness of the initiation event. Trapping of radicals occurs due to the relatively slow motions of chain segments in the noncrystalline phases, in either the solid or molten state. Radicals produced within the crystal lattice by irradiation are caged for longer times and are thus more likely to recombine. In the absence of recombination, radicals can migrate intra- or intermolecularly by hydrogen transfer. Thus, radicals generated within a crystal lattice may migrate until they reach a noncrystalline region, where they

The Chemistry of Polyethylene

undergo reactions with other radicals or with absorbed molecules. Each alkyl radical generated in the initiation step may be responsible for hundreds of subsequent reactions. The hydrogen atoms produced during initiation migrate until they meet and react with other radical species, typically quenching alkyl or alkoxy radicals or bonding with one another to yield molecular hydrogen that diffuses from the polymer.

The initiation of photo-oxidative degradation of polyethylene requires the presence of species other than those found in the pure material. Saturated carbon–carbon and carbon–hydrogen bonds cannot absorb ultraviolet radiation themselves. In order to initiate degradation, chromophores that absorb the appropriate wavelengths of light must be present. Such chromophores include carbonyl groups, unsaturated carbon–carbon bonds, dyes, pigments, catalyst residues, and even antioxidant molecules. Absorption of ultraviolet radiation by chromophores converts them to activated species; as they decay to their ground states, their excess energy can be used to cleave carbon–hydrogen bonds.

Polyethylene molecules containing carbonyl groups can degrade according to the Norrish type I or II regimes:

Type I:

$$\sim\sim CH_2-CH_2-CO-CH_2-CH_2\sim\sim$$
$$\rightarrow \sim\sim CH_2-CH_2-CO\bullet + \bullet CH_2-CH_2\sim\sim$$

Type II:

$$\sim\sim CH_2-CO-CH_2-CH_2-CH_2-CH_2\sim\sim$$
$$\rightarrow \sim\sim CH_2-COH=CH_2 + CH_2=CH-CH_2\sim\sim$$

The acyl radical can further decompse with the loss of carbon monoxide:

$$\sim\sim CH_2-CH_2-CO\bullet \rightarrow \sim\sim CH_2-CH_2\bullet + C=O$$

while adjacent hydroxyl and vinyl groups can rearrange to form a ketone:

$$\sim\sim CH_2-COH=CH_2 \rightarrow \sim\sim CH_2-COCH_3$$

b. Propagation. Propagation is the process by which alkyl radicals are converted to hydroperoxides. This consists of two sequential reactions, as shown in Figure 1. The alkyl radical reacts rapidly with absorbed molecular oxygen to form a peroxy radical that subsequently abstracts a hydrogen atom from an adjacent polymer molecule to yield a hydroperoxide and another alkyl radical.

Propagation and all subsequent reactions are limited to the noncrystalline regions through which absorbed oxygen is free to migrate. In the absence of absorbed oxygen, the alkyl radicals migrate until they meet and react with other radical species. Propagation is considered to be autocatalytic, because it regenerates an alkyl radical. The second reaction of the propagation process is quite slow, making it the rate-limiting reaction for the entire autoxidation process.

   c. *Branching.* Branching consists of numerous reactions that involve the generation of two new radical species from each hydroperoxide. Some of the key reactions are shown in Figure 1. Hydroperoxides may cleave homolytically to yield alkoxy and hydroxy radicals, each of which may abstract a hydrogen atom from a polyethylene chain to generate an alcohol, water, and more alkyl radicals, which reenter the cycle at the propagation stage. Alternatively, more complex reactions can take place that involve direct reaction with polymer chains, as shown in the left-hand side of the branching step in Figure 1. (A reaction path in which peroxide molecules react via a six-membered transition state to abstract a hydrogen atom from adjacent carbon atoms on the backbone to generate vinylene unsaturation, water, and an alcohol has been proposed [1].) The net result of branching is an increase in the concentration of alkyl radicals, which accelerates the overall oxidation process.

   d. *Termination.* Termination, as shown in Figure 1, involves the quenching of alkoxy and alkyl radical species by reaction with one another or with atomic hydrogen. When there is a dearth of absorbed oxygen, the probability of such reactions increases. When alkyl radicals from adjacent chains quench one another, the result is covalent cross-linking.

   e. *Chain Scission.* β-Chain scission occurs when isolated peroxy or alkoxy radicals decompose:

$$\sim\sim CH_2-CHO_2\bullet-CH_2-CH_2-CH_2\sim\sim$$
$$\to\ \sim\sim CH_2-CH=O + CH_2=CH_2-CH_2\sim\sim\ +\ \bullet OH$$
$$\sim\sim CH_2-CHO\bullet-CH_2-CH_2-CH_2\sim\sim$$
$$\to\ \sim\sim CH_2-CH=O + \bullet CH_2-CH_2-CH_2\sim\sim$$

Peroxy radicals yield terminal aldehydes, terminal unsaturation, and hydroxy radicals. Alkoxy radicals yield terminal aldehydes and terminal alkyl radicals. When these types of reactions are dominant over cross-linking, there is an overall decrease in the average molecular weight of the material.

   f. *Course of Degradation.* The precise reaction scheme encountered during polyethylene degradation and the rates of the various reactions depend on a host of interrelated factors. The principal factor that influences the course of degradation, i.e., whether cross-linking or oxidation predominates, is the relative abundances of alkyl radicals and absorbed oxygen. When the concentration of absorbed oxygen is low or the number of initiating events is very high, cross-linking predominates. When oxygen is abundant, propagation occurs after the initiation step, leading to autoxidation. Kinetic rate-controlling factors in the autoxidation process are the generation of alkyl radicals in the initiation step and hydroperoxide formation, the latter being the overall rate-limiting step. The rates of the individual reactions depend on the concentrations of their starting materials, their activation energy, and the temperature. Other factors that must be taken into account include the molecular and morphological structure of the sample, the presence or absence of stabilizers of different types, processing history, and catalyst residues. All in all, the kinetics of degradation are extremely complicated. The effects of some of the principal variables are briefly discussed in the following paragraphs. Those readers wishing to learn more about this complex field are directed to the works of Geuskens, Potts, and Scott listed in the bibliography at the end of the chapter.

The reactions that generate alkyl radicals in the initiation step fall into two categories, those that result in the formation of atomic hydrogen and those that do not. The former is exemplified by the incidence of high energy radiation, the latter by the attack of peroxy radicals generated by homolytic cleavage of organic peroxides. In the first case, the free hydrogen radical split from the polymer may recombine with the alkyl radical, effectively terminating the process before it can get started. Elevated temperatures increase the rate of molecular motion within the sample, thereby reducing the likelihood of caging the alkyl and hydrogen radicals in close proximity and increasing the probability of chain generation. Caging is far more likely to occur when cleavage takes place within the crystalline lattice. Therefore the probability of chain propagation in irradiated samples increases as the degree of crystallinity decreases. Peroxy radicals abstract a hydrogen atom from the polyethylene molecule to form an organic alcohol; the radical remaining on the polymer cannot recombine with its lost hydrogen atom, so it must react with other species.

When incident light is the source of initiation energy, the abundance of chromophores with the appropriate absorption range influences the rate of carbon–hydrogen cleavage. Naturally, the flux of the irradiation is important. When thick or pigmented samples are irradiated, the intensity of radiation may be attenuated as a function of depth, resulting in a reduction of the concentration of alkyl radicals the further the radiation has to penetrate. This is typically encountered in plasma or electron beam treatments.

The rate of propagation is strongly dependent upon the concentration of alkyl radicals and the amount of oxygen available to them. In the case of high crystallinity samples where both the concentration of absorbed oxygen and its rate of diffusion are low, the rate of the initial propagation reaction will be retarded versus that found for low crystallinity samples. In the second propagation reaction, the ease with which a hydrogen atom can be extracted from a polyethylene molecule by the peroxy radical depends on chemical structure. Hydrogen atoms are most readily abstracted from tertiary carbons, followed by secondary and then primary carbons. Thus, branched polyethylene resins are more susceptible to this step than linear resins. The rate of propagation may be increased by raising the temperature of the sample.

The rate of branching depends primarily on the rate of homolytic cleavage of the hydroperoxide group. Such cleavage can be promoted by an increase in the temperature or by the presence of a catalyst (such as the metal complex remaining from the polymerization catalyst). The other reactions involved with the branching process are also promoted by increased temperatures.

The termination step largely depends on the rate at which radical species encounter one another. Elevated temperature increases both the rate of molecular motion and inter- and intramolecular migration of the alkyl radicals, thus promoting their movement and increasing the probability that two radicals will meet and have the chance to react. Termination reactions are most likely when there is little absorbed oxygen with which alkyl radicals can preferentially react.

2. Pyrolysis

Pyrolysis is the spontaneous thermal decomposition that occurs when polyethylene is subjected to extremely high temperatures in the absence of oxygen. At temperatures in excess of approximately 400°C, the carbon–carbon bonds of the polyethylene backbone spontaneously break to yield two shorter chains, each of which is furnished with a terminal radical. Due to the stabilization effect of branches, β-scission is favored in branched resins over random scission in linear regions. For this reason, branched polyethylene samples undergo pyrolysis at temperatures approximately 20°C lower than those required by linear resins [2]. Once terminal radicals have been produced they may undergo "backbiting" reactions with backbone bonds a few carbons removed from the terminus. Backbiting results in the emission of low molecular weight alkanes and alkenes. When allowed to progress to its conclusion, pyrolysis results in the complete conversion of polyethylene to gaseous hydrocarbons. Conditions suitable to cause pyrolysis are rarely encountered under normal service conditions.

B. Mechanisms of Stabilization

The stabilization of polyethylene against autoxidation occurs when any step involved in the process is inhibited. Ideally, initiation would be prohibited entirely, but failing that, retarding one or more of the subsequent steps must be accepted. Given the variety of reactions involved in the autoxidation process, it is only to be expected that stabilization to varying degrees can be effected in many different ways. The principal types of stabilization are shielding against irradiation, quenching of activated species, radical scavenging, and decomposition of hydroperoxides. Representative examples from some of the many families of stabilizers are illustrated in Figure 2. Various other types of stabilization can be employed

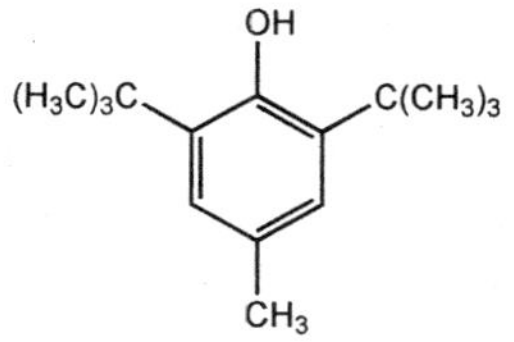

The Chemistry of Polyethylene 383

2,6-t-Butyl 4-methyl phenol
(Butylated hydroxy toluene)

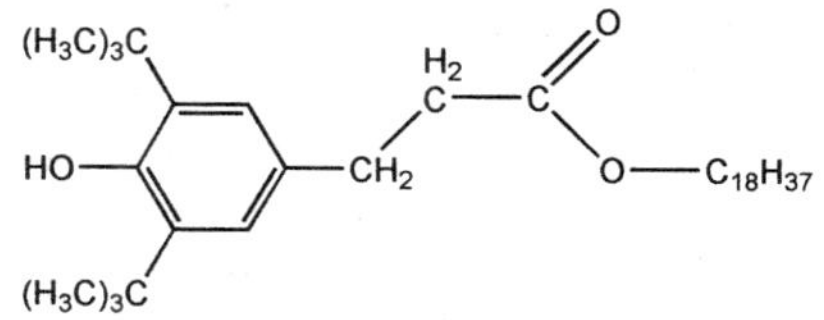

Octadecyl 3-(3,5-di-t-butyl-4-hydroxyl-phenyl) propionate

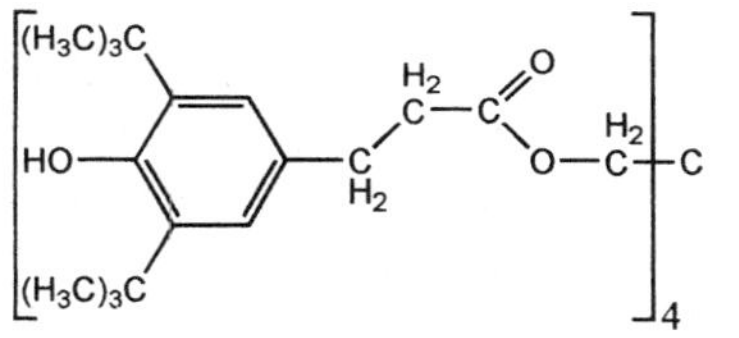

Tetrakis[methylene-3-(3',5'-di-t-butyl-4-hydroxy phenyl) propionate] methane

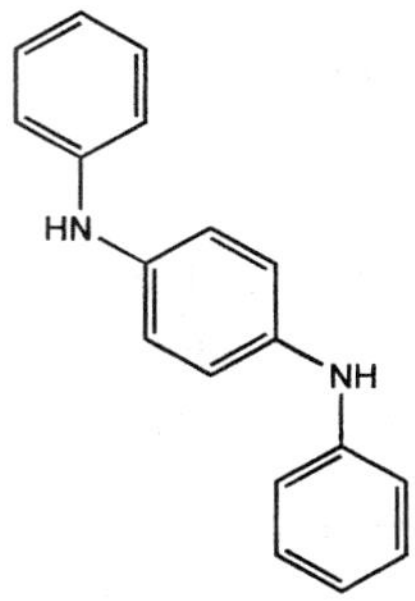

Diphenyl-p-phenylene diamine

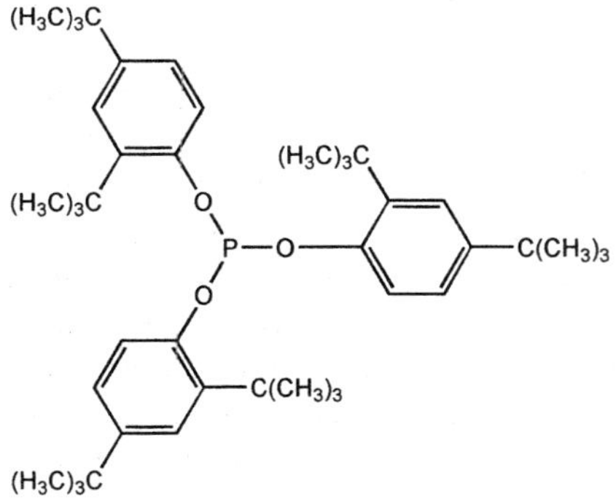

4,4'-Thiobis(6-t-butyl-3-methylphenol)

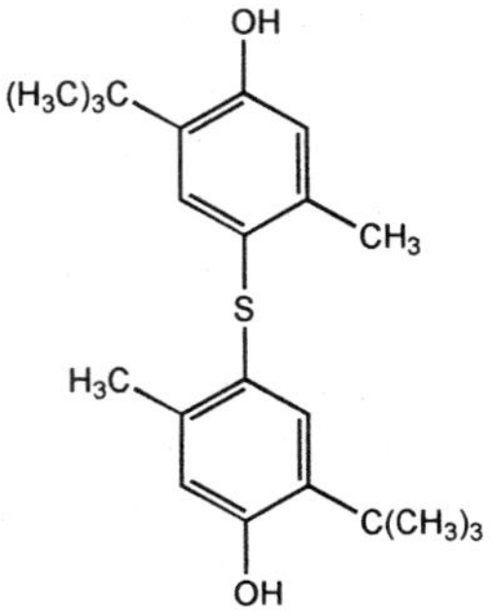

Tri(2,4-di-t-butyl phenyl) phosphite

Figure 2 Examples of some of the families of stabilizers used to inhibit autoxidation of polyethylene.

to guard against specific reactions (such as chelating agents that neutralize the effects of small amounts of metal due to catalyst residues and contamination). The term "antioxidant" is applied specifically to those chemical species that interfere with the propagation and branching of the autoxidation process. A large number of stabilizers and combinations thereof are in commercial use or have been tried in the past. In the space available it is impossible to give anything but a brief overview of this complex topic. Those readers wishing to learn more about this subject are directed to the review articles of Geuskens (1975), Potts (1978), and Scott (1979–1987, 1993) listed in the bibliography.

1. Shielding

Stabilizers that act as shields are primarily effective against the degradative effects of ultraviolet light. The goal of such stabilizers is to prevent ultraviolet radiation from reaching the chromophores that absorb radiation and thereby initiate autoxidation. In doing so they must not transfer the incident energy to the polyethylene molecules in a form that initiates the oxidation process. Ultraviolet shields take two forms: reflectors and absorbers. The first is exemplified by such materials as the rutile form of titanium dioxide, the second by carbon black and various complex organic molecules containing conjugated rings. When incident radiation is absorbed, it is commonly dissipated as heat, fluorescence, or molecular rearrangement.

The usefulness of ultraviolet shields, especially particulate stabilizers, is somewhat limited. Because they are nonspecific, they act upon more than just the ultraviolet portion of the electromagnetic spectrum; in doing so they typically impart a strong color or even opacity to samples. It follows that particulate stabilizers are unacceptable in transparent samples. When transparency is required, much more expensive molecular type ultraviolet absorbers must by employed. The relatively large particle size of pigments and the low concentrations at which they must be used may permit radiation to penetrate a significant distance into a sample before it is blocked. The outer layers of materials are thus not effectively shielded. At a given loading, the smaller the particle size, the more effective is the material as a shield. This lack of surface stabilization may be ameliorated by the use of molecular type shields.

2. Quenching

Ultraviolet quenchers deactivate energetically excited carbonyl bonds in such a way that the absorbed energy is not available to initiate the autoxidation process. Nickel chelating agents are often used for this purpose. The chelating agent forms a transient complex with the excited carbonyl group. When the complex breaks apart, the ultraviolet quencher departs in an excited state, leaving the carbonyl group in its ground state. The quencher subsequently relaxes to its own ground state, dissipating excess energy in the form of heat.

3. Radical Scavenging

Several families of organic molecules contain hydrogen atoms that can be readily abstracted to leave a stable radical species. Such molecules can be used as radical scavengers to inhibit the oxidation process. Stabilization of the radical is achieved by conjugation or interaction with electrophilic species attached to other parts of the molecule. Families of molecules that meet this requirement include hindered phenols and hindered aromatic amines, some examples of which are shown in Figure 2. Such molecules are also known as primary antioxidants, chain-breaking agents, or termination agents. Theoretically, these stabilizers can react with any radical species to hinder the autoxidation process. In practice, the initial propagation reaction of alkyl radicals with absorbed oxygen is so facile that radical scavengers rarely interfere with this step. Their principal reaction is with peroxy radicals that would otherwise participate in the second propagation reaction of the autoxidation process. Stabilizer molecules compete effectively with polymers by sacrificing their hydrogen atoms. Reaction of peroxy radicals with antioxidant molecules slows the production rate of the alkyl radicals that feed the autoxidation cycle. The principal reactions of a commonly used antioxidant [butylated hydroxytoluene (BHT)] are shown in Figure 3.

Each primary antioxidant can deactivate two peroxy radicals. In doing so, hydroperoxides and oxy and methyl radicals are formed. This type of stabilization has two principal drawbacks: The antioxidant is consumed in the process of stabilization, and hydroperoxides are generated that can participate in branching reactions. The degree of autoxidation can be followed by measuring a sample's uptake of oxygen as a function of time. When this is done for a polyethylene stabilized with a radical scavenger, the rate is initially low, and it increases only gradually until the rate of peroxy radical formation overwhelms the capacity of the antioxidant or the antioxidant is exhausted. After this induction period the oxygen uptake rate increases rapidly as the autoxidation process accelerates. This is depicted schematically in Figure 4. The length of the induction period depends on a number of factors, including the antioxidant loading, sample temperature, flux of ultraviolet radiation, and sample processing history. Elevated temperatures decrease the length of the induction period owing to the proliferation of radicals in the system caused by the increased rate of branching reactions.

An undesirable side effect of hindered phenols is the fact that at high temperatures they can react directly with molecular oxygen to yield peroxy radicals that can initiate degradation.

The effectiveness of radical scavengers depends on several interrelated factors. Naturally, those chemical structures that are most capable of stabilizing

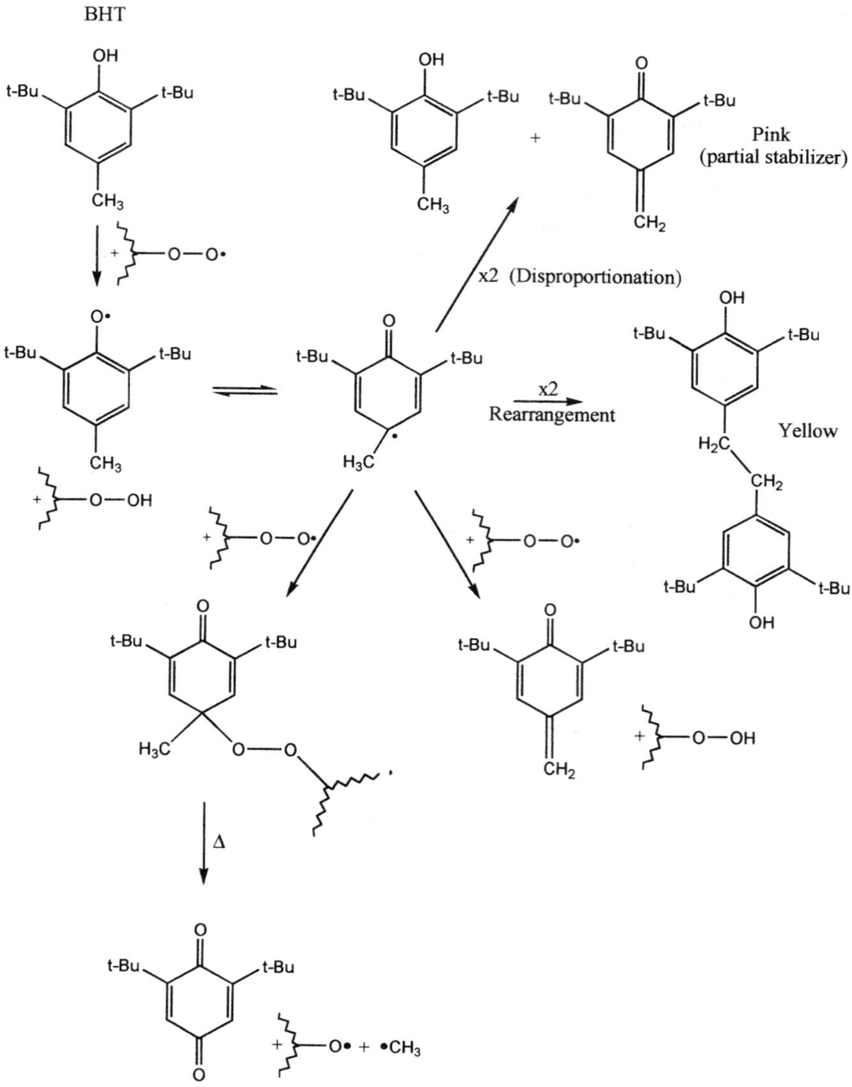

Figure 3 Reaction scheme of butylated hydroxytoluene (BHT) with peroxy radicals.

The Chemistry of Polyethylene

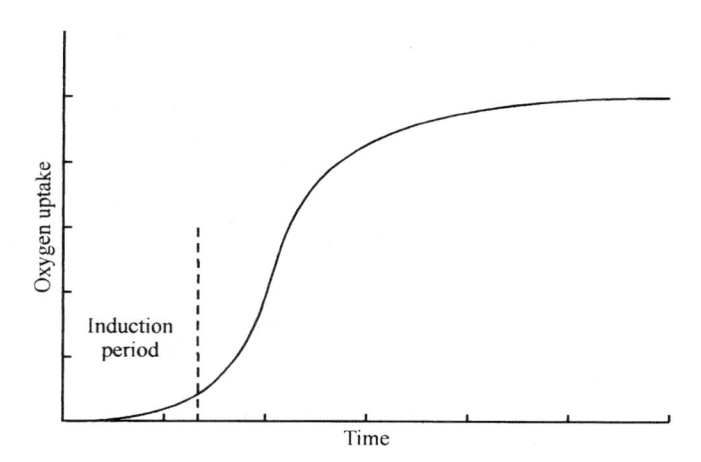

Figure 4 Schematic representation of oxygen uptake as a function of time for a polyethylene sample stabilized with a primary antioxidant.

radicals have the greatest probability of reacting with peroxy radicals. Hindered amines are generally more effective than hindered phenols, but they discolor polyolefins and can stain articles with which they come into contact. Their use is therefore largely restricted to unsaturated elastomers, which are highly susceptible to degradation. The formation of colored reaction products, such as certain hindered quinones, is a distinct disadvantage in certain applications. Relatively small molecules, such as butylated hydroxytoluene, can be quite readily lost due to their volatility at high temperatures, blooming to the surface due to incompatibility or being leached from the polyethylene matrix. More complex hindered phenols, which undergo reactions similar to those of butylated hydroxytoluene, are less likely to be lost. In the extreme it is possible to graft stabilizers directly onto the polymer backbone, thereby ensuring their presence within the material.

It is obvious that although of great importance, primary antioxidants are not the complete answer to limiting autoxidation in polyethylene. In practice, they are commonly used in combination with other types of antioxidants, such as hydroperoxide decomposers, which inhibit other reactions of the autoxidation cycle.

4. Hydroperoxide Decomposition

Stabilizers that decompose hydroperoxides to stable compounds are classified as secondary antioxidants. Families of molecules in this category include organic phosphites and thioesters. By breaking down hydroperoxide molecules, secondary antioxidants reduce the overall concentration of radical species available to participate in the branching reactions of the autoxidation process. Secondary antioxidants are frequently used in combination with primary antioxidants, the overall effect of such stabilizer packages often being synergistic. Because of their synergistic interaction with primary antioxidants, hydroperoxide decomposers are also sometimes known as "synergists."

Most common secondary antioxidants operate along similar lines. Unstable hydroperoxides are reduced to stable alcohols by the removal of an oxygen atom. In the process the stabilizer is oxidized to a higher oxidation state. Organic phosphites are well suited for this purpose. Secondary antioxidants, especially sulfur compounds, may also reduce peroxy radicals to oxy radicals. Ideally, the newly formed oxy radical would abstract a hydrogen atom from a primary antioxidant to yield a stable hydroxyl group. However, due to its highly reactive nature, the oxy radical may also undergo less desirable reactions such as abstracting a hydrogen atom from a polyethylene chain to yield an alkyl radical.

Sulfur-based hydroperoxide decomposers, such as thioesters, suffer from the disadvantage that they may be malodorous or may yield compounds that are. This fact largely precludes their use in many products. For this reason, phosphorus-based compounds are the most widely used type of secondary antioxidants in polyethylene.

C. Biodegradation

The term "biodegradation" refers solely to degradation by biological action, primarily bacterial or fungal attack but also digestion within a living organism either in the alimentary canal or surrounded by body tissue. Biodegradation may take place as part of environmental degradation, but the two terms are not interchangeable. Environmental degradation encompasses a variety of processes, including photodegradation, autoxidation, biodegradation, and abrasion.

To all intents and purposes, chemically pure polyethylene with a molecular weight in excess of approximately 450 is inert to biodegradation. Doubtless, given sufficient time, bacteria and fungi will evolve that can consume polyethylene, but until then any attempt to make polyethylene biodegradable must involve one or more other processes that convert the polymer to a substance that can be digested by presently existing microbiotic life forms [3]. The conversion of polyethylene to a digestible substrate can be accomplished either by reducing its molecular weight substantially or by introducing chemical species that are readily attacked, such as main-chain ester functionality.

It is well known that alkanes of sufficiently short chain lengths can be used as a substrate by a variety of bacteria. Linear alkanes with molecular weights of up to approximately 450 (C_{32}) are readily digested by bacteria, whereas their branched homologues are essentially indigestible [4]. Biodegradation of commercial polyethylene resins is limited to the digestion of the small amounts of low molecular weight chains that are a natural consequence of a broad molecular weight distribution [4,5] or the degradation of additives such as antiblocking agents (for instance, erucamide) or processing aids [4]. For the bulk of a polyethylene sample to undergo biodegradation, it must first be cleaved by a primary process into fragments sufficiently short to be digested. Primary degradation can take the form of photo-oxidation with accompanying Norrish types I and II cleavage or pro-oxidation initiated by chemically active transition metal salts [6], ultraviolet light sensitizers [7,8], or corn oil [9], which reacts with metal salts in soil to form organic peroxides. Once a short chain bearing a terminal hydroxyl group has been generated, it can be attacked by bacterial enzymes. The hydroxyl group is first oxidized to form a carboxylic acid. Subsequent β-oxidation releases two carbon atom based fragments from the end of the chain and regenerates a carboxylic acid group. The liberated fragments enter the citric acid cycle, where they are oxidized to carbon dioxide and water [10].

The introduction of oxygenated species into the polyethylene backbone can improve biodegradability by enhancing primary (nonbiological) degradation of the polymer into digestible fragments or by introducing groups that can be attacked directly by enzymes. The copolymerization of ethylene with carbon monoxide to incorporate ketone functionality falls into the first category [11,12], while copolymerization with 2-methylene-1,3-dioxepane to incorporate ester functionality [13] is of the second type. Copolymers of ethylene and carbon dioxide are used to make loop connectors for six-packs of canned beverages that rapidly photodegrade into a low molecular weight brittle material, the molecules of which are susceptible to biological attack. The incorporation of 10 mol% 2-methylene-1,3-dioxepane results in a product that is readily biodegraded, while that containing 2 mol% is degraded at a far slower rate. The inclusion of sufficient ester functionalities in copolymers to make a significant contribution to biodegradability inevitably results in drastic loss of mechanical properties, especially modulus and yield stress.

An often discussed method that is purported to make polyethylene biodegradable involves blending it with granules of starch [14]. It is hypothesized that when it is composted or buried in a landfill, the starch will dissolve, leaving holes that increase the surface-to-volume ratio of the polymer, making it more susceptible to microbiotic attack. In practice this is not the case; it has been shown that blending with starch alone does nothing to enhance polyethylene biodegradation [15,16].

III. CROSS-LINKING

Cross-linking occurs when adjacent chains become covalently linked, either directly by a carbon–carbon bond or indirectly via a bridging group. Polyethylene can be cross-linked to varying extents, ranging from isolated bonds linking a small proportion of pairs of adjacent chains to multiple links between adjacent chains, which bind the whole sample into a single interconnected network. If each chain is attached to two or more of its neighbors, the whole sample forms a single molecule, each atom being attached to all others via a series of covalent bonds. The portion of the sample incorporated into the network is insoluble, although it can be swollen by hot solvents. The insoluble network is often referred to as a "gel," the remaining material being known as a "sol." Depending upon the extent of cross-linking, the fractional mass of molecules incorporated into the network (the gel content) can vary from 0% to 100%. Products of cross-linking that contain significant proportions of gel are known as cross-linked polyethylene (XLPE). Naturally, the properties of a sample consisting of a single enormous, highly branched molecule are very different from those of a sample composed of a multitude of ordinary polyethylene molecules. The physical properties of a cross-linked network depend upon the average length of the molecular segments between cross-links. The degree of cross-linking is normally described in terms of cross-link density; the shorter the average molecular segment between cross-links, the higher the cross-link density. Even a low cross-link density can have profound effects on a sample's properties, especially in the molten state. Given the pronounced effect of cross-linking on the properties of polyethylene, it may be viewed as either highly advantageous or disadvantageous depending upon the intended use of the material.

A. Mechanisms of Cross-Linking

The methods by which polyethylene can be cross-linked may be categorized into processes that form covalent bonds directly between the carbon atoms of adjacent chains and processes that link adjacent chains via a short chemical bridge. The former category invariably proceeds via a radical process, while the latter generally involves hydrolysis or condensation of species previously grafted onto the polyethylene backbone. Peroxide-initiated and high energy radiation initiated cross-linking mechanisms, which are respectively irradiative free radical and chemical free radical processes, dominate commercial practice. Siloxane-bridged and ultraviolet radiation initiated radical cross-linking processes are each used to a much lesser extent.

1. Radical Cross-Linking

When a hydrogen atom attached to the backbone of a polyethylene molecule is removed, it leaves behind a highly reactive macroradical. In the absence of absorbed oxygen, the macroradical and its associated species can undergo the reactions shown in Figure 5.

Macroradicals can react with each other to form covalent carbon–carbon

The Chemistry of Polyethylene

Crosslinking:	2P·	→ P-P
Reaction with free radical species:	P· + X·	→ P-X
Reaction with free radical scavenger:	P· + AOH	→ PH + AO·
Radical stimulation on adjacent chain:	P'·* + P"	→ P'· + P"· + H·
Intra-molecular hydrogen transfer:	~~CH_2-CH_2-CH·-CH_2~~	→ ~~CH_2-CH·-CH_2-CH_2~~
Inter-molecular hydrogen transfer:	P'· + P"	→ P' + P"·
Abstraction of hydrogen:	~~CH_2-CH_2-CH_2-CH_2~~ + H·	→ ~~CH_2-CH_2-CH·-CH_2~~ + H_2
Disproportionation:	2 ~~CH_2-CH_2-CH·-CH_2~~	→ ~~CH_2-CH_2-CH_2-CH_2~~ + ~~CH_2-CH=CH-CH_2~~

Where:
- P· = Polyethylene macroradical
- P = Polyethylene backbone
- X· = Free radical
- AOH = Anti-oxidant
- P'·* = Excited macroradical

Figure 5 Principal reactions involved in radical-initiated cross-linking of polyethylene.

cross-links or react with other available radical species, or the unpaired electron can migrate, either intra- or intermolecularly, via hydrogen transfer from adjacent carbon atoms. When hydrogen transfer occurs, the newly formed radical can undergo any of the reactions its progenitor could. These reactions can take place in either the molten state or the noncrystalline portion of the solid state. Where there is an abundance of absorbed oxygen relative to the energy flux, significant oxidative degradation can take place in conjunction with cross-linking. Cross-linking under vacuum decreases the consequences of the degradation process. Naturally, the presence of radical scavengers (primary antioxidants) inhibits the cross-linking process. The radicals required for the cross-linking process can be generated directly by cleavage of carbon–hydrogen bonds, either directly under the influence of high energy irradiation or indirectly via hydrogen abstraction by another free radical.

In most cases, removal of hydrogen atoms from the polymer chains occurs approximately randomly throughout the sample. It follows that the probability of any given polymer chain being attacked and the number of events occurring along its length are directly proportional to its length. Thus, high molecular weight chains are more likely to be cross-linked than shorter ones. Polyethylene resins with a high average molecular weight require a lower dose of irradiation or lower concentration of peroxide initiator to effect the same level of gelation (gel content) as resins having a lower average molecular weight. The presence of unsaturated groups, especially terminal ones, strongly influences the cross-linking process. Allylic hydrogen atoms are more readily abstracted than alkyl hydrogen atoms and hence are preferred reaction sites. Many polyethylene resins consist of molecules that each bear a single highly reactive terminal vinyl group. The reaction of two terminal radicals is in effect a chain extension reaction. The reaction of a terminal radical with a radical located in the middle of another molecule results in the introduction of a long-chain branch [17]. The product of this type of reaction is sometimes described as a Y branch.

   a. High Energy Radiation Initiated Cross-Linking. When a polyethylene sample is flooded with high energy photons, one of the effects is to cleave carbon–hydrogen bonds. Cleavage can occur throughout the sample regardless of its semicrystalline morphology or of whether it is in the solid or molten state. The cleavage of carbon–hydrogen bonds liberates hydrogen atoms, leaving behind reactive macroradicals that can take part in the cross-linking process:

$$PH + h\nu \rightarrow H\bullet + P\bullet$$

A liberated hydrogen atom can readily diffuse through the body of the sample until it reacts with another hydrogen atom to yield a hydrogen molecule, reacts with a macroradical to re-form a carbon–hydrogen bond, or diffuses from the surface. If a carbon–hydrogen bond within a crystallite is cleaved and the liberated hydrogen atom diffuses away from the site, the radical on the backbone will migrate until it reaches a noncrystalline region where it will be free to react with other radical species. Cross-linking cannot take place within the crystal lattice [18]. The cleavage of backbone carbon–carbon bonds may also occur, but the radicals so produced are effectively trapped in the vicinity of each other and stand a high chance of reacting to re-form the original bond. The terminal radicals formed by chain scission may be stabilized by disproportionation or by reaction with hydrogen atoms, resulting in molecular weight degradation. Such degradation is of significance only in samples with extremely high levels of crystallinity, such as highly oriented fibers [19].

   The efficiency of high energy radiation induced cross-linking is defined in terms of the number of cross-links formed per unit of radiation. The efficiency of cross-linking solid samples at temperatures in excess of 100°C is increased relative to room temperature, up to a factor of approximately 2 at 130°C [20]. At elevated temperatures, highly crystalline samples exhibit higher levels of gelation than noncrystalline samples subjected to the same radiation dose, presumably due to the concentration of terminal vinyl groups in the vicinity of each other in the relatively small noncrystalline regions [21]. In the case of highly drawn ultra-high molecular weight polyethylene fibers, chain scission predominates within the noncrystalline regions and reduces tensile strength [22], the effect increasing as the draw ratio of the fibers increases.

The shape of the cross-linked polyethylene product required from the high energy radiation process largely dictates the type of radiation that must be employed. The two most common types of radiation used are electron beams (E-beams) and γ rays. The penetrating power of electrons is quite low, the effective electron density decreasing rapidly as a function of penetration distance. In the case of thick specimens, this can result in nonuniform cross-linking. Electron beams are effective when the polymer thickness is less than approximately 0.020 in., e.g., film, thin sheet, thin-walled tubes, wire insulation, fibers, and low density foam. Thicker products require radiation with a higher penetrating power such as the γ rays emitted by cobalt-60. Products requiring high penetrating power include thick sheet, thick-walled tubes, cable insulation, and injection-molded parts.

The efficiency of cross-linking by electron beams is increased in the presence of absorbed acetylene [23]. It is postulated that acetylene diradicals form bridges between adjacent chains. Acetylene is also effective when used as a posttreatment after irradiation in the presence of nitrogen.

  b. Chemical Free Radical Initiated Cross-Linking. Many organic molecules decompose to generate free radicals that can abstract hydrogen atoms from polyethylene to initiate cross-linking. Organic peroxide initiators are the most commonly used class of chemical free radical initiators. Some commonly used peroxide initiators are shown in Figure 6.

Peroxide groups decompose homolytically under the influence of heat to generate a pair of oxy radicals, each of which can abstract a hydrogen atom from a polyethylene molecule.

$$ROOR + \Delta \rightarrow 2RO\bullet$$

$$PH + RO\bullet \rightarrow P\bullet + ROH$$

As it requires the reaction of two unpaired electrons on adjacent chains to form a cross-link, theoretically each peroxide group can generate one cross-link. In practice, an efficiency approaching 100% can be achieved for LDPE [24,25]. The cross-linking efficiency for linear polyethylene is approximately 80% [26]. Factors that affect cross-linking efficiency include the presence of antioxidants, the concentration of absorbed oxygen, the level of branching, and the presence of unsaturation. Primary antioxidants can scavenge peroxy radicals before they have the opportunity to abstract hydrogen atoms from the polymer. The presence of absorbed oxygen leads to chain scission according to the processes outlined in Section II. As the rate of degradation is controlled by the rate at which oxygen can diffuse into the sample, rapid decomposition of the initiator will decrease the probability of chain scission. Low levels of branching appear to enhance cross-linking efficiency [25,27], but high levels promote chain scission as a competing reaction. As the proportion of propylene in an ethylene-*co*-propylene co-

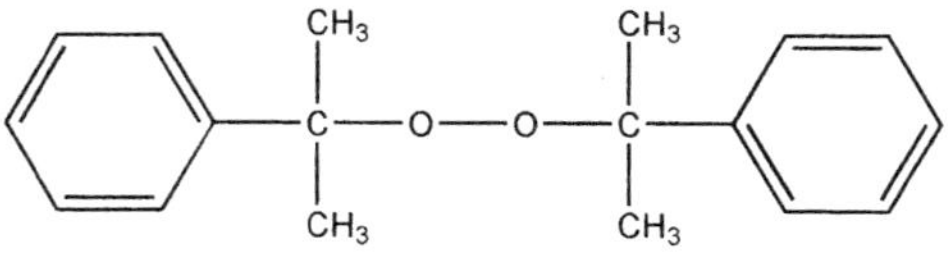

Dicumyl peroxide

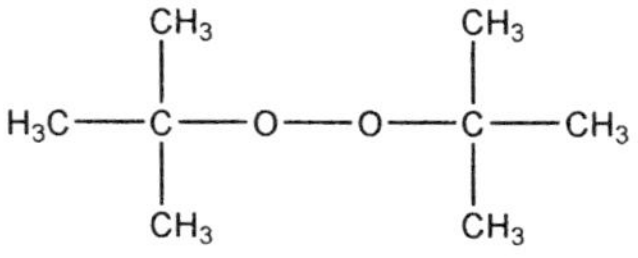

Di-t-butyl peroxide

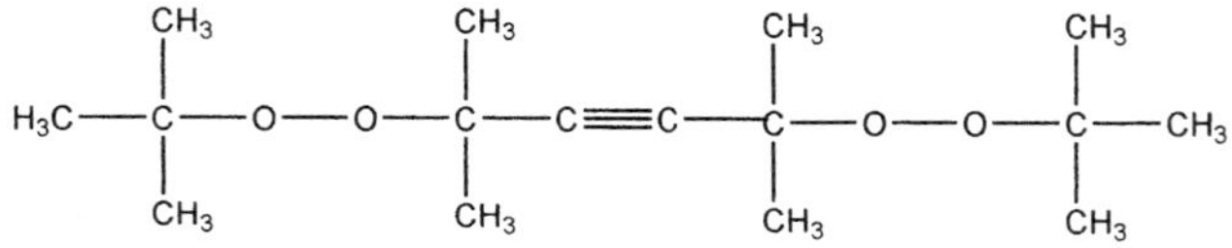

2,5-Dimethyl-2,5-di-(t-butylperoxy) hexyne-3

Figure 6 Examples of commonly used organic peroxide cross-linking initiators.

polymer increases, the tendency for β chain scission adjacent to methyl branches increases due to the stabilization of intermediate radicals [28]. Unsaturation influences the cross-linking efficiency because allylic hydrogen atoms are preferentially abstracted by oxy radicals. When terminal vinyl groups are involved, the result is an initial increase in molecular weight due to end-to-end chain addition, prior to the occurrence of cross-linking [26]. The significance of terminal vinyl groups increases as the molecular weight decreases owing to their relatively lower concentration in the higher molecular weight samples.

The rate at which initiator molecules decompose, and hence the rate of cross-linking, is a function of their chemical stability and the temperature to which they are subjected. The decomposition of organic peroxides is an approximately first-order reaction, its rate increasing exponentially as a function of temperature. The rate of spontaneous decomposition of an initiator at any temperature is typically characterized in terms of its half-life. The approximate half-lives of some commonly used peroxide intiators are listed in Table 1. Initiator molecules for cross-linking polyethylene are selected with respect to the temperature at which the resin must be processed. The goal is to homogenize and mold the resin into the desired shape below the temperature at which peroxide decomposition becomes significant; premature decomposition of the peroxide is known as

Table 1 Approximate Half-Lives of Some Commonly Used Organic Peroxide Cross-Linking Initiators

Initator	Half-life (min) at given temp			
	140°C	160°C	180°C	200°C
Dicumyl peroxide	38	4	0.65	
Di-*t*-butyl peroxide	160	16	2.5	0.35
2,5-Dimethyl-2,5-di-(*t*-butylperoxy)hexyne-3	205	25	3.2	0.5

"scorching." Subsequently the temperature is raised sufficiently for cross-linking to occur within a reasonable period of time. Timing is crucial; if a resin is allowed to reside too long within the barrel of an extruder at too high a temperature, its viscosity may increase until it is unworkable (in extreme cases the extruder screw might even have to be removed for cleaning). On the other hand, economical production requires a high temperature and short residence time at the cross-linking stage. As the curing temperature is increased, the ratio of scission (due to disproportionation of radicals) to cross-linking reactions increases. However, for scission to dominate it is necessary to go to much higher temperatures than those normally encountered during conversion [29].

The efficiency of peroxide cross-linking can be improved with the aid of various coagents that inhibit chain scission and disproportionation [30]. Examples of such additives include pentaerythritol triacrylate and triallyl isocyanurate. Their effect is most pronounced when they are used in combination with low levels of peroxide cross-linking agents.

The relatively long curing step, during which time the shape of the material must be controlled, largely precludes the application of peroxide cross-linking to injection or blow molding processes. Commercially, extrusion forming and rotomolding are the two most important conversion processes. In the extrusion process the resin and initiator (sometimes dissolved in white oil or an alkane) are blended and raised to the forming temperature within the barrel of the extruder. The mixture is shaped within the die and extruded into a curing tunnel, in which the profile of the extrudate is confined while its temperature is raised sufficiently for the initiator to decompose rapidly. The curing time is dictated by the half-life of the initiator. The extrudate passes through the curing process at a rate which permits the decomposition of virtually all the initiator—a cross-linking time equivalent to a minimum of five peroxide half-lives is typical.

This process is used to manufacture such products as thick-walled pipes and the heavily insulated cable used in the electric grid (by coextrusion over a copper core). The polyethylene resin used in rotomolding consists of porous granules that have been previously infused with a solution of peroxide in a volatile solvent. After the solvent is stripped off, an intimate mixture of polymer and initiator remains. A weighed amount of granules is introduced into the mold, which is then tumbled and heated to evenly disperse the granules around its interior. As the heating and tumbling continues, the polymer melts to form a molten layer coating the inside of the mold. Subsequently the temperature is raised and the peroxide decomposes to effect cross-linking. This process is used for the production of large hollow items such as chemical storage tanks and kayaks.

One drawback of peroxide-initiated cross-linking is the presence of residual by-products in the final product. Many of these products are ketones that impart a strong odor to fabricated items, which may or may not be a problem depending upon the intended use.

c. Ultraviolet Radiation Initiated Cross-Linking. The normal effect of ultraviolet radiation on polyethylene is deleterious due to oxidative degradation. However, under appropriate conditions, cross-linking predominates. Ultraviolet radiation alone is sufficient to cross-link polyethylene in the absence of oxygen; however, the length of time required to produce substantial gelation is prohibitive. For optimum efficiency, ultraviolet radiation initiated cross-linking requires the presence of a photoinitiator (sensitizer) and a bridging agent (cross-linker, cross-linking agent). The inclusion of appropriate sensitizers can reduce the reaction time by more than three orders of magnitude [31]. The use of a cross-linking agent in addition to a sensitizer can increase the efficiency of the cross-linking reaction by a further order of magnitude [32]. Examples of photoinitiators include benzophenone, 4-chlorobenzophenone, and sulfuryl chloride [22,33,34]; triallyl cyanurate and triallyl isocyanurate are effective bridging agents [32].

Photoinitiators may be divided into two classes: those that undergo fragmentation to yield radical species, and those that are excited into higher energy states. The products of both types can abstract hydrogen atoms from polyethylene. The newly formed macroradicals can undergo typical radical cross-linking reactions. Excited photoinitiator molecules can also abstract α-hydrogen atoms adjacent to the vinyl groups of bridging agents, such as triallyl cyanurate, to form relatively stable allylic radicals [35]. Allylic radicals can react with the macroradicals on the backbone of polyethylene molecules, grafting the bridging agent onto the polymer. Each triallyl cyanurate molecule can react with up to three polyethylene molecules, linking them together to form a complex cross-linking site. In semicrystalline samples, reactions occur only in the noncrystalline regions that are accessible to initiators and bridging molecules. The factors affecting the efficiency of ultraviolet initiated cross-linking have been thoroughly investigated by Chen and Rånby [36].

From a commercial standpoint, ultraviolet radiation initiated cross-linking appears to have certain distinct attractions. Ultraviolet radiation sources are

The Chemistry of Polyethylene

cheap, present few safety hazards, and are readily available. Highly drawn ultra-high molecular weight polyethylene may be cross-linked without fear of the chain scission that is encountered with the high energy radiation initiated process [22]. On the other hand, ultraviolet radiation does not penetrate deeply into semicrystalline samples. This limits applications to cross-linking in the molten state or to thin solid samples that are essentially devoid of fillers.

2. Silane Bridged Cross-Linking

Organosilane cross-linking of polyethylene is a multistage process involving grafting, blending, molding, and curing [37,38]. The final product consists of polyethylene chains linked to one another by siloxane bridges that can couple two or more chains through a single bridging site.

The first step in the process is the grafting of multifunctional silane groups onto regular polyethylene molecules. This is accomplished by a radical reaction that grafts vinyl siloxane onto polyethylene molecules, as shown in Figure 7. The radical initiator is typically an organic peroxide—often dicumyl peroxide—with vinyl trimethoxysilane being the siloxane of choice. In the presence of moisture, trimethoxysilane groups attached to adjacent chains undergo hydrolysis, eliminating methanol to form siloxane bridges. Each silane group can react with up to three others, generating a complex bridging group that may link several polyethylene chains. Under ambient conditions the cross-linking reaction occurs slowly, limited by diffusion of water into the sample and a low reaction rate. The reaction can be accelerated considerably if the graft copolymer is exposed to steam in the presence of a silanol condensation catalyst such as dibutyltin dilaurate.

In practice, the silane is grafted onto the polymer backbone in high shear compounding equipment, such as a twin-screw extruder, preferably one equipped with a vacuum port to remove the excess silane. Moisture must be excluded from the graft copolymer during subsequent storage to prevent premature, albeit slow, hydrolysis. Because of the practical difficulties of excluding all moisture, the graft copolymer has a limited shelf life. Independently, a masterbatch blend of polyethylene and dibutyltin dilaurate is prepared. The masterbatch is blended into the graft copolymer in the barrel of an extruder as part of extrusion, injection, or blow molding. Care must be taken at this stage that neither of the components is damp. Premature cross-linking, either during storage or in the extruder, can result in drastically lowered processability. In extreme cases cross-linking can progress to such a point that an unprocessable gel is formed, ultimately requiring the shutdown and dismantling of equipment to remove the cross-linked product—which is a major undertaking. Curing takes place in a steam chamber, where the molded parts are kept for about 24 hr. Alternatively, items can be cured by immersion in hot water for a similar period of time. At the end of this period the cross-

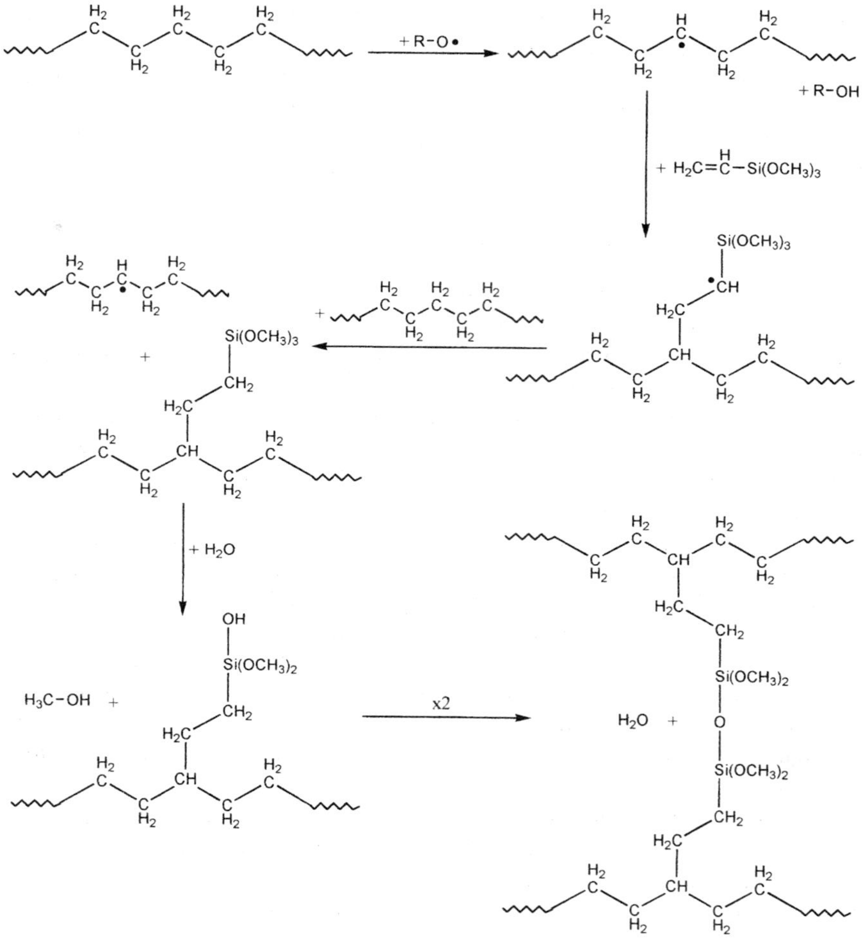

Figure 7 Principal reactions involved in siloxane cross-linking of polyethylene.

linking process is essentially complete and the product is ready for use. Products that are not cured in a steam chamber or in hot water will gradually cross-link over a period of several weeks or months, depending upon ambient conditions.

As a practical process, silane cross-linking has both advantages and disadvantages. On the positive side, the molding and curing stages require little capital investment, allowing short manufacturing runs on a custom basis. The size and thickness of the final product are largely immaterial as the cross-linking agents

The Chemistry of Polyethylene

are dispersed throughout the sample. Molding conditions are less critical than in peroxide-initiated processes. Scrap can be recycled, provided it is reground and returned to the extruder promptly. On the negative side, silane cross-linking is not suited for continuous processes. If large numbers of items are fabricated, the size of the steam chamber required to cure them may become prohibitive. The graft and masterbatch materials are more expensive than the neat resins used in peroxide and high energy irradiation processes. Due to the risk of dimensional instability when the molded products are heated in the steam chamber, it may be advisable to use only high density polyethylene as the base resin.

B. Effects of Cross-Linking

Polyethylene is generally cross-linked for one or more of three purposes: to improve its dimensional stability at elevated temperatures, to improve its impact resistance, or to reduce its propensity to stress crack. In the first instance the goal is to prevent gross deformation, generally above the crystalline melting point, but also under circumstances in which excessive creep below the melting point would be detrimental. Cross-linking influences tensile properties measured at room temperature, but these effects are much less important. The key to the desirable attributes of cross-linked polyethylene lies in the hindered molecular slippage in the noncrystalline regions, either in semicrystalline structures or in the melt state.

Cross-linked polyethylene does not melt in the conventional sense; it does not flow when its temperature is raised above its crystalline melting point. Rather, when its crystalline melting temperature is exceeded, cross-linked polyethylene changes from a ductile semicrystalline solid to a noncrystalline elastomer. The elastic modulus of the noncrystalline state is proportional to its temperature and cross-link density according to classic rubber network theory [39]:

$$E = \frac{n\rho R_g T}{M_c}$$

where

E = elastic modulus
n = a constant ($n = 3$ in the simplest network theory)
ρ = polymer density
R_g = the gas constant
T = absolute temperature
M_c = molecular weight between entanglements

The increase in the elastic modulus as a function of increasing temperature and cross-link density is due to changes in entropy. When the temperature of a rubber is raised, the molecular segments connecting cross-link sites experience an increased force, driving them to adopt a random configuration. It follows that as the temperature increases there is a greater retractive force when the sample is subjected to deformation, i.e., its modulus increases as a function of temperature. The increase in modulus as a function of entanglement density can be understood in terms of the forces required to deform the numerous random coils making up the chain segments between cross-links. As the cross-link density increases, the average end-to-end distance of the random coils between cross-link sites decreases, and the intervening segments must undergo greater alignment to achieve a given overall specimen deformation. The net result of this is a greater decrease in entropy for a given dimensional change for a highly cross-linked sample relative to a lightly cross-linked one. This manifests itself as an increase in elastic modulus as a function of cross-link density. From a practical point of view, the elasticity of cross-linked polyethylene is rarely exploited. Cross-linking mainly serves to maintain the integrity of a sample when its crystalline melting temperature is exceeded. The enhanced dimensional stability of cross-linked polyethylene can be exploited during the simultaneous foaming and cross-linking of polyethylene. A low degree of cross-linking aids in the formation of a narrow distribution of cell sizes by reducing the probability of bubbles bursting.

Cross-linked polyethylene displays thermal memory. When a lightly cross-linked sample is deformed at room temperature it does so in a ductile fashion. If it is subsequently heated above its crystalline melting temperature it will retract to its original dimensions. This property is exploited to encapsulate connections in electrical devices. An expanded tube of cross-linked polyethylene is slipped over a connection and subjected to moderate heat, whereupon it shrinks to form a tight seal around the joint.

At temperatures below the melting point of a sample's crystallites, cross-linking serves to limit the gradual slippage of chains associated with stress cracking and creep. Above a certain degree of cross-linking, environmental stress cracking can be virtually eliminated. The critical degree of cross-linking depends on the intrinsic molecular characteristics of the neat resin. The greater the original resin's resistance to stress cracking, the less cross-linking is required to eliminate it entirely. Creep in polyethylene samples, especially at elevated temperatures, can be greatly reduced by cross-linking, owing to reduced segmental movement in the noncrystalline regions between crystallites.

The effect of cross-linking on a polyethylene resin's room temperature mechanical properties depends to some extent on the conditions under which the cross-linking reaction takes place. When the reaction occurs in the molten state, cross-links are inserted homogeneously throughout the sample. Solid-state cross-linking results in the formation of cross-links only in the noncrystalline regions between crystallites.

For the most part, as cross-link density increases, a sample's degree of crystallinity and crystallite thickness decrease [40–43]. In the case of samples cross-linked in the molten state, this is due to the reduction in the degree of freedom that any given molecular segment has to form part of a crystallite. When cross-linking occurs in the noncrystalline regions of the solid state, the decrease in crystallite thickness is presumably due to strains imposed on the interfacial zones. Concomitant with the decreased degree of crystallinity and crystallite thickness, the Young's modulus, yield stress, elongation at yield, elongation at break, and peak melting temperature of the sample all decrease. When taken to the extreme, all vestiges of crystallinity can be removed from a cross-linked sample, resulting in the creation of a brittle glass [40]. In some cases, irradiation in the solid state leads to an increase in the degree of crystallinity of the sample relative to that of the base resin [44–46]. It is believed that cleavage of taut or entangled tie chains (either directly or by oxidative degradation) permits the thickening of preexisting crystallites. Other effects associated with cross-linking include improved ultimate tensile strength, abrasion resistance, and impact resistance [46–48].

The semicrystalline morphology of polyethylene changes gradually as its degree of cross-linking increases. The changes undergone depend on whether the sample is cross-linked in the solid or molten state. When solid samples are irradiated, the number, location, and long-range ordering of the original crystallites remain intact, but their individual thicknesses are decreased. When cross-linked in the molten state, both the thickness and long-range ordering of crystallites change. For a sample that originally crystallizes with a spherulitic morphology, the nucleation density increases systematically with peroxide concentration [49]. The spherulitic morphology first degenerates to sheaves, then to isolated bundles of lamellae, and finally to isolated micellar crystallites.

IV. CHEMICAL MODIFICATION

Chemical modification of polyethylene can occur within the body of a sample or on its surface. Examples of the former include cross-linking and the neutralization of acid side groups to form ionomers. Commercial chemical modification within the bulk of polyethylene results in significant changes to mechanical properties. Surface modification generally takes the form of oxidation, either with an energetic gas or a corrosive liquid, but chemical grafting is also possible. Commercial surface modification is routinely practiced, with the goal of improving the adhesion of inks and paint.

A. Bulk Modification

Bulk chemical modification is practiced on a commercial basis, with the goal of significantly altering the overall physical properties of a sample, in either the solid or molten state. Examples of commercial bulk modification include crosslinking, the hydrolysis of ester side groups to their alcohol derivatives, and the neutralization of acid functions in ethylene-*co*-methacrylic acid and ethylene-co-acrylic acid copolymers to form ionomers. Chemical modification may be used to impart specific chemical functionality, including binding antioxidant species directly to the backbone [50]. Stabilizers grafted directly onto macromolecules or copolymerized into them have the advantage that they cannot be leached out. However, the high price of monomeric antioxidants and the expense of grafting make it more economical to add higher than normal concentrations of regular stabilizers to counteract leaching.

B. Surface Treatment

In its virgin state, pure polyethylene has a low energy surface that few substances will wet or adhere to. Under many circumstances this detracts from its otherwise desirable properties. This is especially true in packaging applications when information regarding contents must be conveyed or surface decoration is required for aesthetic reasons. Chemical modification of the surface of polyethylene is widely practiced to improve its performance as a substrate for printing and painting. In addition, increased surface energy is beneficial in such applications as metal coating, polyethylene fiber reinforced composites, filters, and medical prostheses or where antistatic or antifogging behavior is required. Three principal benefits derive from surface modification: improved wetting, increased bonding, and surface roughening.

A variety of methods are available for the surface modification of polyethylene, but only a few are widely practiced. From a commercial standpoint the most important methods are corona discharge treatment (CDT), flame treatment, and, to a lesser extent, chemical etching. Corona discharge is routinely applied to films. Flame treatment is used on bottles or other thick molded parts with a smooth exterior profile. Chemical etching is applied to irregularly shaped moldings and interior surfaces. Each of these methods rapidly oxidizes the outer layers of the polymer. Flame and corona treatment take less than a second; chemical etching is over within a few seconds. Other, less practicable methods of surface oxidation include plasma treatment, ozonation, photo-oxidation, ultraviolet light irradiation, and pressing against aluminum foil. Specific chemical functionality can be introduced by various grafting techniques.

The effects of surface modification may be studied by X-ray photoelectron spectroscopy (XPS), also known as electron spectroscopy for chemical analysis (ESCA) [51–53], liquid contact angle measurement [54,55], and chemical derivatization (whereby chemical species are specifically labeled with a molecular tag that confirms the presence of the target species) [56,57].

The Chemistry of Polyethylene 403

1. Corona Treatment

Corona discharge treatment is principally used to oxidize the surface of polyethylene films prior to printing. It is performed by passing the film over a grounded and chilled metal roller above which is closely positioned a bar electrode spanning the width of the film. A high voltage is applied between the treatment bar and the grounded roller, generally in the range of 5–50 kV. The current flow may be either direct or alternating at frequencies of the order of 10–20 kHz. The separation between the electrode and the film is approximately 1 mm, and the current is generally less than 1 A. The intervening gap may be flooded with an "active" gas such as air or oxygen or an "inert" gas such as hydrogen, nitrogen, or argon. The pressure in the corona is generally at or slightly below atmospheric (when a reduced pressure is used, the technique is termed "hybrid plasma treatment").

Ions and electrons generated in the corona readily attack the surface of the film, oxidizing it homogeneously to a depth of approximately 50 Å within a few seconds [56]. The precise nature of the reactions that occur is a matter of some discussion, but it is clear that large numbers of macroradicals are generated when carbon–hydrogen bonds are broken. These macroradicals can participate in autoxidation or cross-linking reactions as described in preceding sections. Even when an "inert" gas is used, oxygen is incorporated into the surface, presumably arising from reactions involving adsorbed oxygen or subsequent reactions when the film is exposed to air. The net result is a surface that is cross-linked, contains unsaturation, and is rich in oxygenated species such as ketones, alcohols, aldehydes, and esters. The increased polarity raises the surface energy, thus facilitating wetting by inks and enabling specific interactions such as hydrogen bonding which enhance adhesion. Enolizable carbonyl groups have been identified specifically as improving hydrogen bonding [57,58]. The effectiveness of surface treatment can be demonstrated by the "adhesive tape test": When ink is applied to an untreated polyethylene surface, the pressure-sensitive adhesive on tape will generally lift it right off; this is not the case for corona-treated surfaces. When a corona-treated film is allowed to age at room temperature or is washed with water, the concentration of polar groups at the surface is observed to decrease [56]. From this it may be inferred that oxygenated species are often attached to relatively mobile lower molecular weight species (possibly produced by chain scission reactions).

2. Flame Treatment

Flame treatment is primarily used to oxidize the surface of blow-molded bottles prior to printing; it is especially well suited to treating round bottles. Bottles are supported on a pair of parallel rollers between which is located a row of gas jets. The surface of the bottle is exposed to the flames as it rotates upon the rollers. Variations on this process can be used to treat noncircular bottles and injection-molded products. Flame treatment is not much used on films due to the risk of overheating them. For best results objects should be exposed at a distance of 0.25–0.5 in. beyond the inner cone of a flame that is slightly oxygen-rich. The exposure duration may be as short as approximately 0.02 seconds [59].

The oxygenated species incorporated by flame treatment are similar to those created during corona discharge treatment. This suggests a common series of reactions following the initial creation of macroradicals by interaction with the ions, radicals, free atoms, excited species, and electrons that compose the flame. The profuse concentration of activation events is such that antioxidants have little impact on the oxidation reactions. X-ray photoelectron spectroscopy reveals that flame treatment modifies the surface to a depth of 40–90 Å [53]. In addition to oxygenated species there is also an increase in the level of nitrogen within the surface layers of the sample.

3. Chemical Etching

Chemical attack with strong acids may be used to simultaneously oxidize and physically etch the surface of polyethylene. The methods used have much in common with the surface preparation techniques described in the section on scanning electron microscopy in Chapter 6. A widely used etchant consists of potassium dichromate, water, and sulfuric acid, the reaction of which produces chromic acid that preferentially attacks the noncrystalline regions between lamellae and at spherulitic boundaries [51]. The etching process introduces oxygenated species and small amounts of sulfonation. Chemical etching is too slow to be of great commercial significance, but it does find a minor role in preparing surfaces with complex shapes for metal coating [60] and improving the adhesion of epoxy resins [51,61,62]. Oxygenated species improve chemical adhesion to the surface, and the microscopically roughened surface presents a larger surface area, the crevices of which can participate in mechanical ''keying.''

4. Miscellaneous Oxidative Methods

Plasma treatment consists of exposing specimens to a low pressure (0.01–10 torr) gas that is excited with radio-frequency (1.5–50 MH_z) or microwave (150–10,000 MHz) electromagnetic radiation. Under the influence of radiation, some of the gas molecules are promoted to an excited state from which they may decay, emitting ultraviolet light; decompose into their component atoms; react directly with the polymer; or lose electrons to form positive ions. Surface oxidation is initiated mainly by the ultraviolet radiation and reactions involving the neutral energetic species. The low pressure gas can be selected to incorporate specific atomic species into the substrate. The gases used include oxygen, nitrogen, argon, ammonia, and hydrogen. In the absence of oxygen, the macroradicals—generated principally by ultraviolet radiation—react to cross-link the skin of the sample, by a process known as CASING (cross-linking by activated species of inert gases)

[63]. Plasma treatment, like other oxidative processes, improves the wetting and adhesion characteristics of polyethylene. When fibers are exposed to plasma for prolonged periods (30–300 sec), microscopic cracks appear in their surface layers that enhance adhesion by mechanical keying [55,64,65]. Although not practicable on a large scale, plasma treatment has the potential for improving the biocompatibility of medical prostheses.

Ozonation can be used to improve the adhesion of polyethylene to aluminum foil during extrusion coating [66]. As the molten polymer leaves the extrusion die it is subjected to a continuous stream of ozone immediately prior to making contact with the aluminum foil. The adhesive strength of the polymer to the metal increases as the level of oxygen incorporated into its surface increased.

Crystallizing polyethylene in contact with aluminum foil results in the incorporation of oxygenated species into the surface of the polymer [52,67]. When the aluminum is dissolved with sodium hydroxide solution, X-ray photoelectron spectroscopy reveals the presence of numerous oxygenated species. The required oxygen is thought to come from air trapped between the surface of the polymer and the metal.

Exposure of polyethylene to ultraviolet or visible light results in surface oxidation, but the process is too slow and nonspecific with regard to depth to be of commercial interest.

5. Graft Modification

The surface characteristics of polyethylene can be specifically modified by grafting various chemical species onto it. For the most part, grafting proceeds via the creation of macroradicals on the polyethylene backbone in the presence of the graft comonomer. Macroradicals can be generated by peroxide decomposition, ultraviolet light irradiation (in the presence of a sensitizer), plasma treatment, or ceric ion initiation. Currently graft modification of surfaces is not practiced on a large scale, but it has potential for small-scale applications such as the production of surface-modified prostheses where biocompatibility is an issue. Examples of chemical substances that have been grafted onto the surface of polyethylene include methyl methacrylate [68], methacrylic acid [68,69], acrylic acid [68,70,71], acrylamide [71–73], 2-methylenepentane-1,5-dicarboxylic acid diethyl ester [74], 2-hydroxyethyl methacrylate [75], sulfonic acid functionality [76,77], and a variety of biologically active substances including heparin and antibodies via glycidyl methacrylate [78].

V. COPOLYMERS OF ETHYLENE WITH POLAR MONOMERS

Monomers containing polar functions are incorporated into polyethylene for two main purposes: to influence the crystallization process or to act as polar ''cross-links." The addition of increasing amounts of polar comonomers serves to reduce the overall degree of crystallinity of the resin, with an accompanying decrease in stiffness. Polar comonomers also inhibit the formation of spherulites, resulting in clearer products due to reduced scattering of light. Polar substituents can interact with one another directly via hydrogen bonds or indirectly, in conjunction with cations, to form ion pairs, "multiplets," or "clusters" that link two or more adjacent chains. Such "cross-links" are heat-sensitive, weakening sufficiently at high temperatures to permit melt flow and re-forming when the resin cools. Copolymers that form polar cross-links with the aid of cations fall into the general category of "ionomers." Polar cross-links stiffen the noncrystalline regions of ionomers, sustaining their moduli at useful levels even when samples contain negligible crystallinity.

The two principal polar monomers copolymerized with ethylene are vinyl acetate (VA) and methacrylic acid (MA). A wide variety of other polar vinyl monomers, including acrylic acid, methyl acrylate, and methacrylonitrile, may also be copolymerized with ethylene but are of less commercial significance. Copolymerization of vinyl acetate with ethylene produces clear, soft materials with high tensile strength which are used primarily in packaging and adhesive applications. Ethylene-*co*-vinyl alcohol copolymers are hydrolyzed derivatives of ethylene-*co*-vinyl acetate copolymers. They are stiffer than their precursors, are somewhat less clear, and have good oxygen barrier properties. Ethylene-*co*-vinyl alcohol copolymers are used as barrier layers in multilayer packaging films. Ethylene-*co*-methacrylic acid and ethylene-co-acrylic acid copolymers find little use in their original state. Neutralization of their acid substituents yields "ethylene ionomers" that are soft and clear, have high tensile strength, and are abrasion- and oil-resistant.

A. Polyethylene-*co*-Vinyl Acetate

Ethylene can be readily copolymerized at high pressure with a variety of vinyl ester comonomers. From a commercial standpoint, vinyl acetate is the most important. Polyethylene-*co*-vinyl acetate is better known as ethylene-vinyl acetate copolymer and is customarily referred to as EVA. The structure of the vinyl acetate branch is shown in Figure 8.

Copolymerization with vinyl acetate has long been used to produce polyethylene resins with lower levels of crystallinity than could be obtained economically by high pressure polymerization of ethylene alone or the copolymerization of ethylene and α-olefins by Ziegler–Natta catalysis. (Recent advances in catalyst design permit the commercial production of linear low density polyethylene resins that have degrees of crystallinity similar to those of ethylene-*co*-vinyl acetate copolymers.) The reactivity ratios of ethylene and vinyl acetate are similar in the high pressure process, resulting in a random distribution of acetate branches along

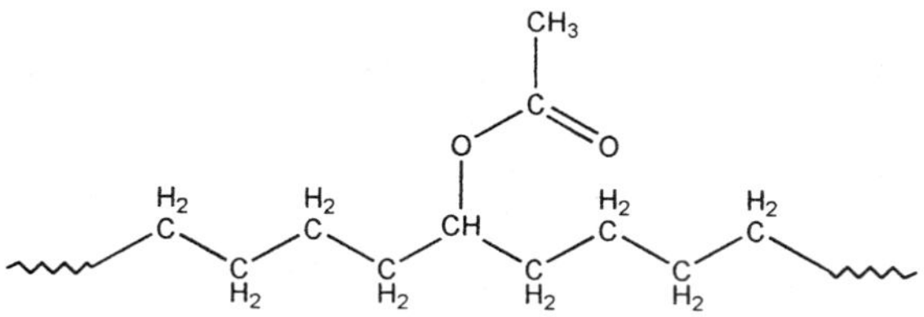

The Chemistry of Polyethylene

Figure 8 Chemical structure of ethylene-*co*-vinyl acetate copolymer.

the length of the polymer chain [79]. In addition to acetate branches, commercial ethylene-*co*-vinyl acetate copolymers also contain short-chain (ethyl and butyl) and long-chain branches as a consequence of their high pressure synthesis.

Commercial ethylene-*co*-vinyl acetate copolymers are available with vinyl acetate concentrations of up to approximately 27 mol% (55 wt%), copolymers containing in excess of approximately 25 mol% being essentially amorphous [79]. The melting range and degree of crystallinity of a series of samples are shown in Table 2. When the total concentration of branches in ethylene-*co*-vinyl acetate resins is taken into account, it is found that they follow the same relationships regarding degree of crystallinity and melting point as a function of branch content as do ethylene-*co*-α-olefin copolymers [80,81]. Investigation of ethylene-*co*-vinyl acetate copolymers by transmission electron microscopy reveals a general degradation of lamellar ordering with increasing vinyl acetate content [82]. This is manifested as increasing lamellar curvature, decreasing thickness, reduced lateral dimensions, and segmentation, indicative of strained unit cells. The observed lamellar thicknesses are of the same order as those of linear low density polyethylene resins having a similar overall branch content. Above 4 mol% branches, lamellae cease to be the preferred crystalline habit.

The modulus of ethylene-*co*-vinyl acetate resins decreases as the co-unit

Table 2 Melting Range and Degree of Crystallinity of a Series of Ethylene-*co*-Vinyl Acetate Copolymers

Vinyl acetate content (mol%)	Melting range from thermal analysis (°C)	Degree of crystallinity from X-ray diffraction (%)
4.3	83–103	27.4
7.6	72–98	19.9
16.8	61–77	~8
27.0	41–44	Noncrystalline

content increases. Ethylene-*co*-vinyl acetate copolymers and linear low density polyethylene resins follow the same relationship of modulus as a function of increasing total branch content and decreasing degree of crystallinity [81,83,84]. Decreases in hardness [84] and yield stress [81] are also observed. The moduli of ethylene-*co*-vinyl acetate copolymers and low density polyethylene samples are found to converge at a temperature of approximately $-70°C$ regardless of their composition [79]. The tensile impact and slow puncture strength of ethylene-*co*-vinyl acetate copolymer samples increases with co-unit content, the highest values being observed at temperatures of approximately $0°C$ and -15 to $-10°C$, respectively [85]. Vinyl acetate incorporation improves the environmental stress cracking resistance of ethylene-*co*-vinyl acetate copolymers relative to that of low density polyethylene resins [84].

B. Polyethylene-*co*-Vinyl Alcohol

Polyethylene-*co*-vinyl alcohol resins are produced indirectly by the hydrolysis (saponification) of ethylene-*co*-vinyl acetate copolymers. The structure of the vinyl alcohol branch is shown in Figure 9. Hydrolysis can be achieved by the action of alcoholic sodium hydroxide or potassium hydroxide in an organic solvent solution at high temperature [84,86]. Ethylene-*co*-vinyl alcohol copolymers are often referred to as EVAL (pronounced as a single word) or by the abbreviation EVOH. Like their ethylene-*co*-vinyl acetate precursors, ethylene-*co*-vinyl alcohol copolymers also contain short- and long-chain branching.

Unlike the acetoxy branches from which they are derived, hydroxy branches can be incorporated into the polyethylene unit cell without causing major disruption. At levels of hydroxy incorporation up to approximately 20 mol% ethylene-*co*-vinyl alcohol copolymers share a common unit cell with polyethylene [87]. The copolymer exhibits a slight increase in average unit cell dimensions relative to pure polyethylene [79]. Thus the degrees of crystallinity and melting temperatures of ethylene-*co*-vinyl alcohol copolymers are higher than those of their ethylene-*co*-vinyl acetate copolymer precursors [79,84]. At low comonomer levels, the melting points of ethylene-*co*-vinyl alcohol copolymers are similar to those of low density polyethylene resins. The melting point and degree of crystallinity of a series of ethylene-*co*-vinyl alcohol copolymers are listed in Ta-

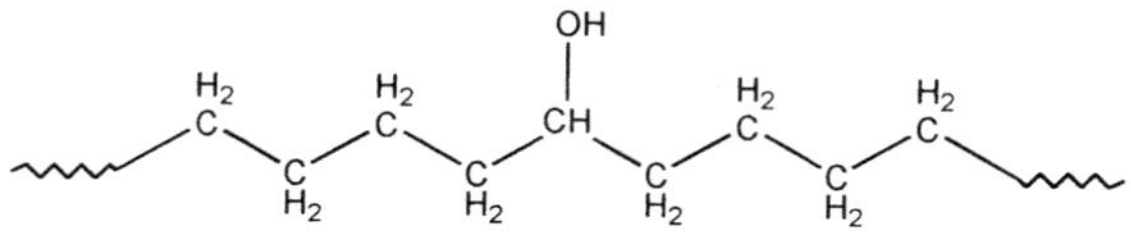

Figure 9 Chemical structure of ethylene-*co*-vinyl alcohol copolymer.

Table 3 Melting Range and Degree of Crystallinity of a Series of Ethylene-co-Vinyl Alcohol Copolymers

Hydroxy content (mol %)	Melting point from thermal analysis (°C)	Degree of crystallinity from X-ray diffraction (%)
4.3	112	38.1
7.6	113	34.9
16.8	107	28.2
27.0	110	23.2

ble 3. (Values for the respective ethylene-co-vinyl acetate precursors are listed in Table 2.) At comonomer incorporation levels in excess of approximately 30 mol%, the melting point of samples increases, reflecting their increasing poly(vinyl alcohol) character [88].

The hydroxyl groups of ethylene-co-vinyl alcohol copolymers display a pronounced polar character. They participate in strong hydrogen bonds that stiffen the noncrystalline regions, significantly altering the bulk properties of the material. The overall stiffness of ethylene-co-vinyl alcohol resins is markedly higher than that of their ethylene-co-vinyl acetate precursors. In common with other ethylene copolymers, increased comonomer content results in reduced modulus and yield stress [84,88]. The Vicat softening temperature of ethylene-co-vinyl alcohol copolymers is also higher than that of ethylene-co-vinyl acetate copolymers, reflecting their higher melting temperatures and stiffness [84]. The glass transition temperature rises as the level of comonomer incorporation increases, increasing by approximately 100°C between 0 and 10 mol% comonomer, thereafter increasing at a much decreased rate [89]. A raised brittleness temperature reflects the higher glass transition temperature [79,84].

C. Ethylene Ionomers

The term "ionomer" is used to describe a variety of copolymers consisting primarily of hydrocarbon backbones to which are attached a relatively small number of ionizable branches. This family of polymers includes two groups that are ethylene-based, one based on ethylene-co-methacrylic acid copolymer and the other on sulfonated ethylene-propylene-diene terpolymer. The latter is an essentially noncrystalline rubber that is beyond the scope of the semicrystalline materials covered in this work (the synthesis, properties, and structure of sulfonate ionomers are reviewed in Ref. 90). Ionomers based on partially or completely neutralized ethylene-*co*-methacrylic acid (EMA) and ethylene-co-acrylic acid (EAA) copolymers are often referred to generically as ethylene ionomers.

The copolymers upon which ethylene ionomers are based are synthesized by the copolymerization of methacrylic acid or acrylic acid with ethylene at high pressure under conditions similar to those used to produce ethylene-*co*-vinyl acetate copolymers. The incorporation of polar branches occurs essentially at random. The structure of the methacrylic acid branch is shown in Figure 10. Short- and long-chain branches are also present as a consequence of the high pressure polymerization process. Ethylene-*co*-methacrylic acid copolymer in its virgin state has little or no commercial significance (its mechanical properties are reported in Ref. 81). Ionomers are produced from the base resin by reaction in the molten state of their acid functionalities with the hydroxides of such metals as calcium, lithium, cesium, sodium, and zinc—the latter two being used most frequently in commercial practice—the product being a metal salt. Due to the hydrocarbon nature of the backbone that accounts for the majority of the copolymer, the cations (also known as counterions) and anions composing the salt functionalities remain associated. Typically, the maximum concentration of acid in the base copolymer is 6 mol%.

The highly polar nature of the salt functionalities has a controlling influence on the morphology, and therefore the properties, of ionomers. At low salt concentrations the polar groups associate as ion pairs, coupling adjacent hydrocarbon chains via a "polar cross-link." As the salt concentration increases, ion pairs phase separate into ionic domains known as "multiplets" that consist entirely of polar substituents. Multiplets form because ion pairs are dipolar, attracting one another by Coulombic interactions. Steric factors limit the size of multiplets to eight ion pairs [91]. Each multiplet is entirely covered by a "skin" of hydrocarbon chains. Multiplets act as complex cross-links, binding several hydrocarbon chains together. At even higher salt concentrations, multiplets aggregate to form "clusters" comprising both polar and hydrocarbon constituents. The very broad band associated with cation vibrations in the infrared spectrum is indicative of the wide range of environments in which cations are found [92]. Numerous

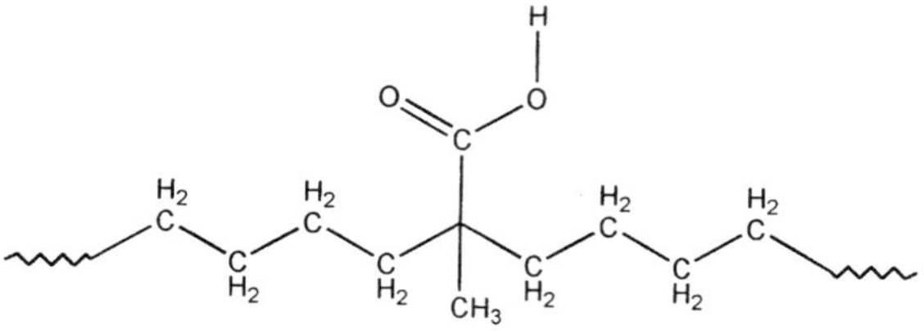

Figure 10 Chemical structure of ethylene-*co*-methacrylic acid copolymer.

The Chemistry of Polyethylene 411

hypotheses proposing to explain the structure of clusters in terms of various models have been advanced. The most significant models are briefly reviewed by Eisenberg et al. [93], who also propose a model purporting to explain all the observed phenomena associated with ionomers in general. It is not necessary to understand the precise nature of clusters and multiplets in order to explain the general physical properties of ethylene ionomers. It mainly boils down to the fact that multiplets and clusters restrict the motions of the polyethylene chains within and around them, acting as fillers that reinforce the noncrystalline regions.

The degree of crystallinity of ethylene-*co*-methacrylic acid copolymers is low to begin with, and neutralization to form ionomers reduces it further [94]. The process of neutralization destroys all evidence of lamellae and spherulites, replacing the original crystalline morphology with one of small, isolated crystallites [95]. Endothermic peaks obtained from differential thermal analysis are quite broad, up to 50°C in width, indicative of a wide range of crystallite sizes. Melting peaks are found in the range of approximately 50–100°C, depending on the concentration of methacrylic acid in the base resin, the extent to which it is neutralized, and the length of time after molding. Changes in the melting characteristics of ionomers can take place for up to 500 days after rapid quenching [96]. As would be expected, the glass transition temperature of ionomers increases with increasing ionic content.

Ethylene ionomers in the solid state behave like rubbers that are lightly cross-linked. As the temperature is increased, the polar cross-links degenerate, permitting viscous flow in the molten state. Upon cooling, the cross-links re-form, and the rubberlike properties are reestablished. These characteristics place ethylene ionomers in the category of "thermoplastic elastomers" (TPE). As the comonomer content or degree of neutralization increases, the modulus, tensile strength, impact resistance, and abrasion resistance are all improved, while the draw ratio at break falls somewhat. Ionomers show a high degree of strain recoverability after deformation, a property defined as "low permanent set." The physical properties of ionomers depend more on their degree of neutralization than on the cation type [97], although some cation effects are observed [98,99]. The low degree of crystallinity of ionomers and their lack of spherulites endows them with exceptional optical clarity. The constrained nature of the noncrystalline regions and specific polar interactions makes ionomers resistant to the diffusion of oils, greases, and fats.

The broad melting range of ionomers makes them readily heat-sealable, a property which, in conjunction with their oil resistance, fits them for meat packaging. The melt viscosity of ionomers increases with the degree of neutralization. The high concentration of oxygenate species on the surface of ionomers facilitates their adhesion to many substances, including glass and metal. The combination of good adhesion to glass, high tensile strength, good puncture resistance, and good abrasion resistance suits ionomers for the encapsulation of chemical reagent bottles to help reduce the hazards associated with fragmentation of broken bottles. Golf ball covers are commonly made from ionomers taking advantage of their high abrasion resistance.

BIBLIOGRAPHY

Bazuin CG, Eisenberg A. Ind Eng Chem Prod Res 20:2671, 1980.
Geuskens G, ed. Degradation and Stabilisation of Polymers. New York: Wiley, 1975.
Longworth R. In: L Holliday, ed. Ionic Polymers. New York: Wiley, 1975, pp. 69–172.
MacKnight WJ, Earnest TR. J Polym Sci Macromol Rev 16:41, 1981.
Potts JE. In: HHG Jellinek. Aspects of Degradation and Stabilization of Polymers. New York: Elsevier, 1978.
Scott G, ed. Developments in Polymer Stabilisation. Vols 1–8. G Scott, London: Elsevier Applied Science, 1979–1987.
Scott G. In: G Scott, ed. Atmospheric Oxidation and Antioxidants. Amsterdam: Elsevier, 1993, pp 141–218.
Wu S. Polymer Interfaces and Adhesion. New York: Marcel Dekker, 1982, Chap. 9.

REFERENCES

1. F Gugumus. Polym Degrad Stabil 27:19, 1990.
2. LA Wall, S Strauss. J Polym Sci 44:313, 1960.
3. G Scott. Polym Age 6:54, 1975.
4. JE Potts, RA Clendinning. In: J Guillet, ed. Polymer Science and Technology, Vol 3. New York: Plenum, 1973.
5. A-C Albertsson, ZG Bánhidi. J Appl Polym Sci 25:1655, 1980.
6. GJL Griffin. Br Patent Int Publ WO88/09354, 1988.
7. A-C Albertsson, S Karlsson. J Appl Polym Sci 35:1289, 1988.
8. GM Ferguson, M Hood, K Abbot. Polym Int 28:35, 1992.
9. J-C Huang, AS Shetty, M-S Wang. Adv Polym Technol 10:23, 1990.
10. A-C Albertsson, SO Anderson, S Karlsson. Polym Degrad Stabil 18:73, 1987.
11. JE Guillet, Y Anerik. Macromolecules 4:375, 1971.
12. E Dan, JE Guillet. Macromolecules 6:48, 1973.
13. WJ Bailey, B Gapud. Am Chem Soc Symp Ser 280:423, 1985.
14. GJL Griffin. Pure Appl Chem 52:399, 1980.
15. E Chiellini, R Solaro, A Corti, G Picci, C Leporini, A Pera, G Vallini, P Donaggio. Chim Ind 73:656, 1991.
16. A-C Albertsson, C Barenstedt, S Karlsson. Polym Degrad Stabil 37:163, 1992.
17. JC Randall, FJ Zoepfl, J Silverman, J Makromol. Chem Rapid Commun 4:149, 1983.
18. GN Patel, A Keller. J Polym Sci Polym Phys Ed 13:303, 1975.
19. PG Klein, JA Gonzalez-Orozco, IM Ward. J Polym 32:1732, 1991.
20. R Kitamaru, L Mandelkern. J Am Chem Soc 86:3529, 1964.

21. T Okada, L Mandelkern, R Glick. J Am Chem Soc 89:4790, 1967.
22. PV Zamotaev, I Chodak. Angew Makromol Chem 210:119, 1993.
23. PG Klein, DW Woods, IM Ward. J Polym Sci Polym Phys Ed 25:1359, 1987.
24. EM Dannenberg, ME Jordan, HM Cole. J Polym Sci 31:127, 1958.
25. D Simunkova, R Rado, O Mlejnek. J Appl Polym Sci 14:1825, 1970.
26. AJ Peacock. Polym Commun 28:259, 1987.
27. EM Dannenberg, ME Jordan, HM Cole. J Polym Sci 31:127, 1958.
28. JH O'Donnell, AK Whittaker. Polym 33:62, 1992.
29. T Bremner, A Rudin. J Appl Polym Sci 49:785, 1993.
30. KJ Kim, YS Ok, BK Kim. Eur Polym J 28:1487, 1992.
31. G Oster, GK Oster, H Moroson. J Polym Sci 34:671, 1959.
32. Q Yan, WY Xu, B Rånby. Polym Eng Sci 31:1561, 1991.
33. G Oster. J Polym Sci 22:185, 1956.
34. BJ Qut, B. Rånby. Appl Polym Sci 48:701, 1993.
35. YL Chen, B. Rånby. J Polym Sci Polym Chem Ed 27:4077, 1989.
36. YL Chen, B. Rånby. J Polym Sci Polym Chem Ed 27:4051, 1989.
37. TR Santelli. US Patent 3,075,948, 1963.
38. HG Scott. US Patent 3,646,155, 1972.
39. LRG Treloar. The Physics of Rubber Elasticity. New York: Oxford Univ Press, 1949.
40. A Charlesby, NH Hancock. Proc Roy Soc Lond. A218:245, 1953.
41. AJ Peacock. PhD Thesis, University of Southampton, 1984.
42. PJ Hendra, AJ Peacock, HA Willis. Polymer 28:705, 1987.
43. JH O'Donnell, AK Whittaker. Radiat Phys Chem 39:209, 1992.
44. C Birkinsaw, M Buggy, S Daly, M O'Neill. J Appl Polym Sci 38:1967, 1989.
45. L Minkova, M Mihailov. Colloid Polym Sci 268:1018, 1990.
46. P Kurian, LE George, D J Francis. J Elast Plast 25:12, 1993.
47. KJ Kim, BK Kim. J Appl Polym Sci 48:981, 1993.
48. I Chodák. Prog Polym Sci 20:1165, 1995.
49. RM Gohil, PJ Phillips. Polymer 27:1696, 1986.
50. D Munteanu, C Cusnderlik, I Tincul. Chem Bull Polytech Inst "Traian Vuia" Timisoara 35:101, 1990.
51. D Briggs, DM Brewis, MB Konieczo. J Mater Sci 11:1270, 1976.
52. D Briggs, DM Brewis, MB Konieczo. J Mater Sci 12:429, 1977.
53. D Briggs, DM Brewis, MB Konieczo. J Mater Sci 14:1344, 1979.
54. AR Blythe, D Briggs, CR Kendall, DG Rance, VJI Zichy. Polymer 19:1273, 1978.
55. S Gao, Y Zeng. J Appl Polym Sci 47:2065, 1993.
56. LJ Gerenser, JF Elman, MG Mason, JM Pochan. Polymer 26:1162, 1985.
57. JA Launauze, DL Myers. J Appl Polym Sci 40:595, 1990.
58. DK Owens. J Appl Polym Sci 19:265, 1975.
59. RL Ayres, DL Shofner. SPE J 28:51, December 1972.
60. DM Brewis, D Briggs. Polymer 22:7, 1981.
61. MS Silverstein, O Breuer. J Mater Sci 28:4153, 1993.
62. FPM Mercx, A Benzina, AD van Langeveld, PJ Lemstra. J Mater Sci 28:753, 1993.
63. RH Hansen, H Schonhorn. J Polym Sci B-2:203, 1966.
64. S Gao, Y Zeng. J Appl Polym Sci 47:2093, 1993.

65. DW Woods, IM Ward. Surf Interface Anal 20:385, 1993.
66. D Briggs, DM Brewis, MB Konieczo. Eur Polym J 14:1, 1978.
67. H Schonhorn, FW Ryan. J Polym Sci A2:231, 1968.
68. Y Ogiwara, M Kanda, M Takumi, H Kubota. J Polym Sci Polym Lett Ed 19:457, 1981.
69. H Kubota, Y Hata. J Appl Polym Sci 40:1071, 1990.
70. GH Hsiue, CC Wang. J Polym Sci Polym Chem Ed 31:3327, 1993.
71. C Batich, A Yahiaoui. J Polym Sci Polym Chem Ed 25:3479, 1987.
72. ZP Yao, B Rånby. J Appl Polym Sci 40:1647, 1990.
73. K Kildal, K Olafsen, A Stori. J Appl Polym Sci 44:1893, 1992.
74. L Lavielle, H Balard. Angew Makromol Chem 147:147, 1987.
75. S Edge, S Walker, WJ Feast, WF Pacynko, J Appl Polym Sci 47:1075, 1993.
76. S Sugiyama, S Tsuneda, K Saito, S Furusaki, T Sugo, K Makuuchi. React Polym 21:187, 1993.
77. D Fischer, HH Eysel. J Appl Polym Sci 52:545, 1984.
78. K Allmér, J Hilborn, PH Larsson, A Hult, B Rånby. J Polym Sci Polym Chem Ed 28:173, 1990.
79. IO Salyer, AS Kenyon. J Polym Sci A-1, 9:3083, 1971.
80. R Alamo, R Domszy, L Mandelkern. J Phys Chem 88:6587, 1984.
81. AJ Peacock, L Mandelkern. J Polym Sci Polym Phys Ed 28:1917, 1990.
82. IG Voigt-Martin, R Alamo, L Mandelkern. J Polym Sci Polym Phys Ed 24:1283, 1986.
83. M Saito, H Tada, Y Kosaka. J Polym Sci Part A-1, 8:2555, 1970.
84. RJ Koopmans, R Van der Linden, EF Vansant. Polym Eng Sci 23:306, 1983.
85. K Porzucek, JM Lefebvre. J Appl Polym Sci 48:969, 1993.
86. MM Coleman, X Yang, H Zhang, PC Painter. J Macromol Sci-Phys B32 295, 1993.
87. K Nakamae, M Kameyama, T Matsumoto. Polym Eng Sci 19:572, 1977.
88. H Yoshida, J Kanbara, N Takemura, Y Kobayashi. Sen'i Gakkaishi 39:T512, 1983.
89. Y Mori, H Sumi, T Hirabayashi, Y Inai, K Yokota. Macromolecules 27:1051, 1994.
90. JJ Fitzgerald, RA Weiss. JMS—Rev Macromol Chem Phys C28:99, 1988.
91. A Eisenberg. Macromolecules 3:147, 1970.
92. AT Tsatsas, JW Reed, WM Risen. J Chem Phys 55:3260, 1971.
93. A Eisenberg, B Hird, RB Moore. Macromolecules 23:4098, 1990.
94. WJ MacKnight, LW McKenna, BE Read. J Appl Phys 38:4208, 1967.
95. R Longworth. In: L Holliday, ed. Ionic Polymers. New York: Wiley, 1975, pp. 69–172.
96. M Kohzaki, Y Tsujita, A Takizawa, T Kinoshita. J Appl Polym Sci 33:2393, 1987.
97. TC Ward, AV Tobolsky. J Appl Polym Sci 11:2403, 1967.
98. E Hirasawa, Y Yamamoto, K Tadano, S Yano. J Appl Polym Sci 42:351, 1991.
99. H Tachino, H Hara, E Hirasawa, S Kutsumizu, S Yano. Macromolecules 27:372, 1994.

8
Orientation of Polyethylene

I. MORPHOLOGY OF ORIENTED POLYETHYLENE

Orientation in polymeric items involves the preferential alignment of molecular segments at various levels. Individual chain segments in the noncrystalline regions may be chain extended and aligned, while nonlinear chain segments may follow a general directional trend. Within crystalline regions the c axes of the unit cells may be aligned with—or preferentially oriented toward—a common orientational axis. Fibrils may be formed which share a common crystalline axis, or lamellar crystallites may be stacked with their lateral planes preferentially arrayed normal to the overall orientation direction. The precise morphology of an oriented sample is a function of many factors, including the structure prior to deformation and the process by which orientation was achieved. The molecular alignment developed during orientation strongly influences the properties of the resulting products; higher levels of orientation magnify the anisotropic response to external influences.

The production of highly oriented polyethylene specimens is a topic that has captured the imagination of many investigators and evoked a great deal of research. The ultimate goal is a structure that is so highly aligned that its properties approach those predicted theoretically for a perfect uniaxial structure. The principal driving force behind this research is the fact that the theoretical modulus of perfectly oriented polyethylene is extremely high—greater than that of steel on the basis of weight. In the drive to produce polyethylene specimens with ever higher degrees of orientation, a variety of esoteric preparative methods have been developed, very few of which have any practical relevance. On a more pragmatic basis, other researchers have attempted to determine the relationships between molecular characteristics, processing parameters, semicrystalline morphology, and the physical properties of oriented polyethylene specimens produced by more mundane, but commercially feasible, techniques.

A. Structures Generated from the Condensed Phase

Oriented solid-state structures are principally generated by one of two methods or by some combination thereof. The first method is solidification from an oriented melt; the second involves solid-state deformation below the melting temperature, i.e., "cold drawing." Many commercial fabrication processes involve both orientation methods. Thus, for example, fiber spinning involves solidification from the oriented melt followed by cold drawing of the as-spun fibers into thinner filaments. A wide range of oriented morphologies may be generated, depending upon the initial melt orientation, the length of time the melt is allowed to relax between orientation and solidification, and the temperature and rate at which solid-state deformation occurs. Naturally, the response of any resin to orientation conditions depends very much on its molecular characteristics. As a general rule, linear polyethylene resins can be oriented more effectively than branched ones. During solidification from the oriented melt, higher levels of orientation can be generated more readily in high molecular weight resins than in low molecular weight ones. Conversely, lower molecular weight isotropic melt-crystallized samples can usually be cold drawn to a greater extent than higher molecular weight samples. Structures generated from materials in which the inherent entanglement density has been reduced fall into a distinct category, which is discussed separately in the following section.

Given the wide range of polyethylene resin molecular characteristics and the variety of conditions and methods by which orientation can be achieved, it is not possible to make blanket statements regarding the morphology of oriented polyethylene. However, as a general rule, two distinct types of morphologies are observed at high levels of orientation: stacked lamellae, which generally result from the crystallization of highly oriented melts, and fibrillar morphologies, which are the result of cold drawing.

1. Structures Generated from Solidification of Oriented Melts

The semicrystalline morphology found in polyethylene crystallized from an oriented melt reflects the melt structure immediately prior to solidification (assuming that there is no subsequent solid state deformation). Therefore, the oriented structures so produced are largely controlled by the rheological response of the molten resin to the applied stress, which in turn is controlled by molecular characteristics. The principal molecular characteristics controlling melt orientation are molecular weight, molecular weight distribution, long-chain branching, and degree of comonomer incorporation. The average trajectory of molecules in the solid state closely matches that of the molten state immediately prior to solidification. The alignment of chain axes with the primary orientation direction is invariably greater in the crystalline regions than in the noncrystalline zones.

In the case of high density polyethylene resins that have a broad molecular weight distribution, the higher molecular weight chains, with their larger number of entanglements and long relaxation times, will become more aligned than shorter chains. When chance brings a number of such highly oriented chain segments into immediate proximity, the entropy barrier to crystallization is reduced relative to that of the surrounding melt. As the temperature of the melt falls, crystallization will begin with these ordered bundles, which form stable microfibrillar crystallites, the axes of which lie parallel with the elongational stress. These microfibrillar crystallites act as nuclei upon which lamellar crystallization from the less well aligned regions takes place as the temperature drops further. As the microfibrillar nuclei are aligned parallel with one another it follows that the lamellar overgrowths will also initially be well ordered, their lateral dimensions lying normal to the stress with their c axes parallel with the microfibrillar nuclei. Such stacks of lamellae are referred to as a "cylindrites" or "row nucleated structures." Cylindritic morphologies are commonly encountered in blown and cast films, in fibers that have not been cold drawn, and near the surface of some injection-molded items. Cylindrites are radially symmetric, irrespective of where they occur [1]. The generation of cylindrites is illustrated schematically in Figure 1.

Cylindrites consisting of scores of stacked lamellae, which may extend for thousands of angstroms, have been observed by electron and atomic force microscopy. The central microfibril is invariably too small to be observed, its existence being inferred on the basis of the observable semicrystalline morphology and independent experiments in which lamellar overgrowths of a similar nature are observed to occur on needlelike nucleating agents.

If the microfibrillar nuclei of cylindrites are widely separated, the lamellar overgrowths will twist as they grow outward, losing their precise alignment with respect to the orienting stress. In such cases the cylindritic cores are well aligned, but the outlying reaches of their associated lamellae may be randomized. The number of microfibrillar nuclei increases as a function of increasing orientation and molecular weight, reducing the distance between them. At extremely high degrees of orientation the lamellar overgrowths of adjacent cylindrites impinge upon each other to form a highly ordered space-filling structure.

Cylindritic morphologies are most commonly seen in linear polyethylene samples, but they also occur in linear low density polyethylene. In the case of traditional linear low density polyethylene, a small portion of the resin consists of high molecular weight, low comonomer content molecules that can be extended to form microfibrillar nuclei. The lower molecular weight, more highly branched molecules crystallize to form lamellar overgrowths when the temperature drops sufficiently. The thickness of the lamellae that comprise the overgrowth reflects the molecular characteristics of the crystallizing molecules and

(a)

(b)

Orientation

(c)

Microfibrillar nucleus

(d)

Lamellar overgrowths

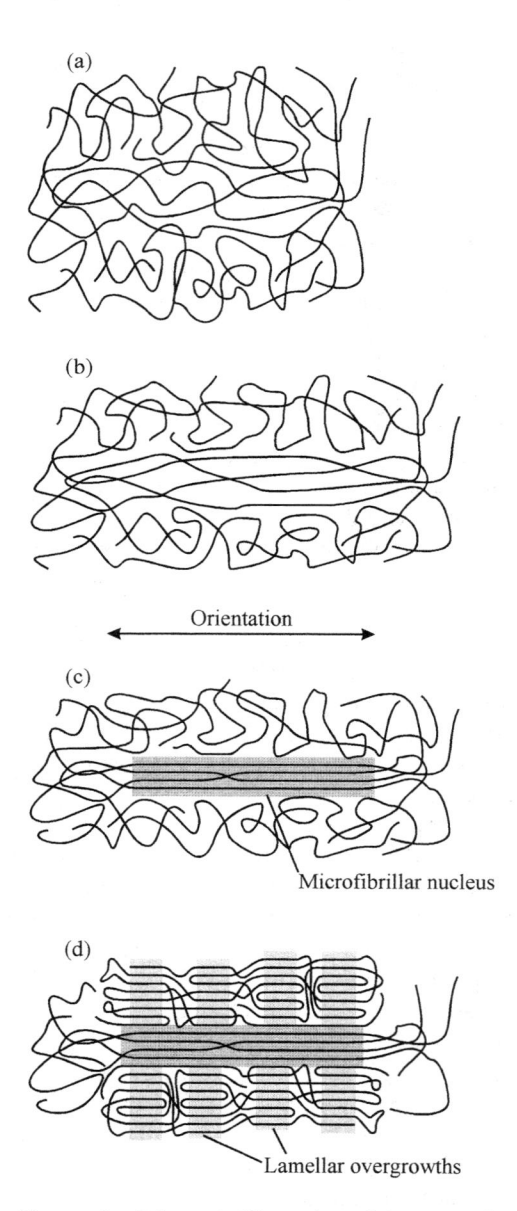

Figure 1 Schematic illustration of the generation of a cylindrite. (a) Random melt; (b) oriented melt; (c) microfibrillar nucleus; (d) lamellar overgrowth form.

the temperature at which crystallization occurred. Thus, the lamellae grown from linear resins are invariably thicker than those grown from branched resins. In branched resins the lamellar thickness reflects the distribution of crystallizable sequence lengths between branches; the higher the degree of branching, the smaller the average lamellar thickness. As melt orientation increases, lamellar stacks become more highly ordered; the individual lamellae show less tendency to twist and become more uniform in thickness.

Throughout a sample, the degree of orientation of individual chain segments depends upon their location. Segments within the microfibrillar nuclei will be highly aligned with respect to the orienting force, as will segments in those parts of the lamellae closest to the nuclei. The degree of orientation decreases as the distance from the nucleus increases, sometimes varying sinusoidally as the lamellar overgrowth twists. As melt orientation increases, the degree of alignment of the segments within the crystalline regions increases rapidly, then levels off as it approaches perfect alignment with the applied stress. Alignment within the noncrystalline regions is much poorer than in the crystallites and increases monotonically with melt orientation. For a given set of processing conditions, alignment within branched sample is lower than for a linear sample having a similar molecular weight distribution. This is mainly due to the lower crystallization temperature of the branched sample, which allows the molecules a longer time to relax prior to solidification.

The elastic modulus of samples crystallized from the melt increases with melt orientation, but even in the most favorable circumstances it is an order of magnitude lower than that predicted for a perfectly aligned sample. This is indicative of the large number of defects (principally entanglements and chain ends) trapped within the sample, which limit the overall level of ordering that can be developed.

As melt orientation prior to crystallization increases, the entropy of the sample decreases due to better alignment of the molecules. One result of this improved alignment is a general increase in the degree of crystallinity as a function of increased orientation, which reflects the lower energy barrier to crystallization. This effect is most noticeable for linear polyethylene samples, especially those with high molecular weights.

When highly branched polyethylene samples, either (dendritic) low density polyethylene or (comblike) very low density polyethylene, crystallize from oriented melts they do not form cylindrites because they contain insufficient linear chain segments to generate microfibrillar nuclei. In such cases the relatively slow crystallization kinetics and low crystallization temperature permit the molecules a relatively long time to relax prior to solidification. The lamellae that form under these circumstances are well separated from one another and do not share a common axis. The resultant semicrystalline morphology is similar to that of low density samples crystallized from an isotropic melt.

2. Structures Generated by Deformation of the Solid State

In the context of this discussion, solid-state deformation will encompass any orientation that takes place at temperatures below the final melting temperature of the polymer. Such deformation may be imposed on samples that are initially isotropic or anisotropic. During commercial forming processes, such deformations are usually taken to the point at which a stable morphology is formed, i.e., beyond the yield point. For a general description of the macroscopic phenomena associated with solid-state deformation, the reader's attention is directed to the section on mechanical properties in Chapter 5.

A major difference between solid-state orientation and melt orientation is the greatly reduced capacity for relaxation of molecules in the former. In oriented melts, molecules try to return to their equilibrium state, i.e., random coils. In the solid state, such retraction is clearly impossible; molecular segments that become oriented in the noncrystalline regions normally remain so. A limited degree of motion (dependent upon temperature) permits molecular segments that are reasonably well aligned in the noncrystalline regions to crystallize in the oriented state. It follows that the alignment of molecular segments comprising the noncrystalline regions in samples prepared from the solid state is higher than in samples drawn in the melt to a similar overall degree of orientation.

Solid-state deformation normally results in the destruction of the crystallites of the original morphology, followed by reordering to form new crystallites. Newly formed crystallites are themselves subject to disruption at higher orientation levels, being replaced by a fibrillar morphology. The proposed mechanisms of solid-state deformation are discussed separately in a subsequent section.

At relatively low levels of deformation induced by shear, compression, or tension (deformation being halted after the yield point but before the onset of strain hardening), a lamellar morphology is formed in which the observed lamellar thickness is independent of the original lamellar thickness, being solely dependent upon the temperature at which the deformation occurred [2–4]. In such morphologies the unit cell c axes are preferentially aligned in the stress direction, while the lateral planes of the lamellae lie approximately normal to the aligning force. It is not unusual to observe a bimodal distribution of c axis and lamellar plane orientations as shown schematically in Figure 2. Such morphologies give rise to small-angle X-ray diffraction patterns exhibiting four maxima rather than the two that would be expected if the lamellar planes were perpendicular to the stress [5]. As an added complication, the c axes within the lamellae are not precisely normal to the lamellar plane but may subtend an angle of up to 50° to it [6]; thus a four-point structure may also be observed in the wide-angle X-ray diffraction pattern.

At medium and high levels of orientation, at draw ratios in excess of approximately 10, a fibrillar morphology is generated. Low to medium molecular

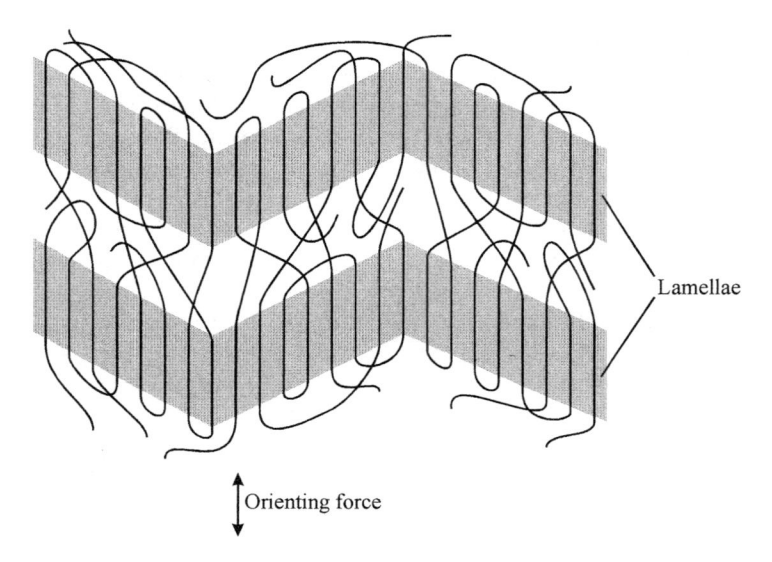

Figure 2 Schematic illustration of bimodal distribution of lamellae with respect to the orienting force in solid-state deformed samples.

weight ($\overline{M}_w$ = 50,000–200,000) high density polyethylene samples are converted directly from an isotropic to a fibrillar morphology during the yielding process when drawn at room temperature, the fibrils remaining in evidence during subsequent strain hardening. As the molecular weight increases, higher drawing temperatures are required to form fibrils. For ultrahigh molecular weight linear polyethylene resins ($\overline{M}_w$ > 1,000,000), it may not be possible to generate a fibrillar morphology by simple tensile drawing at temperatures below the crystalline melting point. (To produce ultradrawn, ultrahigh modulus fibrillar samples, it is invariably necessary to employ a multiple step process beginning with a reduced entanglement sample.) Low density and linear low density polyethylene resins, with their poorly defined yield regions and low draw ratio at break, generally do not form a fibrillar structure prior to failure, even when drawn at elevated temperatures [7].

Fibrillar samples consist of oriented crystallites arranged into needlelike structures of various sizes. There is little consensus of opinion in the literature regarding the nomenclature or sizes of the features that comprise the fibrillar morphology (see, e.g., Refs. 8–11). However, it is clear that a hierarchy of sizes exists; "macrofibrils" consist of bundles of "microfibrils," which in turn are composed of bundles of "nanofibrils." Typical macrofibrils (which are visible to the naked eye) may be up to several hundred micrometers thick with a length that may exceed 1 cm. Microfibrils may be approximately two orders of magnitude smaller than macrofibrils, with nanofibrils approximately two orders of magnitude smaller than microfibrils. It is generally agreed that the aspect ratio of the structures comprising fibrillar morphologies increases as a function of orientation.

Much has been written about fibrillar morphology, and several morphological models have been advanced to explain their observed features and physical properties. The vast majority of this research has centered on high density polyethylene, often drawn at temperatures substantially above ambient that favor high draw ratios. The various morphologies proposed may be rationalized into the hierarchy of structures illustrated in Figure 3, which incorporates the most important features reported by different authors [9,10,12–17].

Nanofibrils consist of stacks of crystallites separated by thin noncrystalline "plates," portions of which are spanned by "intercrystalline bridges," as shown schematically in Figure 4. Each crystallite consists of extended linear segments of a length comparable to the thickness of lamellae formed during melt crystallization. The c axes of the crystallites and the intercrystalline bridges are highly aligned with the fibrillar axis. The misorientation of the c axes in a high modulus ultrahigh molecular weight polyethylene fiber has been estimated to be in the range of 0–13° [12]. Chain entanglements and chain ends are concentrated in the noncrystalline regions. The linear segments that make up the intercrystalline bridges are continuous with the crystalline blocks they link. These extended linear segments are longer than the thickness of the crystalline blocks by a factor slightly greater than the number of blocks that they traverse, most commonly 2. With

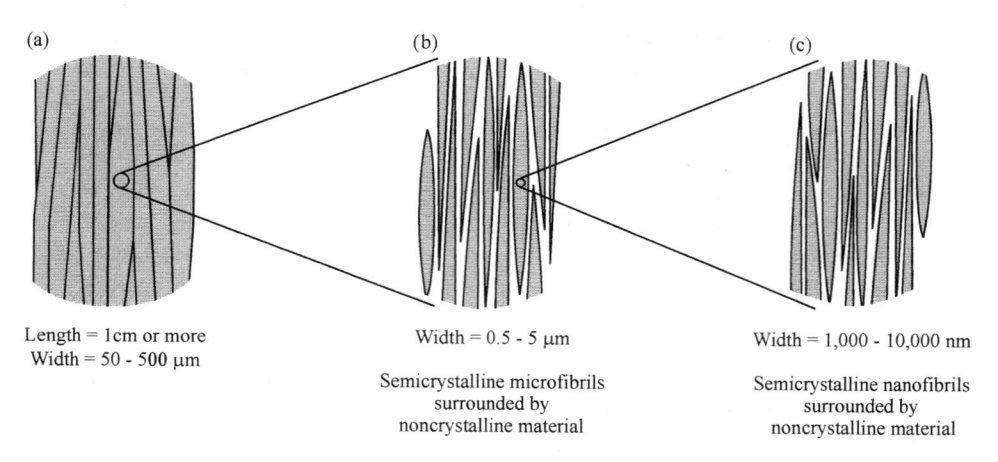

Figure 3 Hierarchy of structures comprising a fibrillar morphology. (a) Macrofibrils; (b) microfibrils; and (c) nanofibrils.

Orientation of Polyethylene

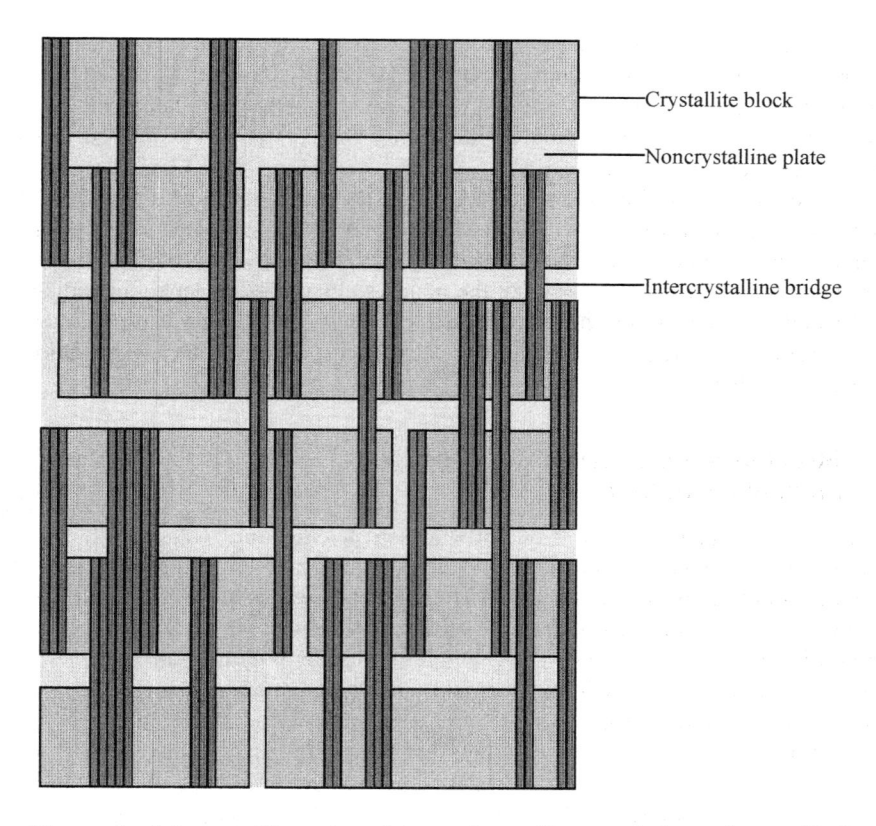

Figure 4 Schematic illustration of the semicrystalline morphology of a nanofibril.

increasing elongation, the noncrystalline plates become thinner and denser, while the number and thickness of the intercrystalline bridges increase.

The thickness of the crystallite blocks in nanofibrils can be readily determined from the frequency of the Raman-active longitudinal acoustic mode. If the intervening plates between crystallite blocks are relatively thick and have a density typical of a noncrystalline phase, there will be sufficient density contrast for small-angle X-ray diffraction analysis to determine the discrete periodicity of the crystalline blocks, which will be slightly greater than the crystallite thickness determined from Raman spectroscopy. In cases in which large numbers of intercrystalline bridges are present, there may be insufficient regularity to yield a small-angle X-ray diffraction peak. If the noncrystalline plates are very thin and densely packed (akin to the partially ordered zones that separate crystalline from disordered regions in melt-crystallized samples), there will be insufficient density contrast between the crystalline and noncrystalline regions for the crystallite blocks to be distinguishable by small-angle X-ray diffraction, and hence the observed repeat distance will be much greater than the crystallite thickness determined from Raman spectroscopy.

The modulus of fibrillar samples can be explained in terms of a fiber-reinforced composite model, in which needlelike nanofibrils are surrounded by a minority matrix of noncrystalline material [15–17]. The modulus is highly dependent upon the shear characteristics of the noncrystalline regions separating adjacent nanofibrils, according the the "shear lag" theory [18]. Improved modulus as a function of increasing orientation is attributed to the increase aspect ratio of the crystalline fibers.

B. Structures Generated from a Reduced Entanglement State

A reduced entanglement state is one in which the average molecular weight between entanglements is substantially greater than that found in the quiescent melt state or a solid sample crystallized therefrom. Such materials are created directly during gas-phase polymerization; the nascent chains making up the granules formed in the reactor are unable to entangle during the polymerization process (which occurs below the melting temperature of the polymer). Alternatively, the dissolution of polyethylene in a large excess of solvent reduces the number of interchain interactions and hence the number of entanglements. Such solutions and gels can be deformed directly, or the solvent can be removed and the resulting solid subsequently deformed. To limit the extent to which the molecules comprising solution-grown crystals, dry "gels," or nascent granules can reentangle, it is necessary to deform them in the solid state. Samples with fewer entanglements have the potential for deformation to much greater extents than molten or melt-crystallized samples. All processes used for making ultrahigh modulus fibers and tapes start from a reduced entanglement state.

Crystallization of polyethylene from flowing solutions or from deformed wet gels yields "shish kebab" structures, which are somewhat akin to cylindrites. The core of a shish kebab consists of an elongated microfibrillar nucleus upon which grow perpendicular lamellae. The most noticeable difference between cylindrites and shish kebabs is that the lamellar overgrowths on the latter have no polymer between them and are not linked to their neighbors by tie chains. An electron micrograph of a shish kebab is shown in Figure 36 of Chapter 4. Shish kebabs are most commonly grown from dilute solutions subject to very high shear fields. Their formation is similar to that of cylindrites: The microfibrillar core develops first, and lamellar overgrowth subsequently occurs. The ratio of the core to overgrowth can be raised by increasing the crystallization temperature [19].

Solid-state deformation of reduced entanglement specimens results in the formation of fibrillar morphologies. Such fibrillar morphologies are similar to those formed by solid-state deformation of melt-crystallized samples to the same extension ratio [20]. Naturally, the ultimate extension ratio of samples with few entanglements is potentially much greater than that of melt-crystallized samples. At very high extension ratios—inaccessible to the melt-crystallized state—an increasingly smaller fraction of the material is required to accommodate chain folds, entanglements and chain ends, and other defects. Such structures may be considered to be essentially crystalline, with a minor component of small noncrystalline regions randomly distributed throughout them. Gel spun fibers of ultrahigh molecular weight polyethylene have been found to contain small amounts of the hexagonal crystalline phase, as evidenced by a secondary melting peak at approximately 156°C [21]. The biaxial drawing of solution-cast films or dry gels results in the formation of uniaxial fibrils, which are randomly distributed within the plane of orientation [22].

II. PROPERTIES OF ORIENTED POLYETHYLENE

Orientation of polyethylene introduces anisotropy with respect to virtually every physical property. At extreme levels of orientation, the degree of anisotropy developed surpasses that attainable by any other polyolefin and is unmatched by other organic polymers with the exception of carbon fibers. Most deliberate attempts to orient polyethylene to high degrees are made with the intent of improving mechanical properties, especially tensile modulus. It is therefore no surprise that the majority of literature references to highly drawn polyethylene detail the effects of orientation on such mechanical properties as elastic modulus, tensile strength, and draw ratio at break.

A. Tensile Properties

11. Elastic Modulus

It has long been appreciated that polyethylene molecules in their extended all-trans configuration have an extremely high elastic modulus when tested parallel with the chain axis. Over the last 40 years many attempts have been made to predict the modulus of perfectly aligned molecules, and assemblies thereof, using a variety of calculation methods. Table 1 reflects the wide range of values so calculated.

The highest reported experimental values of polyethylene elastic modulus attained to date have been in the range of 230–264 GPa [23]. This equals or exceeds some of the values calculated for perfectly aligned polyethylene mole-

Table 1 Calculated Elastic Modulus for Perfectly Oriented Polyethylene Fibers

Predicted elastic modulus (GPa)	Basis for calculation	Ref.
182	Bond stretching, angular deformation	130
340	Complex force field	131,132
235	X-ray diffraction	133
324	Lattice dynamics	134
285	Analogy to diamond	135
300	Molecular orbital	136
290	Raman spectroscopy	137,138
213–229	X-ray diffraction	139
267	Raman spectroscopy	140
360 at 0 K 304 at 300 K	Molecular dynamics	141
209, 186	Molecular orbital	142

cules but is still well short of the most optimistic predictions. Suffice it to say that there appears to be much room for improvement, especially with regard to products of commercially feasible processes.

The reason for the extremely high modulus of oriented polyethylene in comparison to typical specimens (~1 GPa for injection-molded high density polyethylene [24]) is simple. The tensile deformation of isotropic samples at low strains principally involves the distortion of molecules whose trajectories approximate to a random coil, which is largely accommodated by bond rotation. This requires much less force than the molecular stretching required to extend the all-trans configuration, which involves bond elongation and an increase in the C—C—C dihedral bond angle.

As a general rule, the elastic modulus of polyethylene samples increases monotonically with the effective extension ratio. High molecular orientation, as opposed to a high measured draw ratio, is necessary for a high modulus. Factors that favor high molecular orientation include an absence of branches, high molecular weight, reduced entanglement density, and elevated temperatures (not exceeding the crystalline melting temperature).

For samples drawn in the solid state, the elastic modulus is a strong function of the draw ratio. In such cases the molecular orientation closely follows the measured draw ratio. At draw ratios between approximately 10 and 40, the combined data from numerous investigative teams reveal an essentially linear relationship of tensile modulus with draw ratio [25], as shown in Figure 5. The data

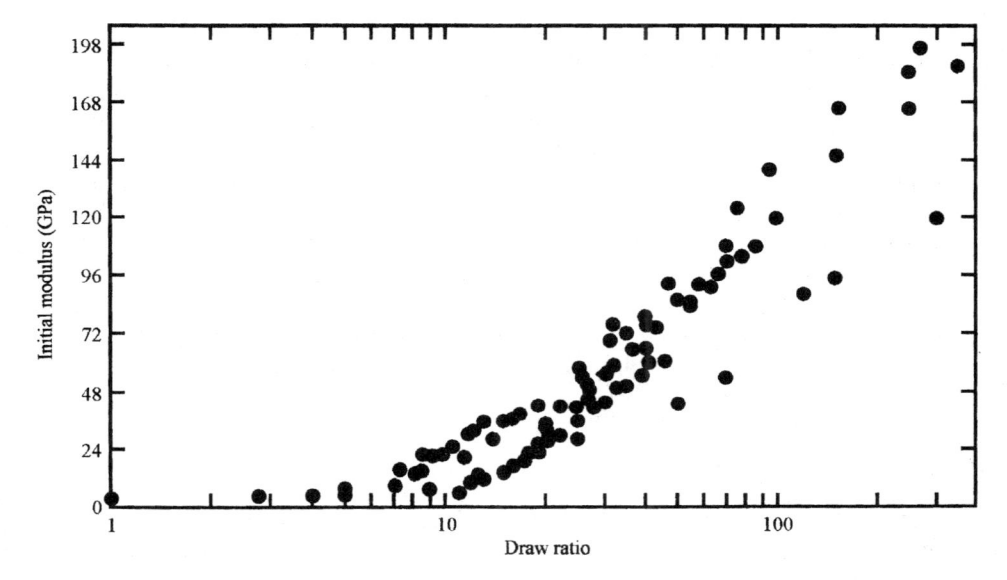

Figure 5 Tensile modulus as a function of draw ratio for various samples oriented in the solid state. (From Ref. 25.)

are quite scattered (which is only to be expected given the multitude of sources), but the vast majority fall within a well-defined envelope. The break in the data corresponding to a draw ratio of approximately 10 appears to correspond to the onset of fibrillar morphology. The break occurs at slightly higher draw ratios for melt-crystallized samples than for reduced entanglement samples. At draw ratios exceeding 40 the data are more scattered, different sets of data following separate relationships.

The modulus that may be achieved by deformation of solid-state samples is clearly a function of the maximum draw ratio attainable. For melt-crystallized samples it has been proposed that the theoretical maximum achievable draw ratio is dependent upon the entanglement density and the square root of the molecular weight of the resin [26], according to

Maximum draw ratio = $M^{1/2}/20.8$

where M = molecular weight. In order to approach the values estimated from this equation, it is necessary to employ ever higher draw temperatures as the molecular weight increases. However, when the deformation temperature reaches a value at which molecular relaxation occurs faster than the orientation process, the modulus of the drawn sample falls [27]. This value depends on the rate at which drawing is performed and the molecular weight of the sample. Faster orientation and a higher molecular weight counteract the effects of relaxation.

For dried gel samples the attainable draw ratio is a function of the molecular weight of the resin and the initial concentration of polymer in the solvent [28,29]. The higher the molecular weight of the sample, the lower the concentration at which the maximum draw ratio is achieved. Theoretically, the optimum gel concentration required to attain maximum draw is inversely proportional to the square root of the molecular weight of the sample. Samples with molecular weights in excess of 1,000,000 require initial gel concentrations less than 1.6 g/100 mL [29]. Given the appropriate preparation conditions, the attainable draw ratio increases with molecular weight, up to a value of approximately 400 for ultrahigh molecular weight polyethylene, yielding fibers having tensile moduli in excess of 150 GPa.

2. Tensile Strength

The tensile strength of oriented samples initially increases linearly with the degree of molecular orientation but levels off or passes through a maximum at higher draw ratios. The degree of orientation at which the maximum tensile strength occurs depends on the molecular characteristics of the resin and its degree of entanglement. For melt-crystallized ultrahigh molecular weight polyethylene samples drawn at 120°C, a tensile strength plateau is achieved at a draw ratio of approximately 9 [30]. At lower molecular weights, the maximum occurs at higher draw ratios: 20–35 for a linear polyethylene resin with a weight-average molecular weight of 224,000 and 30–40 for a linear polyethylene resin with a weight-average molecular weight of 115,000 [31]. For samples drawn in the molten or reduced entanglement states, leveling off occurs at draw ratios of 30 or more. It has been proposed that the leveling off of tensile strength as a function of draw ratio occurs when chain slippage during the drawing process becomes dominant [32]. The maximum values of tensile strength that can be achieved are largely a function of molecular weight parameters of the sample and the initial degree of entanglement. High tensile strength is favored by high molecular weight, narrow molecular weight distribution, reduced entanglement density, and high draw ratio [31,33].

Tensile strengths as high as 9.9 GPa have been reported for individual hot drawn fibers of ultrahigh molecular weight polyethylene. These thin fibers (~17 μm diameter) were highly susceptible to reduction of their strength due to kink bands and microcracks on their surface resulting from general handling [34]. Solution-spun fibers of ultrahigh molecular weight polyethylene have been reported to have tensile strengths of up to 3.0 GPa [33]. Typical values for melt-crystallized samples drawn in the solid state are 0.8–1.3 GPa for samples with draw ratios of 15 and 20 [31]. This compares to engineering tensile strengths of injection-molded high density polyethylene, which typically do not exceed 0.05 GPa.

The tensile strength of a fiber is not uniquely correlated with its tensile modulus. At a given modulus, cold-drawn samples have higher tensile strengths than melt-spun or melt-drawn samples with similar molecular weight characteristics. For a given modulus and fixed number-average molecular weight the tensile strength increases as the weight-average molecular weight increases [33]. The molecular weight distribution plays a significant but less important role than the average molecular weight. For samples with similar weight-average molecular weights the tensile strength at a given modulus increases as a function of the number-average molecular weight, i.e., as the molecular weight distribution narrows [33]. The tensile strength of linear polyethylene melt drawn to a given draw ratio has been shown to depend on the sum of the intrinsic strengths of its component molecules [35].

3. Elongation at Break

The elongation at break of polyethylene fibers is inversely proportional to their orientation. Ultraoriented fibers typically exhibit elongations at break of less than 10% at room temperature. Gel-spun fibers of ultrahigh molecular weight polyethylene that have not been subjected to hot drawing exhibit an elongation at break approaching 500%, but this rapidly falls as the hot draw ratio is increased. The elongation at break stabilizes at a value of approximately 5% for samples with a hot draw ration of 50 and above [36]. This relationship is illustrated in Figure 6. The elongation at break of highly drawn fibers is temperature-dependent; an increase from 5% at $-175°C$ up to 16% at $100°C$ has been reported [37].

B. Miscellaneous Properties

1. Crystallinity Effects

Increased molecular alignment in polyethylene specimens has a positive effect on their degree of crystallinity, average crystallite thickness, and melting temperature. The effects are more pronounced in linear resins than in branched resins and are more pronounced in high molecular weight samples than in low molecular weight samples. The observed melting point is a function of many factors, both intrinsic and extrinsic, including the branch content, molecular weight characteristics, method of deformation, and method of peak melting point determination. The effect of draw ratio on the peak melting temperature of a linear polyethylene resin is shown in Figure 7 [38]. The increase in peak melting temperature with

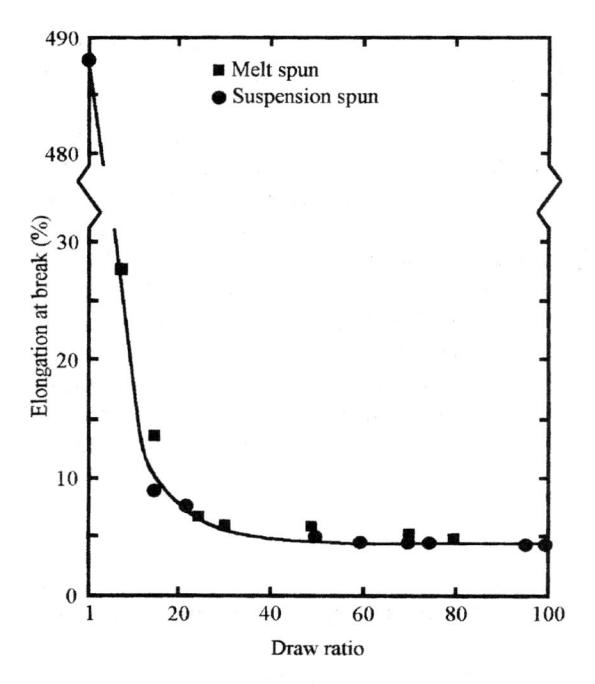

Figure 6 Elongation at break as a function of draw ratio of hot drawn gel-spun and suspension-spun fibers. (From Ref. 36.)

orientation is largely due to an increase in the crystallite c axis dimension. The peaks observed by differential scanning calorimetry also show a distinct reduction in their width, indicative of a narrower distribution of crystallite thicknesses. Analogous observations have been reported for samples of various molecular weights oriented by a variety of methods [7,39,40]. In the extreme case of ultrahigh modulus fibers, a secondary melting peak may be observed at higher temperatures due to the melting of a minor hexagonal phase [21].

The degree of crystallinity of oriented samples follows a pattern similar to that of the melting temperature, increasing with orientation and leveling off at higher degrees of alignment [7,40]. Higher degrees of crystallinity can be obtained for linear resins than for branched resins, the highest values being obtained for ultrahigh molecular weight polyethylene samples prepared from reduced entanglement states. In extreme cases, such as samples prepared by solid-state extrusion of dried mats of solution-grown crystals, the degree of crystallinity can approach 100%. An example of the effect of solid-state extrusion on the degree of

Orientation of Polyethylene

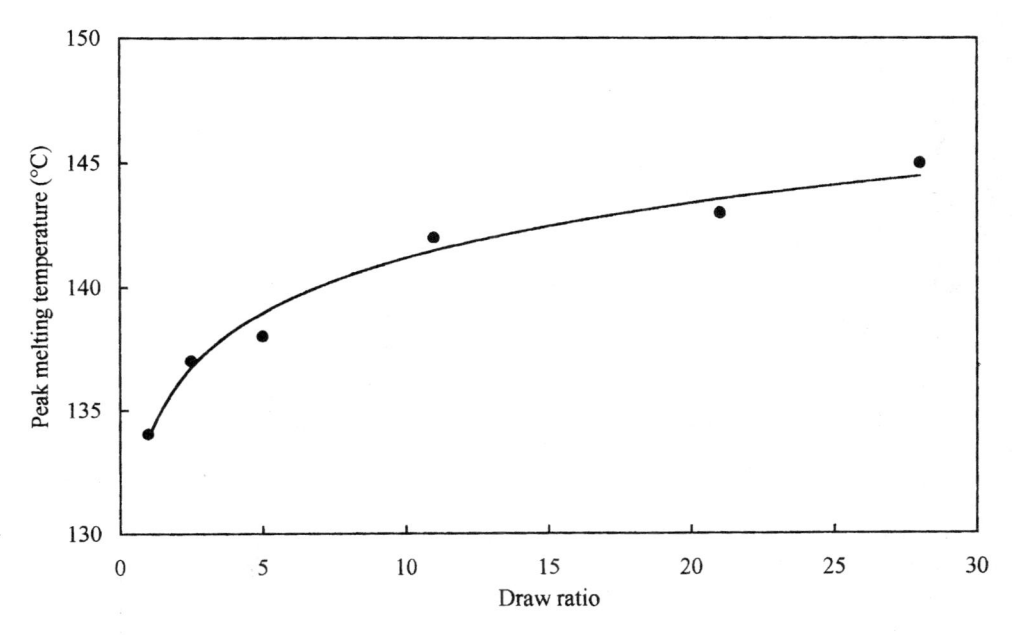

Figure 7 Peak melting temperature (from differential scanning calorimetry) as a function of draw ratio of linear polyethylene with a viscosity-average molecular weight of 300,000 drawn at 100°C. (From Ref. 38.)

crystallinity of melt-crystallized linear polyethylene resins of various molecular weights is shown in Figure 8.

2. Thermal Conductivity

The thermal conductivity of oriented polyethylene measured parallel with the alignment direction increases with orientation. The increase stems from improvements in molecular segment orientation, degree of crystallinity, and connectivity between crystallites, all of which facilitate the transfer of vibrational energy along the sample. The highest value reported is approximately 30 W/(m · K), this being for ultrahigh modulus fibers grown from solution [41]. This value is the highest reported for any polymeric material but is still an order of magnitude lower than that predicted theoretically for perfectly aligned polyethylene. Observed values of thermal conductivity for highly drawn samples can be explained quantitatively on the basis of the Takayanagi model, in which a crystalline region and an amorphous region are arranged in series with one another to form a phase that is arranged parallel with a second crystalline region [42].

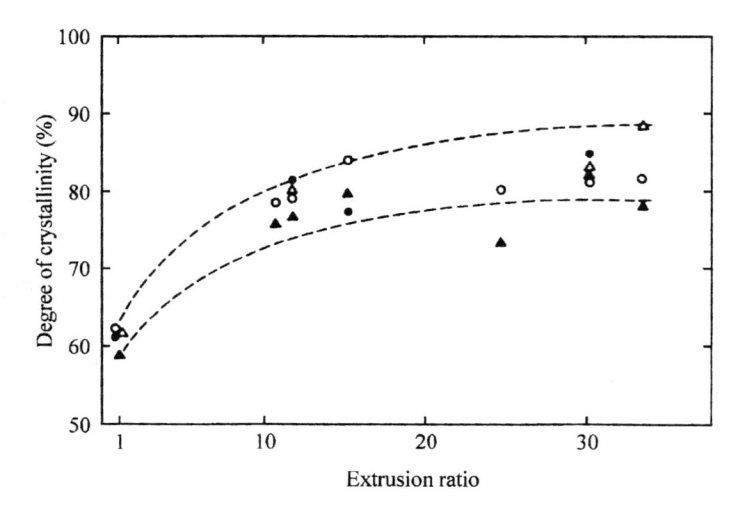

Figure 8 Degree of crystallinity, calculated from differential scanning calorimetry) for solid-state extruded samples of linear polyethylene with average molecular weights of (○) 59,000, (△) 71,000, (●) 92,000, and (▲) 147,000. (From Ref. 39.)

3. Thermal Expansivity

Thermal expansion parallel with the alignment direction decreases with increased orientation, eventually becoming negative at sufficiently high degrees of orientation. Expansivity perpendicular to the orientation direction is greater than that of isotropic samples, increasing with the degree of orientation [39,43,44]. The rate of change of the expansion coefficient diminishes in highly oriented samples. The expansion coefficient parallel with the alignment axis is asymptotic with that of the c axis of the unit cell, while perpendicular to the orientation direction it is asymptotic with the average value of the a and b axes of the unit cell [45].

4. Slow Crack Growth

Slow crack growth in oriented samples is substantially retarded perpendicular to the orientation axis relative to that in isotropic samples. However, parallel with the orientation direction the rate of crack growth is markedly enhanced. For drawn linear polyethylene samples, the decrease in perpendicular crack growth rate has been reported to be an exponential function of the draw ratio [46].

5. Transparency

Solid-state deformation of polyethylene samples in an unconstrained manner normally results in a distinct loss in transparency due to the formation of voids that scatter light. Solid-state extrusion does not permit the formation of voids, and the resulting samples exhibit enhanced optical clarity.

III. ROUTES TO HIGH MODULUS SAMPLES

Orientation may be present in a polyethylene sample for two reasons: either as a consequence of conversion practices that inadvertently generate anisotropy or as a deliberate end in itself, the principal aim of which is to introduce orientation for the sake of improving specific material properties. Most commercial fabrication methods involve some of each. In this section emphasis is placed on methods specifically aimed at obtaining high orientation rather than on commercial practice.

The fabrication of polyethylene products with ultrahigh tensile modulus (i.e., those with a tensile modulus in excess of approximately 100 GPa) generally involves two or more steps. In the first stage the concentration of entanglements is reduced to a manageable level. Subsequent steps deform the reduced entanglement material in such a way that orientation is maximized while relaxation is minimized. The first stage may be omitted if the starting material consists of as-polymerized polyethylene granules or films that inherently contain very few entanglements. The reduction of entanglement concentration involves solvent treatment in some form, ranging from swelling with hot solvent to complete dissolution. The reduced entanglement material can be deformed before or after solvent removal. Deformation of this material in the solid state often takes place in a constrained geometry that prevents the formation of voids. The spinning of solutions or wet gels takes place in an unconstrained manner. The complexity of most of the deformation processes that lead to ultrahigh modulus materials is such that they are not suited to commercial practice. Only the combination of gel spinning followed by hot drawing lends itself to the production of ultrahigh modulus fibers on a commercial scale.

A. Wet Spinning

Wet spinning encompasses the processes of solution and gel spinning. In both cases the entanglement density of the polymer is reduced by solvent dilution prior to drawing. Solvent may be removed before and during the spinning of fibers, which are then typically drawn at elevated temperatures, during which stage additional solvent is removed. Any remaining solvent is finally removed under reduced pressure.

Solution and gel spinning have much in common [47–50]; the outlines of both processes are illustrated in Figure 9. The initial step in both processes is the dissolution of ultrahigh molecular weight polyethylene in a large excess of

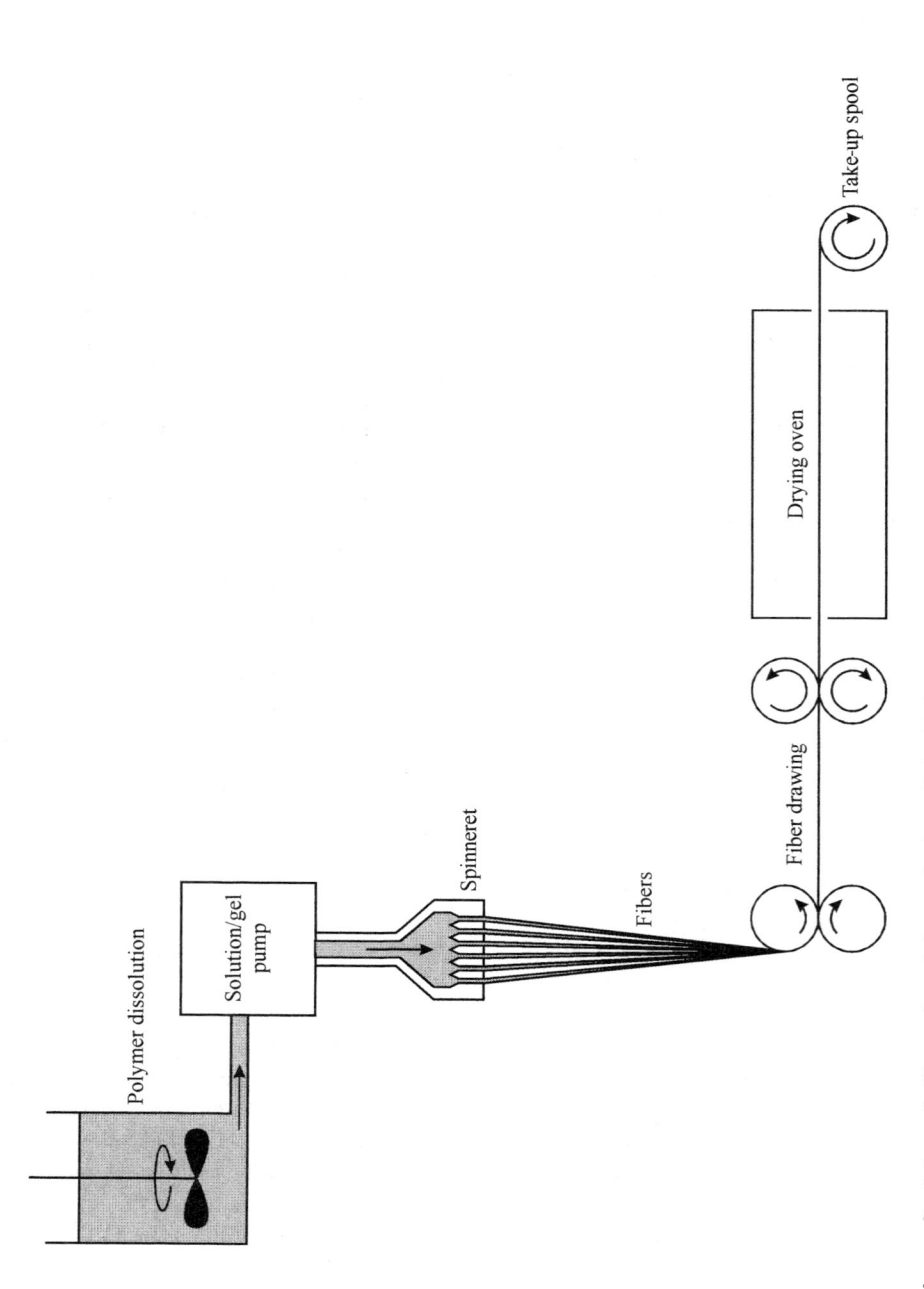

Figure 9 Schematic illustration of the production of ultrahigh modulus fibers by gel or solution spinning followed by hot drawing.

solvent at a temperature in excess of the polymer melting point, typically 140–170°C. Effective gel and solution spinning require polyethylene that has a molecular weight in excess of 300,000 [51]. Concentrations less than or equal to approximately 6% w/v are normally used, with decalin being the preferred solvent. Paraffin oil may also be employed as the solvent, but this typically requires higher dissolution and spinning temperatures. Dissolution of the polymer can be effected by prolonged stirring of the polymer in the solvent or by the extrusion mixing of a suspension of polyethylene granules in a solvent. In the case of solution spinning, the solution is pumped through the holes of the spinneret plate at a temperature sufficiently high that the polymer remains in solution. The strands emanating from the spinneret are cooled below the gelation point of the solution to form gel filaments in which the polymer consists of an entangled network swollen by an excess of solvent. During gel spinning the temperature of the solution is reduced below the gelation point prior to extrusion through the spinneret plate. In both cases the solvent-swollen filament is subsequently hot drawn to a high degree (up to a factor of 20 or more) between two or more pairs of rollers moving at increasingly higher speeds. During the drawing stage much of the solvent exudes from the fibers and evaporates. The remaining solvent is removed by a combination of high temperature and reduced pressure. The tensile modulus of the resulting fibers depends on the extent to which the gel filament can be drawn and the temperature at which drawing occurs [52,53]. Improved draw ratios are attainable by using polyethylene resins with higher molecular weights, which require lower concentrations of polymer in the solvent. Drawing temperatures in excess of 110°C are required to yield fibers with ultrahigh tensile modulus.

As a variation on the process outlined above, complete solvent removal may take place prior to hot drawing [36,49,54,55]. In this case the solvent, which is typically paraffin oil, is extracted with another organic solvent of lower molecular weight, such as methanol or hexane. Final drying to yield porous filaments is achieved under vacuum. The porous filaments may be hot drawn to yield ultrahigh modulus fibers. Dried porous filaments cannot be drawn to the same extent as wet gel filaments produced under identical spinning conditions. However, at a given draw ratio the tensile modulus of the fibers is identical irrespective of whether they were dried prior to hot drawing.

B. Solid-State Extrusion

Polyethylene may be extruded below its melting point if sufficiently high pressures are exerted, in which case the process is known as solid-state extrusion. If the starting material is a reduced entanglement precursor, such as solution-grown crystals [56] or as-polymerized granules [57], fibers and tapes with ultrahigh tensile moduli can be produced. When the precursor is melt-crystallized polyethylene, the attainable tensile modulus may be as high as 46 GPa [58].

Solution-grown crystals are obtained by allowing a very dilute solution of polyethylene (typically <0.5% w/v) to crystallize slowly in the quiescent state. Solution crystals precipitate to form a layer on the bottom of the crystallization vessel. After filtration or decanting of the supernatant liquid, the wet crystal mat is dried under vacuum. The dry product is friable at room temperature. The reduced entanglement state intrinsic to nascent polyethylene granules is readily suited to solid-state extrusion. Thus, granules of ultrahigh molecular weight polyethylene produced by slurry or gas-phase polymerization may be used directly.

In solid-state extrusion, a compacted reduced entanglement powder or a melt-crystallized billet of polyethylene is loaded into the cylinder of a capillary rheometer. High pressure is exerted at an elevated temperature to extrude a fiber from a capillary die with a shallow entrance angle. Extrusion pressures of approximately 2000 atm have been reported for reduced entanglement extrusion at temperatures in the range of 90–128°C [56,57]. Fibers with tensile moduli of up to 60 GPa have been produced by this method. When applied to melt-crystallized high density polyethylene of moderate molecular weights, filaments with tensile moduli of up to 25 GPa have been produced at a pressure of 2000 atm and an extrusion temperature of 136°C [57]. In both cases the fibers emanating from the die were transparent.

Hydrostatic extrusion is a variation of solid-state extrusion in which a billet of melt-crystallized polymer is surrounded by a film of oil that reduces die friction. Filaments with tensile moduli of up to 46 GPa have been produced by this method [58].

Solid-state coextrusion (also known as "split billet" extrusion) is the process by which a film of polymer sandwiched between the two halves of a split polymer billet is extruded through a conical die [59]. As long as the temperature remains below the melting point of the film and the billet, the deformed pieces may be readily separated after they emerge from the die. This process has been applied to compacted solution-grown crystal mats and nascent granules. Multiple passes of a reduced entanglement precursor through the die are needed to generate the effective deformation ratios required to produce samples with ultrahigh moduli [60]. Solid-state coextrusion may be used to form a precursor suitable for subsequent solid-state drawing, which yields ultrahigh modulus products [60–62].

C. Solid-State Drawing

Solid-state tensile drawing to yield highly oriented polyethylene products may be performed on melt-crystallized samples and reduced entanglement precursors such as solution-grown crystal mats, dried gels, as-polymerized films, and solid-state coextruded tapes. Drawing is performed at temperatures between ambient and the crystalline melting point, which in the case of reduced entanglement samples may approach 140°C. Reduced entanglement precursors, especially solution-grown crystal mats, are readily damaged by rough handling; therefore, great care must be taken to grip them gently to prevent premature failure at a point of contact. High degrees of orientation may be achieved by drawing in hot oil [63], water [64], water vapor [65], and heated air [63] or by drawing the sample across a heated shoe [63,66]. Fibers and tapes with tensile moduli in the range of 100–150 GPa can be produced by drawing isotropic reduced entanglement precursors [64,66]. Values in excess of 200 GPa have been reported when a coextruded reduced entanglement film was drawn. Solid-state drawing of melt crystallized samples can produce fibers with tensile moduli of approximately 50 GPa [67,68].

Solution-cast dried gels are obtained by evaporating the solvent from a swollen network of polyethylene in solvent. Polymer is dissolved in a solvent at high temperature to form a dilute solution (typically 0.5–5% w/v). When the solution is allowed to cool in a flat-bottomed crystallization dish, it gels to form an entangled polymer network swollen with solvent. Evaporation of the solvent at room temperature yields a flexible dry gel film that may be readily deformed. Reduced entanglement polyethylene films suitable for drawing to high degrees can be produced directly by polymerizing.

D. Fiber Growth from Sheared Solutions

Highly oriented fibers can be grown from dilute solutions subjected to high shear. Such fibers are termed ''shish kebabs,'' reflecting their characteristic morphology. When observed under the electron microscope, such fibers appear as a series of approximately parallel plates (lamellar crystallites) strung like beads upon a central fiber. A representative electron micrograph of this morphology is shown in Figure 36 of Chapter 4. The proposed molecular arrangement within a shish kebab is illustrated in Figure 37 of Chapter 4. The principal methods used to generate the required shear levels are high speed stirring [69,70], high flow rates [71], and impinging liquid jets [72]. The various methods are reviewed by Barham and Keller [73].

Stirring-induced crystallization from dilute solutions occurs when shear fields generated by vortices are sufficient to align the dissolved molecules. In addition, the concentration and temperature of the solution must be such that crystallization would not occur in the quiescent state but can occur when parallel molecular segments become highly aligned. When these conditions are met, chain segments crystallize on the surface of the stirrer to form fibers. The uncrystallized segments of the molecules composing these fibers are drawn through the solution rapidly. Such segments can crystallize in one of two ways: Highly aligned segments may crystallize with their neighbors to extend the length of the crystalline fiber nucleus, while others may crystallize on the surface of the fibrous nucleus to form platelet overgrowths. As stirring continues over several hours, the length and number of the shish kebabs increase until the surface of the stirrer is wrapped in an aggregate of highly oriented polyethylene. Medium or high molecular weight linear polyethylene can be used as the starting material, at concentrations in the range of 0.5–5 wt%. Crystallization temperatures between 90 and 125°C may be used, depending upon the molecular weight of the resin and the concentration of the solution. The ratio of fiber core diameter to lateral platelet dimension increases with temperature. At high temperatures the fiber core makes up virtually the entire shish kebab, with only a small amount of lamellar overgrowth. The tensile modulus of shish kebabs increases with crystallization temperature, which reflects the proportion of the structure that is load-bearing [74]. At sufficiently high temperatures it is possible to produce fibers with ultrahigh tensile modulus. The relationship between growth temperature and tensile modulus is illustrated in Figure 10.

A continuous highly oriented fiber may be grown from dilute polyethylene solution subjected to shear flow in a Couette apparatus [70,74]. This device consists of a pair of coaxial cylinders, the inner one of which can be rotated rapidly. A dilute polyethylene solution introduced into the gap between the cylinders is subjected to shear flow. When the appropriate conditions of shear, temperature, and concentration are met, the polymer will crystallize on to a seed fiber held in the flow field. As the extended polyethylene fiber grows, it is wound up at a rate equal to its extensional growth rate, the tip of the growing fiber thus remaining at a fixed position. The resulting fiber consists of a bundle of shish

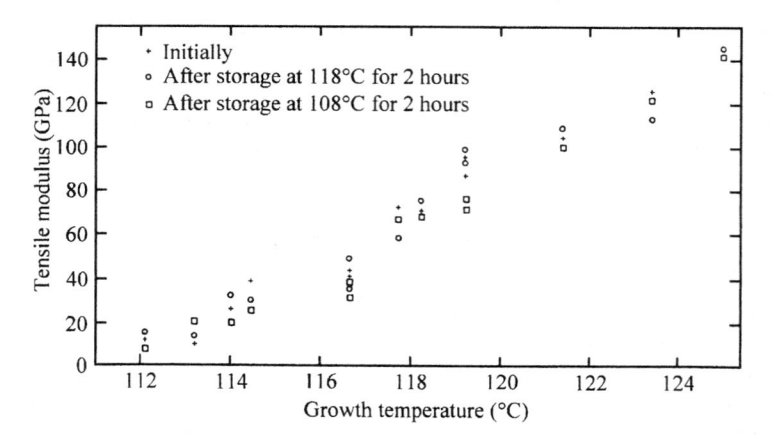

Figure 10 Tensile modulus of "shish kebabs" grown from stirred solution as a function of growth temperature. (From Ref. 74.)

Orientation of Polyethylene 439

kebab fibers with mutually parallel fiber axes. Fibers with tensile modulus of approximately 140 GPa can be produced by this method [74].

A continuous highly oriented fiber can also be produced with the apparatus illustrated in Figure 11 [71]. A solution of ultrahigh molecular weight polyethylene is maintained in the reservoir at a temperature just above that at which it crystallizes in the quiescent state. A stream of this solution is allowed to flow

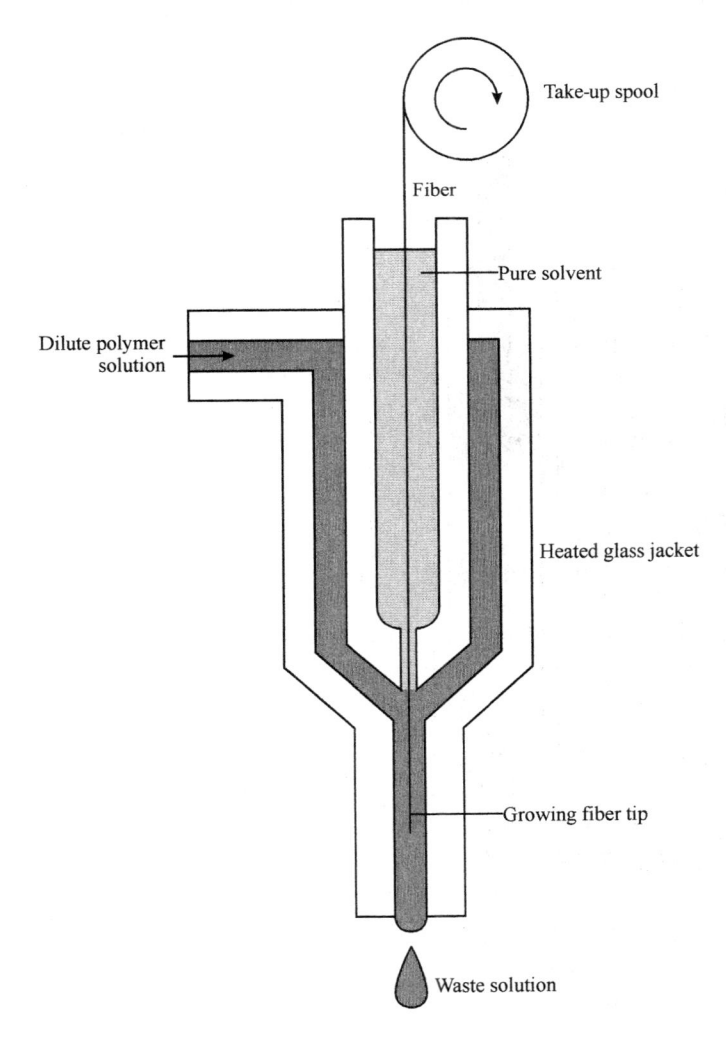

Figure 11 Schematic diagram of apparatus used to produce a continuous shish kebab fiber by Poiseuille flow.

past a growing fiber that is drawn up through the tube down which the solution is flowing. Initially the process is seeded with a polyethylene fiber produced by another method; alternatively, a linen fiber may be used. The rate of fiber growth is controlled by varying the temperature, concentration, and rate of flow of the solution and the rate at which the fiber is wound up.

Highly oriented fibers may be crystallized from solution when two streams of dilute polyethylene solution impinge upon one another at appropriate speeds,

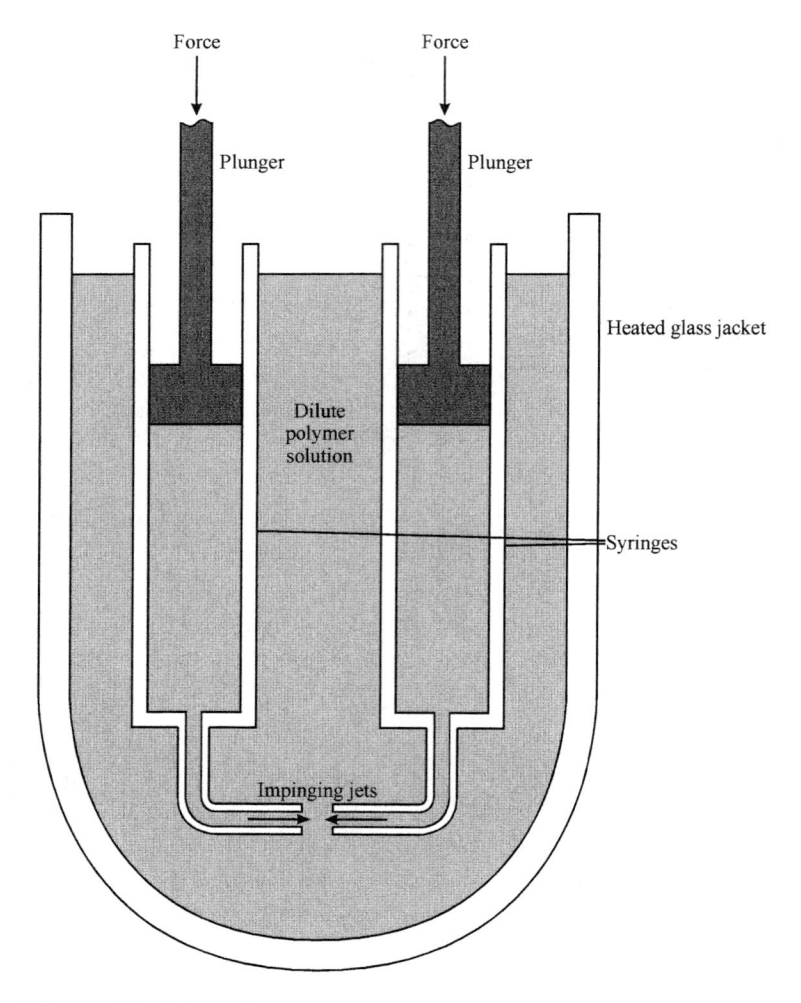

Figure 12 Schematic diagram of apparatus used to produce oriented material by impinging flow.

concentrations, and temperatures [72]. The apparatus used for this experiment is illustrated in Figure 12. Dilute solutions of linear polyethylene in xylene (0.5–5 wt%) at temperatures up to 112°C are forced from the small mutually opposed orifices so that the emerging jets strike one another head on. On the plane of symmetry lying between the two apertures a disk of polymer crystallizes for the duration of solution flow. When flow ceases, the disk collapses. Alternatively, polyethylene solution is rapidly drawn into the two orifices, resulting in the crystallization of fibers along the symmetry axis connecting the apertures. These fibers remain intact when flow ceases. An analogous procedure can be applied to molten polyethylene [75].

E. Miscellaneous Methods

In this section some of the more esoteric methods of producing highly oriented polyethylene are outlined.

1. Blocked Plug Crystallization

Blocked plug crystallization is the process by which a high modulus fiber of polyethylene may be produced by initiating the formation of microfibrils within a capillary die, then plugging up its outlet and increasing the applied pressure [76]. Initially, molten polyethylene is extruded from the capillary rheometer at a temperature approximately equal to its equilibrium melting temperature. Extrusion under these conditions initiates the formation of a small population of microfibrils within the capillary die. When the die outlet is plugged and the pressure is raised the molten polymer crystallizes on to the microfibrillar nuclei. The result is a space-filling shish kebab structure in which lamellae emanating from neighboring microfibrous nuclei interpenetrate, their thicknesses decreasing with distance from the nucleating fiber. Such fibers have tensile moduli of up to 90 GPa.

2. Radial Compression

Relatively thick filaments of highly oriented polyethylene may be produced by applying a radial compressive force to an isotropic rod at elevated temperature [77]. This is achieved by tightly wrapping a rod of polyethylene with numerous windings of a highly stretched elastomeric fiber at approximately 80°C. When the compressive pressure is sufficiently high, the polyethylene rod begins to extrude from both ends of the overwrapping. When the overwrapping is removed, the polyethylene rod is found to be waisted, with a clear portion in the middle where solid-state uniaxial elongation occurred. The effective draw ratio is controlled by the amount of elastomeric fiber that is applied. For a given effective draw ratio, the tensile modulus, elongation at break, and tensile strength are similar to those obtained by conventional solid-state extrusion of the same resin. Tensile moduli of up to 62 GPa have been achieved by using this process.

3. Roller Drawing

Highly oriented films can be produced by drawing a sheet of polyethylene between a pair of freely rotating heated rollers [78,79]. Orientation occurs by two processes: isothermal thickness reduction in the nip between the rollers (75–120°C) and nonisothermal post-roller solid-state drawing. As in many other orientation processes, the tensile strength and modulus of the product increase linearly with the draw ratio. Fibers with tensile moduli of 43 GPa and tensile strengths of 0.67 GPa have been obtained by this method.

4. Swell Drawing

Melt-crystallized polyethylene tapes that have been swollen in hot solvent can be drawn to a much greater extent than the starting material [80]. This is so regardless of whether the solvent is removed before or after drawing. Drawing takes place at temperatures between 90°C and the melting point of the polymer. The extent to which the tapes can be drawn, and consequently the tensile modulus of the final product, are inversely proportional to the concentration of polymer in the swollen tape. Tapes with tensile moduli of up to 102 GPa and tensile strengths of 2.4 GPa have been produced from ultrahigh molecular weight polyethylene precursors. It is possible to envisage a continuous fabrication process whereby polyethylene could be extruded into tapes that are subsequently swollen, drawn, and dried to form uninterrupted lengths of ultrahigh modulus fibers and tapes.

5. Two-Stage Drawing

It is possible to generate highly oriented polyethylene tapes using a two-stage solid-state drawing procedure [81]. The initial drawing is nonisothermal, an approximately ninefold extension taking place in a small region of tape that is locally heated to approximately 120–125°C. The intermediate product is then subjected to isothermal drawing in an oven held at a temperature of 70–80°C. The final product is a tape with a tensile modulus of 37 GPa and a tensile strength of 1.4 GPa.

IV. MECHANISMS OF SOLID-STATE DEFORMATION

The precise mechanisms by which polyethylene deforms in the solid state are not rigorously understood. Various schemes have been proposed to explain the observed macroscopic effects in terms of molecular rearrangement, but the fact remains that it is not possible to observe individual molecules in the static state, let alone track them during deformation. Thus, all proposed deformation mechanisms must be extrapolated from microscopic and macroscopic observations. Hypothetical mechanisms may be constructed from continuous observation of average macroscopic properties or static observation of specimens, whose deformation has been interrupted at known values. Experimental observations are interpreted in light of an understanding of semicrystalline morphology to generate a hypothetical model of deformation. It follows that proposed models for the deformation process have evolved over the years as our knowledge of semicrystalline morphology has improved; however, many discrepancies remain between competing models.

In this section emphasis is placed on the morphological response of polyethylene to tensile deformation, as this is the principal mode by which polyethylene products deform in use. Peterlin's model of polyethylene deformation, which has been highly influential, is discussed first. Subsequently, current views of the yield process are outlined, followed by a general discussion of various aspects of deformation. The macroscopic aspects of polyethylene deformation are dealt with in Chapter 5 in the section discussing mechanical properties.

A. Peterlin's Model of Tensile Deformation

From the mid-1960s to the early 1970s, Anton Peterlin was highly influential in explaining the observed deformational characteristics of polyethylene in terms of molecular and crystalline deformation processes. In 1965 he proposed a mechanism purporting to explain the deformation of polyethylene in terms of slippage along major crystalline planes in lamellae, followed by the disruption of lamellae into crystallite blocks that are rearranged to form fibrils [82]. This mechanism was based on the contemporary view of semicrystalline polyethylene morphology that assumed an overwhelming predominance of tight fold adjacent reentry of chain stems at lamellar surfaces. Over the course of several years Peterlin refined his model on the basis of additional experimental evidence [10,83–85]. A review of his model, the postulates upon which it was based, and the supporting evidence was published in 1971 [86]. Peterlin's model of polyethylene deformation is illustrated schematically in Figure 13.

In Peterlin's model the initial morphology is one of lamellar stacks arranged into spherulites. Each lamella is connected to its neighbors by a relatively small number of tie chains. Initial strain causes twinning and martensitic phase deformation within the lamellae, transforming the orthorhombic lattice into a monoclinic one. This can account for the initial 5–15% elongation, which is largely reversible. Subsequently, crystalline chain segments slip (shear) relative to one another, tilting with respect to the lamellar surface as they do so, causing entire

| (a) | (b) | (c) | (d) |

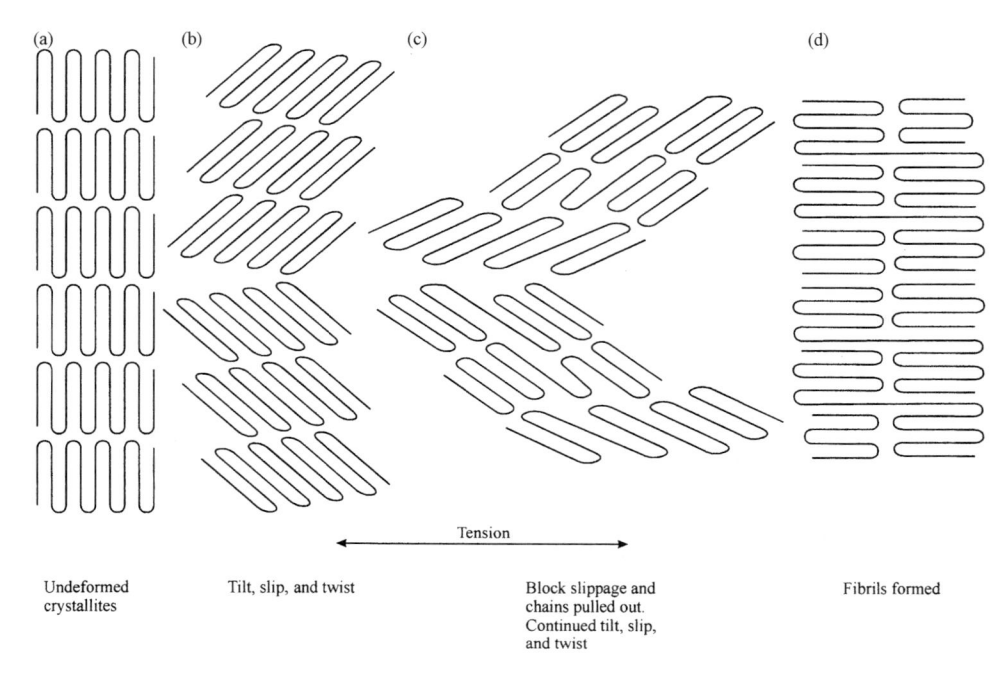

Undeformed crystallites   Tilt, slip, and twist   Block slippage and chains pulled out. Continued tilt, slip, and twist   Fibrils formed

←——— Tension ———→

Figure 13 Peterlin's model of molecular slip, lamellar tilt, and fracture of lamellae into crystalline blocks converting a lamellar morphology into a fibrillar one.

lamellae to rotate. Chain tilt and slippage are responsible for a further 100% elongation. In the next stage, mosaic crystallite blocks of the tilted lamellae are broken off, drawn by partially unfolded chains, and incorporated wholesale into fibrils. The average all-trans segment length in the newly formed fibrils match that of the original lamellae. A combination of continued tensile stress and localized sample heating transforms this primary fibrillar morphology into a "pseudomelt" phase that is akin to a liquid crystal. This mesomorphic phase recrystallizes upon cooling to form a secondary fibrillar morphology, the average all-trans segment length of which depends on the drawing temperature. The degree of chain alignment within the fibrils increases upon further drawing due to the unfolding of chain segments. Within the fibrillar structure itself only a small proportion of chain segments are load-bearing. When the critical stress on an individual chain is exceeded, it breaks, and the load it previously supported is transferred to others in the same plane. Macroscopic rupture occurs when there are no longer sufficient load-bearing chains in a given plane to sustain the applied stress. Rup-

Orientation of Polyethylene

ture then spreads across the plane by a process of "self-accelerating damage accumulation."

Although revolutionary when it was developed, Peterlin's model was limited by the contemporary understanding of the solid-state morphology of polyethylene. Possibly its most significant limitation is the concept of mosaic crystallite blocks being reassembled en masse to form fibrils; this would require a much higher level of adjacent reentry of chains than is currently acknowledged. This said, other components of Peterlin's model are still valid and are widely accepted. Tilting and slippage of chains is thought to occur in the initial phases of compressive and shear deformation of polyethylene; shear dislocation of adjacent blocks within lamellae is considered by many to be a key step in tensile yielding, while others consider that a phenomenon akin to pseudomelting plays an important role in the yielding process.

B. Models Describing Tensile Yield

Currently two competing models exist that purport to describe the tensile yielding mechanism of polyethylene. Each model has ardent supporters of some note who claim that there is convincing evidence to support their viewpoint. The first model, on a chronological basis, is that espoused by Young in the early 1970s. It involves the shear slippage of adjacent crystalline blocks comprising lamellae. The second model was proposed by Flory and Yoon in the late 1970s and was subsequently formalized by Harrison and his coworkers in the 1980s. Proponents of this model believe that applied stress is sufficient to cause isothermal "mechanical melting" of crystallites during the yielding process.

1. Screw Dislocation Model of Yield

In 1973 Young et al. [87] proposed that polyethylene crystallites deform by a combination of three mechanisms—fibrillar slip, lamellar slip, and chain slip, which are illustrated in Figure 14. These mechanisms are based on a morphological model that assumes a high degree of adjacent reentry of chains at lamellar surfaces. Fibrillar slip involves the sliding of adjacent crystallite blocks parallel with the c axis of the unit cell. Lamellar slip is the process by which lamellae slide over the surface of one another. Chain slip is the incremental displacement of all chain stems relative to their neighbors within a section of a crystallite. Young et al. proposed that deformation of the crystalline elements in polyethylene principally involved chain slippage, with minor contributions from the other two mechanisms. Later a fourth deformation mechanism, involving slippage along planes parallel with the a and b axes of the unit cell, was also included in their analysis of crystalline polyethylene deformation [88]. All these mechanisms were

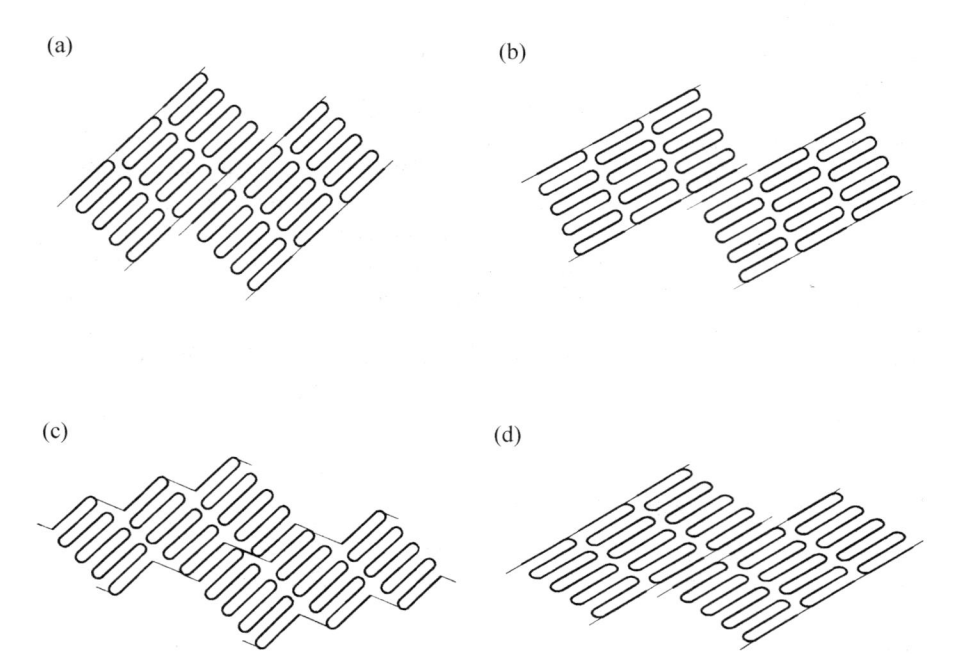

Figure 14 Crystallite deformation mechanisms. (a) Original crystallites; (b) fibrillar slip; (c) lamellar slip; (d) chain slip.

postulated on the basis of evidence from wide-angle X-ray diffraction studies of compressive deformation of high density polyethylene.

Young [89,90] subsequently proposed a yielding mechanism involving the shearing of crystallite blocks past one another along slip planes originating at screw dislocations, as shown in Figure 15. Such screw dislocations are activated by applied stress. The theory supporting this model predicts that the observed yield stress should be directly proportional to lamellar thickness and that increasing temperature should reduce the yield stress, both of which are qualitatively correct. Similar calculations by Crist [91,92] produced a reasonable match between predicted and observed yield stress over a temperature range from approximately $-100°C$ to $90°C$.

Although the screw dislocation model is attractive in its simplicity and the semiquantitative agreement of observed yield stress with theoretical values, major conceptual problems remain. In a tensile configuration, stress is transmitted to crystallites via taut tie chains. It is difficult to conceive how randomly distributed tie chains could apply the critical shear stress necessary to activate screw disloca-

Orientation of Polyethylene 447

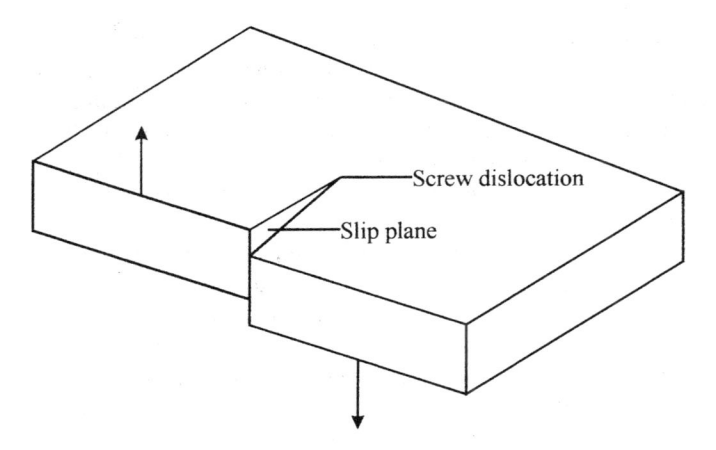

Figure 15 Schematic illustration of the shearing of adjacent crystallite blocks at a screw dislocation.

tions. The appropriate deformation conditions are more likely to pertain to compressive and shear configurations, wherein load is applied more evenly across the surfaces of lamellae. The theoretical calculations of this model are based upon the existence of straight shear planes, which are unlikely to exist in melt-crystallized samples where tight fold adjacent reentry of chains at lamellar surfaces is not prevalent. This problem will be exacerbated as molecular weight, short-chain branching, and crystallization rate increase. This screw dislocation model is also limited in that it does not go far enough to explain tensile phenomena after the initial shear dislocation has occurred.

2. Mechanical Melting Model of Tensile Deformation

In 1978, Flory and Yoon [93] suggested that cold drawing of semicrystalline polyethylene required melting of the original crystallites followed by recrystallization of the deformed material in the oriented state [93]. This view was based upon consideration of the entangled nature of semicrystalline polyethylene, assuming the existence of negligible tight fold adjacent reentry at lamellar surfaces. This model was, to some extent, presaged in 1967 by Sakaoka and Peterlin [94] who suggested that tensile stress may be sufficient to raise the temperature of a sample locally to its melting temperature. Harrison and his coworkers [95–97] proposed a formalized version of this model, viewing yielding as a "stress activated phase transition," in which crystallites melt isothermally to form a noncrystalline phase that recrystallizes during the drawing process. This model has become known as the "mechanical melting model."

According to the mechanical melting model, when a polyethylene sample is stretched it initially undergoes largely reversible deformation, storing energy in the strained noncrystalline regions. Yielding occurs when sufficient stress is applied to initiate isothermal melting of crystallites, i.e., at the ambient draw temperature. Once the crystallites in a given cross section have melted, the resultant molten phase undergoes viscoelastic drawing, during which recrystallization occurs. Thus the original semicrystalline morphology is converted to a fibrillar one by a process of melting, drawing, and recrystallization. Evidence to support this model comes from both experimental and theoretical sources.

The mechanical melting model (alternatively known as the "premelting" model) predicts that the yield stress of a polyethylene sample should be proportional to its heat of fusion and hence its degree of crystallinity, which is found to be the case experimentally [98–100]. The observed increase in yield stress as a function of lamellar thickness can be explained by the approximately linear relationship between lamellar thickness and degree of crystallinity [98].

Calculations by Liu and Harrison [97] and Gent and Madan [101] indicate that the mechanical energy absorbed by a polymer sample during the yielding process is sufficient to melt its crystalline component. The observed reduction of yield stress as a function of increasing temperature is explained by reduced energy input required to disrupt the crystalline phase.

Further supporting evidence comes from careful measurement of local temperature in a sample during the deformation process, by means of an embedded thermocouple [102]. In the initial, approximately elastic, stage of deformation, the temperature remains constant or drops very slightly. A distinct exotherm is detected as the neck envelops the thermocouple, reflecting the heat evolved as the drawn melt recrystallizes. Similarly, the yield stress maximum in high density polyethylene is observed prior to the onset of visible necking [103]. These observations are in accord with stress induced melting followed by recrystallization of the oriented melt.

When a polyethylene specimen is drawn to its natural extension ratio, the thickness of the resulting crystallites is a function of the deformation temperature, irrespective of the original crystallite thickness [2,104]. The resulting thickness is identical to that obtained by crystallization from the quiescent melt at the same temperature. This is strong evidence supporting the proposition that each section of the sample passes through a molten phase, destroying the original semicrystalline morphology.

More evidence in support of chains passing through a molten phase comes from small angle neutron scattering studies by Wignall and Wu [105]. When a partially segregated melt-crystallized blend of deuterated polyethylene and hydrogenous polyethylene is drawn past the yield point, the deuterated molecules become dispersed randomly throughout the hydrogenous polyethylene. This ran-

Orientation of Polyethylene

domization can be accomplished only if the original semicrystalline morphology is totally destroyed and the mixture passes through a molten phase.

3. Double Yield Phenomena

Many low density and linear low density polyethylene samples exhibit what is commonly known as "double yielding." In such cases there is evidence that more than one physical process is active in the region of the force versus elongation curve associated with yielding. Double yielding is not a uniquely defined phenomenon; two principal variants are observed. The first consists of a well-defined yield region consisting of two closely spaced peaks (often overlapping so closely that only one maximum is observed). The second comprises a relatively well-defined maximum (or inflection) followed by a diffuse peak. The former is more typical of polyethylene samples with medium densities (0.92–0.94 g/cm^3), while the latter is more often observed at lower densities (0.90–0.93 g/cm^3). Examples of force versus elongation curves that exhibit double yielding are shown in Figure 5 of Chapter 5. The relative heights of the two maxima are not fixed; they may be of equal magnitude, or one may dominate the other. The magnitude of the peaks has in some cases been observed to vary as a function of time, a minor peak becoming dominant at the expense of the major peak over a period of 24 hr after molding [106]. Phenomenologically, the first peak has been associated with the onset of plastic strain that is slowly recoverable, while the second peak marks the onset of nonrecoverable deformation [107].

The root cause of double yielding is not clear; two explanations have been proposed based upon the competing models of yield. Seguela and Rietsch [108] postulate that the two yielding peaks are due to independent thermally activated slip processes, one involving homogeneous slip of chains within the crystal, the other involving shear of crystalline blocks past one another. These processes correspond to Young's chain slip and fibrillar slip, respectively. Lucas et al. [109] suggested that the first peak is related to mechanical melting of the original lamellae, while the second peak is due to mechanical melting of a second population of oriented lamellae that crystallize from the strained molten phase after the initial yield process. When the two yield peaks are close together, it is conceivable that different lamellar populations comprising a bimodal distribution of lamellar morphologies yield at slightly different strain levels.

C. Strain Hardening

"Strain hardening" is the term generically used to describe the upsweep observed in the force versus elongation curve after the occurrence of yielding and necking. In the case of highly crystalline samples, in which a neck propagates through the length of the sample, it may begin at elongations of several hundred percent. In samples of low crystallinity, strain hardening begins immediately after homogeneous yielding. The elongation and stress increase associated with postyield deformation varies greatly, depending upon the molecular characteristics of the resin and the processing it has undergone. Oriented samples drawn parallel with their initial orientation exhibit steeper strain hardening slopes than when they are drawn perpendicular to the initial orientation. The effects of molecular weight, comonomer content, and comonomer type on the strain hardening characteristics of isotropic samples are quite complex. Mandelkern and coworkers [110] attempted to systematically correlate molecular architecture with the strain hardening characteristics of a series of linear low density polyethylene samples.

There are two principal deformation mechanisms active during the strain hardening process: fibrillar slippage and plastic flow. Fibrillar slippage involves the sliding of adjacent crystallite blocks parallel with the principal axis of deformation. Plastic flow involves local rearrangement of the chain segments comprising crystalline and noncrystalline regions, which may or may not involve the destruction of crystallites. The former is more characteristic of highly crystalline and oriented samples drawn parallel with their principal orientation axes, while the latter is more prevalent in samples of lower crystallinity and oriented samples drawn perpendicular to their principal orientation axes.

During fibrillar slip, crystallite blocks comprising already well oriented fibrils are drawn past one another [111,112]. As this occurs, they draw out some of the intervening noncrystalline chain segments, forcing them into better alignment. In addition, the noncrystalline strata separating crystalline blocks become thinner and denser, with defects such as chain ends and entanglement points becoming more concentrated within them. With improved overall alignment of the sample, the degree of crystallinity increases. As the crystalline domains become larger and effective chains connecting entanglement points become stretched to their limit, it becomes increasingly difficult to deform the sample without rupturing chains. Consequently, the force required to elongate the sample increases and the strain hardening slope increases. There finally comes a point where large numbers of bonds begin to break, and catastrophic rupture of the sample ensues. The overall energy absorbed during the tensile process is largely dissipated by deformational processes, with chain rupture accounting for only a small percentage of the energy absorbed.

In samples that do not attain a fibrillar morphology, i.e., those of lower density (branched or high molecular weight linear samples), plastic flow accounts for much of the postyield deformation. The precise mechanisms involved are not well defined, but it appears that molecular slippage plays a key role. The extension of such a system can be likened to the drawing of a cross-linked network, with entanglement points behaving as cross-links that allow limited slippage. As an increasing proportion of the effective chains linking entanglement points are

Orientation of Polyethylene

stretched taut, the force required to extend the sample increases. The more readily chains can slide past one another, the further the sample can be elongated. Increases in molecular weight, concentration of short-chain branches, and branch size all decrease the rate of molecular slippage, as reflected by an increase in the slope of the strain hardening portion of the force versus elongation curve [110,113].

Samples that were originally isotropic that have been drawn almost to the point of failure return to essentially their original dimensions when subsequently heated above their crystalline melting point. Mills et al. [114] have attempted to explain the strain hardening behavior of low density and linear low density polyethylene samples in terms of rubber network theory. They found that a Mooney–Rivlin plot of tensile deformation yielded a straight line characteristic of a cross-linked rubber. These observations are strongly indicative that plastic flow involves no significant chain disentanglement.

Termonia and Smith [115–119] attempted to predict the force versus elongation curves of polyethylene by computer modeling. They envisage three processes by which samples deform: breakage of van der Waals bonds, slippage of entanglements, and chain breakage. The likelihood of any process occurring at a given point within the specimen is based on the Eyring activation rate theory. Overall their model fits observed data remarkably well, accurately predicting the shapes of force versus elongation curves. In particular it correctly predicts the effects of molecular weight and short-chain branching on strain hardening and elongation at break, providing indirect corroboration of the importance of chain slippage in this region of the force versus elongation curve. Deformation rate, temperature, and polydispersity are also taken into account, the predicted trends mimicking those found experimentally.

D. Rupture Phenomena

All samples break when the appropriate forces are applied; how they do so depends on the molecular characteristics of the resin, specimen morphology, and the external forces. Resins that ordinarily behave in a ductile manner, failing at several hundred percent elongation, can be made to fail in a brittle manner by changing either the test protocol or specimen preparation conditions.

1. Brittle Failure

"Brittle" failure, in which there is no macroscopic orientation prior to break, occurs in polyethylene samples stressed below their glass transition temperature or when a specimen's degree of crystallinity exceeds a critical value at a given molecular weight.

Only at sub-glass transition temperatures does failure occur in a microscopically brittle fashion. At such temperatures the rate of chain motion is so slow that plastic deformation is essentially eliminated. When sufficient stress is applied, a brittle crack propagates rapidly across a specimen, taking the path of least resistance. Examination of fracture surfaces by scanning electron microscopy reveals cleavage of crystallites in the planes containing chain c axes, with chain rupture presumably occurring in the noncrystalline regions.

"Brittle" failure above the glass transition temperature occurs only in well-crystallized samples [100,110]. Careful examination of the fracture surface of such samples reveals the presence of localized ductile failure. In such cases a stable neck is not established, and rupture ensues before macroscopic deformation can occur. In linear polyethylene samples, the transition between ductile and brittle failure as a function of molecular weight is well defined for any given set of crystallization conditions. As the crystallization rate increases, the molecular weight at which the ductile-to-brittle transition occurs decreases. At weight-average molecular weights below 25,000, isotropic melt-crystallized samples are invariably brittle. At molecular weights in excess of approximately 150,000, it is very difficult to generate sufficiently high levels of crystallinity for brittle failure to occur.

A possible explanation of brittle failure in highly crystalline polyethylene specimens involves the ratio of tie chains connecting crystalline lamellae to the degree of crystallinity. When a sample is initially stressed, the load is carried by the lamellae and the taut tie chains connecting them. For a sample to deform in a ductile manner the sample must first yield, which involves the destruction of crystallites in some manner. Each taut tie chain can transmit only a finite load prior to rupture. If there are insufficient taut tie chains in a given cross-sectional plane to sustain the stress required to destroy the crystallites, then the tie chains themselves will break prior to ductile yielding. The flux of tie chains required depends on either the degree of crystallinity or the thickness of the lamellae (depending upon the model of yielding to which one ascribes). Thus, if the ratio of number of tie chains to degree of crystallinity (or lamellar thickness) is lower than a critical value, brittle failure will ensue. For this reason, factors such as increased molecular weight and short-chain branching reduce the likelihood of brittle failure either by increasing the concentration of tie chains or by decreasing the degree of crystallinity.

2. Tensile Rupture Following Ductile Deformation

The principal molecular factors that control the extent to which an isotropic melt-crystallized sample can be drawn prior to failure are those that primarily control its rate of crystallization. High molecular weight and the presence of branches inhibit drawing. In addition, a sample's initial orientation (imposed during preparation) and the rate and temperature at which deformation occurs also influence its extensibility. A sample's initial morphology has little if anything to do with its final draw ratio [98,99].

The extent to which a sample draws is a function of the separation between its entanglement points, the rate of molecular slippage past the entanglements, and the ratio of the molecular relaxation rate to the deformation rate. Samples with a high molecular weight between entanglements (such as those crystallized from solution) can be drawn to very high degrees because the effective chains between entanglements can be highly extended before they are pulled taut [120,121]. In melt-crystallized samples, which have a much lower molecular weight between entanglements, the effective chains are pulled taut at much lower draw ratios. Once the effective chains have been pulled taut, any further orientation requires chain slippage. If chain slippage cannot take place on a reasonable time scale, then chain rupture will ensue. Chain slippage is inhibited by increasing molecular weight and branching.

The draw ratio at break of isotropic melt-crystallized samples passes through a maximum as a function of deformation rate [117]. If chain relaxation takes place on a time scale faster than sample deformation, it is possible for molecules to disentangle faster than they can be aligned. In such instances the orientation at break is low. These conditions are met when room temperature samples are deformed at very low strain rates, the critical strain rate varying inversely with the molecular weight. When the deformation rate greatly exceeds the rate of chain slippage, then taut chain segments rupture before they have time to slip and relieve the applied stress. In such cases, orientation prior to rupture is limited by the immediate extensibility of the effective chains linking entanglements. At intermediate strain rates, molecular slippage can take place sufficiently fast to relieve excessive stress on chain segments, but not so fast that widespread disentanglement occurs. In this range of strain rates, the maximum draw ratio is obtained.

A maximum of draw ratio as a function of deformation temperature is also observed for many samples [117,122]. Initially, increased temperature facilitates the relief of excessive strain on individual chain segments by molecular slippage, but above a critical temperature the disentanglement rate overtakes the deformation rate. The temperature at which this occurs in linear samples increases as a function of molecular weight [123]. In ultrahigh molecular weight polyethylene samples, the maximum draw ratio is not obtained until the deformation temperature exceeds the melting temperature of the sample.

3. Low Stress Brittle Failure

Low stress brittle failure is the process by which a crack propagates across the thickness of a specimen at very low stress levels. The macroscopic process and the effect of molecular parameters and failure conditions on the rate of failure are discussed in Chapter 5. From a morphological standpoint, the process of slow crack growth has been explained in terms of tie chains and the rate of molecular disentanglement [124–126].

Low stress brittle failure can occur when the overall applied stress level is so low that the rate of molecular disentanglement exceeds the creep rate and when one or more stress concentrators are present. Stress concentrators take the form of structural inhomogeneities such as scratches, voids, and nonpolymeric inclusions. The first event is the formation of a craze at the point of highest stress. This involves localized yielding of the sample to form load-bearing fibrils, which occurs when the concentrated stress field equals or exceeds the local yield stress [127]. The fibrils consist of alternating layers of noncrystalline regions and lamellae whose lateral planes are perpendicular to the applied stress. For the crack to propagate, the fibrils that compose the craze must rupture. The rate of rupture is controlled by the rate of disentanglement (pullout) of the load-bearing tie chains that connect the lamellae. Disentanglement is slowed by increased molecular weight and short-chain branching. The longer the molecules, the more lamellar stacks are likely to be encompassed by the trajectory of a single chain. Effectively, the higher the molecular weight, the more entangled will be the molecules in the fibril. Short-chain branches inhibit disentanglement by acting as "anchors" that cannot be readily pulled through the crystalline matrix. The more numerous and the larger the branches, the greater the hindrance to chain disentanglement. When load-bearing fibrils rupture, the local stress ahead of the craze increases and new fibrils form, driving the craze further into the sample. Finally there comes a point when the cross-sectional area of the remaining ligament is so small that the stress it experiences exceeds the instantaneous yield stress, and failure proceeds in a ductile manner.

4. Environmental Stress Cracking

Environmental stress cracking is the brittle failure of a stressed sample in the presence of a sensitizing agent, with failure occurring more rapidly than when stress is the only factor. Its general progression is very similar to that of low stress brittle failure in an inert environment. A craze comprising oriented fibrils forms at the base of a notch, scratch, or other surface irregularity. The crack advances when load-bearing fibrils at the tip of the crack break and new ones form at the advancing front of the craze. The final stage is ductile failure, which occurs when the stress field between advancing crack and the opposing surface of the sample exceeds the short-term yield stress.

Environmental stress cracking occurs faster than low stress brittle failure when the rate of molecular disentanglement is increased by local interactions between the polymer and the solvent. Swelling of the noncrystalline regions at the notch tip increases the local free volume. This plasticization accelerates segmental motion, which facilitates the formation of fibrils. The degree of plasticization depends upon the interaction between the stress cracking agent and the polymer; the better the match of the solvent's cohesive energy density with the polymer's solubility parameter, the greater will be the swelling [128]. Disengagement of the molecules comprising the fibrils and their subsequent rupture may also be accelerated. This occurs when the (relatively long) time required for diffusion of the stress cracking agent into the crystallites of the fibrils is less than the time required for them to rupture unaided [129].

REFERENCES

1. W Glenz, A Peterlin. J Macromol Sci Phys B4:473, 1970.
2. R Corneliussen, A Peterlin. Makromol Chem 105:193, 1967.
3. T J Bessell, R J Young. J Polym Sci Polym Lett Ed 12:629, 1974.
4. F Ania, F J Baltá Calleja, R K Bayer. Polym 33:233, 1992.
5. S G Burnay, G W Groves. J Mater Sci 12:1139, 1977.
6. T Katayama, T Amano, K Nakumura. Kolloid-Z Z Polym 226:125, 1968.
7. L Araimo, F De Candia, V Vittoria, A Peterlin. J Polym Sci Polym Phys Ed 16:2087, 1978.
8. S N Magonov, S S Sheiko, R A C Deblieck, M Möller. Macromolecules 26:1380, 1993.
9. J Clements, I M Ward. Polymer 24:27, 1982.
10. A Peterlin, R G Snyder. J Polym Sci Polymer Phys Ed 19:1727, 1981.
11. J Clements, R Jakeways, I M Ward. Polymer 19:639, 1978.
12. D T Grubb, K Prasad. Macromolecules 25:4575, 1992.
13. Y K Wang, D A Waldman, R S Stein, S L Hsu. J Appl Phys 53:6591, 1982.
14. G Capaccio, I M Ward. J Polym Sci Polym Phys Ed 20:1107, 1982.
15. R G C Arridge, P J Barham, A Keller. J Polym Sci Polym Phys Ed 15:389, 1977.
16. P J Barham, R G C Arridge. J Polym Sci Polym Phys Ed 15:1177, 1977.
17. R G C Arridge, P J Barham. Polymer 19:654, 1978.
18. H L Cox. Br J Appl Phys 3:72, 1952.
19. A Zwijnenburg, A J Pennings. J Polym Sci Polym Lett Ed 14:339, 1976.
20. P Smith, P J Lemstra, J P L Pijpers, A M Kiel Colloid Polym Sci 259:1070, 1981.
21. Y-L Hsieh, J Ju. J Appl Polym Sci 53:347, 1994.
22. N S J A Gerrits, R J Young, P J Lemstra. Polymer 31:231, 1990.
23. T Kanamoto, R S Porter. In: L A Kleintjens, P J Lemstra eds., Integration of Fundamental Polymer Science and Technology. New York: Elsevier, 1986, pp. 168–177.
24. G Graff, ed. Modern Plastics Encyclopedia. New York: McGraw-Hill, 1993 p. 209.
25. L-H. Wang, R S Porter, T Kanamoto. Polym Commun 31:457, 1990.
26. B Kalb, A J Pennings. J Mater Sci 15:2584, 1980.
27. G Cappaccio, T A Crompton, I M Ward. J Polym Sci Polym Phys Ed 18:301, 1980.
28. P Smith, P J Lemstra, H C Booij. J Polym Sci Polym Phys Ed 19:877, 1981.
29. T Ogita, R Yamamoto, N Suzuki, F Ozaki, M Matsuo. Polymer 32:822, 1991.

30. A Kaito, D Nakayama, H Kanetsuna. Polym. J 14:757 1982.
31. M A Hallam, D L M Cansfield, I M Ward, G Pollard. J Mater Sci 21:4199, 1986.
32. J P Pennings, D J Dikstra, A J Pennings. J Mater Sci 26:4721, 1991.
33. P Smith, P J Lemstra, J P L Pijpers. J Polym Sci Polym Phys Ed 20:2229, 1982.
34. V A Marachin, L P Mjasnikova, D Zenke, R Hirte, P Weigel. Polym Bull 12:287, 1984.
35. M A Hallam, G Pollard, I M Ward. J Mater Sci Lett 6:975, 1987.
36. A.J Pennings, J Smook. J Mater Sci 19:3443, 1984.
37. B Dessain, O Moulaert, R Keunings, A R Bunsell, J Mater Sci 27:4515, 1992.
38. V I Selikhova, L A Ozerina, A N Ozerin, N F Bakeyev. Vysokomol Soyed A28: 342, 1986.
39. S Kojima, C R Desper, R S Porter. J Polym Sci Polym Phys Ed 16:1721, 1978.
40. J Clements, G Capaccio, I M Ward. J Polym Sci Polym Phys Ed 7:693, 1979.
41. B Poulaert, J-C Chielens, C Vandenhende, J-P Issi, R Legras. Polym Commun 31: 158, 1990.
42. A G Gibson, D Greig, I M Ward. J Polym Sci Polym Phys Ed 18:1481, 1980.
43. C L Choy, F C Chen, K Young. J Polym Sci Polym Phys Ed 19:335, 1981.
44. G A Orchard, G R Davies, I M Ward. Polymer 25:1203 1984.
45. A G Gibson, I M Ward. J Mater Sci 14:1838, 1979.
46. Z Zhou, N Brown. Polymer 35:1948, 1994.
47. P Smith, P J Lemstra. J Mater Sci 15:505, 1980.
48. US Patents 4,344,908 Smith P, Lemstra P J, Aug 17, 1982; 4,422,993 Smith P, Lemstra P J, Dec 27, 1983; 4,430,383 Smith P, Lemstra P J, Feb 7, 1984; 4,436,689 Smith P, Lemstra P J, Kirschbaum R, Rijpers J P L, Mar 13, 1984.
49. J Smook, A J Pennings. Polym Bull 10:291, 1983.
50. US Patent 4,668,717. Lemstra P J, Meijer H E H, van Unen L H T, May 26, 1987.
51. C Sawatari, T Okumura, M Matsuo. Polym J 18:741 1986.
52. P J Barham. Polymer 23:1112, 1982.
53. P Smith, P J Lemstra. Polymer 21:1341, 1980.
54. P Smith, P J Lemstra. Makromol Chem 180:2983, 1979.
55. B Kalb, A J Pennings. Polym Commun 21:3, 1980.
56. T Kanamoto, E S Sherman, R S Porter. Polym J 11:497 1979.
57. A E Zachariades, M P C Watts, T Kanamoto, R S Porter. J Polym Sci Polym Lett Ed 17:485, 1979.
58. A G Gibson, I M Ward, B N Cole, B Parsons. J Mater Sci Lett 9:1193, 1974.
59. P D Griswold, A E Zachariades, R S Porter. Polym Eng Sci 18:861, 1978.
60. T Kanamoto, A Tsuruta, K Tanaka, M Takeda, R S Porter. Macromolecules 21: 470, 1988.
61. T Kanamoto, A Tsuruta, K Tanaka, M Takeda, R S Porter. Polym J 15:327, 1983.
62. T Kanamoto, T Ohama, K Tanaka, M Takeda, R S Porter. Polymer 28:1517, 1987.
63. P J Lemstra, N A J M van Aerle, C W M Bastiaansen. Polym J 19:85, 1987.
64. K Furuhata, T Yokokawa, K Miyasaka. J Polym Sci Polym Phys Ed 22:133, 1984.
65. K Ishikawa, K Miyasaka, M Maeda. J Polym Sci A-2, 7:2029, 1969.
66. P Smith, H D Chanzy, B P Rotzinger. Polym Commun 26:258, 1985.
67. G Capaccio, T J Chapman, I M Ward. Polymer 16:469, 1975.
68. G Capaccio, T A Crompton, I M Ward. Polymer 17:644 1976.

69. A Keller, F M Willmouth. J Macromol Sci-Phys B6:493 1972.
70. A J Pennings, A Zwijnenburg, R Lageveen. Kolloid-Z Z Polym 251:500, 1973.
71. A Zwijnenburg, A J Pennings. Colloid Polym Sci 253:452, 1975.
72. F C Frank, A Keller, M R Mackley. Polymer 12:467, 1971.
73. P J Barham, A Keller. J Mater Sci 20:2281, 1985.
74. M J Hill, P J Barham, A Keller. Colloid Polym Sci 258:1023, 1980.
75. M R Mackley, A Keller. Polymer 14:16, 1973.
76. J A Odell, D T Grubb, A Keller. Polymer 19:617, 1978.
77. P D Griswold, R S Porter, C R Desper, R J Farris. Polym Eng Sci 18:537, 1978.
78. A Kaito, N Nakayama, H Kanatsuna. J Appl Polym Sci 30:1241, 1985.
79. M J Shankernarayanan, D C Sun, M Kojima, J H Magill. Int Polym Proc 1:66, 1987.
80. M R Mackley, S Solbai. Polymer 28:1115, 1987.
81. D G Peiffer. Polym Eng Sci 20:167, 1980.
82. A Peterlin. J Polym Sci Polym Symp 9:61, 1965.
83. A Peterlin. J Polym Sci Polym Lett Ed 18:123, 1967.
84. A Peterlin. Kolloid-Z Z 233:857, 1969.
85. A Peterlin. J Polym Sci Polym Phys Ed 7:1151, 1969.
86. A Peterlin. J Mater Sci 6:490, 1971.
87. R J Young, P B Bowden, J M Ritchie, J G Rider. J Mater Sci 8:23, 1973.
88. R J Young, P B Bowden. Phil Mag 29:1061, 1974.
89. R J Young. Phil Mag 30:85, 1974.
90. R J Young. Mater Forum 11:210, 1988.
91. B Crist. Polym Commun 30:69, 1989.
92. B Crist In: E L Thomas, ed. Structure and Properties of Polymers, Vol. 12. Weinheim: VCH, 1993.
93. P J Flory, D Y Yoon. Nature 272:226, 1978.
94. K Sakaoku, A Peterlin. Makromol Chem 108:234, 1967.
95. T Juska, I R Harrison. Polym Eng Rev 2:13, 1982.
96. T Juska, I R Harrison. Polym Eng Sci 22:766, 1982.
97. T Liu, I R Harrison. Polym Eng Sci 28:1162, 1988.
98. R Popli, L Mandelkern. J Polym Sci Polym Phys Ed 25:441, 1987.
99. A J Peacock, L Mandelkern. J Polym Sci Polym Phys Ed 28:1917, 1990.
100. M A Kennedy, A J Peacock, L Mandelkern. Macromol 27:5297, 1994.
101. A N Gent, S Madan. J Polym Sci Polym Phys Ed 27:1529, 1989.
102. M Nakamura, S M Skinner. J Polym Sci 89:423, 1955.
103. C G'Sell, N A Aly-Helal, S L Semiatin, J J Jonas. Polymer 33:1244, 1992.
104. C J Farrell, A Keller. J Mater Sci 12:966, 1977.
105. G D Wignall W Wu. Polymer Commun 24:354, 1983.
106. A J Peacock. Personal observation.
107. N W Brooks, R A Duckett, I M Ward. Polymer 33:1872 1992.
108. R Seguela, F Rietsch. J Mater Sci Lett 9:46, 1990.
109. J C Lucas, M D Failla, F L Smith, L Mandelkern, A J Peacock. Polym Eng Sci 35:1117, 1995.
110. M A Kennedy, A J Peacock, M D Failla, J C Lucas, L Mandelkern. Macromolecules 28:1407, 1995.

111. W W Adams, D Yang, E L Thomas. J Mater Sci 21:2239 1986.
112. B M Ginzburg, S H Tuichiev. J Macromol Sci Phys B31:291, 1992.
113. G Capaccio, I M Ward. J Polym Sci Polym Phys Ed 22:475, 1984.
114. P J Mills, J N Hay, R N Haward. J Mater Sci 20:501 1985.
115. Y Termonia, P Smith. Macromolecules 20:835, 1987.
116. Y Termonia, P Smith. Macromolecules 21:2184, 1988.
117. Y Termonia, S R Allen, P Smith. Macromolecules 21:3485 1988.
118. Y Termonia, P Smith Colloid. Polym Sci 270:1085, 1992.
119. Y Termonia, P Smith. Macromolecules 26:3738, 1993.
120. P Smith. Macromolecules 16:1802, 1983.
121. T M Malik, P J Carreau, H P Schreiber, A Rudin, W Tchir. Int Polym Proc, Vol. V, 42, 1990.
122. R Seguela, F Rietsch. Polymer 27:703, 1986.
123. A J Peacock, L Mandelkern, R G Alamo, J G Fatou. J Mater Sci 33:2255, 1998.
124. X Lu, X Wang, N Brown. J Mater Sci 23:643, 1988.
125. Y-L Huang, N Brown. J Polym Sci Polym Phys Ed 39:129, 1991.
126. J J Lear, P H Geil. Int J Polym Mater 15:147 1991.
127. X Wang, N Brown. Polymer 30:1456, 1989.
128. H R Brown. Polymer 19:1186, 1978.
129. A L Ward, X Lu, Y Huang, N Brown. Polymer 32:2172 1991.
130. L R G Treloar. Polymer 1:95, 1960.
131. T Shimanouchi, M Ashina, S Enomoto. J Polym Sci 59:93, 1962.
132. R F Schantele, T Shimanouchi. J Chem Phys 47:3605, 1967.
133. I Sakurada, T Ito, K Nakamae. J Polym Sci C15:75, 1966.
134. G Wobser, S Blasenbrey. Kolloid-Z Z 241:985, 1970.
135. F C Frank. Proc Roy Soc Lond A319:127, 1970.
136. D S Boudreaux. J Polym Sci Polym Phys Ed 11:1285, 1973.
137. J L Koenig, D L Tabb. J Macromol Sci Phys B9:141, 1974.
138. G R Strobl, R Eckel. J Polym Sci Polym Phys ed 14:913, 1976.
139. J Matsuo, C Sawatari. Macromolecules 19:2036, 1986.
140. R P Wool, R S Bretzlaff, B Y Li, C H Wang, R H Boyd. J Polym Sci Polym Phys Ed 24:1039 (186.
141. A S Deazle, B J Howlin, C Lekakou, G J Buist, J R Jones, J M Barton. J Chem Cryst 24:17, 1992.
142. A Kurihara, H Chuman, S Enomoto. In: M Doyama, J Kihara, M Tanaka, R Yamamoto eds., Computer Aided Innovation of New Materials II. New York: Elsevier, 1993, 1443–1446.

9
Use and Fabrication of Polyethylene Products

I. USES OF POLYETHYLENE PRODUCTS

Polyethylene resins, with their wide range of physical properties, find outlets in an extensive array of manufactured goods. The key to the adaptability of polyethylene lies in its tunable semicrystalline morphology, which can be controlled by manipulating molecular and processing variables. Toughness, hardness, clarity, and other physical characteristics can be regulated by altering average molecular weight, comonomer type, and comonomer content. Resins suited to most commercial thermoplastic fabrication processes can be created by controlling average molecular weight, molecular weight distribution, and branching characteristics. Manipulation of the material prior to and during crystallization provides another way of influencing ultimate properties. Polyethylene resins can thus be adapted to many end uses by virtue of both their physical properties and processing their characteristics. From an economic standpoint, the generally low price of polyethylene resins can give them a competitive edge compared to other materials (both polymeric and nonpolymeric) that adequately meet the desired end use requirements.

Table 1 of Chapter 1 summarizes the key properties of the various classes of polyethylene with respect to one another. When perusing this table it must be borne in mind that there is much overlap between the properties of the different classes. Property measurements depend upon the precise molecular characteristics, processing conditions, and testing methods employed.

In the following sections the key attributes of the four main classes of polyethylene are outlined. Examples of applications to which each type is suited are given. The bulk of the chapter is devoted to outlining the most important fabrication and finishing processes and products available from them. The requirements for each application are discussed in terms of how the properties of the different classes of polyethylene fit them to a given end use. The final section provides a breakdown of the principal markets of polyethylene in terms of type, fabrication technique, and use.

A. High Density Polyethylene

High density polyethylene consists of molecules that are essentially linear, typically with fewer than one branch per 200 carbon atoms in the backbone. The linearity of these resins permits the development of high degrees of crystallinity, which endow them with the highest modulus and lowest permeability of all the classes of polyethylene. This combination makes them suitable for certain medium- and large-scale liquid holders such as drums, tight-head pails, and chemical storage tanks. On a smaller scale, a combination of stiffness, low permeability, and high environmental stress crack resistance makes high density polyethylene admirably fit for bottles to contain household, industrial, and automotive chemicals such as liquid detergent, bleach, motor oil, and antifreeze. Such attributes are also valuable in thin-walled food containers such as milk bottles and margarine tubs. The low glass transition temperature of polyethylene suits it for freezer applications such as ice cream containers. High crystallinity has the drawback of causing opacity and imparting a matte surface finish; therefore, high density polyethylene bottles and moldings are often pigmented to render them more attractive. Low permeability, corrosion resistance, and stiffness are desirable pipe attributes; water, sewer, and natural gas pipes are the principal outlets. High density polyethylene's good tensile strength suits it for short-term load-bearing film applications such as grocery sacks, general merchandise bags, and commercial trash can liners. Its acceptable stiffness, durability, and lightness make it suitable for a variety of household and commercial low-load-capacity applications, including crates, pallets, and pails. Other domestic applications include pharmaceutical bottles, cosmetic containers, and general storage containers. Reasonable toughness coupled with good rigidity make high density polyethylene the resin of choice for many toys. High density polyethylene's low glass transition temperature and satisfactory heat deflection temperature and its combination of reasonable stiffness and acceptable toughness suit it for nonstructural exterior uses such as lawn furniture, playground equipment, and trash cans. When very large polyethylene parts are to be produced, high density resins are often selected because of their superior stiffness, which gives them the ability to retain their shape over broad expanses, such as in municipal garbage cans, storage tank covers, and canoes. An added advantage in such applications is high abrasion resistance. Fabricated items may be cross-linked to further improve their resistance to chemical and physical abuse in such applications as chemical storage tanks and whitewater kayaks. The chemical resistance and low moisture permeability of high density polyethylene sheeting are exploited in its use as a sheet liner for liquid and solid waste containment pits. All in all, high density polyethylene finds use in an enormously broad range of applications in many aspects of everyday life.

B. Low Density Polyethylene

Because of the radical nature of the high pressure process by which low density polyethylene resins are made, they contain many long and short branches. The numerous short-chain branches effectively reduce the degree of crystallinity far below that of high density polyethylene, resulting in a flexible product with a low melting point. Long-chain branches confer desirable non-Newtonian rheological characteristics, high melt strengths combined with relatively low viscosities at the shear rates encountered during processing. These rheological characteristics eminently suit low density polyethylene to the film-blowing process. Blown film is the principal outlet of this class of polyethylene, accounting for more than half of all usage. Low density polyethylene's low crystallinity results in films that are transparent, soft to the touch, and moderately tough. The ready deformability and high creep of such films makes them unsuitable for high load applications or situations in which prolonged low level stress is encountered. Principal uses of low density polyethylene films include commercial and retail packaging applications. Other applications include diaper backing, shrink-wrap, dry cleaning bags, moisture barriers in construction, agricultural groundcover, and greenhouse skins. Low density polyethylene can be drawn down to form very thin films that may be coated directly onto cardboard. The resulting product is a waterproof and heat-sealable composite that is widely used in juice and milk cartons. Minor uses of low density polyethylene include wire and cable insulation and flexible pipe. Injection- and blow-molded items made from this resin are flexible and reasonably tough, suiting them for such applications as squeeze bottles and food storage containers. The admirable rheological characteristics of molten low density polyethylene make it valuable as an additive to improve the processing characteristics of other less tractable polyethylene resins such as high density and linear low density polyethylene.

C. Linear Low Density Polyethylene

The generic classification of linear low density polyethylene covers a spectrum of ethylene-*co*-1-alkenes, ranging from clear materials with stiffnesses similar to that of low density polyethylene to rigid opaque materials that share many of the characteristics of high density polyethylene. It is the modulus of a resin that largely determines its suitability for a given application. Modulus depends on the degree of crystallinity, which is controlled by the level of comonomer incorporation and is reflected in the resin's density. The majority of linear low density polyethylene resins fall within the density range encompassed by low density polyethylene, and thus they share many of the same markets. In the realm of film, which is its largest outlet, linear low density polyethylene distinguishes itself by superior toughness, as measured by tear strength, impact resistance, and puncture resistance. Linear low density polyethylene films are used in many packaging and nonpackaging applications, including grocery sacks, fresh produce packages, stretch-wrap, domestic trash can liners, and scientific balloons. It is also extruded to form wire and cable insulation, pipes, and sheet for use where the stiffness of high density polyethylene is not required. Injection molding is used to convert linear low density polyethylene into such items as food container lids and toys, where flexibility combined with toughness is needed. On a larger scale, it is used for food processing containers, storage tanks, and highway barriers.

Very low density polyethylenes are a subset of linear low density polyethylene in which the comonomer content is so high that crystallization is largely suppressed. Such materials are also known as ultralow density polyethylene or plastomers. The limited crystallinity of these resins results in low levels of stiffness and high clarity. These materials are used where transparency, softness, strain recovery, and toughness are at a premium; such applications include medical tubing, meat packaging, and diaper backing.

D. Ethylene-Vinyl Acetate Copolymer

The numerous short-chain hydrocarbon branches and acetate branches in ethylene-vinyl acetate copolymers limit the development of crystallinity. The resulting materials have low modulus and good clarity. Long-chain branches endow these copolymers with non-Newtonian rheological characteristics similar to those of low density polyethylene. The bulky acetate side groups inhibit the sliding of chains past one another during deformation, resulting in good strain recovery compared to low density polyethylene. The high branch content of these polymers results in low lamellar thicknesses, which translate to low melting and processing temperatures. Ethylene-vinyl acetate copolymers are used primarily in packaging films. In such applications, their flexibility, toughness, elasticity, and clarity are desirable attributes. Outlets for such products include meat packaging and stretch-wrap. The other main use of ethylene-vinyl acetate copolymers is as a component of adhesives.

II. PRINCIPAL FABRICATION PROCESSES

The goal of all polyethylene fabrication processes is to convert pellets or powder into a usable solid-state product via manipulation of the fluid state. The majority of conversion processes include the steps of melting, homogenizing, shaping, and

Use and Fabrication of PE Products 463

cooling. The most important processing technique is extrusion, which is used to create products directly via the continuous processes of film blowing, film casting, profile forming, and substrate coating. Extrusion is also incorporated as a preliminary step in the major fabrication processes of blow molding and injection molding. The other principal polyethylene conversion process is rotational molding. Subsequent finishing steps may involve localized melting, manipulation, and cooling. The subsidiary processing techniques of cross-linking and foaming can be applied to many of the primary fabrication processes. Sintering and gel spinning, which account for a tiny fraction of polyethylene use, are most often applied to ultrahigh molecular weight polyethylene.

In this section the principal fabrication processes pertaining to polyethylene are outlined. End use products are discussed in terms of how the attributes of various polyethylene resins suit them to particular conversion processes and applications. It is not intended to provide a detailed discussion of the theory or operation of processing equipment. Those wishing to inquire further into these aspects of polymer processing should consult the bibliography at the end of the chapter.

A. Extrusion

An extruder is basically an Archimedean screw designed to handle granular solids and viscous melts at high temperatures. Extruders serve three purposes: They melt, homogenize, and transport resin. The end result is a supply of molten resin suitable for forming into useful products.

1. Principles of Extrusion

The principal components of a single-screw extruder are shown in Figure 1. Polyethylene pellets (mixed with appropriate additives, antioxidants, pigments, etc.) are fed by gravity from the hopper into the rear (feed section) of the extruder barrel. This zone may be cooled to prevent premature melting of the polymer, which could block the throat of the hopper. Rotation of the screw pushes the resin forward into the compression zone. Here it is melted by a combination of mechanical shearing and heat from the barrel heaters. The screw in this zone is often internally cooled to facilitate the transport of molten polymer adhering to the barrel wall. In the compression zone the depth of the screw channel decreases, ensuring consolidation of the molten polymer and the exclusion of trapped air. Polyethylene resins obtain most of the energy required to melt them from mechanical work. The processing of resins that exhibit relatively little shear thinning, such as linear low density polyethylenes (especially those made with metallocene catalysts), may involve the input of so much mechanical work that cooling, rather than heating, of the outer jacket is required. The shearing and compaction

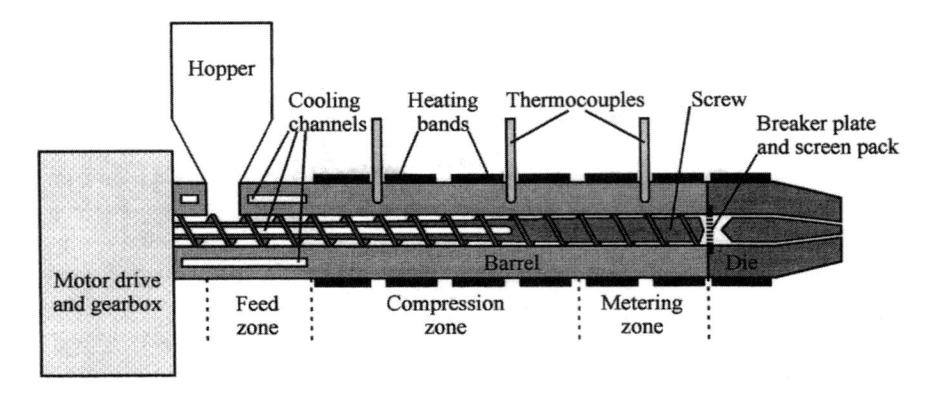

Figure 1 Principal components of a single-screw extruder.

that occur in the compression zone serve to homogenize the melt, ensuring that additives are uniformly dispersed and that molecular composition variation is minimized. When intensive mixing is required, short mixing sections may be incorporated into the screw. Such mixing sections consist of rows of pins, parallel interrupted mixing flights, reverse flights or various other patented or proprietary devices. The final portion of the barrel is the metering zone, which pumps the homogenized resin to its outlet. Ideally, the profile of an extruder screw would be matched to the rheological characteristics of the resin being processed. In practice, this can be achieved only in equipment dedicated to the extrusion of a particular resin under a given set of extrusion conditions. A breaker plate consisting of a perforated metal disk limits the output, increasing pressure in the barrel to ensure adequate mixing. The breaker plate also supports the screen pack, which consists of a stack of metal mesh disks that serves as a filter to remove particulate contaminants.

2. Products of Extrusion

The molten output of an extruder may be shaped in a variety of ways to yield useful end products of indefinite length. Film is formed by bubble blowing and chill roll casting. The products of these processes may also be referred to as tubular and flat film, respectively. Chill roll casting is also used to produce sheeting. Profiles, such as pipes and conduits, are produced by extrusion through cooled dies that have appropriately shaped orifices. Coatings are applied as thin films to substrates such as paper and cardboard. Coating technology is also used to insulate wire and cable.

B. Film Blowing

Film blowing involves the inflation and stretching of a molten tube of polymer. Once reduced to the correct thickness, the tube is cooled, flattened, and wound up onto rolls.

1. Process

The basic equipment used in film blowing is illustrated schematically in Figure 2. The output from an extruder is turned 90° to flow upward through an annular die that produces a tube of molten polymer. The upper end of this tube is pinched flat between a pair of rollers, typically mounted several tens of feet above the die. A constant volume of air trapped within the tube inflates it to form a bubble. As the tube cools, it is inflated and drawn upward simultaneously. The pinch rolls are driven at a speed in excess of the extrusion speed at the die exit, imparting vertical orientation in the material flow direction. The ratio of the bubble diameter to the die diameter, which is known as the "blow-up ratio" (BUR), is controlled by the volume of air entrapped. Bubble inflation applies hoop stress, which imparts transverse orientation perpendicular to the material flow direction. When the molten polymer exits the die it is clear, but this changes as it cools and crystallizes. Cooling is aided by a stream of air applied to its outer surface from an air ring mounted above the die. The newly formed crystallites scatter light, increasing haze noticeably, even as the thickness of the film is drastically reduced. The onset of crystallization is quite sharp, resulting in a clearly defined "frost line," the height of which above the die is known as the "frost line height" (FLH). Once the flattened tube of film leaves the pinch rolls, it passes around a series of rollers designed to maintain constant tension. The product may be wound up directly onto a roll as a flattened tube, or, alternatively, the creases on either side of the tube may be sliced off, yielding two films that may be slit into narrower widths before being wound up independently. Bubbles with diameters ranging from a few inches up to more than 30 ft can be produced by this process.

Many variations on the basic film-blowing process exist. Some of the more common variations include the use of rotating dies and coextrusion. Rotating dies serve to even out thickness variations around the circumference of the film, ensuring that rolls of film have no cumulative thick spots that would cause distortion. Coextrusion is used when it is desired to produce a film made up of layers of different polyethylene resins or of polyethylene in combination with other thermoplastics. In coextrusion, two or more separate extruders feed a single die designed in such a way that their outputs are combined concentrically.

The thickness of a film is controlled by extruder output rate, die gap, blow-up ratio, and take-up rate. The die gap is changed by raising or lowering the conical core of the die. By appropriate manipulation of the blow-up ratio and the take-up rate, it is possible to generate films that have the same thickness but

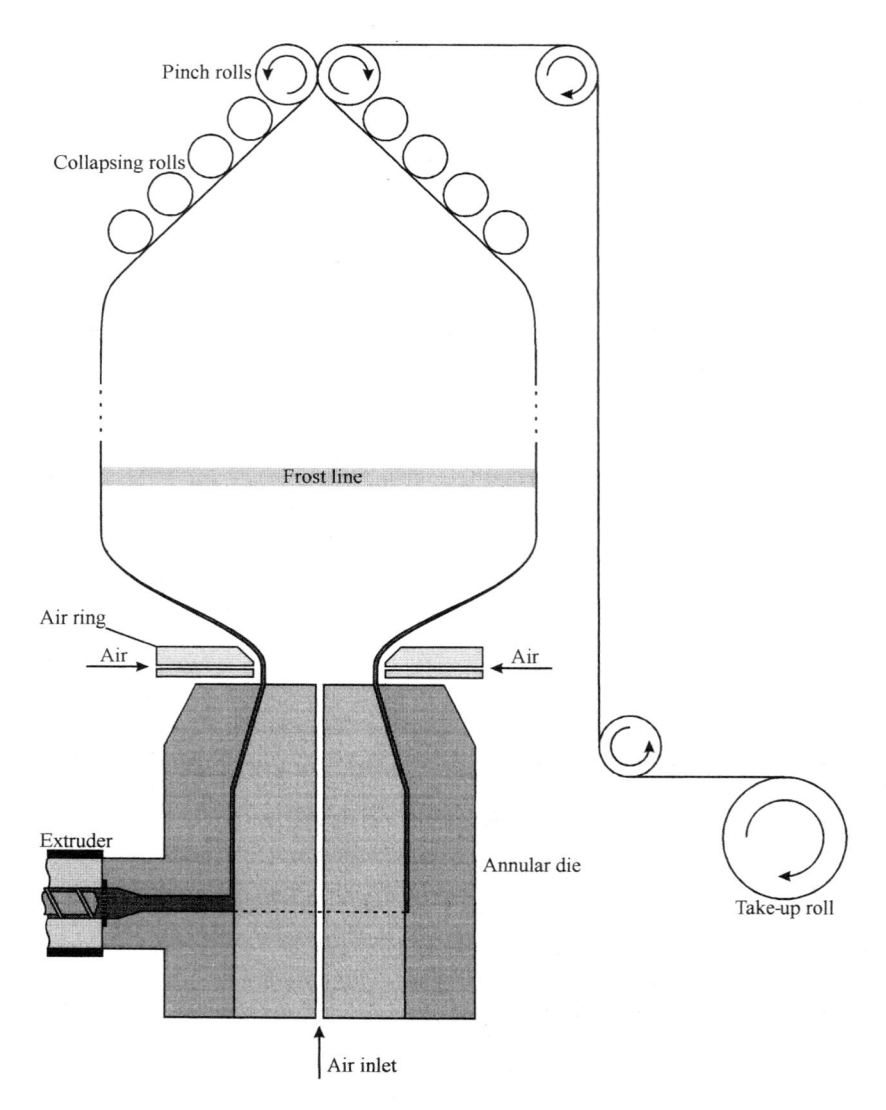

Figure 2 Principal components of film blowing equipment.

Use and Fabrication of PE Products 467

in which preferential molecular orientation is either parallel with or perpendicular to the machine direction of film travel during fabrication. This controls the principal tensile and tear strength directions of the film. (It is convenient to refer to such properties with respect to the "machine direction," which is parallel with the material flow during fabrication. The perpendicular direction is known as the "transverse" or "cross" direction.) If the primary orientation is parallel with the machine direction, the tensile strength will be greatest in the machine direction, but resistance to tear will be greatest in the transverse direction (by convention, tear strength is stated in terms of the direction of rupture propagation). As the orientation level increases, the disparity between the machine and transverse direction properties increases, whereas the impact and puncture resistance decrease.

Orientation is also influenced by the time it takes the film to crystallize. The longer the crystallization time, the greater will be the relaxation of molecular orientation in the molten film. Therefore, a short crystallization time will result in greater film orientation than a longer time. The crystallization time is controlled by the die temperature, the temperature and pressure of the air issuing from the air ring, the ring's height above the die, and the thickness of the molten film. The frost line height may be used as an approximate indicator of the crystallization time—higher for long crystallization times, lower for short times. A resin's rate of relaxation is a function of its molecular characteristics; a high molecular weight or broad molecular weight distribution increases relaxation time and results in increased orientation. A high molecular weight tail is especially efficacious in increasing film orientation.

The rheological characteristics of a resin strongly influence its processability. Increased melt index (decreased average molecular weight) reduces motor load and decreases power consumption. However, reduced molecular weight lowers melt strength (decreasing bubble stability), toughness, and tensile strength. Resins with a melt index between 0.5 and 2 generally give the best balance of film and processing characteristics. Resins that exhibit substantial shear thinning rheological behavior are more readily extruded and have greater bubble stability than those of comparable weight-average molecular weight that are less shear thinning. Shear thinning is enhanced by a broad or bimodal molecular weight distribution and long-chain branching. Thus, low density polyethylene, with its broad molecular weight distribution and long-chain branches, is blown into film more readily than high density or linear low density polyethylene. The new metallocene type of linear low density polyethylene resins, which have narrow molecular weight distributions, are more difficult to extrude and blow into films than conventional linear low density polyethylene resins. The incorporation of long-chain branches into metallocene products improves their processability.

In its virgin state, polyethylene film has a low energy surface that inks do not readily wet or adhere to. Corona treatment (described in Chapter 7) increases a film's surface energy and thus improves its printability.

2. Products

The attributes of a blown film depend upon the interaction between the resin's molecular characteristics and processing conditions. In general, physical properties are influenced by processing conditions within limits established by molecular characteristics. The key physical properties of polyethylene films include machine and transverse direction tear strengths, impact resistance (high speed penetration), puncture resistance (low speed penetration), and machine and transverse direction tensile strengths. Other attributes, such as clarity, gloss, permeability, and cling, may also assume importance, depending upon end use requirements.

The presence of modest amounts of short-chain branching improves the toughness of linear low density polyethylene films relative to those made from high density polyethylene resins having a similar molecular weight and molecular weight distribution. Such films are also tougher than low density polyethylene films having equivalent melt indices and densities. Ethylene-hexene and ethylene-octene copolymer films are tougher than those made from ethylene-butene copolymers. Metallocene-catalyzed linear low density polyethylene films have better clarity and are more impact-resistant than those made from conventional linear low density polyethylene. Low levels of long-chain branching in metallocene-based linear low density polyethylene films improve their tear strength.

Polyethylene film is used in many diverse applications, the most common of which is packaging. Bags of various descriptions are made from low density, linear low density, and high density polyethylene films. High strength-to-weight ratio, toughness, flexibility, general barrier properties (especially impermeability to water), and heat sealability make them ideal for general-purpose applications such as merchandise bags, trash bags, and box liners. When greater load-bearing capacity is required, such as in grocery sacks, more crystalline resins, with their higher yield stresses and greater tensile strengths, are used. For less physically demanding applications, such as dry cleaning bags and newspaper bags, lower density films are quite acceptable. Linear low density polyethylene films are used when the highest degree of toughness is required. In specific applications, other properties of polyethylene come to the fore. Clarity is desirable in many food packaging applications, requiring the use of low crystallinity films in such applications as bread bags and fresh produce packages. In the case of fresh produce packaging, the relatively high oxygen permeability of very low density polyethylene films helps maintain product freshness. In low temperature applications, such as freezer bags, the low glass transition temperature of polyethylene favors

Use and Fabrication of PE Products 469

low temperature flexibility and toughness. The excellent low temperature physical properties of ethylene-vinyl acetate copolymer films suit them for use in ice bags. For packaging dry foods such as rice, dried fruit, and cereal, the heat sealability of polyethylene permits the use of the "form, fill, and seal" process, in which packages are created, filled, and sealed continuously on a single machine. In such applications, a stiffer film permits faster throughput, favoring the use of high density polyethylene, which has the additional advantage of reduced moisture vapor transmission rates. To enhance sealability, a thin layer of ethylene-vinyl acetate copolymer may be coextruded with the high density polyethylene. The melting range of the ethylene-vinyl acetate copolymer is relatively broad, occurring at a temperature approximately 40°C lower than that of high density polyethylene.

Stretch-wrap (cling) films used in commercial, retail, and domestic applications must meet a demanding combination of requirements. They must stretch and conform to their contents, cling to themselves, be strong enough to maintain the integrity of their contents, and resist puncture and tearing. A low molecular weight additive that blooms to the surface (such as polyisobutylene) may be added to improve cling, but increasingly, low crystallinity is relied upon to ensure adequate cling. When wrapping items stacked on pallets, load retention after stretching is paramount; for such applications linear low density polyethylene films are generally used. Ethylene-vinyl acetate copolymer meets the requirements of retail meat wrapping, which requires a clear, tough film that is impermeable to grease, fat, and aqueous liquids.

Shrink-wrap films require good heat sealability and biaxial shrinkage to encapsulate the contents of packages. In addition, they must also display good puncture and tear resistance, especially at the elevated temperature required for shrinking. Low density polyethylene films are generally used in this application. When greater clarity is required, metallocene-based linear low density polyethylene may be used.

In its broader widths, polyethylene film is used as a barrier material in industrial, construction, and agricultural applications. Polyethylene film is used as a moisture barrier in walls and floors. In addition to their barrier properties, such films must be tough and abrasion-resistant in order to avoid holing, which would nullify their barrier properties. Highly stabilized polyethylene film is used in the construction of greenhouses, in which case puncture and tear resistance is a requirement. Carbon black filled films are used as ground cover to limit the growth of weeds under fruits and vegetables. Impermeability to moisture is also advantageous when polyethylene film is used in diaper construction.

On a lighter note, low density and linear low density polyethylene films are used in high altitude scientific balloons. In this application, low temperature ($-80°C$) toughness is a major consideration.

An especially tough film is created from strips slit from polyethylene film on a bias. Two layers of such strips are laid down at an angle to one another, then fused together by heat sealing. Such films are used in express letter envelopes.

C. Chill Roll Casting

Chill roll casting is the process by which a flat molten sheet of polymer is drawn uniaxially and quenched against a metal drum. Both film and sheets are made by this process.

1. Process

The basic equipment used in chill roll film casting is shown schematically in Figure 3. The output from an extruder is turned 90° to flow downward through a slit die, which may approach 20 ft in width. The sheet of molten polymer is drawn into contact with the surface of a large roller, the interior of which is cooled by water. Intimate contact with the chill roll is ensured by the use of an "air knife," which holds the molten polymer against the roller with a wide jet of air spanning the width of the roller. After passing around the chill roll, where it crystallizes, the film is drawn between a pair of nip rolls. At this point the film is somewhat narrower than the width of the die. Thereafter the slightly thickened edges are trimmed from the film, and it is slit into the desired widths before being wound up. An assortment of other rollers are also used to provide extra cooling,

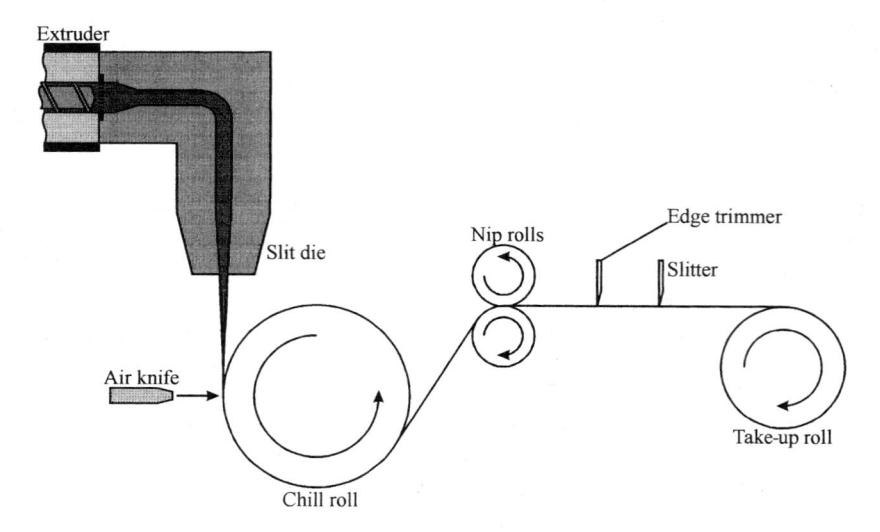

Figure 3 Principal components of chill roll casting equipment.

maintain film tension, or heat treat the film. Coextruded films are produced when the outputs of two or more extruders are fed into the die in such a way that laminar flow maintains separate layers.

The thickness of the cast film is controlled by the extruder output, the die gap, and the amount of draw between the die and the chill roll. The die gap is regulated by a series of adjusters across the width of the die, each of which may be adjusted independently to ensure uniform film thickness. Film thickness is scanned continuously during production, using a noncontacting β-radiation gauging device. The die gap may be adjusted automatically on the basis of feedback from the gauging device. The amount of draw experienced by the molten polymer is controlled by the rate of rotation of the nip rolls and chill roll relative to the die output.

2. Products

The interaction between film casting conditions and resin characteristics is very similar to that encountered during film blowing. Film properties may be tailored by the judicious selection of resin parameters and processing conditions. Cast films are generally made from a lower molecular weight resin than blown films, their melt indices typically falling in the range of 2.5–6.

Films made by chill roll casting may be used in most of the applications described above for blown films. Some differences are encountered, owing to the fact that cast films are inevitably uniaxially oriented in the machine direction whereas blown films are typically biaxially oriented. Faster quenching rates are encountered in chill roll casting than in film blowing, resulting in lower crystallinity films that have greater clarity and lower stiffness.

In addition to film production, chill roll casting yields polyethylene sheeting of nominal thicknesses in excess of 0.012 in. In general, sheeting is not subjected to the extreme drawdown experienced by thin film. Thus, sheets are typically substantially less oriented than cast films. Polyethylene sheeting finds its greatest outlet as geomembranes, which are principally used to line and cover landfill sites and as pond liners in mining and industrial applications. Liners several acres in extent can be created by joining the edges of adjacent sheets. Joints may be created by heat-sealing or welding; the necessary heat is provided electrically, ultrasonically, or by a stream of heated inert gas. The majority of geomembranes have thicknesses in the range of 0.06–0.08 in. In general, the more permanent the installation and the more hazardous the contents, the thicker the liner. In extreme cases, sheets as thick as 0.25 in. may be used for containment purposes. Such sheets are typically made of linear low density polyethylene, which has the flexibility, toughness, and chemical resistance typically required in such cases. Other uses for polyethylene sheeting include pickup truck bed liners, dunnage trays, and garden pond liners, all of which are produced by thermoforming.

D. Profile Forming

Profile forming is the process by which the molten output from an extruder is shaped and cooled by an extrusion die to form a continuous product with a constant cross-sectional profile. In the case of polyethylene, products are predominantly tubular, with relatively thick walls in comparison to their diameter.

1. Process

The basic equipment used in pipe extrusion is illustrated in Figure 4. The output from an extruder is fed into an annular extrusion die, which is typically coaxial with the extruder screw. The die consists primarily of an outer barrel and an inner "torpedo" or mandrel, which is supported by several narrow streamlined radial fins known as a "spider." Molten polymer flows around the torpedo and out the end of the die, its profile conforming to the shape of the annulus. The length of the die must be great enough that the stream of polymer divided by the spider can "heal" before it leaves the die. If this requirement is not met, "weld lines" corresponding to the fins of the spider will weaken the product. When the molten cylinder emerges from the die, its diameter is larger than that of the desired product. To reduce it to the correct dimensions it is subjected to a sizing step during which it is cooled to the solid state. The molten polymer is drawn into the calibrator by hauling on the solidified portion of the pipe with a traction device. Tubing (nominally less than 1 in. in diameter) and flexible pipe is wound onto drums, whereas rigid products are cut to the desired length using a saw that travels with the product.

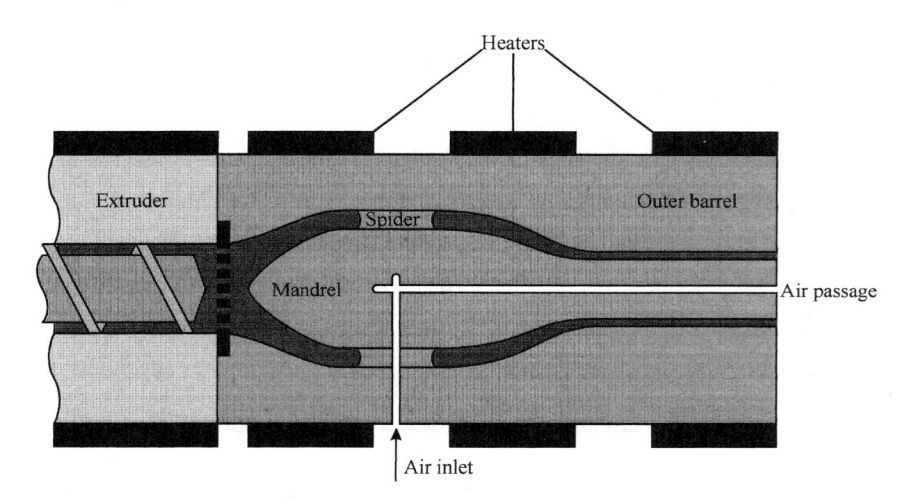

Figure 4 Principal components of pipe extrusion equipment.

Use and Fabrication of PE Products

The use of a spider and the problems associated with weld lines can be avoided if a crosshead or an offset die is used. In these dies the molten polymer flow enters from one side and envelops the mandrel directly. One such extrusion head is illustrated in Figure 5. These types of dies are typically used in the extrusion of small diameter tubing or when an internally cooled extended mandrel is used.

Three variations of size calibration are employed, depending on the end use and the diameter of the product; these are illustrated in Figures 6–8. External calibration is used when the external diameter and outer surface finish are paramount. To prevent the molten cylinder from collapsing, it may be inflated, a floating plug being used to contain the air at the downstream end. Internal calibration, over an extended tapered mandrel, is used when a smooth inner wall is required. For small tubing, the molten output may be drawn through a series of progressively smaller sizing plates. Corrugated pipe is obtained by inflating the molten polymer within a two-part female die that is closed around the product as it emerges from the extrusion die.

The wall thickness of tubes and pipes is controlled by a combination of the dimensions of the die annulus, the extruder output rate, and the haul-off rate. Uniform wall thickness is ensured by adjusting the position of the outer barrel so that it is concentric with the mandrel.

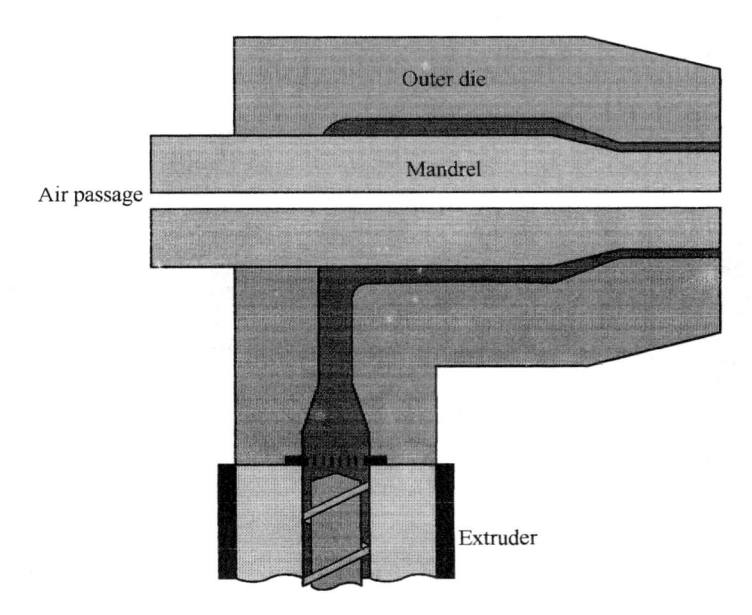

Figure 5 Crosshead type of pipe extrusion die.

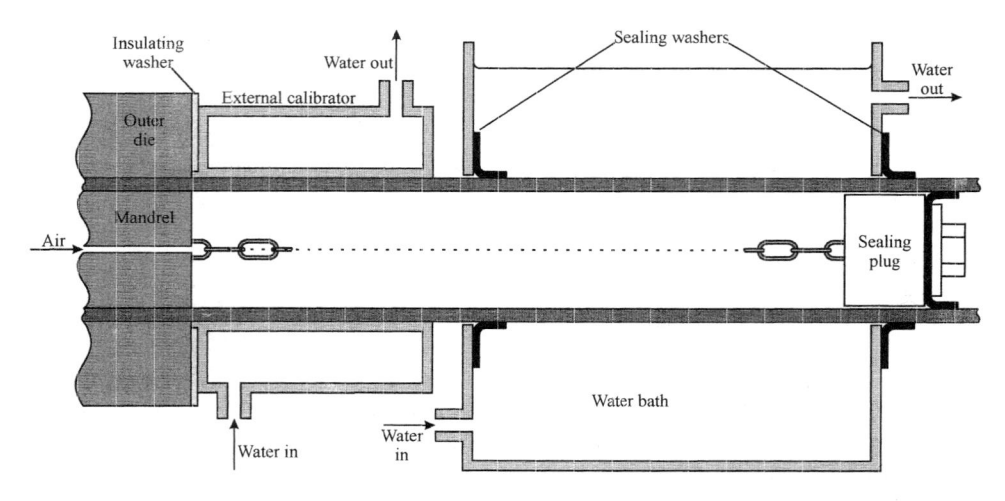

Figure 6 External calibration of large pipes.

The shear rates encountered during pipe extrusion are relatively low, which permits the use of relatively high molecular weight resins with fractional melt indices. High molecular weight favors low creep and high toughness, both of which are desirable in most pipe applications. Tubing, with its thinner walls, requires a lower viscosity resin, preferably one with good shear thinning characteristics, such as low density polyethylene.

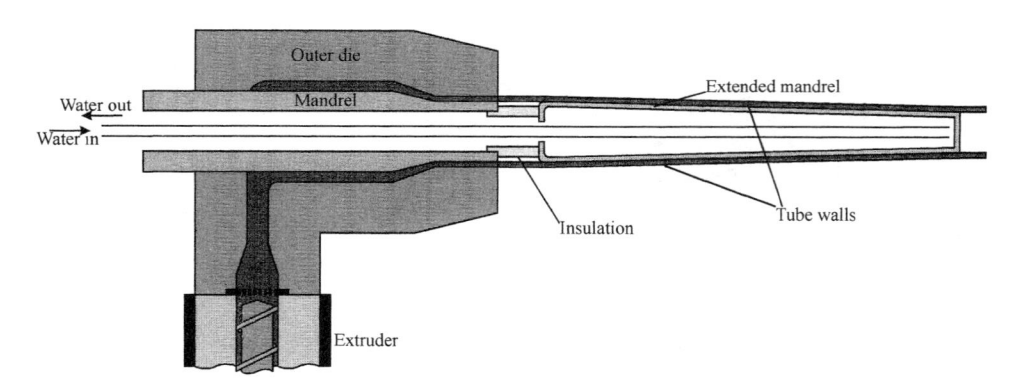

Figure 7 Internal calibration using a tapered mandrel.

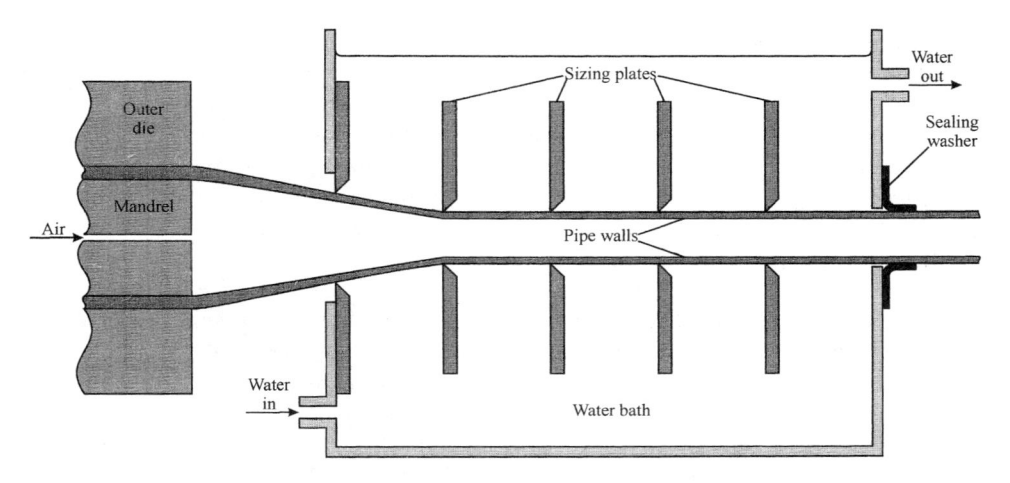

Figure 8 External calibration of tubing using sizing plates.

2. Products

One of the major and growing outlets of polyethylene pipe is natural gas delivery. Currently, approximately one-fourth of all polyethylene pipe goes into this application. It is used in both transmission (long distance, high pressure) and distribution (local, low pressure, typically less than 4 in. diameter) applications. In this field two types of resins are used, high density and "medium density" polyethylene. The latter is a linear low density polyethylene containing sufficient 1-alkene comonomer to reduce its density to approximately 0.935–0.940 g/cm^3. The principal difference between high and medium density polyethylene pipes is their stiffness; the high density product is essentially rigid, while the medium density product is somewhat flexible. The rigidity of high density polyethylene means that pipe walls can be made thinner and hence the product is lighter and cheaper than an equivalent length of medium density polyethylene pipe. In addition, the high density polyethylene pipe has slightly better environmental stress crack resistance. With respect to resistance to slow crack growth (long-term brittle failure), there is some discussion regarding which material is preferable, neither one being consistently superior to the other. High density polyethylene pipes are used in direct burial applications, in which individual lengths of pipe are welded together and laid in a trench. Heat for the welding process may come from a flow of hot inert gas, ultrasonic excitation, an electrically heated plate inserted between the pipe ends, or the friction of spinning one pipe against another. Traditionally, high pressure gas transmission pipes have been made of steel, but high density polyethylene pipe is making inroads into this market, starting at the lower end of the diameter range. Medium density polyethylene pipe, being somewhat flexible and generally of smaller diameter, is available in uninterrupted lengths wound on a drum. It is used to line deteriorating steel transmission lines, by pushing it in from one end. In smaller diameters it can be laid directly by pushing a capped length underground between holes dug tens of yards apart. Lining old steel pipes and the direct pushing of medium density polyethylene pipe are cheaper than the trenching and welding that must be used with rigid pipes.

Potable water delivery is another area in which the use of polyethylene pipe is increasing. Both high density and medium density resins are used; considerations regarding the choice of material are similar to those encountered for gas pipe. Principal applications include municipal transmission lines and private wells. Polyethylene pipe is not used in indoor residential applications due its low creep resistance when exposed to hot water. High density polyethylene transmission pipes are buried directly, while medium density polyethylene pipes are used as liners or are pushed in their smaller sizes. Medium density polyethylene pipe is available in diameters of up to 12 in., while high density polyethylene pipe as large as 48 in. in diameter have been extruded. Large diameter pipes may also be made from thick extruded sheets that are spirally wound around a mandrel and welded at the edges. Irrigation pipes, hoses, and fittings are commonly made from polyethylene. Abrasion resistance, flexibility, and ease of installation are among polyethylene's desirable characteristics in this application.

Mining, industrial, and sewer applications account for approximately one-third of all polyethylene pipe usage. The corrosion and abrasion resistance of polyethylene pipes are the principal characteristics that suit them for such purposes. These applications do not involve high pressures; therefore, medium density polyethylene is often the material of choice. This is particularly true in temporary installations, where the same length of flexible pipe may be used several times in different locations. Industrial and mining applications typically involve the transport of water, aqueous solutions, or slurries. In the oil industry, crude product and brine are the principal liquids transported. Sewer pipe can be of the liner or direct burial type, liners generally being larger in diameter.

Corrugated pipe accounts for a little less than one-fifth of all polyethylene pipe. It contains a large percentage of off-specification and reground resin that is not suitable for high pressure pipe or potable water applications. The corrugations endow it with good crush resistance combined with lateral flexibility. Its principal use is drainage pipe into which slots are cut parallel with its long axis to permit the influx of water. The majority of corrugated pipe is 4 in. in diameter, but product up to 24 in. in diameter is available. In its smaller sizes it is used in perimeter drainage around houses. Highway applications run as high as 15 in. in diameter.

Medium density polyethylene conduit is used to protect communication

Use and Fabrication of PE Products

cables. This relatively thin walled pipe can be pushed underground or snaked within walls. Electrical wires or fiber-optic cables are then pulled through. This installation process is much more economical than conventional cable laying techniques. It is expected that this application will become more important as fiber-optic communication for television and computer needs expands.

Low density polyethylene resins, with their good rheological characteristics, are typically used in the extrusion of small-diameter polyethylene tubes. Such tubes are found in aerosol spray cans and trigger spray bottles. More specialized uses occur in the field of medicine, including the use of corrugated tubing in respirators.

E. Film Coating

Film coating is used to apply a thin layer of polymer to a continuous sheet of substrate. It principally involves the coating of thin cardboard and aluminum foil, both of which are used in food packaging.

1. Process

Film coating employs equipment that has much in common with that used in chill roll film casting. The general layout of a film coating line is illustrated schematically in Figure 9. The output from a slit die is drawn downward into a

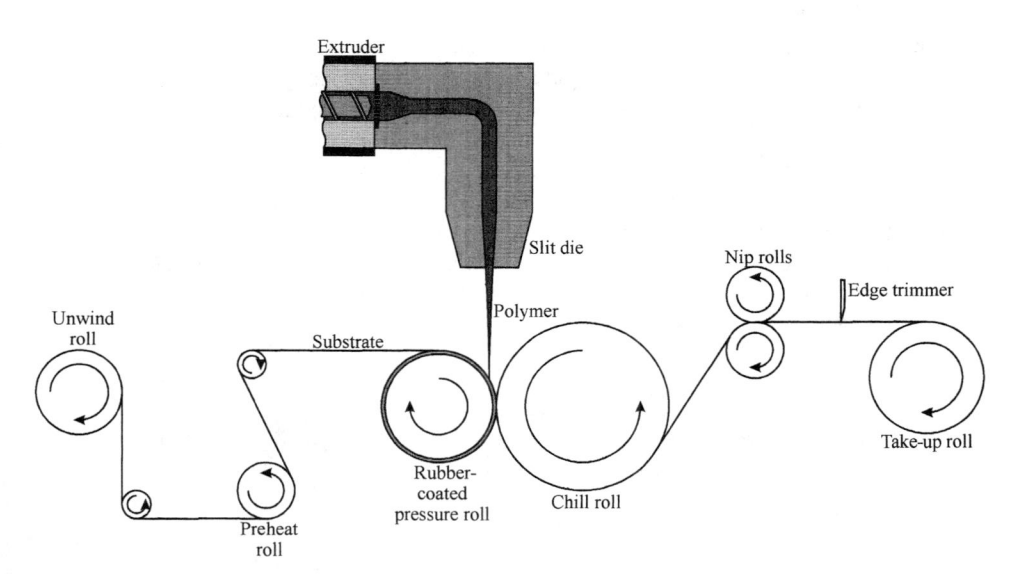

Figure 9 Principal components of a film coating line.

nip consisting of a chill roll and a rubber-coated pressure roll over which the substrate is fed. The molten polymer is simultaneously forced into intimate contact with the substrate and quenched against the chill roll. The substrate may optionally be preheated to improve polymer adhesion. Subsequently the coated product is trimmed to the desired width and wound up on rolls using conventional equipment. When it is desired to coat the substrate on both sides, a second extruder, die, and nip rolls are added downstream prior to slitting and wind-up.

The resin used in film coating generally has a very high melt index, the molten polymer flowing freely from the die as a "curtain" prior to meeting the substrate. Low melt viscosity ensures that intimate contact is made with the substrate in the nip, thus facilitating adhesion. In all cases, the substrate moves faster than the speed at which the polymer leaves the die; the polymer is thus drawn down substantially. The precise thickness of the coating is controlled by adjusting the die gap and the relative rates of the substrate feed and extruder output.

2. Products

The substrate most commonly coated with polyethylene is thin card, the product being termed "milk board," indicative of one of its principal uses. Both sides of the cardboard are coated, with the side destined to become the interior of the package being given the thicker coating. In addition to dairy product packaging, milk board is also used to package other food products such as fruit juices and frozen vegetables. In these applications, the primary role of the polyethylene coating is to the prevent liquid contents or exterior condensation from saturating and softening the cardboard. A second crucial role of the coating is as a heat-sealing layer. Low density polyethylene dominates this market; its excellent flow characteristics and flexibility at low temperatures making it particularly suitable. Aluminum foil used to seal dairy product containers and other food packages is normally coated with polyethylene as a heat seal layer. Ethylene-vinyl acetate copolymer is used preferentially in this application, because it has superior adhesion to aluminum. Photographic paper is coated with an extremely thin layer of polyethylene. Great care is taken to ensure that resins used for this purpose are as free of particulate contaminants as possible in order to ensure defect-free development of the underlying emulsion.

F. Wire and Cable Coating

Wire and cable coating involves the extrusion of a concentric layer of polymer around a continuous wire core.

1. Process

The basic equipment used to coat wire is shown in Figure 10. It consists of an extruder equipped with a crosshead or offset tubular die, the wire being paid out

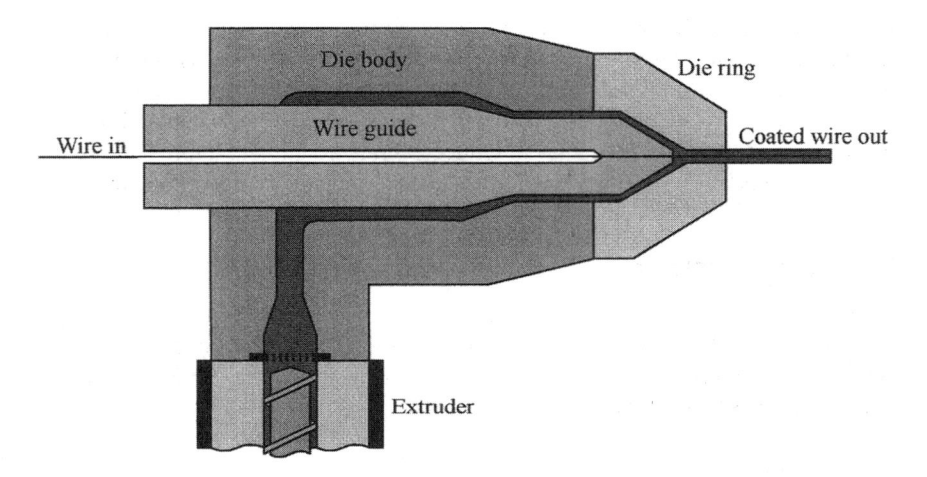

Figure 10 Crosshead type of wire-coating die.

through a hole in the center of the mandrel. To minimize disruption of the melt flow, crosshead dies may subtend an angle of substantially less than 90° to the axis of the extruder. Thin insulation layers cool rapidly upon exposure to air, but thicker layers may require additional cooling in the form of a water bath in order to fully solidify the polymer. The finished product is drawn through the system using a capstan traction system prior to being taken up on spools or drums. Most wire-coating lines are equipped with an electrostatic device to check the insulation for pinholes or discontinuities and an automatic measuring device to ensure uniform thickness.

The thickness of the polymer coated onto the wire is controlled by a combination of the dimensions of the wire and the orifice, the output rate of the extruder, and the wire feed rate. Concentricity of the wire within its insulating jacket is ensured by adjusting the position of the outer die ring. Interchangeable wire guides and outer rings are available that accommodate a range of wire diameters and insulation thicknesses. Coating equipment exists that covers the range from the fine wire used in miniature electronic devices up to the high voltage transmission cable used in the national electricity grid.

2. Products

All the major varieties of polyethylene are used in wire and cable coating. Lower density resins, including ethylene-vinyl acetate copolymers, are used for general-purpose utility applications, the need for high flexibility being a major concern. Higher density resins are used when elevated end use temperatures are likely to be encountered. Cross-linked resins are used to insulate high voltage power transmission cables and other applications where temperatures may temporarily exceed the crystalline melting point of polyethylene. Resins often contain an inert filler, such as calcium carbonate, to impart a measure of fire retardance. Foamed insulation is used in coaxial cables to reduce overall weight and improve flexibility; such resins are typically cross-linked linear low density polyethylene.

G. Injection Molding

Injection molding involves the introduction of molten polymer into a mold cavity, where it cools and solidifies. A wide variety of primarily nonstructual commodity items are produced by this process; products include pails, pallets, crates, housewares, toys, and mragarine tubs.

1. Process

The basic equipment used in injection molding is illustrated in Figure 11. A reciprocating screw extruder is used to melt and homogenize the polymer resin, pumping it into a reservoir ahead of the screw. As the reservoir fills, the screw withdraws to accommodate the charge of polymer building up ahead of it. When sufficient molten resin has accumulated, the screw stops rotating and is thrust forward, injecting the polymer into a closed mold. Once the mold is full, "packing pressure" is maintained as the resin cools and solidifies; this minimizes product shrinkage. After the polymer has solidified, the mold opens and the molding is removed. Removal of the molding is accomplished with the aid of "ejection pins" that protrude from the face of the die when it opens. Meanwhile, the screw recommences turning, accumulating molten polymer in readiness for the next shot. As a finishing step, sprue may have to be removed manually from the molding.

The principal variations of injection molding equipment are found in the mold. In its simplest form, a mold consists of two parts—one fixed and one movable—that define one or more cavities into which the molten polymer is injected. Molds may be opened and closed with mechanical toggles, hydraulic rams, or a combination of the two. The principal requirement of the mold closing system is that it must exert sufficient force to prevent the mold from opening when the resin is injected. Injection molding machines are defined in terms of maximum closing force on the mold, the smallest equipment being rated at a few tons, while the largest can generate several thousand tons of closing force. "Shot size" is limited by the diameter of the screw and its back travel distance.

Molds must be made from an extremely durable material if they are to survive the large forces associated with injection molding over prolonged production runs. Hardened stainless steel is normally used, but more exotic materials

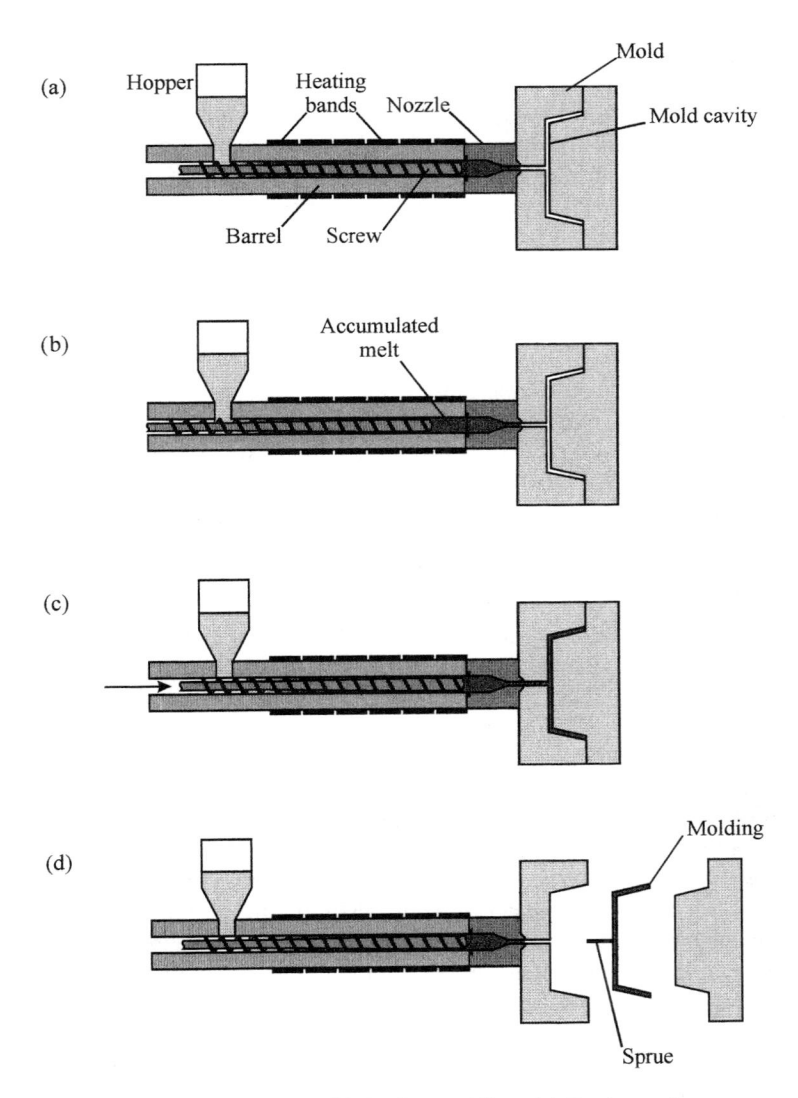

Figure 11 Basic process of injection molding. (a) Reciprocating screw extruder; (b) melt accumulates ahead of screw; (c) injection and cooling of melt; (d) mold opens and ejects molding.

may be employed in exceptional circumstances. Chrome plating of the mold cavity produces parts with a glossy surface finish. Water cooling is used to increase the crystallization rate and hence reduce the overall cycle time. When molten polymer is injected into a die, it compresses and heats the air within the cavity. To prevent superheated air from scorching the surface of the molding, vents are incorporated into the mold. These vents typically take the form of shallow (<0.001 in.) channels machined into the die face where the mold surfaces meet. Molds for more complex items, such as pallets and crates, typically consist of several moving parts, which permits the fabrication of products with recesses, projections, or undercuts that could not be removed from a two part mold. Such molds are typically large and expensive; whenever possible, products are designed in such a way that they can be molded in a two-part die.

Traditionally, molten polymer is injected into the mold cavity via a "cold runner," which consists of an unheated channel within the fixed part of the mold. This has the disadvantage that polymer solidifies within the runner, forming a solid sprue that must be removed from the molding after ejection. Sprues can be reground and fed back to the hopper, but their existence is wasteful in terms of manpower, equipment, and resin. In molds that have multiple cavities, the polymer wasted in the cold runners can account for a significant percentage of each injection shot. Modern multicavity molds use a "hot runner" system to deliver molten polymer to each cavity. In this system the runners leading to each cavity are individually heated, the resin within them remaining molten at all times. Advantages of the hot runner system include reduced waste, more uniform delivery of resin to each cavity, reduced shot volume, and shorter cycle times. Disadvantages include increased complexity and cost.

The location of the gate (or gates) through which molten polymer enters the mold is carefully chosen to provide an even flow of polymer to all parts of the cavity. For simple moldings such as pails and margarine tubs, a single "pin" gate is commonly located at the center of the underside of the bottom. For more complex items or those with large aspect ratios, several gates may be required to ensure uniform filling of the mold. When multiple gates are used, the region where two flows of molten polymer meet, known as a "weld line," may be visible to the naked eye. If the impinging flows are too cold when they meet, they may not fuse adequately and the resulting weld line may be a source of weakness. The use of elongated "edge" gates eliminates the creation of weld lines when molding broad or wide items. The gates on hot runner molds can be equipped with pneumatic valves to mechanically close the gate; this arrangement yields parts with superior surface finish at the point of injection.

One of the characteristics of injection molding is that melt orientation imparted by rapid flow is frozen into the final product. Solidification occurs fastest next to the mold surface, with the result that highly oriented skins are often found overlaying a less oriented core. Orientation may be beneficial or detrimental,

Use and Fabrication of PE Products 483

depending upon its alignment relative to the applied load in end use applications. When orientation is parallel with tensile forces, it can impart added strength but may be a source of weakness when forces are applied perpendicularly. Perhaps the most detrimental effect of orientation is warpage, which may occur immediately upon removal from the mold or at a later date when the product is exposed to temperatures sufficient to soften it. It would be possible to avoid orientation entirely if molds were heated prior to injection and only cooled once the shot were complete; however, this would be uneconomical.

The range of melt indices employed in injection molding is very broad. The melt index of the resin required is principally determined by the aspect ratio and thickness of the molding. Items that have thin cross sections or a very large aspect ratio must be molded from low viscosity resins that flow readily. Melt indices as high as 350 may be used in extreme cases, such as for the lids of dairy product tubs. The low molecular weight of such resins makes them subject to brittle failure. To ameliorate this, the degree of crystallinity must be minimized, this is accomplished by the use of resins with a high degree of branching, such as very low density polyethylene. These low density materials have very little stiffness, making them unsuitable for structural purposes. At the high molecular weight end, resins with melt indices of about 5 are employed to mold parts with thick cross sections. Such items are typically required to be sturdy, which makes them prime candidates for the use of high density polyethylene.

2. Products

Injection-molded polyethylene items are ubiquitous, finding widespread use in commercial and domestic applications. The largest portion is used in containers and packaging of many kinds. Approximately two-thirds of all injection-molded polyethylene items are made from high density resins. Another 20% are linear low density polyethylene, about 12% are low density polyethylene, and the remainder are ethylene-vinyl acetate copolymer. Resins are pigmented in virtually all applications.

Industrial containers such as crates, pallets, and pails which are often reusable or hold heavy loads are generally made from high density resins. The melt index is typically quite low (5–10), which results in products with good toughness. Polyethylene crates are used to distribute dairy and bakery products to stores and to hold and transfer intermediate parts within factories. Polyethylene finds some use in pallets, but these are generally more expensive than wooden ones, so their use is limited. Pails with resealable lids are used to transport food and other products in bulk. To reduce the possibility of food contamination, the low molecular weight component of resins used to mold pails is kept to a minimum. Products transported in pails include pickles, fats, paint, grout, and plaster.

Many liquid or semiliquid foods are retailed in high density, thin-walled polyethylene tubs, dairy products being a major outlet. Other retail food uses include carbonated beverage bottle bases, milk bottle caps, and other closures, especially those with integral spouts.

The low cost of polyethylene resins and their high intrinsic toughness make them prime candidates for use in many household items such as laundry baskets, trash cans, buckets, drinking cups, toys, and garden furniture. The choice of resin is principally determined by the stiffness and toughness requirements. Laundry baskets and trash cans and the like are normally made from low density or linear low density resins, whereas drinking cups and yard furniture are more likely to be made from high density resins. To improve processability, the molecular weight of the resins used is generally as low as possible while still maintaining the requisite toughness.

H. Blow Molding

Blow molding is a versatile conversion process used to produce a wide variety of objects by inflating a molten tube of polymer inside a hollow mold. Two major variations of the method exist: extrusion blow molding and injection blow molding. The former is predominant in polyethylene molding. Common items fabricated by blow molding include bottles, drums, chemical storage tanks, and toys.

1. Process

The principal steps involved in extrusion blow molding are illustrated schematically in Figure 12. The initial step is the extrusion of a free-hanging tube of molten polymer, known as a "parison." Once the parison has reached a predeter-

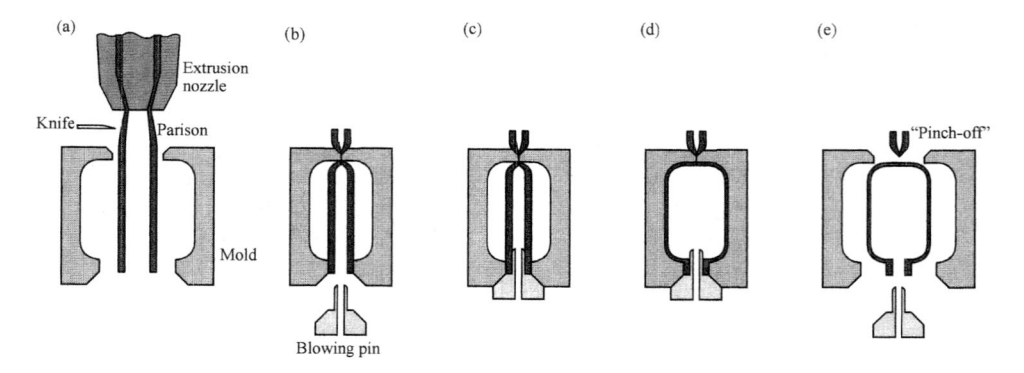

Figure 12 Principal steps of extrusion blow molding. (a) Parison extrusion; (b) encapsulation of parison; (c) insertion of blowing needle; (d) inflation and cooling; (e) ejection.

mined length, a blow mold closes around it, pinching it off from the extrusion die and sealing one end. A blowing needle is then inserted, and a blast of compressed air inflates the parison until it conforms to the interior of the mold. The mold remains closed sufficiently long to permit the polymer to solidify, whereupon it opens and the product is ejected. As a final step, the molding may need to have excess neck and base "pinch-off" removed, either automatically in conjunction with ejection or as a separate manual step. Pinch-off material is ground up and recycled directly to the extruder feed.

Parison extrusion may involve changing the extrusion die annulus to vary the wall thickness of the parison as a function of its length. This is achieved by the programmed vertical adjustment of a conical mandrel within the die during extrusion. The goal of programmed parison thickness control is to ensure that the wall thickness in the final product is as uniform as possible. Factors that must be taken into account include die swell, parison drawdown under its own weight, and the shape of the final product. Thus, those parts of a bottle that have a large diameter are blown from thicker parison sections than those having a smaller diameter. This is illustrated in Figure 13.

Multiple parisons may be extruded simultaneously, especially when small items are being made. Concentric coextrusion of polyethylene and other thermoplastic resins is often practiced in order to yield layered products with a combination of the requisite stiffness, thoughness, and barrier properties.

When large items (such as chemical storage drums) are to be blown, it may not be possible to continuously extrude the parison sufficiently fast to prevent it

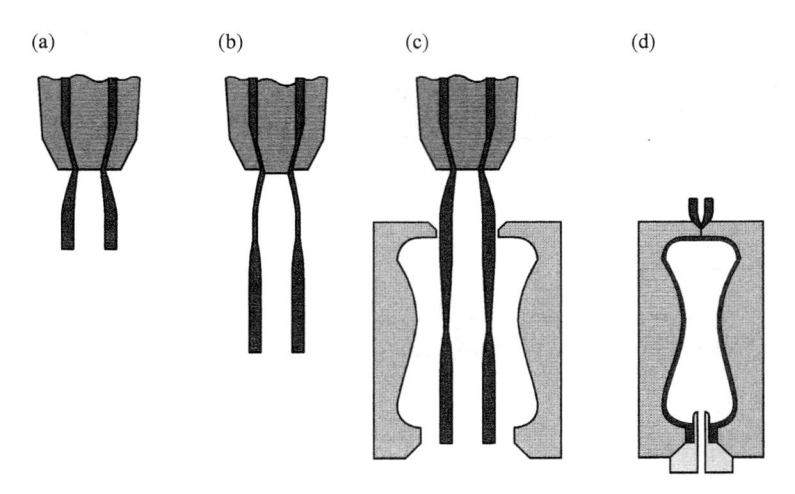

Figure 13 Programmed parison wall thickness to yield even wall thickness in the final product.

from rupturing under its own weight or partially solidifying prior to inflation. Under such circumstances an accumulator may be employed; examples are shown in Figures 14 and 15. In the reciprocating screw method, a reservoir of molten polymer is accumulated ahead of the screw, which gradually moves backward to accommodate the excess melt. When sufficient polymer has accumulated, the screw stops turning and is rammed forward, expelling the polymer rapidly from the die. In the ram accumulator method, the molten polymer is pumped from the extruder into a separate reservoir equipped with a hydraulically actuated ram that ascends as the reservoir fills. The parison is extruded rapidly when the ram descends.

When parisons are being continuously extruded, the blow mold must move aside immediately after it has pinched off the appropriate length. Such is the case typically encountered when molding products of less than a few gallons in volume. The blow mold containing the parison is transferred to a remote blowing

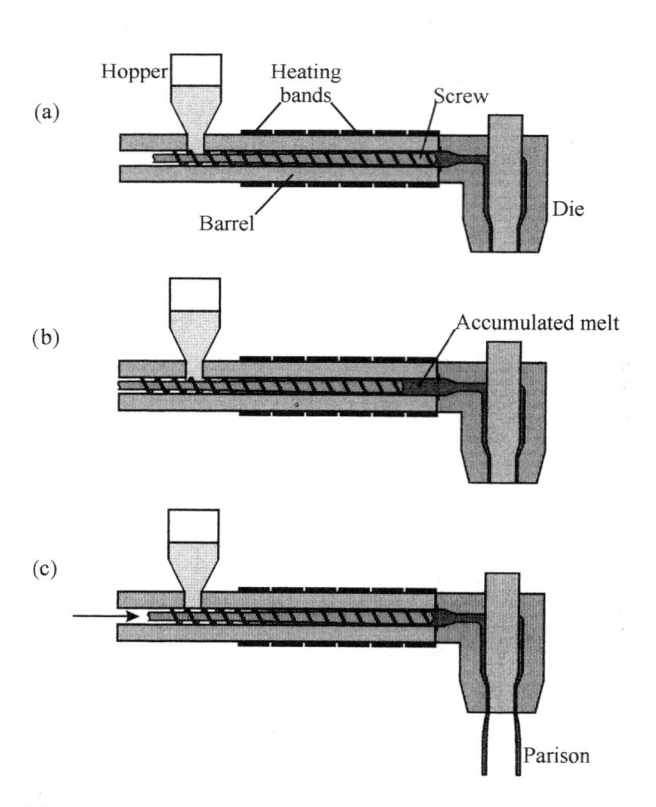

Figure 14 Reciprocating screw accumulator.

Use and Fabrication of PE Products

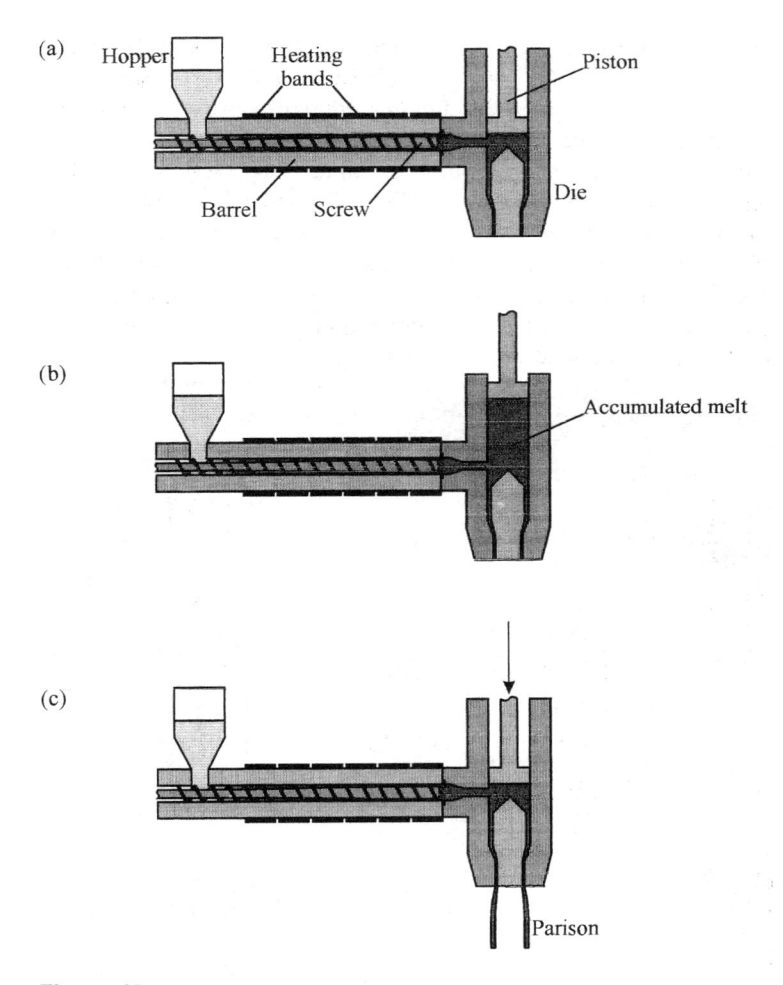

Figure 15 Ram accumulator.

station, where the molding process continues. Often the output of the extruder is greater than can be handled by a single blow mold (or set of molds when multiple parisons are being extruded). In such cases, two or more blowing stations may exist, each provided with its own mold (or set of molds), which pinch off lengths of parison in turn.

"Shuttle-type" blow molding machines translate their molds linearly between the pinch-off point and the blowing station. Alternatively, it is possible to mount a series of blow molds on the circumference of a wheel (either vertical or horizontal). In wheel-type machines, parison pinch-off, blowing needle insertion, inflation, cooling, and ejection take place simultaneously in different molds. Examples of shuttle- and wheel-type blow molding configurations are illustrated schematically in Figures 16 and 17, respectively.

The principal steps of the injection blow molding process are illustrated in Figure 18. The initial step is the injection molding of an inflatable preform. This simultaneously creates a bottle neck with well-defined dimensions and a tube of polymer sealed at one end that will later be inflated. The preform is then transferred to an oven, where the tube is reheated to its softening temperature. The softened preform is then transported to a conventional blow mold, and compressed air is introduced to inflate it.

Injection blow molding has certain advantages over extrusion blow molding in that it produces containers with well-defined mouth and neck dimensions

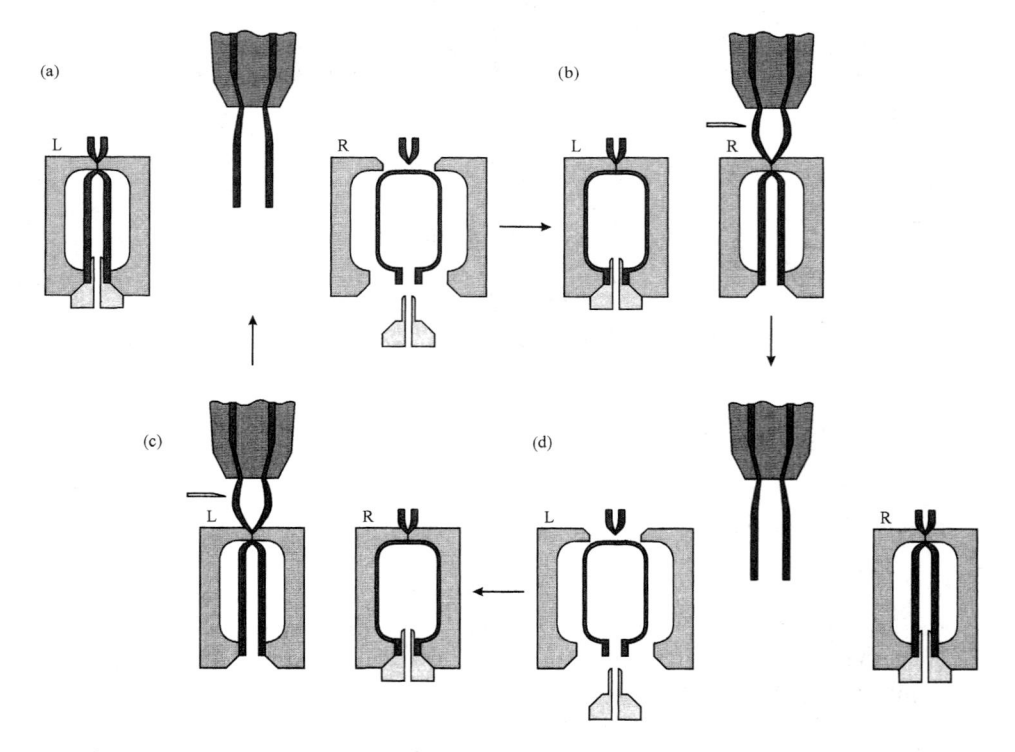

Figure 16 Shuttle type of blow molding process (a) L, blowing pin insertion; R, ejection. (b) L, inflation and cooling; R, encapsulation of parison. (c) L, ejection; R, blowing pin insertion. (d) L, encapsulation of parison; R, inflation and cooling. L = left mold; R = right mold.

Use and Fabrication of PE Products

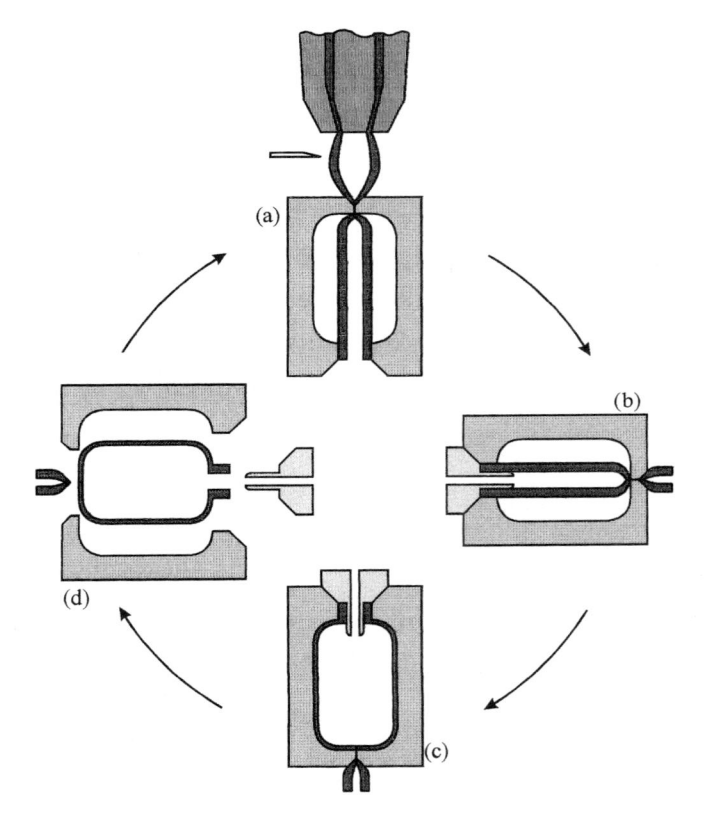

Figure 17 Wheel type of blow molding process. (a) Encapsulation of parison; (b) insertion of blowing needle; (c) inflation and cooling; (d) ejection.

and no potentially weak pinch-off points or thin spots. Additionally, the process is essentially scrap-free. Against this must be weighed the fact that injection blow molding is a more complex (i.e., costly) process than extrusion blow molding. As such it is best suited to making reusable containers or those requiring greater structural integrity. Resins used in injection blow molding are typically of lower molecular weight than those used to make comparable extrusion blow molded items.

Inflation of the parison is normally accomplished with compressed air, but in some cases the vaporization of liquid nitrogen is used instead. If the air between the parison and the mold wall is not vented as the parison inflates, it will prevent the molten polymer from making uniform contact with the mold surface, resulting in poor surface finish. Slit-type vents may be provided at the mold parting line,

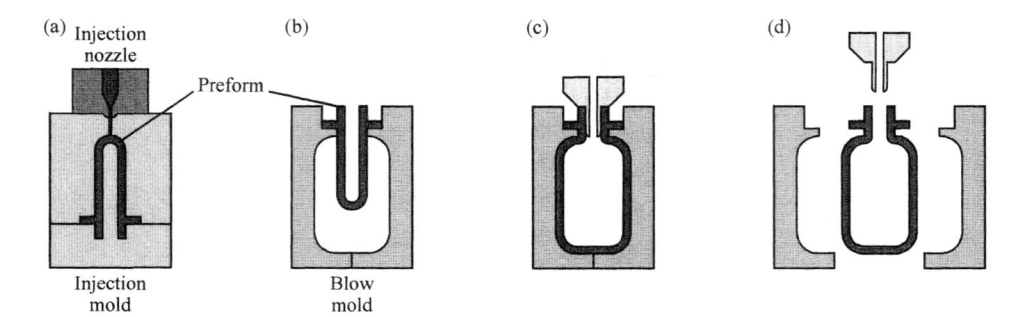

Figure 18 Principal steps of injection blow molding. (a) Injection molding of preform; (b) transfer of preform to blow mold; (c) inflation and cooling; (d) ejection.

or small holes (< 0.01 in. in diameter) may be drilled into the mold in recesses liable to trap air pockets. Porous sintered metal inserts may also be used as vents. Blow molds are typically made of stainless steel, but other materials are used under special circumstances. Hardening of the mold is seldom required, as the loads associated with the blowing process are not excessive. Molds used for prototype testing or short runs may be made from softer metals or even castable polymers such as epoxy or polyester resins. The interior of a blow mold does not need to be highly polished or chrome plated to yield products with good surface gloss; a smooth finish is all that is required. A matte surface finish can be obtained by sand or bead blasting the mold interior. In designing blow molds, special care must be taken to avoid excessive drawing of the parison into corners or recesses that could result in unacceptably thin areas. Other factors that must be considered include shrinkage, especially in higher density resins; general weakness at the pinch-off points; and dimensional stability, especially in the vicinity of the neck.

In practice, the blow molding process is much more complex than the brief outline given above. Much ingenuity has gone into developing machinery capable of producing a wide assortment of products with all manner of shapes and sizes. Those wishing to learn more about this subject should consult the texts listed in the bibliography.

2. Products

More than 90% of all blow-molded polyethylene products are made from high density resins. The majority are bottles that are consumed domestically, such as those used for milk, household chemicals, and cosmetic products. Other major outlets include chemical transportation drums, pails, and fuel tanks.

Milk, fruit juice, vinegar, and water bottles are all blown from similar grades of high molecular weight, high density polyethylene resin. These resins have fractional melt indices and a broad, sometimes bimodal, molecular weight distribution. Such resins have good melt strength, which permits high stretch, resulting in thin-walled, puncture-resistant containers. The high crystallinity of such resins imparts good structural rigidity but limits clarity. To reduce the possibility of food contamination, the low molecular weight component of these resins is kept to a minimum. As with most polyethylene products, the low energy surface of blown bottles is not immediately receptive to printing inks. Adhesive labels are routinely applied to identify and describe the contents. Alternatively, flame treatment may be used to oxidize the surface of bottles prior to screen printing.

Motor oil bottles and household chemical bottles such as those containing liquid detergents, toiletries, bleach, and cosmetics are made from high density polyethylene with a lower molecular weight than that used for milk bottles. As a general rule, the smaller the bottle, the lower the molecular weight. These bottles are intended to withstand the rigors of chemical containment and use over a prolonged period of time and hence have relatively thick walls. Such bottles typically have a textured surface, handles, or a sculpted shape, all of which make them easier to grip. A primary cause of failure of such bottles in the past was environmental stress cracking. This has largely been eliminated by the use of resins with a relatively narrow molecular weight distribution, which effectively limits the level of detrimental low molecular weight chains. Linear low density polyethylene resins are sometimes used when especially active environmental stress cracking agents are to be contained. "Squeezable" bottles that must be flexible and elastic may be made from low density polyethylene, but only when environmental stress cracking is unlikely to be an issue.

Industrial and agricultural chemical storage tanks, fuel tanks, and transportation drums are all made from high molecular weight, high density polyethylene. Such containers are tough, light, and corrosion-resistant. When an added measure of product containment is required, a liner of polyethylene can be blown directly into a steel drum. Barrels of up to 55 gal capacity and storage tanks capable of holding several hundred gallons are produced by accumulator extrusion blow molding. The interior of fuel tanks may be chemically treated to reduce their permeability. Alternatively, a fluorocarbon polymer with low permeability may be coextruded with polyethylene to create a tank with multilayer walls.

Low density polyethylene is the resin of choice when blow molding preschool children's playthings. It is easily molded to form rings, blocks, etc. that are soft, lightweight, resilient, and noninjurious when used as projectiles.

I. Rotational Molding

In rotational molding, free-flowing polyethylene powder is loaded into a cold cavity mold, which is then spun and heated simultaneously, spreading the polymer over the interior of the mold and melting it. The mold is then cooled to solidify the polymer, and the product is removed. This process yields medium to large hollow items, such as chemical storage tanks, garbage containers, and kayaks. This process is also known as "rotomolding" or, occasionally, "rotational casting."

1. Process

Rotational molding is a four-step process: loading the mold with resin, rotating and heating the mold, rotating and cooling the mold, and, finally, removal of the product. The basic process is outlined in Figure 19. A predetermined weight of polyethylene powder is loaded into a cold mold, which is then closed. The mold is then transferred to an oven, where it is rotated and heated. Temperatures in excess of 250°C are commonly used. Rotation typically takes place about two mutually perpendicular axes, which distributes the polymer powder over the interior surface. As the mold heats up, the resin begins to melt and adhere to its interior. Several minutes elapse before all the powder ceases to flow freely. Heating and rotation are continued until the molten resin flows under its own weight to form a void-free coating on the interior of the mold. Rotation is maintained while the mold is cooled with sprays of cold water or by fan-blown air. Once the polyethylene has solidified, the two halves of the mold are unbolted and the molding is removed. Cycle times are quite long, the time required for heating and cooling increasing with wall thickness. For extremely large items, cycle times as long as 1 hr or more can be expected.

The equipment used in rotational molding is relatively unsophisticated and inexpensive compared to other polyethylene-molding techniques. Molds are subjected to low stresses and can therefore be constructed of relatively lightweight and low cost materials such as sheet steel or cast aluminum, the latter being used for smaller items. Temperature and time regulation are not as critical as in other

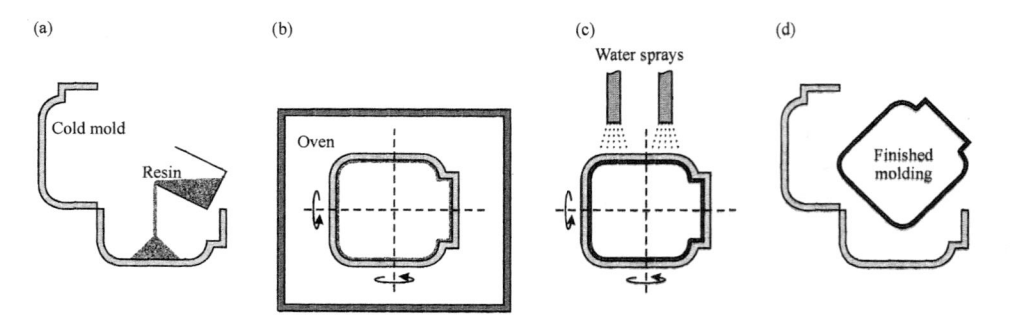

Figure 19 Steps involved in rotational molding. (a) Loading resin powder into the mold; (b) rotating and heating; (c) rotating and cooling; (d) product removal.

techniques, so the use of electronic controllers is minimized. A variety of molding machines are available that are equipped with from one to five arms. Each arm, bearing one or more molds, sequentially advances through the various stages of the process. Shuttle-type machines are equipped with two molds that are alternately run in and out of a single oven from opposite ends. Other machines, such as the carousel type, rotate their arms through loading, heating, cooling, and unloading stations. Clamshell machines have a single station located in a chamber, the upper half of which pivots backward to allow access to the molds for loading and unloading. When closed, the chamber doubles as an oven and cooling compartment. Not all molds are rotated biaxially; rock-and-roll machines oscillate their molds about one axis while rotating them about another. This type of machine is useful for molding products with a large aspect ratio, such as kayaks. Other machines rotate their molds around a single axis, producing tubular products such as oversized drainage culverts.

Rotational molding produces items that are essentially free of orientation. The minor residual stresses that are found arise from quenching effects. The outer skin layer is cooled more rapidly than the interior, resulting in a slight density gradient, with the density increasing toward the inner surface. This imparts a small degree of internal stress that can result in warpage if not controlled. Such problems are more common with resins of higher density. Positive gas pressure within the mold improves both dimensional stability and internal surface finish (which tends to be somewhat grainy). The biaxial rotation of molds within the oven ensures even heating of the mold surface, resulting in uniform wall thicknesses with no weld lines. Wall thickness can be controlled locally by insulating the mold in specific locations or by the confined application of extra heat supplied by hot air piped internally to the desired spot. Multilayered moldings can be created with the aid of a "drop box" within the mold that contains additional polymer. Once a continuous outer skin has formed, the drop box releases its contents, which melt to form an inner layer.

The process of rotational molding is relatively slow and labor-intensive. Not being suitable for high output rates, its principal use is in the fabrication of medium and large items for which there is limited numerical demand. On the positive side, it is possible to change colors without the wasteful purging associated with extrusion-based processes. There is also relatively little scrap.

The range of melt indices that can be used in rotational molding is quite limited. The lower limit is determined by the requirement that the molten polymer must flow and fuse within a reasonable period of time. The upper limit is determined by the toughness required of the finished product. Melt indices as low as 3 may be used, but 5 or 6 is more common, with an upper limit of about 10. Resins with higher melt indices can be used when the need for good surface finish outweighs the requirement of good physical properties. In general, the larger the product, the lower the melt index and the higher the density. For best results, the molecular weight distribution is kept as narrow as possible; low molecular weight tails tend to melt too early, while high molecular weight tails increase melt viscosity.

To counteract the effects of high temperature experienced for extended lengths of time, the resins used in rotational molding must be well stabilized. Stabilizing additives are incorporated into the resin by extrusion blending. The resulting pellets must then be ground to a free-flowing powder prior to rotomolding. A 35 mesh powder, in which more than 95% passes through a 500 µm screen, is commonly used. To further limit oxidative reactions, some processors flush their molds with nitrogen during molding.

When an extremely tough product is required, cross-linked polyethylene is used. Cross-linking can be achieved by molding a low viscosity resin at a temperature insufficient to activate the cross-linking agent. Once the polymer has fused, the oven temperature is raised to decompose the cross-linking agent.

2. Products

Polyethylene accounts for approximately 85% of all products made by rotational molding. Linear low density resins make up almost two-thirds of the polyethylene used, the remainder being split approximately equally between high and low density polyethylene. The process is typically used to manufacture medium to large hollow, double-walled, or intricate parts at relatively low output rates.

Rotationally molded products have walls of uniform thickness that are free of orientation and weld lines. As such they are resistant to environmental stress cracking and are therefore suitable for use in large-scale chemical containment uses. Linear low density polyethylene resins excel in such applications. Storage tanks of up to several thousand gallons capacity are used in agricultural, domestic (septic systems), and industrial applications. On a smaller scale, rotationally molded fuel tanks (chemically treated to reduce diffusion) are used on tractors, cars, and small gasoline-powered appliances.

Civil engineering uses include large diameter pipe for culverts and air ducts and energy-absorbing highway safety devices such as crash barriers filled with polyurethane foam and barrels filled with sand installed around bridge abutments. On a more down-to-earth level, the components of portable privies are frequently rotationally molded.

The good impact resistance of rotationally molded polyethylene is exploited in chemical shipment containers, commercial trash cans, and dumpster lids. Rotational molding is used extensively to manufacture playground equipment and garden furniture, where toughness is also at a premium. Aquatic applications include kayaks, buoys, and the hulls of small boats, any of which may be cross-linked for additional impact resistance.

III. SUBSIDIARY FABRICATION PROCESSES

Various subsidiary processes can be combined with the primary fabrication processes to yield products with modified physical properties. The most important of the subsidiary processes are cross-linking and foaming.

A. Cross-linking

Cross-linking involves the formation of chemical bonds between adjacent chains to create a molecular network. The chemistry of cross-linking, the processes used to accomplish it, and its effect on the physical properties of polyethylene are addressed in Chapter 7.

B. Foaming

Foamed polyethylene exhibits several desirable physical characteristics that make it useful in a variety of relatively small markets. It can be made as either a closed-cell or an open-cell product. In the former case bubbles of gas expand to a limited extent within a molten polymer constrained by pressure. The product comprises cells completely bounded by solidified polymer walls. Open-cell products form when bubbles are allowed to expand in a relatively unconstrained manner. The cells so formed have only partial walls and are thus interconnected like those of a sponge.

Foaming is most often combined with profile forming. Either the blowing agent can be injected into the molten polymer within an extruder as a low boiling point liquid such as butane or carbon dioxide or can it be generated in place by thermal decomposition of a solid or liquid blowing agent, such as azodicarbonamide, blended into the polymer. When the pressure on the molten polymer is relieved as it exits the die, the dissolved gas expands to foam the molten polymer. In doing so it absorbs heat, thus accelerating the polymer's solidification. When the molten polymer exiting the die is constrained in volume it yields a closed cell product. Unconstrained expansion results in open-cell foams.

The principal attributes of foamed polyethylene that are exploited commercially are its insulating characteristics and its shock-absorbing ability. Closed-cell foams are used primarily as insulation, while open-cell foams are used in damping applications. Closed-cell applications include pipe insulation, coaxial cable insulation, and sleeping pads. Open-cell polyethylene foam applications included athletic shoe inserts, athletic pads, and foamed packaging sheeting. In athletic shock-absorbing applications, ethylene-vinyl acetate is the preferred resin type for its good elastic characteristics. When crush resistance is of greater importance, higher density resins are employed.

IV. SECONDARY FABRICATION PROCESSES

A. Thermoforming

Thermoforming is used to convert flat polyethylene sheets into three-dimensional objects. It is used to produce such items as pickup truck bed liners, agricultural feeding troughs, and panels for portable sanitation facilities.

The principal steps of thermoforming are illustrated schematically in Figure 20. In the first step, a cold sheet is clamped into a frame that grips its edges. The sheet is then heated sufficiently to soften but not melt it. The heated sheet is forced to conform to the shape of the mold by vacuum, pressure, or mechanical means or by some combination of any two or all three of them. The shaping force is maintained while the sheet solidifies. Upon removal from the mold, the product is trimmed as necessary.

The attainment of a final product with uniform wall thickness requires correctly designed equipment, the appropriate resin, and carefully chosen processing parameters. This is especially true when products with a deep draw are being fabricated. The molten sheet should have sufficient elasticity that it does not sag and flow unduly under its own weight, but it must be capable of being extended to the desired dimensions without tearing or necking. Thus, extremes of molecular weight are to be avoided. The precise molecular parameters required will

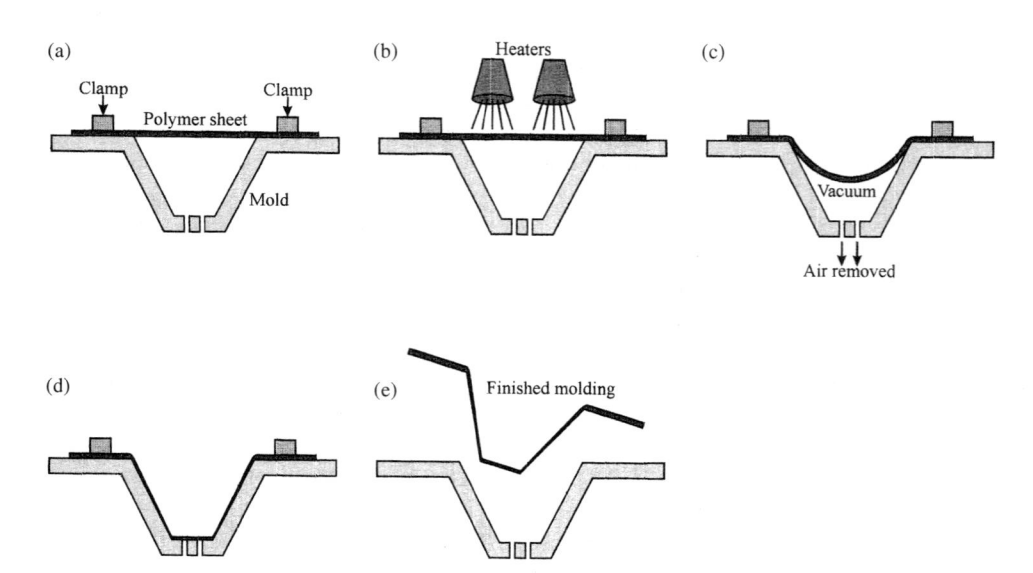

Figure 20 Principal steps of thermoforming. (a) Clamping; (b) heating; (c) forming; (d) cooling; (e) removal.

Use and Fabrication of PE Products

depend on sheet thickness, processing rates, depth of draw, and the processing temperature.

B. Bonding Techniques

The bonding of polyethylene to itself, other polymers, and nonpolymeric substrates involves raising its temperature sufficiently for most or all of the crystalline regions to melt, then holding it in position until it resolidifies. Molecular interdiffusion across the boundary or intimate surface contact creates the bond. The term "sealing" applies to the bonding of thin films, while "welding" is used to describe the bonding of thicker items such as sheet and pipes.

The very low surface energy of polyethylene prohibits the widespread use of adhesive bonding techniques. Even the adhesion of printing ink to polyethylene requires prior chemical modification of the surface by one of the techniques described in Chapter 7.

1. Sealing

The sealing of polyethylene films involves the melting and resolidification of two surfaces in intimate contact with each other. Heat can be applied externally from a heated sealing bar or generated internally by the electromagnetic excitation of polarizable side groups. Sealing is used extensively in the field of packaging to create bags and pouches that are used for all manner of food, medical, and general purposes. More specialized applications of heat sealing include the creation of pond liners and research balloons.

The principles of heat sealing are illustrated schematically in Figure 21. In the first step, the surfaces to be joined are brought into alignment with one another. Next, a heated sealing bar applies sufficient heat and pressure to melt the films and force their surfaces into intimate contact. The temperature, pressure, and dwell time are all carefully controlled to ensure adequate molecular diffusion

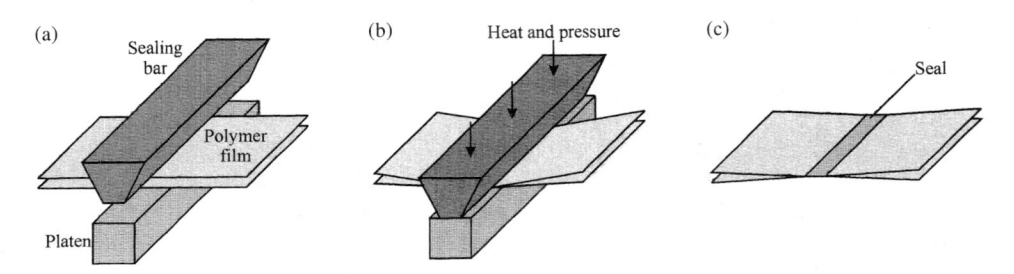

Figure 21 Schematic illustration of the heat-sealing process. (a) Polymer films aligned; (b) sealing bar applies heat and pressure; (c) seal removed from under sealing bar.

without undue oxidative degradation or excessive flow that could result in molten polymer being squeezed away from the bond. In the final step, the heat source and pressure are removed, and the bond is allowed to solidify prior to the application of load. Heat sealing is only applicable to thin films in which heat transfer into and out of both thicknesses of film is very rapid.

Sealing bars come in various shapes and sizes. In their simplest form they consist of an electrical heater within a metal bar that is covered with a nonstick coating such as polytetrafluoroethylene. Impulse heat-sealing bars support a polytetrafluoroethylene-covered nichrome wire that is momentarily pulsed with sufficient electrical current to melt the film between the sealing bar and the platen. When a continuous bond is to be created, such as the edge seal required to form a tube from a flat film, band sealers are used. In this process, metal bands apply pressure and transport the film through heating and cooling zones. The rapid rate at which polyethylene film can be sealed suit it to the requirements of modern high speed packaging processes such as vertical form, fill, and seal. When complex or prototype seals are to be created, it is possible to seal polyethylene films with the aid of a handheld iron, the use of which requires much skill if consistent bonding is to be achieved.

Internal melting of films to effect heat sealing can be accomplished by radio-frequency heating (the technique also being known as dielectric heat sealing and high frequency welding). In this process an alternating electric current (typically with a frequency of 27.12 MHz) is applied between the sealing bar and the platen, which excites polarizable groups sufficiently to raise the temperature of the polymer into its melting range. In the case of polyethylene, this technique is limited to copolymers containing polar comonomers such as vinyl acetate or methacrylic acid. During sealing, the platen and sealing bar remain cold, preventing the outer surfaces of the film, with which they make contact, from melting.

The ability to create an effective heat seal may be limited by surface contamination. During packaging operations, contents in the form of liquids or powders may adulterate the sealing surfaces, preventing interdiffusion of the polymer. Other factors that may detrimentally affect sealing include blooming of additives to the surface and chemical surface modification such as corona treatment.

2. Welding

The welding of polyethylene involves the melting of surfaces that are subsequently brought into contact. Two principal variations exist, the direct bonding of adjacent surfaces and a more traditional welding approach in which an independent bead of polyethylene heated by hot air is used to create the bond. Direct bonding is widely applied to pipes, while hot air welding is used to fabricate tanks and the like from polyethylene sheet.

a. Direct Bonding. Lengths of polyethylene pipe can be joined end to end by the techniques of butt or socket welding, and junctions can be created by saddle welding.

In butt welding (also known as butt fusion and hot plate fusion), a heated metal plate is placed between the squared and cleaned ends of adjacent lengths of pipe. Slight pressure is applied to maintain good contact between the pipe ends and the plate. After a preset time the plate is rapidly withdrawn and the molten pipe ends are brought into contact with sufficient pressure to produce intimate contact and create interior and exterior circular beads of polymer. The pressure is then reduced and the joint is allowed to solidify. In smaller pipes (up to approximately 2 ft in diameter) this process can be automated, which produces more consistent joints than can be attained manually. Control variables include temperature, dwell time, initial contact pressure, and holding force during solidification. The application of too much pressure to the molten pipe ends will result in thin bonds that are weak. Conversely, with too little pressure the interdiffusion of molecules across the interface will not be adequate.

In socket fusion the cleaned ends of adjacent lengths of pipe are inserted into a socket that is heated to create the bond. The temperature of the socket is commonly raised by the resistive heating of wires embedded within it. The applied heat melts the interior of the socket and the exterior of the pipe ends to form the weld. Socket fusion involves less melt flow than butt fusion, so it is important that the surfaces to be joined be free of contaminants such as dirt, grease, and oxidation. As such joints are frequently made outdoors, rain and blowing dust can be troublesome. This problem can be alleviated to a great extent by the use of temporary welding shelters.

b. Hot Air Welding. Hot air welding involves melting a polyethylene welding rod and the surfaces to be bonded with a hot air jet. Before welding takes place, the workpieces are clamped in the desired position, with a V-shaped gap defining the welding channel. Typically the welding rod, which may be triangular, is automatically fed concentrically through the hot air jet of the welding gun. The hot air simultaneously melts the welding rod and the surfaces to be joined. To ensure adequate joint strength the substrates must have thicknesses exceeding one 1/16 in. Naturally, the joint surfaces and that of the welding rod must be kept scrupulously clean.

V. FABRICATION AND USES OF ULTRAHIGH MOLECULAR WEIGHT POLYETHYLENE

Ultrahigh molecular weight polyethylene (MW > 1,000,000) has many desirable solid-state characteristics, including excellent abrasion resistance, high toughness, and extremely high fiber modulus. However, the ultrahigh molecular weight that endows these resins with such desirable solid-state properties also endows them with melt viscosities so high that they cannot be processed by any standard technique involving melt flow. The fabrication of useful products from ultrahigh molecular weight polyethylene requires either extremely high pressures or dissolution. Compression molding and ram extrusion are used to process the melt at high pressures, compacted powder billets can be sintered, and gel spinning is used to produce highly oriented fibers.

A. High Pressure Melt Forming

The high pressure melt forming of ultrahigh molecular weight polyethylene is essentially a forging process in which resin powder is subjected to heat and extreme pressure within a die. Under these conditions, the resin particles amalgamate to form a void-free mass, molecules from adjacent particles entangling at their interfaces. High pressure melt forming processes have very low production rates.

In compression molding, an appropriate weight of resin powder is placed in the cavity of a robust mold, where it is heated above its melting point and compressed by a hydraulically actuated piston. Pressure and heat are maintained for sufficient time that molecular relaxation relieves local internal stresses caused by the deformation of particles. To prevent warpage, pressure is maintained while the molding is cooled. Products of compression molding are typically billets or blanks from which finished articles are subsequently machined.

In ram extrusion, pressure is applied directly to heated polymer resin in an extrusion die by a hydraulic piston. The shape of the die orifice determines the profile of the product, which slowly issues from it. Products include rods, bars, tubes, and sheets. The thickness of sheets may be reduced by subsequent rolling.

Ultrahigh molecular weight polyethylene's good balance of abrasion resistance, low coefficient of friction, chemical inertness, and high toughness suit it for a variety of specialty applications. The upper use temperature of ultrahigh molecular weight polyethylene is limited to approximately 100°C, above which unacceptable softening occurs. It is particularly useful as an unlubricated bearing material in applications that are inaccessible for routine maintenance, such as aeronautical cable guides, bushings in business machines, and replacement hip joints. Ultrahigh molecular weight polyethylene's lack of need of lubrication also suits it as a material for bushings in food and pharmaceutical processing equipment, where lubricants could be a source of contamination. In mining and quarrying operations, chutes and hoppers may be lined with ultrahigh molecular weight polyethylene sheeting to reduce friction and prevent wear on underlying metal surfaces. Seals and bushings in hydraulic rock hammers and drills, which are subject to abrasive dust and grit, are further examples of mining applications. Ultrahigh molecular weight polyethylene parts are also used extensively in the textile industry.

B. Sintering

Sintering is the process of heating a compacted preform of ultrahigh molecular weight polyethylene powder above its crystalline melting temperature. This is accomplished most effectively when the average powder particle size is minimized. The particle size of ex-reactor polymer powders may be reduced by using a pulverizing mill; a fluid energy mill is suitable. An average particle size of approximately 100 μm or less results in good final properties; particle shape plays a lesser role. A predetermined weight of polymer powder is placed in a mold at a temperature, below its crystalline melting temperature, where it is subjected to pressures in excess of 2000 psi. Under the influence of high pressure the powder particles deform and pack together to produce a porous solid that can be handled. After removal from the mold, the compacted preforms are heated in an oven to a temperature in excess of their crystalline melting point. At this temperature, individual particles melt and coalesce with their neighbors. The length of time spent in the oven depends upon the sample thickness; it is essential that the crystalline melting temperature be exceeded throughout the part. The extremely high viscosity of ultrahigh molecular weight polyethylene prevents the preforms from sagging or collapsing under their own weight. The density of the final product depends on the packing pressure, particle size, and particle shape. Useful products with densities ranging from approximately 0.88 to 0.93 g/cm^3 can be produced. At the higher end of the density range, applications overlap with those of products of high pressure melt forming processes. At the lower end of the density range, the product is porous and can be used for filtration purposes and in storage batteries.

C. Gel Spinning

The dissolution of ultrahigh molecular weight polyethylene in a large volume of solvent greatly reduces its entanglement density. As the solution cools, it forms a gel that consists of a swollen entangled network. Such gels can be spun into highly oriented fibers by using the process and equipment described in the wet spinning section of Chapter 8. The resulting product consists of highly oriented fibers, the modulus of which is in porportion to their degree of molecular alignment. These fibers are used in a few niche application where a high modulus-to-weight ratio is important. Applications include bulletproof vests and military helmets.

VI. MARKETS

The demand for all types of polyethylene in the United States totaled approximately 26,000 million pounds (Mlb) in 1998 [1]. This accounted for almost 30% of worldwide use. In the United States, demand was broken down into 12,400 Mlb of high density polyethylene, 7300 Mlb of linear low density polyethylene, and 6300 Mlb of low density polyethylene and ethylene-vinyl acetate combined; this translates to approximately 49%, 28%, and 24%, respectively. In terms of conversion processes, extrusion accounts for 52%, blow molding for 16%, injection molding for 12%, and rotational molding for 3%. The major outlets of the different polyethylene types are outlined in this section.

In the figures associated with the following subsections, values are rounded to the nearest whole percentage point and thus may not necessarily add up to 100%.

1. High Density Polyethylene

Approximately 85% of all high density polyethylene used in the United States is converted to end products by blow molding, extrusion, and injection molding, as shown in Figure 22.

Products of blow molding are used as outlined in Figure 23. Bottles, which make up the bulk of high density polyethylene blow-molded products, are largely

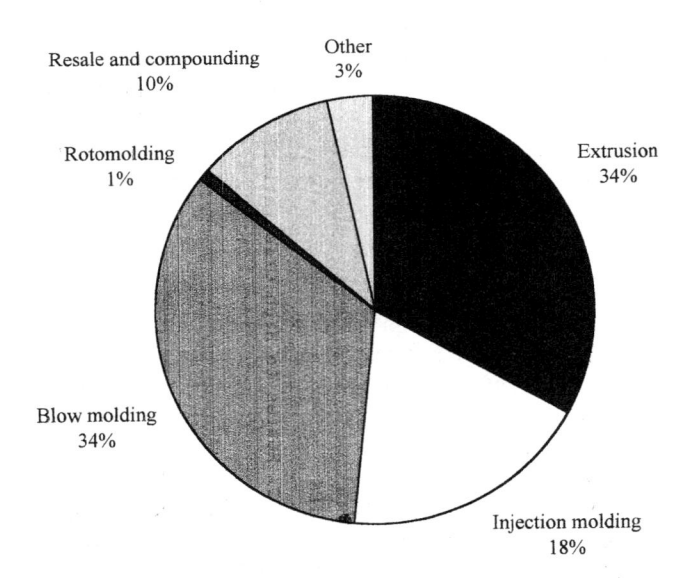

Figure 22 US demand for high density polyethylene.

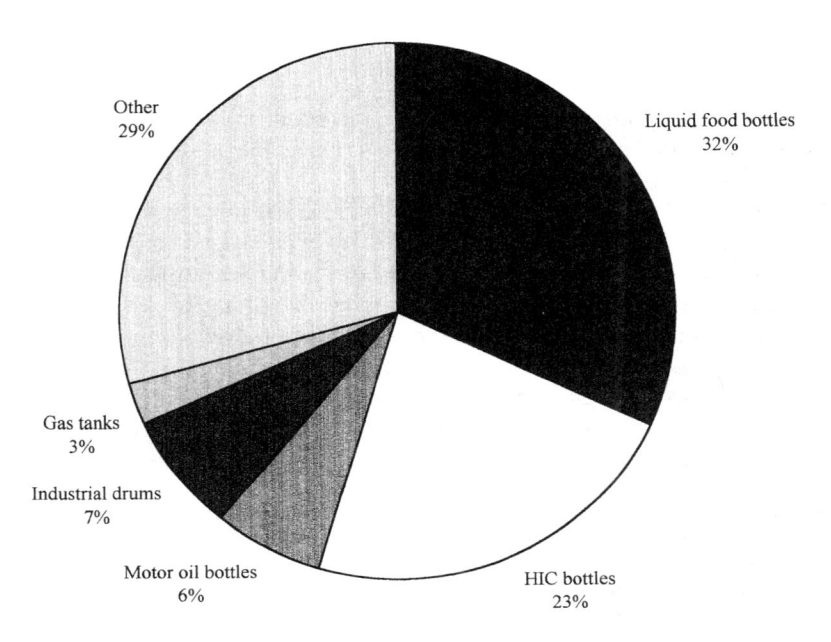

Figure 23 High density polyethylene extrusion products.

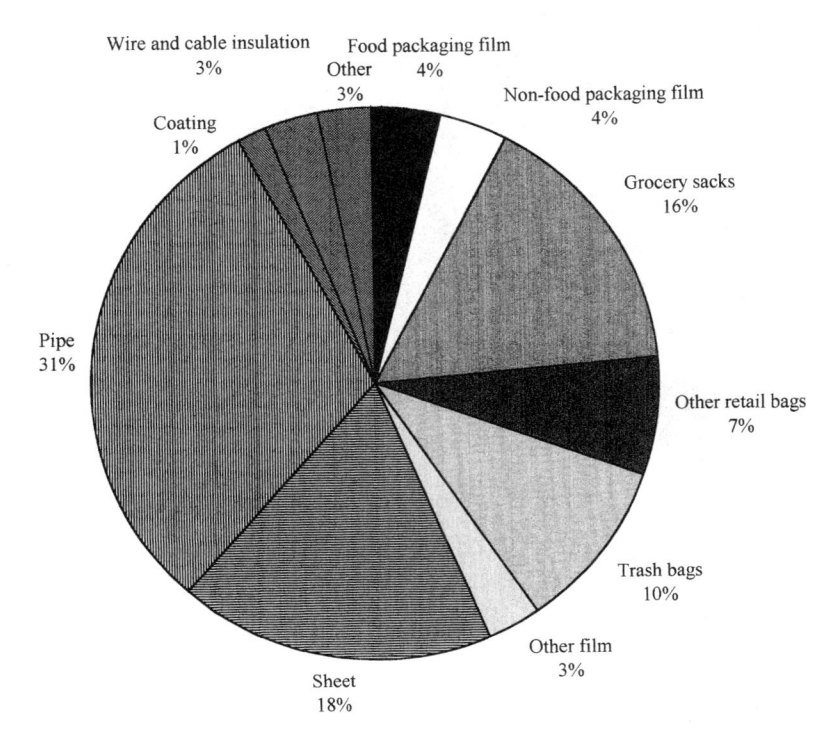

Figure 24 High density polyethylene blow molding products.

used for liquid foods (primarily milk and juice), household and industrial chemicals, motor oil, pharmaceuticals, and cosmetics. Other major outlets include drums, fuel tanks, and housewares.

Film accounts for almost half of all extruded high density polyethylene, as shown in Figure 24. Such films are used primarily to make bags such as grocery sacks, general merchandise bags, and institutional trash can liners. Approximately one-third of extruded high density polyethylene goes into pipe. Sheet is another major outlet, accounting for almost one-fifth of extruded polyethylene.

High density polyethylene is converted into a variety of products by injection molding as shown in Figure 25. Many of these products are used in packaging and transport applications such as food containers, crates, pallets, and pails. Other significant outlets include toys, housewares, and drinking cups.

2. Low Density Polyethylene and Ethylene-Vinyl Acetate Copolymer

The vast majority of low density polyethylene and ethylene-vinyl acetate copolymer is converted into end products by extrusion, as shown in Figure 26. Other conversion processes account for less than 5% each.

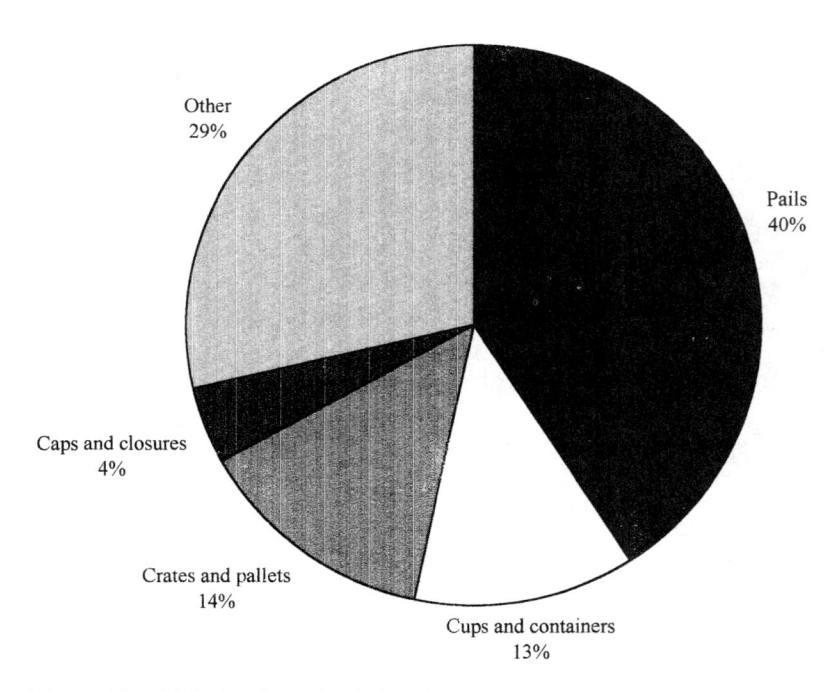

Figure 25 High density polyethylene injection molding products.

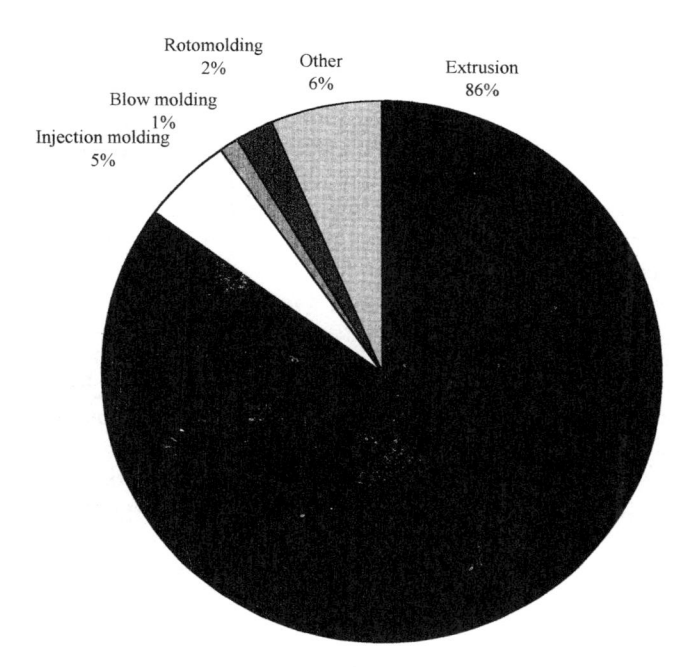

Figure 26 US demand for low density polyethylene and ethylene-vinyl acetate copolymers.

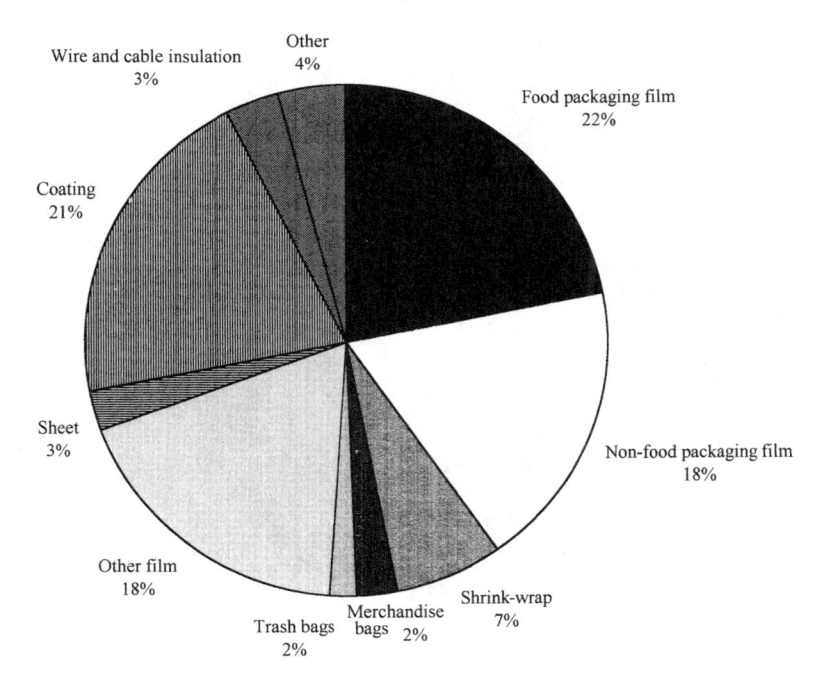

Figure 27 Low density polyethylene and ethylene-vinyl acetate copolymer extrusion products.

The principal uses of extruded low density polyethylene and ethylene-vinyl acetate copolymer products are outlined in Figure 27. Packaging film is the major outlet for low density polyethylene, accounting for approximately 40% of all usage. Another major use is the coating of paperboard and aluminum for food packaging. The former primarily uses linear low density polyethylene, and the latter, ethylene-vinyl acetate copolymer.

3. Linear Low Density Polyethylene

The principal fabrication process used to convert linear low density polyethylene to end products is extrusion, which accounts for approximately 80% of all usage. This is illustrated in Figure 28.

Linear low density extruded products share many of the same markets as low density products, but the proportions are different, as shown in Figure 29. Film dominates extrusion products, with packaging films, stretch-wrap and domestic trash bags consuming much of the product.

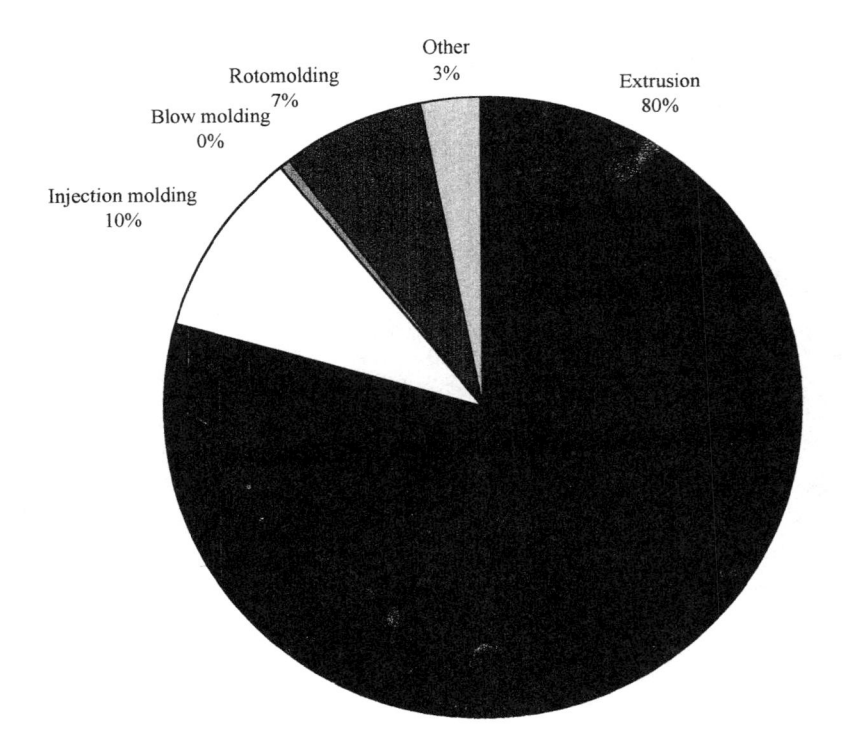

Figure 28 US demand for linear low density polyethylene.

Use and Fabrication of PE Products 507

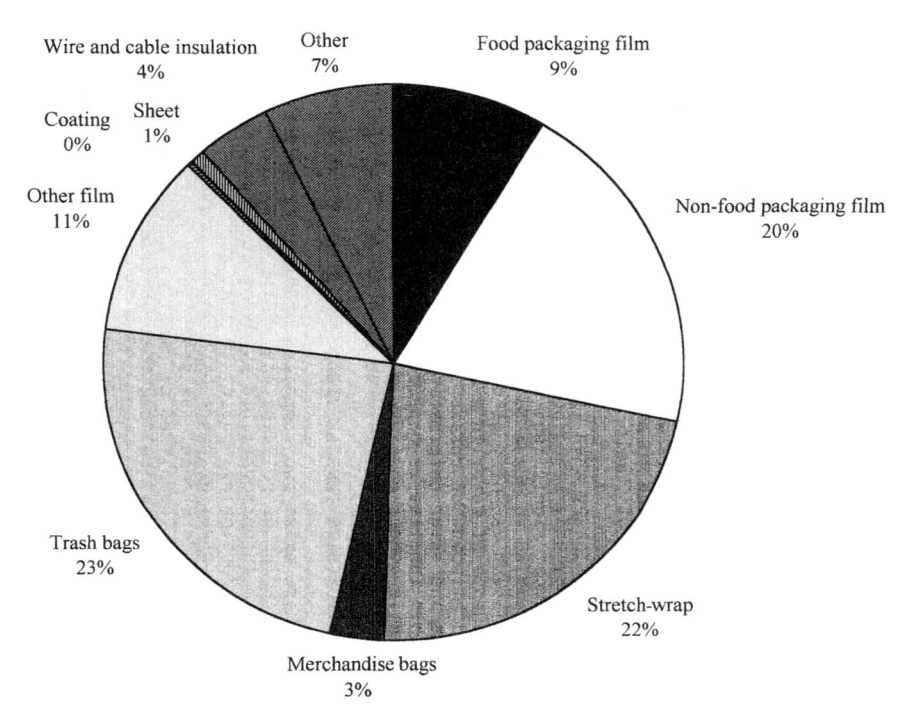

Figure 29 Linear low density polyethylene extrusion products.

Pie chart values:
- Wire and cable insulation 4%
- Other 7%
- Food packaging film 9%
- Coating 0%
- Sheet 1%
- Other film 11%
- Non-food packaging film 20%
- Trash bags 23%
- Merchandise bags 3%
- Stretch-wrap 22%

BIBLIOGRAPHY

Chanda M, Roy S K. Plastics Technology Handbook, 2nd ed. New York: Marcell Dekker, 1993.
Pearson J R A. Mechanics of Polymer Processing. London: Elsevier, 1985.
Rauwendaal C. Polymer Extrusion. Munich: Hanser, 1990.
White J L. Principles of Polymer Engineering Rheology. New York: Wiley-Interscience, 1990.

REFERENCE

1. Modern Plastics, New York: McGraw-Hill, January 1999, p. 72.

10
The Future of Polyethylene

I. DISPOSAL AND RECYCLING

One of the by-products of a highly industrialized society is a massive amount of waste that must be dealt with in one way or another. Although discarded plastic accounts for less than 10% by weight of all waste (much less, for instance, than either paper or grass clippings in the United States), it is perceived by the general public as a major contributor to the overall problem of waste disposal. As such, the disposition of plastic after it has fulfilled its original purpose is a problem that the polymer industry must confront. Consequently there is continuing interest in learning how to recycle plastic waste efficiently and, incidentally, in improving the overall image of "plastic" in the mind of the public.

Many of the issues raised with respect to polymer recycling are nontechnical; public perception, political maneuvering, and economic incentives are just a few of the factors involved. Given the volatility of the subject, only a brief technical review of each of the three principal strategies currently employed in the recycling of polymers in general and polyethylene in particular are presented in this chapter. A fourth method, "quaternary recycling," is a euphemism for incineration; it can only be considered to be recycling in terms of partial energy recovery.

A. Primary Recycling

Primary recycling consists of the preconsumer reuse of a single polymer stream for the purpose for which it was originally intended. This has been practiced for decades by polymer processors, who routinely save scrap, such as injection molding sprues, edge trimmings from film, and sundry malformed moldings, and subsequently regrind or shred them and return them directly to the feed hopper. This type of recycling poses few problems as long as the scrap material is clean and neither cross-linked nor thermally degraded. Minor deterioration of physical properties inevitably occurs when regrind is incorporated. However, as long as the raw material is adequately stabilized and fabrication conditions are not too extreme, as much as 10% scrap can be accommodated.

B. Secondary Recycling

Secondary polymer recycling consists of the collection, cleaning, and refabrication of useful items from the postconsumer waste stream. This is the recycling strategy most visible to the public. The more contaminated the waste stream or the more homogeneous the desired product, the greater will be the number and complexity of steps required to generate a usable resin. Two subcategories of secondary recycling exist: intensive mixing of commingled plastic waste to produce a polymer blend, and complete separation of the polymeric components of the waste stream to yield single-polymer products. Both of these approaches have drawbacks. Commingled plastic products have poor mechanical properties unless the components are suitably compatibilized, and single-polymer recycled resins are expensive to produce and susceptible to contamination.

A mixed waste stream comprising commingled plastic types can be processed to yield a replacement for lumber for use in nonstructural applications. This process involves collecting the waste plastic; sorting, shredding, washing, drying, and dry blending it with additives; melt mixing and extrusion. After the initial collection and delivery of polymeric waste to a recycling facility the next step is manual sorting. Two sorting strategies are employed. In the first method, unusable material such as nonpolymeric trash and identifiably noncompatible polymeric items—principally thermosets—are removed, the assumption being that the remaining polymers are processable. The second sorting strategy involves the positive selection of specific polymer types that are known to be processable, the remaining material being discharged to waste. The selected polymers are mechanically shredded to yield flakes which are then washed and dried. Drying must be thorough when polyesters and nylons are part of the waste stream; the presence of water will result in hydrolytic decomposition of these resins in the compounding extruder. The flakes are dry blended with processing aids, thermal stabilizers, antioxidants, and compatibilizers prior to intensive melt mixing in a twin screw extruder and profile extrusion. It is necessary to add compatibilizers, because the principal polymers found in postconsumer waste are inherently incompatible [1]. Compatibilizers migrate to the interfaces between incompatible polymer phases, where they are soluble to a limited extent in both the polymers, acting as a bridging material. Compatibilizers may consist of diblock copolymers, the opposite ends of which are miscible with different incompatible polymers, or a copolymer that is partially soluble in adjacent incompatible resins. The resulting billets issuing from the extruder may be cut, drilled, and joined by screws much as wood is handled.

The Future of Polyethylene

One scenario for secondary recycling to recover polyethylene as a single product involves collection, sorting, shredding, prewashing, grinding, separation by density, washing, spin drying, hot air drying, extrusion homogenization, and pelletization. This process starts with the consumer, who saves plastic waste and delivers it to a collection facility. Once a sufficient quantity has been amassed, the materials are baled and transported to a recycling plant. Here the items are manually sorted according to polymer type and color prior to shredding. The resulting flakes are washed to remove the majority of the contaminants, including paper from labels, the remains of the original contents, and the assorted debris that inevitably accompanies postconsumer plastic waste. The flakes are then ground into fine particles that are fed to a hydrocyclone, where they are separated on the basis of density. This eliminates the higher density polymeric contaminants such as poly(vinyl chloride) (PVC) and poly(ethylene terephthalate) (PET) but cannot remove polypropylene. The remaining granules are then additionally washed one or more times until acceptably clean, spin dried to remove most of the water, and then hot air dried to remove the remaining moisture. The granules are then dry blended with appropriate stabilizers, melt homogenized in an extrude, and pelletized.

The recovery of a useful polyethylene product from postconsumer mixed waste is fraught with peril. Many factors can reduce the value of the end product or even render it worthless. The two principal concerns are feedstock heterogeneity and the control of molecular polydispersity in the end product. The fact that the molecular composition of virgin resins can be tailored to customer needs may be the undoing of the recycling effort. Virgin polyethylene is marketed on the basis of discrimination of one product from another based upon specific characteristics. When polymers are recycled, their molecular composition and hence their physical properties cannot be controlled to a high degree of precision. Hence they cannot be marketed in terms of finely controlled characteristics, and thus they sell for less than prime virgin resins. Given the low cost of virgin polyethylene, it is difficult to reclaim polyethylene economically. The key to economic secondary recycling is automation; current techniques are labor-intensive, especially the manual sorting step. Unless more effective methods of sorting and cleaning can be devised, only a small fraction of the postconsumer waste stream will be recycled.

The problems associated with the recovery of a high grade recycled polyethylene product start with the quality of the postconsumer waste stream. In addition to polyethylene, the waste stream inevitably contains polymeric and nonpolymeric contaminants. Small amounts of certain thermoplastics can be accommodated in recycled polyethylene, but the presence of even a few tens of parts per million of others can cause significant problems [2]. Particles of thermosets, such as polyurethanes and cross-linked polyesters, that survive the sorting and washing processes can plug screenpacks. Those that make it into the final product can act as stress concentrators, reducing material properties (especially toughness), or they may be unsightly, reducing aesthetic appeal. Thermoplastics with high melting temperatures, such as nylons and poly(ethylene terephthalate), can cause similar problems. Polymeric contaminants, such as polypropylene, which melt in the same temperature range as polyethylene are rarely compatible in the solid state and can seriously reduce the mechanical properties of final products. Poly(vinyl chloride) is an especially unpleasant contaminant in that it is prone to degradation upon extrusion, emitting hydrogen chloride, which damages metal parts, and leaving behind charred fragments which are unsightly and reduce physical properties. Nonpolymeric contaminants include dirt, food residue, oil, paper, aluminum foil, and pigments. The solid contaminants can generally be excluded, but chemicals, especially those containing a hydrocarbon chain, can permeate polyethylene and become virtually impossible to remove. The presence of low molecular weight hydrocarbon chains can detrimentally affect the processing behavior of the product, altering rheological characteristics, "plating out" in extruder dies, and accumulating at die lips, where they degrade to emit smoke. Greases, fats, and other food residues that make it into the final product can impart distinctive and unpleasant odors. Unless sorting yields a single-product waste stream, such as unpigmented milk jugs, uniform coloration will be an issue. When no attempt is made to separate different colors, the resulting polyethylene product from a typical waste stream is invariably some shade of green.

The problems associated with recycled polyethylene are not limited to those caused by extraneous materials. Postconsumer polyethylene is frequently degraded to some extent, especially when it has been exposed to sunlight, such as in the case of agricultural films. Degradation may result in cross-linking or chain scission and the incorporation of carbonyl, vinyl, and hydroxyl functionalities, all of which promote further degradation. The original antioxidants and stabilizers that were partially consumed during fabrication and use are further depleted during the cleaning process. The recycled product therefore requires restabilization, often at a higher level than originally applied, to counteract the deleterious effects of contaminant [3,4].

The polydispersity of a recycled polyethylene resin determines its utility as a feedstock suitable for further fabrication processes. Both the average composition and the distribution of the component molecules must be addressed. Polydispersity comes in the form of broad molecular weight distribution, various comonomer types, and broad composition distribution. Careful sorting can isolate certain products such as high density polyethylene milk jugs, but this is not the general case. The molecular weight distribution of the product will be affected by the average molecular weight and molecular weight distribution of the waste stream components and the level to which they are degraded. Compounding the problem is the fact that many containers are made from polyethylene blends or consist of multiple layers. Waste stream heterogeneity translates into products having wider polydispersities than typical virgin resins. The material characteristics of such products normally render them unsuitable for use as a single feedstock in fabrication processes. Among the problems encountered are low melt strength, poor environmental stress crack resistance, inadequate toughness, and inconsistency of the product from batch to batch. In practice, recycled polyethylene resins are typically used as a minor component in blends with virgin resins.

C. Tertiary Recycling

The reconstitution of low molecular weight organic feedstock by the controlled decomposition of polymeric waste is known as tertiary recycling. In the case of polyethylene, massive degradation is brought about by the application of heat in one of three processes known collectively as thermolysis [5]. Pyrolysis occurs in the absence of oxygen, gasification when a controlled amount of oxygen is supplied, and hydrogenation in the presence of hydrogen and a catalyst. The equipment used includes fluidized beds, rotating kilns, and retorts. Solvolysis, whereby chemical degradation is used to reduce the molecular weight of polymers produced by condensation polymerization, is not feasible in the case of polyethylene.

Pyrolysis yields a variety of products depending upon the nature of the processing equipment, the temperature employed, and the components of the waste stream with which it is fed. The product typically consists primarily of a synthetic crude oil with lesser quantities of bituminous and gaseous products. The higher the processing temperature, the greater will be the proportion of the gaseous products [6]. Pyrolysis of polyethylene at 700°C results in a split of 51% gas, 42% oil, and 6% bitumen residue [7]. Hydrocarbon gases are collected, separated and purified (a fraction being burned to heat the reactor), liquid products can be used as a feedstock in a commercial oil refinery, while the tar can be treated like the nondistillable fraction obtained from crude oil. Catalytic cracking can be combined with heat to yield products that are primarily gases and low boiling hydrocarbons [8].

Gasification is performed at higher temperatures than pyrolysis, in the presence of a controlled amount of oxygen. The result is complete decomposition of the polymer, which in the case of polyethylene yields a syngas consisting primarily of carbon monoxide and hydrogen, which are valued as feeds in a variety of chemical syntheses [5].

Waste polymer hydrogenation is a variant of a similar process used in oil refining. Degradation takes place in the presence of high pressure hydrogen over a bed of catalyst such as a zeolite or silica-alumina [9]. Decomposition and hydrogenation occur simultaneously, to yield products consisting primarily of liquid hydrocarbons that may be used as fuels [10].

D. Future Development of Recycling Strategies

Decisions regarding the future of polyethylene recycling are likely to be driven as much by societal and legislative pressures as they are by technical and economic ones. Primary recycling will likely continue much as it has done in the past, unaffected by external issues. If anything, it is likely to decrease in importance as processing equipment and virgin resins improve to the extent that less scrap is generated in the first place. The importance of secondary, tertiary, and quaternary recycling relative to traditional disposal methods may be decided based in part upon studies that evaluate environmental impact "from cradle to grave," i.e., from the processes used by the oil industry to extract crude oil from the ground all the way through to the final disposition of the polymer. Factors that would be taken into account include resource depletion, energy expenditure, and pollution of the environment [11].

Increased demand for recycled polymer streams will raise their value and hence improve the viability of recycling strategies that are currently marginal or completely uneconomical. However, virgin resins are relatively inexpensive and are likely to stay that way, which will limit the effect of this positive economic driver. Looking at recycling from another point of view, the cost of landfill may increase to the point that it becomes such an expensive disposal method that recycling may start to make more sense in terms of limiting the money spent on postconsumer garbage disposal.

From an energetic standpoint, the most desirable of the postconsumer recycling strategies is secondary recycling to yield single-product streams. The main trouble with this is one of material discrimination. Some polymers, such as the poly(ethylene terephthalate) in soda bottles and the high density polyethylene used for milk jugs, are readily identified, but it is no simple matter to identify many others. Manual sorting is effective, but it has several limitations, not the least of which is its cost. Automated identification and sorting strategies are being developed, but none is widely used. Spectroscopic identification methods are promising but are currently in their infancy [12,13]. As packaging becomes more complex, with the increasing use of controlled atmosphere containers and multilayer films, etc., the problem of material discrimination will only become more complicated.

Due to the added difficulties associated with the handling and cleaning of films relative to compacted materials such as crushed bottles, the recycling of polyethylene is largely limited to blow and injection molded containers [14]. Pilot studies have been performed with regard to recycling various single-source agricultural products such as greenhouse film, mulch layers, and cotton bale overwrap, but these have been hindered by issues of polymer degradation and contamination [15,16].

In addition to the effect of contamination on polyethylene material proper-

The Future of Polyethylene

ties, there is concern regarding contaminants leaching from products made from recycled resin. The principal consideration in this case is one of food contamination. Currently no recycled resins are cleared for use in food contact applications. It is possible to incorporate recycled polyethylene as a core layer in a multilayer package, but this requires the use of multiple extruders, which may not be available on all fabrication equipment and may increase overall costs.

Once all the possible single-product stream materials have been separated, the remaining polymers are presently suitable only for conversion into plastic lumber, which has relatively little value. Concerns regarding the material limitations of plastic lumber (especially in the long term) restrict its use to light load and non-safety-critical applications.

On the face of it, tertiary recycling and incineration should be suitable for dealing with postconsumer waste not acceptable for secondary recycling, but concerns with respect to pollution and economic viability limit their use. One of the principal problems is that of poisonous gaseous by-products such as hydrogen chloride, hydrogen cyanide, and phosgene. Methods are available for dealing with such gases, but they involve increased operational complexity and cost [17]. Disposal of the ash from incineration is also of some concern as it inevitably contains traces of heavy metals that render it toxic. From an energy standpoint, both tertiary recycling and incineration are unfavorable relative to secondary recycling as the energy expended in the polymerization process is lost.

II. METALLOCENE TECHNOLOGY

Metallocene catalysts are used to produce a distinctive range of resins that are less polydisperse than those available from traditional catalyst systems. Each catalyst contains only one type of active site, all of them polymerizing the available monomers in an identical fashion. The net result is a uniform polymeric product that has a narrow molecular weight distribution and homogeneous comonomer incorporation.

Although metallocene types of structures have been known for several decades [18,19], their potential as commercial catalysts remained unrealized until 1980, when Kaminsky and coworkers [20,21] discovered that methylalumoxane improved their catalytic activity dramatically. Since that discovery, massive and intense research programs have been undertaken to bring metallocene products to commercial fruition. Reviews of the development of metallocene catalysts can be found in papers by Horton [22] and Kaminsky [23]. The reason for this enormous interest lies in the ability of metallocene catalysts to provide well-defined polymer products, opening the way to the molecular engineering of resins with properties tailored to the precise needs of the end user. Particular effort has been expended to replace conventional (Ziegler–Natta-type) linear low density polyethylene and very low density polyethylene resins. Elastomeric metallocene-catalyzed very low density polyethylene resins became available commercially in 1993, with metallocene-type linear low density polyethylene (sometimes dubbed mLLDPE) products following in 1995.

Metallocene-type ethylene–α-olefin copolymers are characterized by their narrow molecular weight distribution ($\overline{M}_w/\overline{M}_n \approx 2.0$) and homogeneous composition distribution. This is in contrast to Ziegler–Natta copolymers, which are broadly polydisperse in terms of molecular weight and composition, their longer molecules incorporating a lower percentage of comonomers than their shorter molecules. The difference between the two types of linear low density polyethylene is shown schematically in Figure 1, in which fractional content is plotted as a function of molecular weight and comonomer content. The lack of high and low molecular weight tails in metallocene-type linear low density polyethylene resins has significant effects on their processing characteristics and physical properties. Resins made with so-called constrained geometry catalysts contain a small but significant amount of long-chain branching in addition to the short-chain branching derived from the comonomer [24].

The relatively open active sites of metallocene catalysts permit the copolymerization of nontraditional cyclic comonomers, such as styrene and norbornene, with ethylene. Although such resins are not commercially available at present, they have the potential for exhibiting novel physical characteristics, possibly expanding the use of polyethylene into new markets. Metallocene technology has also been developed for the production of isotactic and syndiotactic polypropylene, copolymers of propylene with other olefins, and syndiotactic polystyrene.

Metallocene-type linear low density polyethylene resins are not as readily processed as conventional resins. The root of the problem lies in the narrow molecular weight distribution of the metallocene product, which results in limited shear sensitivity. The difference between the two types of resins is illustrated in Figure 2, which shows their viscosity versus shear rate characteristics. For a given melt index (or weight average molecular weight), metallocene-type resins exhibit lower viscosities at low shear rates and higher viscosities at high shear rates than their Ziegler–Natta counterparts. This translates to decreased melt strength and elasticity in unconstrained situations like film blowing and increased back pressure, temperature, and motor load in extruders. Low melt strength results in less stable bubbles and webs in film production and consequently more breaks or reduced line speeds. According to one resin producer, the preferred extruder configuration for processing metallocene-type linear low density polyethylene is one of "dual channel, low-shear, barrier mixing screws in smooth-bore extruders" [25]. On the positive side, the absence of low molecular weight species in metallocene-type linear low density polyethylene resins results in less "plate-out" and smoking at the die lips. In addition, metallocene-type linear low density polyeth-

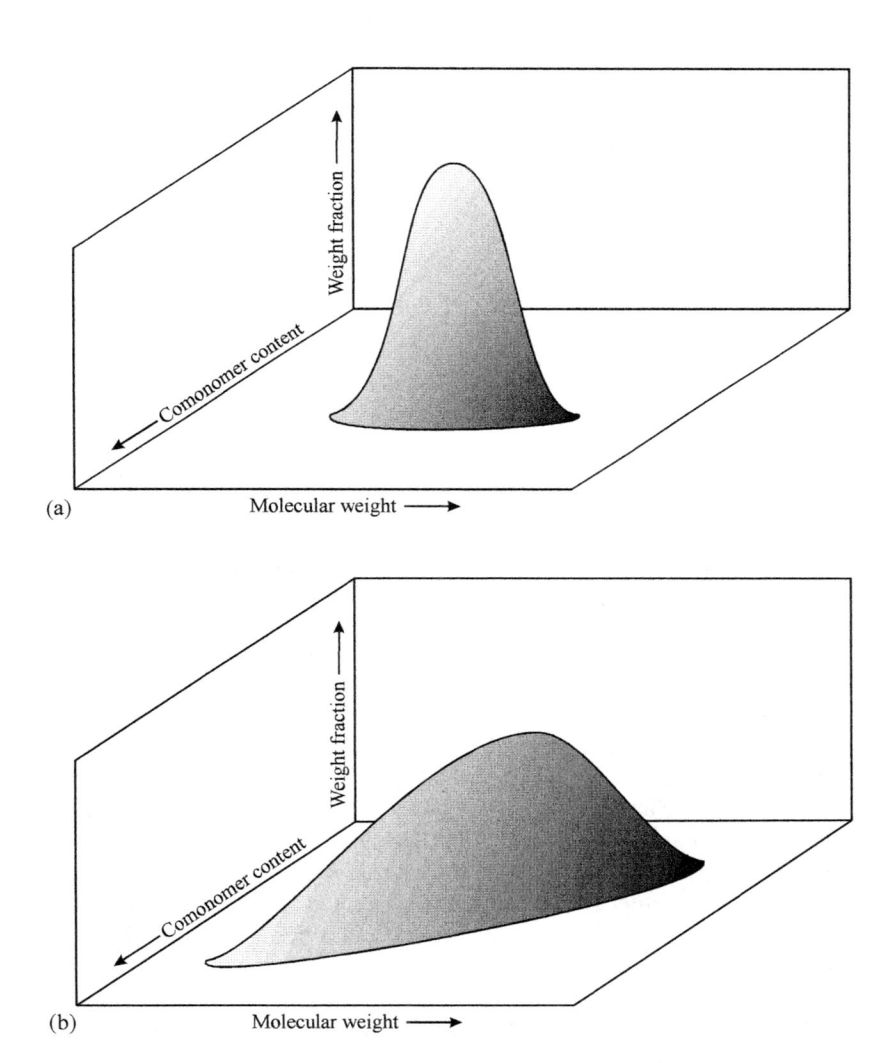

Figure 1 Schematic diagram showing the difference in molecular composition between (a) metallocene-type and (b) conventional linear low density polyethylene.

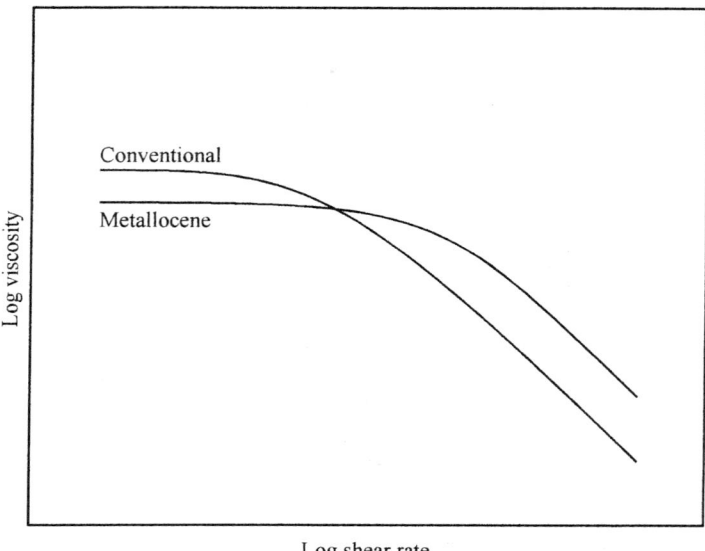

Figure 2 Schematic diagram showing the viscosity versus shear rate characteristics of conventional and metallocene linear low density polyethylene resins.

ylene can be drawn down more than conventional resins, which is beneficial from the standpoint of downgauging to produce thin films.

There are two basic methods by which the processing characteristics of single-site resins can be improved: broadening their molecular weight distribution or incorporating long-chain branches. A broader molecular weight distribution can be achieved by blending components with different molecular weights, creating polymers with different molecular weights simultaneously by running two catalysts in one reactor, or running a pair of reactors in series, employing different polymerization conditions in each reactor. Of the three options, the mixed catalyst approach is the most attractive; few resin producers have the capability to run reactors in series, and postreactor blending is expensive. Long-chain branching can be obtained from constrained geometry catalysis.

The mechanical properties of metallocene-type linear low density polyethylene resins are better than conventional resins in many respects, but there are also deficiencies in certain areas. On the positive side, the impact strength, puncture resistance, and tensile strength of metallocene-type linear low density polyethylene films are all improved by a considerable margin over those of conventional resins—300%, 50%, and 40%, respectively, according to one resin manufacturer

[26]. On the downside, metallocene-type linear low density polyethylene films have lower tear strengths than their conventional counterparts. When tear strength is not a crucial factor, the use of metallocene-type linear low density polyethylene resins permits downgauging, which is always attractive to film converters.

In many other areas the physical characteristics of metallocene-type linear low density polyethylene resins are highly attractive. Metallocene products have excellent optical characteristics; both gloss and clarity are significantly better than those of conventional resins of the same density. The absence of low molecular weight species improves environmental stress cracking resistance and presumably will have a favorable effect on slow crack growth. The lack of highly branched, noncrystallizable, low molecular weight species means that very low density elastomeric resins are not sticky, mitigating a major problem when it comes to material handling. Pellets of metallocene-type very low density polyethylene can be dry blended with polypropylene prior to melt mixing to create rubber-toughened products with higher elastomer loadings than are possible with conventional ethylene-propylene copolymer rubbers. Metallocene-type linear low density polyethylene films, lacking low molecular weight noncrystallizable species, are less likely to contaminate goods with which they come into contact, making them attractive in food packaging applications. Another advantage in this area is increased breathability, which permits gaseous exchange while preventing moisture loss, which is a desirable combination for maintaining the freshness of fruits and vegetables. Metallocene-type linear low density polyethylene resins have lower melting points than conventional resins of similar density, which is desirable when it comes to heat sealing. Metallocene-type resins also exhibit good hot tack and seal integrity, both of which are advantageous when it comes to form, fill, and seal packaging.

Some of the principal outlets of metallocene-type linear low density polyethylene and very low density polyethylene resins are in coatings, tie layers, and seal layers in multilayer film packaging. Contents include fresh produce, meat, coffee, pet food, cheese, and breakfast cereal. The trend is toward the replacement of ionomer and ethylene-vinyl acetate copolymer films in applications where clarity and puncture resistance are desired without the superior grease and oil resistance or heat sealing characteristics of ionomer films. Other film outlets include stretch films and bale wrap for rubbers. The clarity and softness of metallocene-type linear low density polyethylene resins and the absence of extractable species in them make them attractive for certain medical applications such as respirator masks, blood bags, and intravenous fluid bags. The use of metallocene-type linear low density polyethylene as a component of foam is being explored with regard to such applications as footwear, gaskets, and packaging.

The widespread penetration of metallocene resins into commodity markets is currently limited by their higher cost compared to conventional resins and their poor processing characteristics. The cost differential comes primarily from the metallocene catalyst and cocatalyst. This differential has been falling over the years and will continue to be eroded by the development of more active catalyst systems, more efficient production of catalysts, and improvements in polymerization know-how. Processing problems will be mitigated by the development of more sophisticated resins, processing machinery development, familiarity with the product, and blending with more easily processed resins. The industry went through a similar process of refinement when high density polyethylene was introduced after processors had become accustomed to low density polyethylene. Given the many desirable characteristics of metallocene-type resins, it is unlikely that processing difficulties will prevent their adoption in many important applications. Having invested billions of dollars in the development of metallocene products, resin manufacturers have a very strong incentive to make them a commercial success.

III. EMERGING PRODUCTION PROCESSES

Throughout the polyethylene industry there is a sustained effort to develop more efficient methods of resin production. Two principal forces drive process development: reduced production costs and enhanced resin characteristics. Development of production processes takes two forms: improving the efficiency of existing plants and designing new facilities. Enhancing the efficiency of existing reactors is desirable as it extends their useful life and reduces capital expenditure on new equipment. This economic advantage is partially offset by the need to upgrade ancillary equipment such as heat exchangers, finishing extruders, and resin-handling apparatus. In addition to the traditional goal of reducing production costs, there is the added objective of improving the end use properties and processing characteristics of polyethylene resins. With the advent of metallocene catalysis, manufacturers are preparing to take full advantage of the potential for tailoring resins to meet precise customer needs. To these goals must be added the incentive to license new technology to other producers with less proactive development strategies.

Announced upgrades to existing reactor technology have principally come in the area of gas-phase polymerization. The challenge for much research has been the increased dissipation of the large amounts of heat developed within the fluid bed in which polymerization takes place. Three companies have announced improvements in this regard which are variously termed "condensing mode" or "supercondensing" technology. Condensing mode technology—which claims to double the output of gas-phase reactors—operates by separating liquid ethylene from the recycle stream and injecting it directly through special nozzles into the fluid bed, where it instantly vaporizes, absorbing heat as it does so [27,28]. Supercondensing technology is believed to be similar to, but distinct from, condensing mode technology, providing an output increase of threefold or greater

[29]. By some estimates, improvements in reactor technology could increase output by as much as 15-fold over a plant's original capacity [29]. Also in the field of gas-phase technology, manufacturers are developing mixed catalysts that yield resins with bimodal polydispersities, which provide benefits in processing and properties [30]. Both mixed metallocene and Ziegler–Natta with metallocene combinations have been suggested.

The Spherilene process is used to produce spherical reactor particles with a smooth surface, which, it is claimed, provide several advantages [31]. The key to this process lies in careful control of the structure of the magnesium chloride catalyst support, which is replicated as the polymer particles grow. The resulting product comprises particles with diameters in the range of 0.5–4 mm that require no pelletization prior to shipment. The spherical shape of the particles results in a free-flowing product with high bulk density over a wide range of densities and melt indices.

Several companies are using, or are in the process of developing, plants that exploit twin-reactor technology [32–35]. These configurations are variously referred to as "series," "staged," "tandem," or "cascade" reactors. This capability permits manufacturers to produce resins comprising an intimate blend of components that have beneficial processing characteristics and improved solid-state properties. A bimodal molecular weight distribution increases shear thinning at a constant melt index, thereby improving processability, while a bimodal composition distribution can improve toughness, especially when the high molecular weight component has a higher concentration of short-chain branches. Such resins are gaining favor as materials for gas and water pipes [36]. Twin-reactor technology can be used with loop, solution, slurry, or gas-phase reactors in various combinations. The narrow molecular weight distribution and composition distribution of metallocene polyethylene resins lend themseleves to exploitation in this type of plant. It thus seems likely that twin-reactor technology will gain in importance as metallocene technology develops.

BIBLIOGRAPHY

Benedikt GM, Goodall BL. eds. Metallocene-Catalyzed Polymers—Materials, Properties, Processing and Markets. Norwich, NY: Plastics Design Laboratory, 1998.
Heitz W. ed. Recycling of Polymers. Makromol Chem Macromol Symp 57, 1992.
Morse PM. Chem Eng News, July 6, 1998, p 11.

REFERENCES

1. M Xanthos, A Patel, S Dey, SS Dagli, C Jacob, TJ Nosker, RW Renfree. Adv Polym Tech 13:231, 1994.

2. P Breant. Makromol Chem Macromol Symp 57:353, 1992.
3. LT Pearson. In: PN Prasad, ed. Frontiers of Polymer and Advanced Materials. New York: Plenum, 1994.
4. R Pfaendner, H Herbst, K Hoffmann, F Sitek. Angew Makromol Chem 232:193, 1995.
5. G Mackey. ACS Symp Series 609:161, 1995.
6. JA Conesa, R Font, A Marcilla, AN García. Energy & Fuels 8:1238, 1994.
7. G Menges, R Fischer, V Lackner. Int Polym Process 7:291, 1992.
8. R Lin, L White. J Appl Polym Sci 58:1151, 1995.
9. Y Ishihara, H Nanbu, T Ikemura, T Takesue, T Kuroki. Mater Life 2:238, 1990.
10. SJ Ng, H Seoud, M Stanciulescu, Y Sugimoto. Energy & Fuels 9:735, 1995.
11. C Mølgaard, L Alting. ANTEC '94, p 3021.
12. J Graham, PJ Hendra, P Mucci. Plastics Rubber Compos Process Appl 24:55, 1995.
13. TQ Soler. Makromol Chem Macromol Symp 57:123, 1992.
14. RD Leversuch. Mod Plast 70:65, 1993.
15. C Llop, A Pérez. Makromol Chem Macromol Symp 57:115, 1992.
16. RD Leversuch. Mod Plast October 1993, p 17.
17. W Kaminsky. Angew Makromol Chem 232:151, 1995.
18. EO Fischer Angew Chem 22:620, 1952.
19. R Wilkinson, IM Birmingham. J Am Chem Soc 76:5072, 1954.
20. H Sinn, W Kaminsky. Adv Organomet Chem 18:99, 1980.
21. W Kaminsky, M Miri, H Sinn, R Woldt. Makromol Chem Rapid Commun 4:417, 1983.
22. AD Horton. Trends Polym Sci 2:158, 1994.
23. W Kaminsky. Macromol Chem Phys 197:3907, 1996.
24. JC Stevens. Stud Surf Sci Cat 101:11, 1996.
25. Plast News, July 22, 1996, p 10.
26. B Vernyi. Plast News, Sept 18, 1995, p 1.
27. P Layman. Chem Eng News, Jan 23, 1995, p 8.
28. Mod Plast, March 1995, p 13.
29. B Vernyi. Plast News, Mar 6, 1995, p 1.
30. RD Leversuch. Mod Plast, December 1994, p 26.
31. P Galli. Plast Rubber Compos Process Appl 23:1, 1995.
32. RD Leversuch. Mod Plast, August 1995, p 20.
33. B Veryni. Plast News, Oct 23, 1995, p 7.
34. R Colvin. Mod Plast, December 1995, p 21.
35. RD Leversuch. Mod Plast, January 1997, p 33.
36. J Scheirs, LL Böhm, JC Boot, PS Leevers. Trends Polym Sci 4:408, 1996.

Index

Adjacent reentry model of crystallization, 88–90, 93–94 (*see also* Crystallization, mechanisms of)
Applications, 22–25, 459–507, 502–507
  blow molded products, 490–491
  coatings, 478, 479–480
  ethylene-vinyl acetate copolymer, 25, 462, 504–506
  film, 468–470, 471
  foams, 495
  high density polyethylene, 22–24, 460–461, 502–504
  injection molded products, 483–484
  linear low density polyethylene, 24–25, 461–462, 506
  low density polyethylene, 24, 461, 504–506
  pipe, 475–477
  rotationally molded products, 494
  sheet, 471
  very low density polyethylene, 24–25
Arc resistance, 215–216 (*see also* Testing, electrical properties)

Autoxidation, 377–382
  branching, 380
  chain scission, 380
  initiation, 377–379
  propagation, 379–380
  termination, 380

Backbiting, 44–45
Biodegradation, 388–389
Blocking (*see* Testing, physical properties, cling measurement)
Blow molding, 484–491
  process, 484–490
  products, 490–491
Bonding techniques, 497–499
  sealing, 497–498
  welding, 498–499
Bottle blowing (*see* Blow molding)
Branching in low density polyethylene, 44–45
Brittle failure (*see* Rupture phenomena)

Cable coating (*see* Coating, wire and cable)
Capacitance, 212–214 (*see also* Testing, electrical properties)

Catalysis
    chromium oxide, 61–62
    constrained geometry (*see* Catalysis, metallocene)
    metallocene, 40, 60–61
    metal oxide, 61–62, 65–66
    single-site (*see* Catalysis, metallocene)
    uniform-site (*see* Catalysis, metallocene)
    Ziegler-Natta, 53–54, 58–59
Central core model, 93–94
Characterization, 241–370 (*see also* Testing)
    ASTM methods, 366–370
    of cross-linked polyethylene, 305–310
        determination of cross-link density, 308–310
        gel content analysis, 306–308
    melt rheological, 258–269
        capillary rheometry, 260–262
        die swell measurement, 267–268
        drag flow rheometry, 263
        dynamic mechanical analysis, 268–269
        melt elongation, 263–267
        melt indexing, 259–260
        melt strength determination, 265–267
    molecular, 242–258
        calorimetric investigation of composition distribution, 254–256
        composition distribution determination, 252–258
        cross-fractionation, 256–258
        infrared spectroscopy, 251
        light scattering, 249

[Characterization]
        membrane osmometry, 249–250
        molecular weight determination, 242–250
        nuclear magnetic resonance spectroscopy, 251–252
        size elution chromatography, 242–247
        spectroscopy, 250–252
        temperature rising elution fractionation, 253–254
        vapor pressure osmometry, 250
        viscometry, 247–249
    rheological (*see* Characterization, melt rheological)
    solid state, 269–310
        atomic force microscopy, 276
        crystallinity measurement (*see* Degree of crystallinity)
        crystallization half-time determination, 295–296
        crystallization temperature determination, 293–295
        differential scanning calorimetry, 292–296
        dilatometric analysis, 297–298
        electron microscopy, 271–276
        electron microscopy, scanning, 272–274
        electron microscopy, transmission, 274–276
        infrared spectroscopy, 285–288
        melting temperature determination, 293–295
        microscopy, 269–276
        nuclear magnetic resonance, 291

[Characterization]
  optical microscopy, 270–271
  Raman internal mode analysis, 290
  Raman longitudinal acoustic mode analysis, 290–291
  Raman spectroscopy, 288–291
  scattering measurements, 276–285
  small angle light scattering, 282–284
  small angle neutron diffraction, 282
  spectroscopy, 284–291
  thermal analysis, 291–298
  thermogravimetric analysis, 296–297
  vibrational spectroscopy, 285–291
  X-ray diffraction, 278–281
  X-ray diffraction, small angle, 281
  X-ray diffraction, wide angle, 278–281
Chemical modification, 401–405
  of the bulk, 401–402
  of surfaces, 402–405
    corona treatment, 402–403
    etching, 403
    flame treatment, 403–404
    graft modification, 405
    oxidative methods, 404–405
Chemical reactions, 375–412
Cilia, 68–69
Cling, 352–354
Coating, 477–480
  film, 477–478
    process, 477–478
    products, 478

[Coating]
  wire and cable, 478–480
    process, 478–479
    products, 479–480
Composition
  distribution, 10
  molecular, 5–10 (*see also* Structure, molecular)
Connections, interlamellar (*see* Tie chains)
Copolymers
  ethylene with 1-alkenes (*see* Linear low density polyethylene; Very low density polyethylene)
  ethylene with polar monomers, 405–412 (*see also* Ethylene vinyl ester copolymer; Ethylene vinyl acetate copolymer; Ionomers)
  ethylene with vinyl acetate (*see* Ethylene vinyl acetate copolymer)
  ethylene with vinyl alcohol copolymer (*see* Ethylene vinyl alcohol copolymer)
  ethylene with vinyl ester (*see* Ethylene vinyl ester copolymer; Ethylene vinyl acetate copolymer)
  ionomers (*see* Ionomers)
Corona treatment, 402–403
Creep, 152–158
Cross-linked polyethylene, 5–6, 399–401
  discovery, 36–37
  properties, 20
Cross-linking, 389–401
  effects of, 399–401
  radical, 390–397
    chemical decomposition induced, 393–396

[Cross-linking, radical]
  high energy induced, 392–393
  peroxide induced, 393–396
  ultraviolet radiation initiated, 396–397
  silane bridged, 397–399
Crystallites, 14, 76–80
  lamellar, 14, 77–80
  thickness, 110–111
Crystallization, 76–119
  adjacent reentry model, 88–90
  Flory model, 88–89, 91–94
  kinetics, 98–107
    molecular composition effects, 102–107
    orientation effects, 100–102
    temperature effects, 100–101
  mechanisms of, 83–98
  from the molten state, 114–119
    effects of orientation, 119 (*see also* Crystallization, from oriented melts)
  nucleation, 84–87
    heterogeneous, 86
    homogeneous, 85–86
    primary, 84–87
    secondary, 87
    self, 87
  primary, 87–98
  from oriented melts, 416–424
  from oriented solutions, 424–425
  rate (*see* Crystallization, kinetics)
  regimes, 94–98
  secondary, 98
  from solution, 109–113
  thermodynamics of, 83
Crystals
  single, 77, 110–111 (*see also* Crystals, solution)

[Crystals]
  solution, 77–78, 109–113
    gel spun, 112
    shish kebabs, 111–112, 424
    thickness, 110–111
Cylindrites, 417–419

Deformation of the solid state
  mechanisms of, 442–451
    double yield phenomena, 449
    mechanical melting model of tensile deformation, 447–449
    Peterlin's model, 443–445
    screw dislocation model of yield, 445–447
    strain hardening, 449–451
    yield, 445–449
  structures generated by, 420–424
Degradation, 376–382 (*see also* Biodegradation)
Degree of crystallinity
  comparison of methods of determination, 303–305
  determination, 298–303
    by density, 298–300
    by differential scanning calorimetry, 300–301 (*see also* Characterization, solid state)
    by nuclear magnetic resonance spectroscopy, 302–303
    by ultrasonic measurement, 303
    by vibrational spectroscopy, 302 (*see also* Characterization, solid state)
    by wide angle X-ray diffraction, 301 (*see also* Characterization, solid state)

Index

[Degree of crystallinity]
  effect of orientation on, 429–431
Density, 2–3, 124–127
  and crystallinity, 2, 124–127
  measurement, 298–300
Differential scanning calorimetry, 292–296
Diffusion (*see* Properties, barrier)
Discovery of polyethylene, 28
Distribution
  composition, 10
  molecular weight, 7–10 (*see also* Molecular weight)
Draw ratio at break (*see* Properties, tensile, elongation at break)

Elastic modulus, 130–133 (*see also* Properties, oriented; Testing, physical properties)
Elongation at break (*see* Properties, tensile)
Environmental stress cracking, 165–170 (*see also* Environmental stress crack resistance)
Environmental stress crack resistance, 337–338
Ethylene vinyl acetate copolymer, 3, 406–408
  applications, 462, 504–506
  development, 40
  properties, 19
Ethylene vinyl alcohol copolymer, 408–409
Ethylene vinyl ester copolymer, 3
Extrusion, 463–464
  pipe, 472–474
  principles of, 463–464
  products, 464, 475–477

Fabrication processes, 462–502
  blow molding, 484–490
  cross-linking, 389–401
  extrusion, 463–464
    profile and pipe, 472–474
  film blowing, 465–468
  film casting, 470–471
  film coating, 477–478
  foaming, 495
  high pressure melt forming, 500–501
  injection molding, 480–483
  profile forming, 472–476
  rotational molding, 492–494
  sintering, 501
  spinning, gel and solution, 433–435
  thermoforming, 496–497
Fibers
  gel spun, 112
  shish kebab, 111–112
Fibrils, 420–424 (*see also* Morphology, oriented)
Film blowing, 465–470
  process, 465–468
  products, 468–470
Film casting, 470–472
  process, 470–472
  products, 472
Film coating (*see* Coating, film)
Flame treatment, 403–404
Flory model of crystallization, 88–89, 91–94 (*see also* Crystallization, mechanisms of)
Foaming, 495
Free radical polymerization, 43–48
Fringed micelles, 76–77

Gel permeation chromatography (*see* Size elution chromatography)

Gloss, 205, 210 (*see also* Testing, physical properties, reflective property determination)

Haze (*see* Properties, optical)
Heat distortion temperature, 175–177 (*see also* Testing, physical properties, tensile heat distortion, flexural heat distortion)
High density polyethylene, 2
  applications, 22–24, 460–461, 502–504
  discovery, 34–35
  market development, 37–38
  properties, 17
High modulus samples (*see* Oriented morphologies)
High pressure polymerization, 43–53
History, 27–41

Injection molding, 480–484
  process, 480–483
  products, 483–484
Insulation (*see* Coating, wire and cable)
Interface, 69–70
Interfacial region (*see* Interface)
Ionomers, 4–5, 409–412
  discovery, 40, 410–412
  properties, 19–20

Lamellae (*see also* crystallites), 14, 77–80
  loosely connected, 111
  melt crystallized, 78–80, 114–119
  thickness, 114
Linear low density polyethylene, 3
  applications, 24–25, 461–462, 506
  discovery, 39

[Linear low density polyethylene]
  market development, 39–40
  properties, 18
Loops, 68–69
  entangled, 107
Low density polyethylene, 2
  applications, 24, 461, 502–504
  discovery, 28
  market development, 28–34
  polymerization, 43–53
  properties, 17–18
Low pressure polymerization, 53–60

Market development, 28–41
Markets (*see* Applications)
Melt index, 224–226
  measurement, 259–260
Metallocene
  catalysis, 40, 60–61
  products, 40 (*see also* Metallocene, technology)
  technology, 515–520
Micelles, fringed, 76–77
Microscopy (*see* Characterization, solid state)
Microfibrils, 417, 421–422 (*see also* Morphology, oriented)
Modulus, initial, 130–133 (*see also* Testing, physical properties, elastic modulus)
Molecular weight
  determination, 242–250 (*see also* Characterization, molecular)
  distribution, 7–10
  manipulation, 41
  number-average, 7–8
  peak, 9
  viscosity average, 9–10
  weight-average, 8–9
  z-average, 9

Index

[Molecular weight]
  $z+1$-average, 9
Morphology, 10–15, 67–119, 415–425
  amorphous, 70
  noncrystalline, 68–70
    connections, interlamellar (*see* Tie chains)
    oriented, 415–425
      cylindrites, 417–419
      fibrils, 420–424
      from melts, 416–423
      microfibrils, 417, 421–422
      nanofibrils, 421–424
      shish kebabs, 111–112, 424
      from solutions, 424–425
    semicrystalline, 10–11, 67–71
    three phase, 67–71, 94

Nanofibrils, 421–424 (*see also* Morphology, oriented)

Oligomers, 1
Orientation, 415–456
  effect on crystallization from the melt state, 119
Oriented morphologies (*see also* Morphology, oriented)
  high modulus samples, routes to 433–442
    blocked plug crystallization, 441
    fiber growth from sheared solutions, 437–441
    radial compression, 441–442
    roller drawing, 442
    solid state drawing, 436–437
    solid state extrusion, 435–436
    spinning, gel and solution, 433–435
    swell drawing, 442
    two stage drawing, 442
Oxidation (*see* Autoxidation)

Permeability (*see* Properties, barrier)
Peroxide induced cross-linking, 393–396
Phillips process, 61–66
  discovery, 34
Pipe extrusion, 472–474
Polymerization:
  free radical, 43–48
  high pressure, 43–53
  Ziegler-Natta, 53–60
Polymethylene, 27
Production processes, 43–66
  emerging processes, 520–521
  high pressure, 43–53
  low pressure, 53–60
  medium pressure, 61–66
Profile forming, 472–474
Properties, 123–237
  abrasion resistance, 195–200 (*see also* Testing, physical properties)
    effect of external parameters, 198–200
    effect of molecular characteristics, 197–198
  barrier, 185–193
    effect of external conditions, 192–193
    effect of morphology on, 190–192
    compared to other polymers 20–22
  compressive, 145–147 (*see also* Testing, physical properties)
  creep, 152–158 (*see also* Testing, physical properties)
  creep rupture (*see* Properties, low stress brittle failure)
  cross-linked polyethylene, 20, 399–401

[Properties]
    electrical, 210–220
        arc resistance, 215–216 (*see also* Testing, electrical properties)
        bulk resistance, 211–212
        capacitance, 212–214 (*see also* Testing, electrical properties)
        dielectric constant, 212–213
        dielectric strength, 214–215 (*see also* Testing, electrical properties)
        dissipation factor, 213–214
        effect of molecular characteristics on, 216–217, 220
        permittivity (*see* Properties, electrical, dielectric constant)
        power factor, 214
        resistance, 211–212 (*see also* Testing, electrical properties)
        surface resistance, 212
        treeing, 217–220 (*see also* Testing, electrical properties)
        water trees, 217–219
    elongational (*see* Properties, tensile)
    elastic modulus (*see* Properties, tensile)
    ethylene-vinyl acetate copolymer, 19
    environmental stress cracking, 165–170 (*see also* Testing, physical properties)
    flexural modulus, 147–148 (*see also* Testing, physical properties)
    fracture (*see* Properties, rupture)

[Properties]
    friction, 200–204 (*see also* Testing, physical properties)
        effect of external conditions, 203–204
        effect of molecular characteristics, 202–203
    hardness, 146–147 (*see also* Testing, physical properties)
    high density polyethylene, 17
    impact resistance (*see* Properties, rupture)
    initial modulus (*see* Properties, tensile, elastic modulus)
    intrinsic, 15–17
    ionomers, 19–20
    linear low density polyethylene, 18
    long term brittle failure (*see* Properties, low stress brittle failure)
    low density polyethylene, 17–18
    low stress brittle failure, 159–165, 432 (*see also* Rupture phenomena)
    mechanical, 127–170
        long term, 151–170
    melt (*see* Properties, rheological)
    microhardness (*see* Properties, hardness)
    microindentation hardness (*see* Properties, hardness)
    modulus:
        elastic, 130–133 (*see also* Properties, oriented; Testing, physical properties)
        flexural, 147–148
    optical, 204–210
        gloss, 205, 210 (*see also* Testing, physical properties)

[Properties]
    haze, 204–205, 206–209 (*see also* Testing, physical properties)
    haze, internal, 206–208 (*see also* Testing, physical properties)
    haze, surface, 206, 208 (*see also* Testing, physical properties)
    refractive index, 210
    transparency, 205, 432–433 (*see also* Testing, physical properties)
  oriented, 425–433
    elastic modulus, 425–428
    elongation at break, 429
    tensile strength, 428–429
  puncture resistance (*see* Properties, rupture)
  rheological, 220–237
    die swell, 230–233
    effect of molecular characteristics, 237
    melt elasticity, 230–237
    melt fracture, 233–235
    melt index, 224–226
    melt orientation, 233–237
    melt strength, 236–237
    melt viscosity, 222–229
    melt viscosity as a function of shear rate, 226–229
    solution viscosity, 229–230
    viscoelasticity, 221
    zero shear viscosity, 222–224
  rupture, 148–151, 159–170 (*see also* Rupture phenomena; Testing, physical properties, impact and puncture resistance)
  slow crack growth (*see* Low stress brittle failure)

[Properties]
  solvent stress cracking (*see* Environmental stress cracking)
  stress relaxation, 158–159
  surface contact, 193–204
  tear resistance (*see* Properties, rupture)
  temperature effects, 141–144
    toughness, 149–151
    ultimate tensile stress, 141
  wear resistance (*see* Properties, abrasion resistance)
  yield phenomena, 133–137 (*see also* Testing, physical properties)
  tensile, 128–145
    draw ratio at break (*see* Properties, tensile, elongation at break)
    elastic modulus, 130–133 (*see also* Properties, oriented; Testing, physical properties)
    elongation at break 138–141 (*see also* Properties, oriented; Testing, physical properties)
    elongation rate effects, 144–145
    strength, 141 (*see also* Properties, oriented *and* Testing, physical properties)
  thermal, 170–185
    α transition, 184
    β transition, 184–185
    expansion, 180–181, 432
    γ transition (*see* Properties, thermal, glass transition)
    glass transition, 182–183
    heat capacity, 178–180
    heat conduction, 178, 431

[Properties]
    heat distortion temperature, 175–177 (*see also* Testing, physical properties, tensile heat distortion, flexural heat distortion)
    heat of fusion, 177–178
    melting range, 170–175
    transitions, 182–185
    very low density polyethylene, 19
Pyrolysis, 382 (*see also* Recycling, tertiary)

Random coil, 12
Recycling, 509–515
    future development, 514–515
    primary, 509
    secondary, 510–513
    tertiary, 513
Refractive index, 210
Resistance (*see* Properties, electrical)
Rheology (*see* Properties, rheological)
Rotational molding, 491–494
    process, 492–494
    products, 494
Rotomolding (*see* Rotational molding)
Rupture phenomena, 451–455 (*see also* Properties, rupture)
    brittle failure, 451–452
    ductile failure, 452–453
    environmental stress cracking, 454–455
    low stress brittle failure, 453–454

Sealing, 497–498
Sheaves, 114–115
Shish kebabs, 111–112, 424

Siloxane cross-linking, 397–399
Single crystals (*see* Crystals, solution)
Single-site catalysis (*see* Metallocene, catalysis)
Size elution chromatography, 242–247
Small angle light scattering, 282–284
Spherulites, 14–15, 114–119
Spinning, gel and solution, 433–435
Stabilization, 382–388
    hydroperoxide degradation, 387–388
    quenching, 384–385
    radical scavenging, 385–387
    shielding, 384
Standard Oil process, 35
Stress relaxation, 158–159
Structure
    crystallite, 14
    molecular, 1–6
        cross-linked polyethylene, 5–6
        ethylene-vinyl acetate copolymer, 3
        high density polyethylene, 2
        ionomer, 4
        linear low density polyethylene, 3
        low density polyethylene, 2
        very low density polyethylene, 3
    noncrystalline, 12
    sheaves, 114–115
    spherulites, 14–15, 114–119

Tear strength, 323–325
Temperature rising elution fractionation, 253–254

Index

Tensile strength (*see* Properties, tensile)
Testing, 241–370 (*see also* Characterization)
  ASTM methods, 366–370
  of electrical properties, 358–366
    arc resistance, 362–363
    breakdown voltage, 361–362
    capacitive properties, 360–361
    corona discharge resistance (*see also* Testing, electrical properties, partial discharge resistance)
    dielectric strength measurement, 361–362
    partial discharge resistance, 364–365
    resistance, 358–360
    tracking resistance, 362–364
    treeing, 365–366
  of physical properties, 310–358
    abrasion resistance, 355–357
    birefringence measurement, 350–351
    cantilever beam testing, 321
    Charpy impact test (*see* Testing, physical properties, impact beam testing)
    cling measurement, 352–354
    compressive testing, 319–320
    creep measurement, 326–327
    dart drop test (*see* Testing, physical properties, film puncture resistance)
    dynamic mechanical analysis, 334–337
    elastic modulus, 314–316
    elongation at break, 318–319
    Elmendorf tear strength (*see* Testing, physical properties, tear propagation resistance)

[Testing]
    environmental stress crack resistance, 337–338
    film puncture resistance, 333–334
    flexural heat distortion temperature, 322–323
    flexural testing, 321–323
    force versus deformation measurements, 310–327
    four point bending, 321
    friction, coefficient of, 354–355
    gloss, 347–350
    hardness, 319–320
    haze determination, 346–347
    impact beam testing, 327–330
    impact plaque testing, 330–332
    impact and puncture resistance, 327–334
    initial tear resistance, 323–324
    Izod impact test (*see* Testing, physical properties)
    optical characterization, 346–351
    organoleptic analysis, 345
    permeation, 342–345
    puncture resistance, 333–334
    refractive index measurement, 350–351
    reflective property determination, 347–350
    shear modulus determination, 320–321
    shear strength, 320
    shear testing, 320–321

[Testing]
  stress crack resistance, 337–339
  tear propagation resistance, 324–325
  tear strength, 323–325
  tensile impact testing, 332–333
  tensile measurements, 314–319
  tensile strength, 317–318
  tensile heat distortion temperature, 319
  thermal stress crack resistance, 339
  three point bending, 321
  transparency determination, 350
  water vapor transmission rate, 342–345
  weathering, artificial, 341–342
  weathering, outdoor, 340–341
  weathering resistance, 340–342
  yield phenomena, 316
Thermal analysis, 291–298
Thermoforming, 496–497
Tie chains, 68–69, 107–109
Transparency, 205, 432–433 (*see also* Testing, physical properties)

Ultrahigh molecular weight polyethylene:
  fabrication and uses, 499–501
  high pressure melt forming, 500–501

[Ultrahigh molecular weight polyethylene]
  sintering, 501
  spinning, 501 (*see also* Oriented morphologies, high modulus samples, routes to)
Unit cell, 12–13, 71–76
  hexagonal, 75–76
  monoclinic, 74–75
  orthorhombic, 12–13, 72–74
  triclinic (*see* Unit cell, monoclinic)
Uses (*see* Applications)

Variable cluster model, 93–94
Very low density polyethylene, 3
  applications, 24–25
  properties, 19

Water trees, 217–219
Welding, 498–499
Wire coating (*see* Coating, wire and cable)

X-ray diffraction (*see* Characterization, solid state)

Yield phenomena, 133–137 (*see also* Testing, physical properties)

Ziegler-Natta
  catalysis, 53–54, 58–59
  polymerization, 53–60
Ziegler polymerization, 34–35